AF581659

Nutrition and Cardiovascular Health

Nutrition and Cardiovascular Health

Special Issue Editors

Paramjit S. Tappia
Heather Blewett

MDPI • Basel • Beijing • Wuhan • Barcelona • Belgrade • Manchester • Tokyo • Cluj • Tianjin

Special Issue Editors
Paramjit S. Tappia
St. Boniface Hospital
Canada

Heather Blewett
Agriculture and Agri-Food Canada
Canada

Editorial Office
MDPI
St. Alban-Anlage 66
4052 Basel, Switzerland

This is a reprint of articles from the Special Issue published online in the open access journal *International Journal of Molecular Sciences* (ISSN 1422-0067) (available at: https://www.mdpi.com/journal/ijms/special_issues/nutrition_cardiovascular_health).

For citation purposes, cite each article independently as indicated on the article page online and as indicated below:

LastName, A.A.; LastName, B.B.; LastName, C.C. Article Title. *Journal Name* **Year**, *Article Number*, Page Range.

ISBN 978-3-03928-887-8 (Pbk)
ISBN 978-3-03928-888-5 (PDF)

Contents

About the Special Issue Editors

Paramjit S. Tappia received his B.Sc. (Honors) in Pharmacology from the University of Sunderland, U.K. in 1985 and a Ph.D. in Biochemistry from the University of Wolverhampton, U.K. in 1992. He received postdoctoral training in the Institute of Human Nutrition, University of Southampton, U.K. from 1992–1995. In 1996, he joined the Institute of Cardiovascular Sciences, St. Boniface Hospital Research Centre, Winnipeg, Canada as a Research Associate. He is currently a Principal Investigator and Clinical Scientist in the Asper Clinical Research Institute, St. Boniface Hospital, Winnipeg, Canada. As a clinical trialist, Dr. Tappia has conducted several trials in the area of diabetes, heart failure, gut health, and cancer diagnostics, as well as nutrition and human health. To date, he has published 115 full-length papers in scientific journals and 28 book chapters in the area of cardiovascular sciences and nutrition with H-index of 34 (Google Scholar). He has co-edited 3 books in the area of human health and serves on the Editorial Boards of 3 international journals. Dr. Tappia also serves as the Editor of the CV Network, which is the official news bulletin of the International Academy of Cardiovascular Sciences; a global organization for the promotion of prevention of cardiovascular disease.

Heather Blewett (nee Hosea) completed her Ph.D. under the supervision of Dr. Carla Taylor in the Department of Human Nutritional Sciences at the University of Manitoba in 2006. Her Ph.D. thesis explored the effects of zinc deficiency on T-cell maturation and function in growing rats. She went on to work with Dr. Catherine Field at the University of Alberta (2006–2008). During this time, Dr. Blewett expanded both her nutritional and immunological training by studying the importance of amino acids and fatty acids for intestinal and immune defense against *E. Coli* using a novel piglet model. She also investigated the effect of a "natural" trans fat called vaccenic acid on the pro-inflammatory tendency of immune cells in a rat model of obesity. After leaving the University of Alberta, Dr. Blewett joined Peter Zahradka's lab at the Canadian Centre for Agri-Food Research in Health and Medicine (CCARM) to gain experience using clinical trials to test the efficacy of Canadian crops in the treatment of vascular disease (2009–2011). Dr. Blewett was hired into her dream job as a Research Scientist with the Government of Canada in the Department of Agriculture and Agri-food Canada in May 2011 and is located at the St. Boniface Research Centre in Winnipeg, Manitoba. She uses her expertise in human nutrition to perform the clinical trials that are needed to substantiate food health claims for Canadian crops including LDL cholesterol lowering, reduction in postprandial glucose response, and increased satiety. Her background in immunology also allows her to focus on the effect of Canadian food products on immune function in the context of chronic diseases (i.e., cardiovascular disease, diabetes, obesity).

International Journal of
Molecular Sciences

Editorial

Nutrition and Cardiovascular Health

Paramjit S. Tappia [1,*] and Heather Blewett [2]

1 Asper Clinical Research Institute & Office of Clinical Research, St. Boniface Hospital, Winnipeg, MB R2H2A6, Canada

2 Agriculture and Agri-Food Canada, Winnipeg, MB R2H2A6, Canada; hblewett@sbrc.ca

* Correspondence: ptappia@sbrc.ca

Received: 23 March 2020; Accepted: 24 March 2020; Published: 26 March 2020

There is unequivocal experimental, epidemiological and clinical evidence demonstrating a correlation between diet and increased risk of cardiovascular disease (CVD). While nutritionally poor diets can have a significant negative impact on cardiovascular health, dietary interventions with specific nutrients and/or functional foods are considered cost-effective and efficient components of prevention strategies. It has been estimated that nutritional factors may be responsible for approximately 40% of all CVD [1]. Indeed, in one of the seminal studies conducted on modifiable risk factors and heart health (the INTERHEART study), >90% of all myocardial infarctions were attributed to preventable environmental factors with nutrition identified as one of the important determinants of CVD [2]. There is increasing public interest and scientific investigation into establishing dietary approaches that can be undertaken for the prevention and treatment of CVD. This Special Issue provides an insight into the influential role of nutrition and dietary habits on cardiovascular health and disease as well as the therapeutic and preventive potential of novel nutraceuticals and specific nutrients. Fourteen outstanding papers, from experts in the field, provide a broad range of contributions detailing various aspects of nutrition and cardiovascular health as well as highlighting possible mechanisms of beneficial action.

Casas et al. [3] have reviewed the role of overall nutrition, specific nutrients, foods and dietary practices in relation to cardioprotection and prevention of CVD. This review Furthermore describes some of the mechanisms in the cardioprotective properties of individual nutrients, foods and nutritional patterns.

This Special Issue contains two reviews on the effects of long-chain polyunsaturated omega-3 fatty acids on CVD. Innes and Calder [4] provide a review of the literature on the use of marine derived omega-3 (n-3) fatty acids (eicosapentaenoic acid and docosahexaenoic acid) with a focus on primary and secondary prevention of CVD, along with a discussion of the potential mechanisms for their effects. Goel et al. [5] discuss the inconsistencies regarding the cardioprotective effects of fish and fish oils. Although many experimental studies and some clinical trials have documented the benefits of fish oil supplementation in decreasing the incidence and progression of atherosclerosis, myocardial infarction (MI), heart failure (HF), arrhythmias and stroke, recent large-scale clinical studies have failed to demonstrate any benefit on cardiovascular outcomes and mortality. This is an area of investigation that needs some refinement in order to fully understand the beneficial effects of fish oils and n-3 polyunsaturated fatty acids in general. Indeed, we have Furthermore reported inconsistencies in outcomes of CVD management in some trials with n-3 fatty acids [6,7].

Micronutrient deficiency is present in HF and is associated with adverse clinical signs and symptoms. Indeed, a pathophysiological role as well as prognostic value has been ascribed to micronutrient deficiency. In this regard, Cvetinovic et al. [8] review evidence that demonstrate a correction/normalization of micronutrient status is linked to a concomitant improvement in physical performance and quality of life in HF. The Special Issue Furthermore includes articles on specific micronutrients. Globally, vitamin D deficiency is highly prevalent and has been linked to many non-communicable diseases, but the role of vitamin D in the pathogenesis of HF remains to be defined. Roffe-Vazquez [9] discuss the molecular mechanisms involved in inflammatory processes, remodeling, fibrosis and atherosclerosis in humans due to a deficiency in vitamin D, through in vitro and animal experimentation. Furthermore, by employing human cardiac microvascular endothelial cells and a rat model of isoproterenol-induced fibrosis, Lai et al. [10] specifically report that vitamin D may, in fact, reduce the development of fibrosis. Although more research is required, vitamin D could be part of a prevention strategy for individuals at risk of HF.

The role of magnesium (Mg) deficiency as a risk factor of arterial hypertension is not completely known. Kostov and Halecheva [11] describe the many benefits of Mg and discuss the role of Mg deficiency in increasing the risk of atherosclerosis, endothelial dysfunction and arterial stiffness, which all contribute to hypertension. Thus, healthy dietary practices that include the recommended amounts of Mg may constitute a nutritional approach for normal blood pressure.

Nutraceuticals and functional foods have established efficacy for the treatment and/or prevention of adverse human health. Among these natural health products is American Ginseng. Parikh et al. [12] describe the cardioprotective effects of American Ginseng root extract (GBE) using a rat model of congestive heart failure due to MI induced by coronary artery ligation. Although treatment with GBE did not improve cardiac remodeling and function, attenuation of oxidative stress and TNFα was seen. The reduction of TNFα to below baseline levels was suggestive of the use of GBE as a prophylactic or as a preventive adjunct for cardiovascular disease. Jakovljevic et al. [13] have examined the potential health benefits of Aronia melanocarpa extract (SAE) in a rat model of metabolic syndrome (MetS). SAE was found to lower blood pressure and improve cardiac function. It is noteworthy that SAE improved glucose tolerance, liver damage and reduced oxidative stress. From the evidence provided, nutritional supplementation with SAE may potentially exert cardioprotective effects in patients with MetS. Cocaine is a potent stimulant drug that disrupts the electrical signals of the heart, increases blood pressure, heart rate, and the occurrence of heart-related fatal events including sudden cardiac death. In the review by Kim and Park [14] the cellular and molecular mechanisms of cocaine on the cardiovascular system are described. While there is unequivocal information on the adverse acute effects of cocaine on the heart (electrophysiological abnormalities, arrhythmia, and acute MI), the data on the chronic effects of cocaine on the vascular system (coronary artery disease (CAD) and/or subclinical atherosclerosis) are less clear and inconsistent. However, interestingly, chronic effects of cocaine are more likely in individuals with higher CAD risk and with deleterious health choices and behaviour.

Extracellular matrix (ECM) remodeling and fibrosis are key players in HF. Although some experimental studies have reported that nutraceuticals can diminish cardiac fibrosis, clinical data are conflicting. Jahan et al. [15] detail the molecular mechanisms involved in the control of fibroblast activation. Specifically, the role of Zeb2 transcription factor which could serve as an effective target for the attenuation/prevention of cardiac fibrosis is discussed and the potential of specific nutraceuticals described. Lim et al. [16] describe the current understanding of α-Klotho, a protein with anti-aging properties, as a potential therapeutic against age-associated vascular abnormalities. From the evidence presented in this review, it is conceivable that Klotho-based interventional trials could be initiated and yield important data for the effective clinical use of α-Klotho in several pathophysiological conditions including chronic kidney disease, cancer, diabetes and HIV infection where age-related vascular alterations are implicated. High sodium intake has been shown to have a positive relationship with death [17]. Paczula et al. [18] discuss the role of endogenous cardiotonic steroids in mediating salt-induced hypertension and organ damage. Compelling lines of evidence are provided in the link between high-salt diet and organ damage and thus it can be inferred that control of dietary salt is of critical importance in the prevention and/or risk reduction. Irisin has been characterized as a myokine that has been linked to insulin resistance, obesity and other non-communicable adult diseases. However, information regarding the role of irisin in childhood and early adulthood is inconsistent. In this regard, Elizondo-Montemayor [19] discusses the potential role of irisin in cardiovascular and metabolic health and disease in the pediatric population.

Taking all contributions into account it is clear that nutrition plays an important role in cardiovascular health and disease. In general, nutrients exhibit a diverse range of properties including anti-oxidant effects, anti-inflammatory actions, modification of signal transduction mechanisms, as well as metabolic, molecular and membrane actions. Although the cardioprotective effects may not be due to a single nutrient, but a balanced and varied diet of food items that can provide different benefits may prove to be key to cardiovascular health. We are grateful to all contributors, who are highly regarded and well-recognized in their respective field of interest. They have helped to generate a valuable issue of the journal that is very much special. It is hoped that experts within the field and those that have a general interest in nutrition and human health will find the information presented in this Special Issue, educational and insightful to prompt further investigation and advancement in understanding into the essential role of nutrition in cardiovascular health and disease treatment and/or prevention.

Acknowledgments: Infrastructural support for this work was provided by the St. Boniface Hospital Research Foundation, Winnipeg, Canada.

References

1. Willet, W.C. Potential benefits of preventive nutrition. In *Preventive Nutrition: The Comprehensive Guide for Health Professionals*; Bendich, A., Deckelbaum, R.J., Eds.; Humuna Press: Totowa, NJ, USA, 1999; pp. 423–441.
2. Yusuf, S.; Hawken, S.; Ounpuu, S.; Dans, T.; Avezum, A.; Lanas, F.; McQueen, M.; Budaj, A.; Pais, P.; Varigos, J.; et al. Effect of potentially modifiable risk factors associated with myocardial infarction in 52 countries (the INTERHEART study): Case-control study. *Lancet* **2004**, *364*, 937–952. [CrossRef]
3. Casas, R.; Castro-Barquero, S.; Estruch, R.; Sacanella, E. Nutrition and cardiovascular health. *Int. J. Mol. Sci.* **2018**, *19*, 3988. [CrossRef] [PubMed]
4. Innes, J.K.; Calder, P.C. Marine Omega-3 (N-3) Fatty Acids for Cardiovascular Health-an Update for 2020. *Int. J. Mol. Sci.* **2019**, *20*, 1362. [CrossRef] [PubMed]
5. Goel, A.; Pothineni, N.V.; Singhal, M.; Paydak, H.; Saldeen, T.; Mehta, J.L. Fish, fish oils and cardioprotection: Promise or fish tale? *Int. J. Mol. Sci.* **2018**, *19*, 3703. [CrossRef] [PubMed]
6. Shaikh, N.A.; Tappia, P.S. Why are there inconsistencies in the outcomes of some omega-3 fatty acid trials for the management of CVD? *Clin. Lipidol.* **2015**, *10*, 27–42. [CrossRef]
7. Xu, Y.J.; Gregor, T.; Dhalla, N.S.; Tappia, P.S. Is the jury still out on the benefits of fish, seal and flax oils in cardiovascular disease? *Ann. Nutr. Disord. Ther.* **2014**, *1*, 40.
8. Cvetinovic, N.; Loncar, G.; Isakovic, A.M.; von Haehling, S.; Doehner, W.; Lainscak, M.; Farkas, J. Micronutrient depletion in heart failure: Common, clinically relevant and treatable. *Int. J. Mol. Sci.* **2019**, *20*, 5627. [CrossRef] [PubMed]
9. Roffe-Vazquez, D.N.; Huerta-Delgado, A.S.; Castillo, E.C.; Villarreal-Calderón, J.R.; Gonzalez-Gil, A.M.; Enriquez, C.; Garcia-Rivas, G.; Elizondo-Montemayor, L. Correlation of vitamin D with inflammatory cytokines, atherosclerotic parameters, and lifestyle factors in the setting of heart failure: A 12-month follow-up study. *Int. J. Mol. Sci.* **2019**, *20*, 5811. [CrossRef] [PubMed]
10. Lai, C.-C.; Juang, W.-C.; Sun, G.C.; Tseng, Y.-K.; Jhong, R.-C.; Tseng, C.-J.; Wong, T.-Y.; Cheng, P.W. Vitamin D attenuates loss of endothelial biomarker expression in cardio-endothelial cells. *Int. J. Mol. Sci.* **2020**, *21*, 2196. [CrossRef]
11. Kostov, K.; Halacheva, L. Role of magnesium deficiency in promoting atherosclerosis, endothelial dysfunction, and arterial stiffening as risk factors for hypertension. *Int. J. Mol. Sci.* **2018**, *19*, 1724. [CrossRef] [PubMed]
12. Parikh, M.; Raj, P.; Yu, L.; Stebbing, J.A.; Prashar, S.; Petkau, J.C.; Tappia, P.S.; Pierce, G.N.; Siow, Y.L.; Brown, D.; et al. Ginseng berry extract rich in phenolic compounds attenuates oxidative stress but not cardiac remodeling post myocardial infarction. *Int. J. Mol. Sci.* **2019**, *20*, 983. [CrossRef] [PubMed]
13. Jakovljevic, V.; Milic, P.; Bradic, J.; Jeremic, J.; Zivkovic, V.; Srejovic, I.; Nikolic Turnic, T.; Milosavljevic, I.; Jeremic, N.; Bolevich, S.; et al. Standardized *Aronia melanocarpa* extract as novel supplement against metabolic syndrome: A rat model. *Int. J. Mol. Sci.* **2018**, *20*, 6. [CrossRef] [PubMed]
14. Kim, S.T.; Park, T. Acute and chronic effects of cocaine on cardiovascular health. *Int. J. Mol. Sci.* **2019**, *20*, 584. [CrossRef] [PubMed]

15. Jahan, F.; Landry, N.M.; Rattan, S.G.; Dixon, I.M.C.; Wigle, J.T. The functional role of zinc finger E box-binding homeobox 2 (Zeb2) in promoting cardiac fibroblast activation. *Int. J. Mol. Sci.* **2018**, *19*, 3207. [CrossRef] [PubMed]
16. Lim, K.; Halim, A.; Lu, T.S.; Ashworth, A.; Chong, I. Klotho: A major shareholder in vascular aging enterprises. *Int. J. Mol. Sci.* **2019**, *20*, 4637. [CrossRef] [PubMed]
17. Cook, N.R.; Appel, L.J.; Whelton, P.K. Sodium intake and all-cause mortality over 20 years in the trials of hypertension prevention. *J. Am. Coll. Cardiol.* **2016**, *68*, 1609–1617. [CrossRef] [PubMed]
18. Paczula, A.; Wiecek, A.; Piecha, G. Cardiotonic steroids-A possible link between high-salt diet and organ damage. *Int. J. Mol. Sci.* **2019**, *20*, 590. [CrossRef] [PubMed]
19. Elizondo-Montemayor, L.; Mendoza-Lara, G.; Gutierrez-DelBosque, G.; Peschard-Franco, M.; Nieblas, B.; Garcia-Rivas, G. Relationship of circulating irisin with body composition, physical activity, and cardiovascular and metabolic disorders in the pediatric population. *Int. J. Mol. Sci.* **2018**, *19*, 3727. [CrossRef] [PubMed]

International Journal of
Molecular Sciences

Review

Nutrition and Cardiovascular Health

Rosa Casas [1,2], Sara Castro-Barquero [1], Ramon Estruch [1,2] and Emilio Sacanella [1,2,*]

1 Department of Internal Medicine, Hospital Clinic, Institut d'Investigació Biomèdica August Pi i Sunyer (IDIBAPS), University of Barcelona, Villarroel, 170, 08036 Barcelona, Spain; rcasas1@clinic.cat (R.C.); sacastro@clinic.cat (S.C.-B.); restruch@clinic.cat (R.E.)

2 CIBER 06/03: Fisiopatología de la Obesidad y la Nutrición, Instituto de Salud Carlos III, 28029 Madrid, Spain

* Correspondence: esacane@clinic.ub.es; Tel.: +34-93-2275745; Fax: +34-93-2275758

Received: 19 November 2018; Accepted: 6 December 2018; Published: 11 December 2018

Abstract: Cardiovascular disease (CVD) is the leading cause of death in Western countries, representing almost 30% of all deaths worldwide. Evidence shows the effectiveness of healthy dietary patterns and lifestyles for the prevention of CVD. Furthermore, the rising incidence of CVD over the last 25 years has become a public health priority, especially the prevention of CVD (or cardiovascular events) through lifestyle interventions. Current scientific evidence shows that Western dietary patterns compared to healthier dietary patterns, such as the 'Mediterranean diet' (MeDiet), leads to an excessive production of proinflammatory cytokines associated with a reduced synthesis of anti-inflammatory cytokines. In fact, dietary intervention allows better combination of multiple foods and nutrients. Therefore, a healthy dietary pattern shows a greater magnitude of beneficial effects than the potential effects of a single nutrient supplementation. This review aims to identify potential targets (food patterns, single foods, or individual nutrients) for preventing CVD and quantifies the magnitude of the beneficial effects observed. On the other hand, we analyze the possible mechanisms implicated in this cardioprotective effect.

Keywords: Mediterranean diet; cardiovascular disease; inflammation; nutrients; polyphenols; MUFA; PUFA; bioactive compounds; phytosterols; dietary pattern

---

1. Introduction

Data obtained in 2013 showed that the leading global cause of death in Western countries is cardiovascular disease (CVD), accounting for 17.3 million of all deaths worldwide per year (or 31.5% of all global deaths), despite steadily decreasing during the past 10 years [1,2]. One in three deaths in the United States and one in four deaths in Europe are caused by CVD [3]. So, in 2035, 45.1% (>130 million adults) of the US population are projected to have clinical expression of CVD [1,4]. CVD describes a range of disorders that affect the heart and blood vessels, such as hypertension, stroke, atherosclerosis, peripheral artery disease, and vein diseases [4]. The probability of developing CVD is associated with unhealthy dietary patterns (i.e., excessive intake of sodium and processed foods; added sugars; unhealthy fats; low intake of fruit and vegetables, whole grains, fiber, legumes, fish, and nuts), together with a lack of exercise, overweight and obesity, stress, alcohol consumption, or a smoking habit (Figure 1) [5–7]. Additionally, CVD often coincides with multiple co-morbidities, such as obesity, diabetes, hypertension, or dyslipidemia, which represent four of the 10 greatest risk factors for all-cause mortality worldwide [8]. Furthermore, the rising incidence of CVD over the last 25 years has become a public health priority, especially the prevention of CVD (or cardiovascular events) through lifestyle interventions [9]. On the one hand, a large body of scientific evidence has reported that nutrition might be the most preventive factor of CVD death [10], and could even reverse heart disease [8]. On the other hand, diet seems to play an important role in the management of other risk factors, such as excess weight, hypertension, diabetes, or dyslipidemia [8]. In this sense, the identification and classification of nutrients, foods, or dietary patterns that can enhance CVD prevention is a priority.

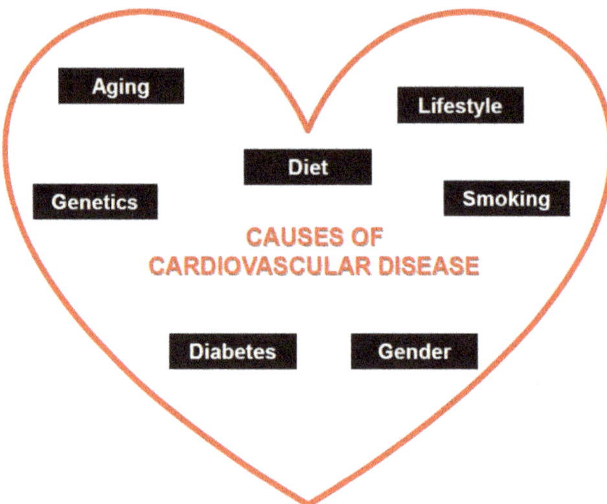

Figure 1. Unhealthy dietary patterns, together with a lack of exercise, overweight and obesity, aging, gender, genetics, or a smoking habit, among others, might lead to the development of cardiovascular disease (CVD).

Atherosclerosis is an inflammatory disease that contributes to major incidence and mortality of CVD. Oxidative stress and systemic inflammation are modifiable by nutrition [10–13], with an excess energy intake and physical inactivity as contributors of pro-inflammatory cytokines' secretion [14]. Inflammatory processes involve the sub-endothelial area of the arterial wall, accumulating lipids and lipid-laden macrophages among other cell types [15,16]. Current scientific evidence shows that chronic inflammation plays a key role in the pathogenesis of coronary artery disease (CAD), including the initiation and progression of atheroma plaque and rupture, and post-angioplasty and restenosis [17]. The main mediators of CAD development are C-reactive protein (CRP), interleukin (IL)-1, IL-6, IL-8, IL-1β, IL-18, monocyte chemoattractant protein (MCP)-1, and tumor necrosis factor (TNF)-α, among others. Moreover, those mediators are considered potential inflammation biomarkers and their expression may correlate with CAD severity [17–19].

As such, current evidence indicates that Western dietary patterns compared to healthier dietary patterns, such as the "Mediterranean diet" (MeDiet), leading to an excessive production of proinflammatory cytokines associated with a reduction of the synthesis of anti-inflammatory cytokines [20–23]. Therefore, the intake of fruits, vegetables, whole grains, nuts, seeds, and legumes is associated with lower inflammation [24–28], whereas red meat intake has been correlated with a higher inflammatory level [24,29–31]. Consequently, increased adherence to healthier dietary patterns, characterized by higher intake of fruits, vegetables, legumes, nuts, and whole grains, may mitigate low grade inflammation, preventing CVD [32–35].

In addition, microbiota has been linked to intestinal health, the immune, system and bioactivation and metabolism of nutrients, such as vitamins B and K and bioactive compounds. Recent clinical studies suggest a correlation between elevated plasma trimethylamine N-oxide (TMAO), which is produced by gut bacteria metabolism of dietary components, such as L-carnitine, betaine, and choline, and a higher risk of diabetes, hypertension, and atherosclerosis [36–38]. Therefore, it has been well studied that diet affects the composition and activity of gut microbiota and situations of gut microbiota dysbiosis may be involved in the development of CVD.

This review aims to identify potential targets (food patterns, single foods, or individual nutrients) for CVD prevention, quantify the magnitude of the beneficial effects observed, and analyze the mechanisms implicated in these cardioprotective effects. Besides, studies were limited to humans with no time restriction. Relevant studies, systematic reviews, and meta-analyses were searched to obtain the reference lists. The Medical Subject Headings search terms included: Inflammation, oxidative stress, inflammatory markers, IL, CRP, TNF-α, IL-6, dietary pattern, Mediterranean diet, Dietary Approach to Stop Hypertension (DASH diet), atherosclerosis, fruits and vegetables, olive oil, nuts, wine, fiber, micronutrients, vitamins, minerals, omega-3 fatty acids, lycopene, phytosterols, and polyphenols.

2. Atherosclerosis

Early stages of atherosclerosis are involved the internalization of lipids in the intima, mainly low-density lipoproteins (LDL), which is translated to endothelial dysfunction [39]. The disruption of the endothelial function promotes the inflammatory response, thrombus formation, and multiple pathological consequences, such as calcifications, stenosis, rupture, or hemorrhage [15,40].

The inflammatory response is enhanced by the infiltration of low-density lipoproteins (LDL) particles in the extracellular matrix (EM) while circulating monocytes are attached to the endothelium and transformed into macrophages infiltrating into the sub-endothelial area. The retention of LDL in EM is mediated by proteoglycans, which facilitate retention in the intima [41]. The LDL particles attached in the intima are susceptible to oxidative modifications by reactive oxygen species (ROS) and enzymatic modification released from inflammatory cells. Macrophages are converted into foam cells after oxidized LDL (oxLDL) particles are absorbed by them. Additionally, endothelial dysfunction enhances platelets' adhesion, which secretes chemotactic substances and growth factors, promoting plaque progression [42]. Vascular smooth muscle cells (VSMC) are also involved in plaque progression. Foam cells' growth factors and cytokines' secretion enhance VSMC migration to the intima where they contribute to the formation of the fibrous cap [43]. If the progression of lipid accumulation persists, foam cells and macrophages' apoptosis is induced jointly with pro-thrombotic molecules' secretion [44,45]. Atherosclerotic plaque progression and plaque disruption, promoted by pro-thrombotic agents, initiate platelet activation and aggregation, which leads to the coagulation cascade and, consequently, thrombus formation [46]. The clinical manifestations of advanced atherosclerosis are coronary heart disease, ischemic stroke, peripheral artery disease, heart failure, or sudden death [47].

3. Oxidative Stress and Inflammation

Oxidative stress has been related to the pathogenesis of atherosclerosis [48]. ROS and reactive nitrogen species (RNS) are mainly produced through mitochondrial activity and other pathways, such as nitric oxide (NO) synthase, and oxidase enzymes, such as Nicotinamide adenine dinucleotide phosphate (NADPH) oxidases (Nox), xanthine oxidase (XO), lipoxygenase, myeloperoxidase, uncoupled endothelial nitric oxide synthase (eNOS), and the mitochondrial respiratory chain via a one-electron reduction of molecular oxygen. Note the role of Nox in oxidative stress, as upregulated and overactive Nox enzymes contribute to oxidative stress and CVD. Several signaling pathways regulate inactivation and degradation of ROS and RNS, including catalase, glutathione peroxidase, and superoxide dismutase among others. An excess of ROS and RNS leads to oxidative stress, promoting cell proliferation, migration, autophagy, necrosis, DNA damage, endoplasmic reticulum stress, endothelial dysfunction, and higher levels of oxLDL [49,50]. Moreover, ROS activate the inflammatory response that directly affects plaque progression and endothelial function, increasing the levels of inflammatory cytokines, such as interleukins (IL-6, IL-8), TNF-α, and MCP-1, and adhesion molecules, such as intercellular adhesion molecule 1 (sICAM-1) and vascular cell adhesion molecule (sVCAM-1) [51]. Simultaneously, transcription factors activation, mainly nuclear factor kappa B

(NF-κβ) and nuclear factor (erythroid-derived 2)-like 2 (Nrf2), and signal transduction cascades result in a high production of inflammatory cytokines and inducible nitric oxide synthase [52]. NO has important anti-inflammatory, antihypertensive, and antithrombotic actions due to its strong vasodilator activity and anti-platelet aggregation. Additionally, anti-inflammatory effects are enhanced by the ability of NO to inhibit NF-κβ expression and the subsequent adhesion molecules [53]. Oxidative stress contributes to endothelial eNOS dysfunction [54,55]. Dysfunctional eNOS generates superoxide anions instead of NO, which is translated to a higher ROS production and contributes to atherogenesis [56]. In the case of inducible NOS (iNOS), which is expressed in cells after cytokines or bacterial lipopolysaccharide stimulation, an excessive and sustained production of NO has been linked with inflammatory diseases and septic shock [57]. Therefore, a decrease of NO production by eNOS leads to endothelial dysfunction while an excessive NO production by iNOS may induce pro-inflammatory and pro-atherogenic factors.

The causes and risk factor of atherosclerosis and oxidative stress are not well defined. However, certain health conditions and habits may contribute to atherosclerosis development, such as high total cholesterol and low-high-density lipoprotein cholesterol (HDL-c) levels, hypertension, type 2 diabetes mellitus (T2DM), obesity, and physical inactivity. Additionally, healthy dietary patterns and lifestyle modifications are potential strategies for atherosclerosis and oxidative stress prevention.

4. Dietary Patterns

Several studies correlate healthy dietary patterns with lower plasmatic concentrations of pro-inflammatory markers [58], whilst a Western-type diet (meat-based dietary pattern) is associated with higher levels of low-grade inflammation [31]. For that reason, CVD guidelines recommend a healthy diet. Dietary intervention allows a better combination of multiple foods and nutrients. Therefore, healthy dietary patterns support a greater magnitude of beneficial effects than the potential effects of a single nutrient supplementation, because of the synergistic health effects among them. The current body of evidence shows that healthy dietary patterns share similarities, such as a high intake of fiber, antioxidants, vitamins, minerals, polyphenols, monounsaturated, and polyunsaturated fatty acids (MUFA and PUFA, respectively); low intake of salt, refined sugar, saturated, and trans fats; and carbohydrates of low glycemic load [59]. This translates to a high intake of fruits, vegetables, legumes, fish and seafood, nuts, seeds, whole grains, vegetable oils (mainly, extra virgin olive oil [EVOO]), and dairy foods together with a low intake of pastries, soft drinks, and red and processed meat [59,60].

Mediterranean and DASH dietary interventions are well studied for CV outcomes. Both dietary patterns may reduce the incidence CVD through the down-regulation of low-grade inflammation and better control of body weight, which also improve other risk factors, and are correlated with lower numbers of clinical events [59,60]. Thus, this will be the focus of this research.

4.1. Mediterranean Diet

Until now, the main benefits of the Mediterranean diet (MeDiet) against CVD (Figure 2) have been associated with a better control of risk factors to improve blood pressure (BP), lipid profile, glucose metabolism, arrhythmic risk, or gut microbiome [59]. Also, some authors have suggested that MeDiet may exert an anti-inflammatory effect (Table 1) in the vascular wall as a possible mechanism to explain the link between MeDiet and low CVD prevalence [61]. Interestingly, MeDiet seems to modulate the expression of pro-atherogenic genes as cyclooxygenase-2 (COX-2), MCP-1, and low-density lipoprotein receptor-related protein (LRP1) [62], reducing plasmatic levels of plaque stability and rupture related molecules as MMP-9, IL-10, IL-13, or IL-18 [63,64].

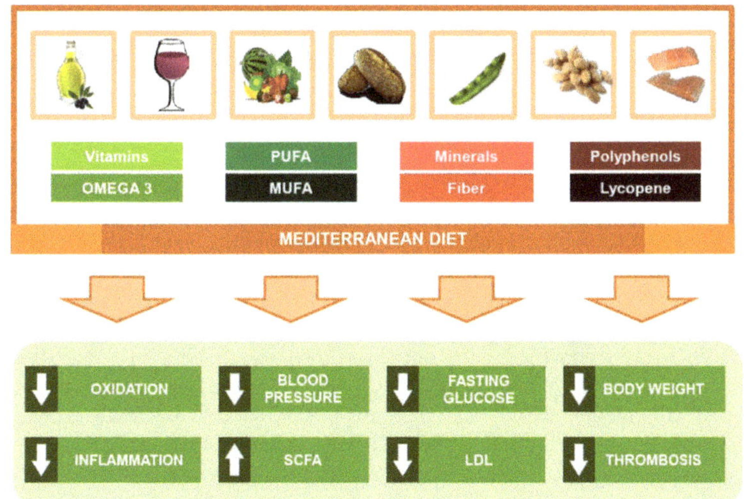

Figure 2. Main protection mechanisms of the Mediterranean diet against cardiovascular disease.

Table 1. Potential inflammatory effects of Mediterranean and DASH diet on CVD.

	Pro- and Anti-Inflammatory Markers and Genes	Leukocyte Expression	Oxidative Stress Markers
MeDiet	sVCAM-1, sICAM-1, RANTES, MIP-1β, TNF-α, TNFR-60, IL-1β, IL-6, IL-7, IL-10, IL-12p70, IL-13, IL-18, MMP-9, VEGF, CRP, TCF7L2, APOA2, CETP, COX-2, MCP-1, LRP1	Lymphocytes: CD11a, CD49d, CD40 Monocytes: CD11a, CD11b, CD49d, CD40	MDA, oxLDL
DASH diet	sICAM-1, IL-6, CRP, PAI-1	-	-

APOA2: Apolipoprotein A2; CETP: Cholesteryl ester transfer protein plasma; COX-2: Cyclooxygenase-2; CRP: C-reactive protein; IL: Interleukin; LRP1: Low-density lipoprotein receptor-related protein; MCP-1: Monocyte chemoattractant protein; MDA: Malondialdehyde; MMP-9: Metallopeptidase-9; oxLDL: Oxidized LDL; PAI-1: Plasminogen activator inhibitor 1; sICAM-1: Soluble intercellular adhesion molecule 1; sVCAM-1: Soluble vascular cell adhesion molecule; TNF-α: Tumor necrosis factor; TNFR: Tumor necrosis factor receptor; TCF7L2: Transcription factor 7-like 2; VEGF: Vascular endothelial growth factor.

The observational study, ATTICA, evaluated the link between MeDiet and the incidence of metabolic syndrome (MetS) in 1514 men and 1528 women (>18 y) without clinical evidence of CVD or any other chronic disease [65] during 10 years. Authors found that an increase of 10% in the MeDiet adherence score was associated with a 15% lower odds for CVD incidence. Nevertheless, the inflammatory factors studied (adiposity, CRP, IL-6), whose components are associated with a higher likelihood of CVD, showed a higher incidence (29%) in those subjects away from the MD [65]. Also, the Multi-Ethnic Study of Atherosclerosis (MESA) investigated if a dietary quality score based in a MeDiet pattern was related with regional adiposity [66]. Authors studied 5079 individuals free of CVD (61 ± 10 years) and found that a high quality dietary pattern was associated with less regional adiposity and a lower body mass index (BMI), CRP, and insulin resistance. Thus, Lahoz et al. [67] conducted a cross-sectional analysis of 1411 subjects of the Screening PRE-diabetes and type 2 DIAbetes (SPREDIA-2) study (mean age 61 years, 43.0% males) to assess whether the 14-point Mediterranean Diet Adherence Screener (MEDAS) was associated with serum CRP levels. After adjusting for confounders, the authors showed an inverse correlation between the adherence to MeDiet and CRP levels ($p = 0.041$). Also, a substudy of the MOLI-SANI cohort (6879 women and 6892 men) found that men with a higher adherence to a healthy high-antioxidant diet (HAC), vitamins, and phytochemicals enriched diet, inside a MeDiet pattern, were more protected against hypertension and inflammation than those with a healthy low-antioxidant diet [68]. Authors found HAC was associated with a significant decrease in CRP levels (β = 0.03, p = 0.03). Finally, Sureda et al. [69] conducted a study of two cross-sectional nutritional surveys with men and women (219 males and 379 females) aged among 12–65 years old, who lived on the Balearic Islands. Results showed that the male adult population with a higher adherence to the MeDiet showed lower concentrations of pro-inflammatory biomarkers, such as TNF-α and hs-CRP. Also, in this population, lower levels of leptin or plasminogen activator inhibitor 1 (PAI-1) were observed, while adiponectine levels were increased. Moreover, females (young and old), with a higher adherence to the MeDiet, showed lower hs-CRP levels. Lower leptin levels were showed only in the young female group, while PAI-1 reduction was only observed in female adults.

On the other hand, the PREDIMED (Prevención con Dieta Mediterránea) study, the largest interventional study about MeDiet, which included 7447 subjects (55 to 80 years of age, 57% women) at high CV risk, without CVD at baseline, showed a lower prevalence of CV events in participants assigned to a MeDiet supplemented with extra-virgin olive oil (EVOO) or nuts than those assigned to a low-fat diet after five years intervention [70]. Focusing on the PREDIMED study, the MeDiet has reported an anti-inflammatory effect on the expression of adhesion molecules in leukocytes, but also improvements in the circulating levels of soluble adhesion molecules (sVCAM-1, sICAM-1, E- and P-selectin, cytokines (IL-1, IL-6, IL-8, IL-12p70, CRP, TNF-α, tumor necrosis factor receptor (TNFR)-60 and 80, etc.), chemokines (MCP-1, Regulated on Activation, Normal T Cell Expressed and Secreted [RANTES], macrophage inflammatory proteins [MIP-1β], etc.) and molecules related with vulnerability atheroma plaque (IL-10, IL-13, IL-18, or Matrix metallopeptidase-9 [MMP-9]) after three months, one, three, and five years intervention in those participants that followed a MeDiet supplemented with EVOO or nuts [63,64,71–74]. On the one hand, the results obtained support that MeDiet exerts an important immunomodulatory effect, reducing proinflammatory biomarkers, especially those related to atheroma stability plaque. On the other hand, these anti-inflammatory effects seem to appear in the short and medium-term (three months, one year) and later is maintained in the long-term (three and five years). In a Swedish randomized crossover study [75], the adherence to a Medierranean-style diet and its correlation with inflammatory biomarkers (CRP, IL-6), vasoregulation, vascular endothelial growth factor (VEGF), and serum phospholipid fatty acid composition was investigated. The 22 subjects free of CVD (10 women) received a Mediterranean-type diet or Swedish diet for four weeks. No changes were observed for CRP or IL-6 although the MeDiet group showed significant reductions in leukocytes and platelets levels (by 10% and 15%, respectively) and in VEGF levels (15%). Esposito and et al. [76] assessed the effects of a Mediterranean dietary pattern on endothelial function and vascular inflammatory markers in patients with the MetS during two years (90 patients/intervention group). Compared with controls, the MeDiet group showed lower serum concentrations of high-sensitivity ((hs)-CRP, IL-6, IL-7, and IL-18 ($p \leq 0.04$; all)) and improved the endothelial function score ($p < 0.001$), defined as a measure of BP and platelet aggregation response to l-arginine. These results are in agreement with Azzini et al. [77], who also reported improvements in the CV risk profile and modulation of inflammatory levels (IL-10 and TNF-α), and a reduction in oxidative stress (malondialdehyde [MDA]).

The influence of polyphenols' content of a MeDiet pattern was also studied in the PREDIMED study [78]. Total polyphenol excretion (TPE) in urine was analyzed in 1139 participants homogenously and randomly in one of the three groups. Authors observed that MeDiet (supplemented with EVOO or nuts) significantly increased their TPE after one year of dietary intervention, decreasing inflammatory biomarkers compared with baseline (sVCAM-1, sICAM-1, IL-6 MCP-1, TNF-α).

Epigenetic studies have also reported similar results. Thus, Arpón et al. [79,80] conducted a substudy on 36 participants of the PREDIMED cohort after five years of intervention, where the main results were that a MeDiet supplemented with EVOO or nuts may influence the methylation status of peripheral white blood cell (PWBCs) genes. These changes were mainly observed in genes related to intermediate metabolism, diabetes, inflammation, and signal transduction. However, interactions among MeDiet and COX-2, IL-6, apolipoprotein A2 (APOA2), cholesteryl ester transfer protein plasma (CETP), and transcription factor 7-like 2 (TCF7L2) gene polymorphisms have been demonstrated [81–84].

4.2. DASH Diet

A large body of evidence supports that adherence to a DASH dietary pattern is linked to improvements in BP [85], body weight [86], glucose-insulin homeostasis [87], blood lipids and lipoproteins [88], inflammation grade [89,90], endothelial function [91,92], the gut microbiome [93,94], CVD risk [95,96], and total mortality [97,98]. The DASH diet is characterized by a high intake of fruits and vegetables, legumes, low fat dairy, whole grain products, nuts, fish, and poultry; a reduced intake of saturated fat, red meat and processed meats, and sweet beverages; and a low intake of sodium and refined grains [89,99].

Focusing on inflammatory markers and oxidative stress, several studies have shown the protective effect of the DASH diet on CVD (Table 1). A recent systematic review and meta-analysis of randomized trials [89], which included six randomized control trials (RCT) with 451 participants who were followed for 3–24 weeks, studied the effect of the DASH diet on inflammatory biomarkers. Results showed that the DASH diet significantly reduced high hs-CRP concentrations (mean difference (MD) = −1.01, 95% confidence interval (CI): −1.64, −0.38; I-squared (I2) = 67.7%) compared to other diets. When the follow up of participants was longer, the reduction of hs-CRP serum levels was greater. However, when the DASH diet effect was compared to other healthy diets, no significant changes were observed. Besides, a meta-analysis conducted by Neale et al. [100] about 17 RCTs observed that following a healthy diet (MeDiet, Nordic diet, Tibetan diet, and DASH diet) was associated with a significant reduction of CRP levels (−0.75, 95% confidence interval (CI): −1.16, −0.35; p = 0.003). No changes were found for the other biomarkers (TNF-α, total adiponectin, high-molecular-weight adiponectin, adiponectin:leptin ratio, resistin, or retinol binding protein 4). Eichelmann et al. [101] studied the link between plant-based diets (Nordic diet, MeDiet, vegetarian diet, plant-based diet, Paleolithic diet, and DASH) and obesity-related pro-inflammatory markers (CRP, IL-6, TNF-α, sICAM-1, leptin, adiponectin, and resistin) on 29 interventional trials with a total of 2689 participants. Results showed improvements in obesity-related inflammatory profiles after following plant-based diets: CRP (−0.55 mg/L), IL-6 (−0.25 ng/L), and sICAM-1 (−25.07 ng/mL). No significant changes were observed for TNF-α, resistin, adiponectin, and leptin.

In a cross-sectional analysis of 1493 men and women (aged 50–69 years), potential associations between dietary quality (through the DASH dietary quality score), adiposity, and biomarkers of glucose metabolism, lipid profile, and inflammation were assessed [102]. Results showed that a higher adherence to the DASH dietary pattern was associated with improvements in adiposity measures (BMI, $p < 0.05$, waist circumference, $p < 0.001$), and lower concentrations of TNF-α, IL-6, CRP, WBC, and PAI-1 ($p < 0.05$; all), such as pro-inflammatory, pro-thrombotic, and pro-atherogenic markers. Improvements in lipoprotein profile parameters (LDL-c, HDL-c, and lower large very low density lipoprotein [VLDL] particles, $p < 0.001$ all) and glucose homeostasis biomarkers (HOMA, insulin and glucose, $p < 0.05$ all) were also shown.

With respect to interventional studies, the DASH diet has reported improvements on insulin resistance, inflammation, and oxidative stress in women with gestational diabetes [103]. The RCT was performed with 32 pregnant women diagnosed with gestational diabetes (GD) at 24- to 28-week gestation. All of them were randomly assigned to a DASH diet group or a control group (16 participants/group) and were followed up for four weeks. The DASH diet compared with the control showed significant reductions in serum insulin levels (−2.62 μIU/mL, p = 0.03), fasting plasma glucose (FPG) (−7.62 mg/dL, p = 0.02), and the homeostasis model of assessment-insulin resistance (HOMA-IR) score (−0.8, p = 0.03). Furthermore, Kawamura et al. [104] also reported significant reductions in the BMI, BP, FPG, and fasting insulin level ($p \leq 0.003$; all) after analyzing 58 Japanese participants with untreated high-normal BP or stage 1 hypertension (30 men and 28 women; mean age 54.1 ± 8.1 years), who followed a modified DASH diet (salt 8.0 g/day) during two months. Finally, the DASH diet also increased the plasma total antioxidant capacity (45.2 mmol/L, $p < 0.0001$) and total glutathione levels (108.1 μmol/L, $p < 0.0001$). Neither group showed changes in serum hs-CRP levels. Saneei et al. [105] conducted a cross-over study, which examined the effects of the DASH diet on markers of systemic inflammation in 60 post-pubescent girls with MetS (aged 11–18 years and weight mean was 69 kg). Participants were randomized into two groups: The DASH diet or usual dietary advice (control group) and were followed for six weeks. Results did not show significant changes on TNF-α, IL-2, IL-6, and adiponectin levels, whereas hs-CRP levels were significantly lower (−0.09 mg/L, $p = 0.002$) in those participants with higher adherence to a DASH diet compared to the control group.

5. Foods

5.1. Fruits and Vegetables

The European Society of Cardiology (ESC) and American Heart Association Nutrition Committee strongly endorse the daily consumption of multiple servings of both fruits and vegetables in order to reduce CVD risk [106,107]. These recommendations are based upon epidemiological studies and meta-analysis, mainly [106–113]. A recent meta-analysis [108] with 83 studies (71 clinical trials and 12 observational studies) showed that a higher intake of fruit or vegetable was significantly inversely associated with CRP and TNF-α levels ($p < 0.05$; both) and directly associated with an increased proliferation of γδ-T cell populations ($p < 0.05$). Also, Corley et al. [111] studied, in 792 participants aged 70 years from the Lothian Birth Cohort 1936, the association between biomarkers of systemic inflammation (such as CRP and fibrinogen) and specific single foods (fruits and vegetables). The dietary intake was measured using a 168-item Food Frequency Questionnaire (FFQ). Authors described that a higher fresh fruit intake was associated with lower CRP levels (≤3 mg/L) (β = 0.100, 95% CI 0.82, 0.99). No significant association was found between vegetables and CRP. Similar results ($p < 0.05$) were found between fibrinogen levels and fruit intake (β = 0.083) or combined fruits and vegetables intake (β = 0.084).

Also, in the cross-sectional study conducted by Holt et al. [112], in 285 healthy adolescent boys and girls aged 13 to 17 years, it was found that serum CRP levels were inversely associated with fruit intake (r = −0.19; $p = 0.004$), while IL-6 was inversely associated with fruit and vegetable intake and TNF-α only with vegetable consumption ($p < 0.05$; both). The HELENA Cross-Sectional Study [113], which aimed to demonstrate that a healthy diet might reduce adiposity and systemic inflammation, found that fruits and nuts were negatively linked with IL-4 (all subjects, $p < 0.05$; both) and TNF-α (only girls, $p = 0.036$). Contrastingly, vegetables showed only significant inverse correlations with sE-selectin (all subjects, $p \leq 0.0012$; both). This study was carried out in 464 adolescents (13–17 years) of the European HELENA cohort. In a cross-sectional analysis [114], in 1005 Chinese women aged 40 to 70 years, the association between vegetable intake and inflammatory and oxidative stress markers was studied. Results showed that a higher intake of cruciferous vegetables was associated with lower concentrations of TNF-α (p trend = 0.001), IL-1β (p trend = 0.004), and IL-6 (p trend = 0.02). Additionally, the mean difference of concentrations among the highest and the lowest quintiles of cruciferous vegetables intake were 12.66% for TNF-α, 18.18% for IL-1β, and 24.68% for IL-6. Any association was observed between the consumption of cruciferous vegetable and oxidative stress markers (F2-isoprostanes and 2,3-dinor-5,6-dihydro-15-F2t-IsoP).

Finally, in a sub-study from the PREDIMED study, Urpí-Sardà et al. [77] found that the participants who increased more than 62.7 g/day of their consumption of vegetables after one year decreased their plasma concentration of TNFR60 from 1.7 μg/L to 1.5 μg/L ($p < 0.05$), as shown in Table 2.

Table 2. Potential inflammatory effects of different foods on CVD.

	Pro- and Anti-Inflammatory Markers and Genes	Leukocyte Expression	Oxidative Stress Markers
Fruits & vegetables	TNF-α, TNFR-60, IL-1β, IL-4, IL-6, γδ-T cell, fibrinogen, sE-selectin	-	F2-isoprostanes, 2,3-dinor-5,6-dihydro-15-F2t-IsoP
Olive oil	sVCAM-1, sICAM-1, RANTES, MIP-1β, TNF-α, TNFR-60, IL-1β, IL-6, IL-7, IL-10, IL-12p70, IL-13, IL-18, MMP-9, CRP, MCP-1, NT-proBNP, NF-κβ	-	plasma antioxidant capacity, antioxidant enzymes-catalase, and glutathione peroxidase
Nuts	CRP, IL-6, TNF-α, TNF-β, TNF-R2, sICAM-1, fibrinogen, PF4, resistin	-	oxLDL
Wine and beer	IL-1α, IL-5, IL-6, IL-6r, IL-8, IL-10, IL-15, IL-18, CRP, MDC, sVCAM-1, sICAM-1, E-selectin, fibrinogen, CD40 ligand, MCP-1, factor VII, PAI-1, IFN-γ, RANTES, TNF-β	Lymphocytes: LFA-1 Monocytes: LFA-1, MAC-1, VLA-4, CCR2, CD36, CD15	SOD, MDA

CCR2: C-C chemokine receptor type 2; CD15: Sialil-Lewis X; CRP: C-reactive protein; IL: Interleukin; LFA: Lymphocyte function-associated antigen 1; MAC-1: T-lymphocytes and macrophage-1 receptor; MCP-1: Monocyte chemoattractant protein; MDA: Malondialdehyde; MDC: Macrophage-derived chemokine; MMP-9: Metallopeptidase-9; NT-proBNP: Pro-brain natriuretic peptide; oxLDL: Oxidized LDL; PAI-1: Plasminogen activator inhibitor 1; PF4: Platelet factor 4; sICAM-1: Soluble intercellular adhesion molecule 1; SOD: Superoxide dismutase; sVCAM-1: Soluble vascular cell adhesion molecule; TNF: Tumor necrosis factor; TNFR: Tumor necrosis factor receptor; IFN-γ: Interferon gamma.

5.2. Olive Oil

Several studies and meta-analyses have demonstrated the anti-inflammatory effects of olive oil (OO) rich diets [115–117]. Bioactive components of EVOO, the main key food of the MeDiet, have demonstrated improvements in inflammatory status, oxidative stress, and endothelial dysfunction [115]. A recent meta-analysis conducted by Schwingshackl et al. [118] in 30 RCT (3106 participants and daily consumption of 1 mg and 50 mg OO) found a significant decrease in CRP (−0.64 mg/L, $p < 0.0001$, n = 15 trials) and IL-6 (−0.29 mg/L, $p < 0.04$, n = 7 trials) compared to controls. Also, the flow-mediated dilatation (FMD) value was significantly increased in subjects with highest OO intake (0.6%, $p < 0.002$).

Several studies from the PREDIMED study and others related to the MeDiet have reported that a MeDiet supplemented with EVOO leads to decreased levels of N-terminal pro-brain natriuretic peptide (NT-proBNP) [119], the progression of intima media thickness (IMT) in those with elevated baseline IMT [120,121], improved systolic and diastolic BP in both hypertensive and non-hypertensive patients [122–124], and decreased expression and concentration of circulating inflammatory biomarkers related to atherosclerosis [63,64,71–74]. Urpí-Sardà et al. [72] reported that those participants with a higher adherence to a MeDiet + EVOO and whose intake of VOO was higher than 24 g/day showed lower plasma concentrations of TNFR60 after one year of intervention. Moreover, Camargo et al. [125] observed that, after isolation of peripheral blood mononuclear cells (PBMCs), the MeDiet exerted an inhibitory effect on the expression of genes related to plaque progression, such as MMP-9, NF-κβ or MCP-1, by increasing IκB kinase (IκBα) expression after the intake of a MeDiet + EVOO compared with two other diets ($p < 0.05$; all). IκBα stabilizes the NF-κβ molecule in the cytoplasm, maintaining it in an "unactivated" state. Widmer et al. [126] observed that daily consumption of 30 mL of OO or 30 mL of OO supplemented with epigallocatechin 3-gallate (EGCG) in 82 subjects (≥18-y) with early atherosclerosis showed improvements of endothelial dysfunction, independently of EGCG supplementation, after four months of intervention. However, the OO + EGCG group showed significant reductions in inflammatory parameters, such as sICAM-1 ($p \leq 0.001$), white blood cells ($p < 0.05$), monocytes, and lymphocytes ($p < 0.05$; both). Additionally, Oliveras-López et al. [127] showed an increase in plasma antioxidant capacity, antioxidant enzymes-catalase, and glutathione peroxidase, as well as an improvement in the superoxide dismutase (SOD) expression after analyzing

45 healthy adults (age: 21–45 years old, mean BMI: 21.4 ± 0.5 kg/m^2) who ingested 50 mL of EVOO for 30 days. Along with the other studies, the intake of EVOO seems to have positive effects on endothelial function. Finally, in the VOLOS (Virgin Olive Oil Study), participants with mild dyslipemia were randomized in two groups of intervention (daily 40 mL of EVOO with 166 mg/L of hydroxytyrosol vs. refined OO with 2 mg/L of hydroxytyrosol) during seven weeks. Results showed significant reduction of thromboxane B2 (TXB2) levels of 20% in the EVOO groups [128], as shown in Table 2.

5.3. Nuts

Nuts, specifically peanuts and walnuts, have been demonstrated to reduce the CVD morbidity and mortality in numerous large prospective cohort studies [129,130]. Mente et al. [131] predicted that nut intake might offer a preventative risk reduction on heart disease (RR = 0.67 [95% CI: 0.57–0.77]). Also, nut intake is associated with weight loss improvements [132], lower LDL-c levels [71,133,134], hypertension risk [133,135], and T2DM, improving hyperglycemia and insulin resistance [136], and inflammatory and oxidants mediators [134,137–139].

In a recent extensive meta-analysis, 23 RCTs were evaluated [139] to investigate the effects of nut intake over some inflammatory biomarkers (CRP, sICAM-1, sVCAM-1, IL-6, E-selectin, TNF-α). Authors found significant reductions of sICAM-1 (−0.17 ng/mL, p = 0.01) after nut intake. No changes were observed among the others inflammatory markers. Similar results were found in the meta-analysis conducted by Neale et al. [138]. No significant differences were shown after analyzing a wide number of inflammatory biomarkers, such as CRP, adiponectin, IL-6, sICAM-1, sVCAM-1, and FMD, in a total of 32 RCT studies. A significant improvement in FMD after nut intake was observed.

On the one hand, Yu et al. [137] conducted a cross-sectional study to investigate if nut intake was correlated to inflammatory biomarkers levels (CRP, IL-6, and tumor necrosis factor receptor 2 (TNFR2)) from 5013 participants in the Nurses' Health Study (NHS) and Health Professionals Follow-Up Study (HPFS) who were non-diabetic. Results showed that a higher nut intake showed lower inflammatory biomarkers levels (CRP: RR = 0.80 [95% CI: 0.69, 0.90, p-trend = 0.0003]; and IL-6: RR = 0.86 [95% CI: 0.77, 0.97, p-trend = 0.006]).

However, a randomized, parallel-group study on 50 patients with MetS and supplemented with 30 g/day of raw nuts (15 g walnuts, 7.5 g almonds, and 7.5 g hazelnuts) showed significant reductions of plasma concentrations of IL-6 (−1.1 ng/L; p = 0.035) compared with a control diet [140]. These results are according with the data showed by Hernández-Alonso et al. [141], who reported, in a randomized cross-over study in 54 participants, significant reductions of fibrinogen, oxLDL, and platelet factor 4 levels in the pistachio-supplemented group compared with control group ($p < 0.05$; all) after four months of intervention. Additionally, the pistachio-supplemented group showed lower IL-6 (−9%) and resistin gene expression (−6%) ($p < 0.05$; both). Similar results (−10.3% for IL-6 and CRP, −15.7% for TNF-α) were showed in a randomized, crossover study for 12-weeks carried out by Liu et al. [142] in 20 Chinese subjects with T2DM with mild hyperlipidemia (nine men and 11 women, mean age of 58 years, and BMI of 26 kg/m^2). See Table 2.

5.4. Wine and Other Fermented Alcoholic Beverages

Nowadays, there is enough evidence from both epidemiologic studies and RCTs to conclude that regular moderate consumption of fermented alcoholic beverages, mainly red wine and beer, has cardioprotective effects and can exert a positive effect on CV risk factors [143–145].

5.4.1. Wine

Wine or wine-derived phenolic compounds that exert effects through mechanisms on atherosclerosis are clearly identified. On the one hand, wine and their phenolic compounds decrease oxidation of LDL-c and oxidation stress, and increase in NO, improving endothelial function. Also, ethanol increases HDL-c levels and inhibits platelet aggregation, promotes fibrinolysis, and reduces systemic inflammation [144,146].

Janssen et al. [147] investigated the relationship of wine consumption and CV risk markers (CRP, fibrinogen, factor VII, and PAI-1) in a multi-ethnic sample of 2900 healthy women of middle-age, who were followed up for seven years. Authors concluded that moderate wine consumption may protect against CVD, after observing lower concentrations of CRP ($p < 0.001$), fibrinogen ($p < 0.001$), factor VII ($p < 0.01$), and PAI-1 ($p < 0.05$) compared to abstainers or women that drink little wine.

On the other side, Estruch et al. [148] reported that both red wine and gin have anti-inflammatory properties in the atherosclerotic process through the reduction of fibrinogen levels (−9%) and IL-1α (−21%), as well as lower plasma hs-CRP (−21%), sVCAM-1 (−17%), and sICAM-1 (−9%) levels. Moreover, monocyte expression was significantly reduced by 27–32% (LFA-1, MAC-1, VLA-4). In another randomized, crossover consumption trial in 67 males at high risk of CVD, Chiva-Blanch et al. [149] investigated the effects of ethanol and phenolic compounds of 30 g alcohol/day of red wine, and the equivalent amount of dealcoholized red wine and gin (30 g alcohol/d) for four weeks on the expression of inflammatory biomarkers related to atherosclerosis. Thirty g/day of alcohol of red wine showed an increase in plasma concentrations of IL-10 and decreased macrophage-derived chemokine (MDC). On the other hand, sICAM-1, E-selectin, and IL-6 were reduced by phenolic compounds of red wine. Phenolic compounds also inhibited the expression of LFA-1 in T-lymphocytes and MAC-1, and CCR2 expressions in monocytes. Concentrations of CD40 antigen, CD40 ligand IL-16, MCP-1, and sVCAM-1 were downregulated in both groups: Ethanol and phenolic compounds of red wine. A current study conducted by Roth et al. [150] found that aged white wine presents a greater ability to repair and maintain endothelial integrity than gin. In this randomized, controlled, crossover study, 38 high-risk male volunteers (55–80 years), who received 30 g ethanol/day as aged white wine or gin for three weeks, were evaluated. After intervention, T-lymphocytes' expression of CD31 and CD40 and monocytes expression of CCR2 and CD36 were lower after consumption of aged white wine. Additionally, for aged white wine, a significant reduction was observed in plasma concentrations of IL-8 and IL-18, sICAM-1 and sVCAM-1. Both beverages showed significant reductions in LFA-1, MAC-1, VLA4, CD40, and CD31 expression, as well as lower concentrations of interferon gamma (IFN-γ). Finally, Estruch et al. [151], in a new study, where wine and gin were compared again, found that wine intake significantly decreased plasma SOD activity [8.1 U/gHb (95% CI: 138, 25; $p = 0.009$)] and MDA levels [11.9 nmol/L (95% CI: 21.4, 2.5; $p = 0.020$)] compared to the gin group, as shown in Table 2.

5.4.2. Beer

Among fermented beverages, beer has a moderate polyphenol content that confers grater cardioprotective effects than spirits and distilled beverages [145]. So, De Gaetano et al. [152] described in a consensus document the effects of moderate beer consumption on health and disease, where they concluded that epidemiological studies showed that low-moderate doses of beer intake protect against CV risk and its effect is comparable to that reported for moderate red wine consumption.

In 1999, Wannamethee et al. [153] studied 7735 British men during 17 years and found that regular beer intake was associated with lower total mortality (HR = 0.84 (CI: 0.71–10.01)).

Finally, Chiva-Blanch et al. [154] performed a randomized, crossover controlled clinical trial with 33 subjects to evaluate the effects of three beverages types: A non-phenolic alcoholic beverage, such as gin; beer, which is a moderate phenolic alcoholic beverage; and a non-alcoholic beer, with the same amount of phenolic compounds. Beer and gin showed improvements in the HDL cholesterol levels (around 5%) and adiponectin (around 7%) compared to the non-alcoholic beer intervention. Related to circulating inflammatory biomarkers, IL-1α levels increased (around 24%) and IL-5 levels decreased around 14% after beer and gin intervention. However, non-alcoholic beer showed significant improvements in homocysteine concentration (decreased by around 6%) and improved folic acid levels around 9%. Related to inflammatory biomarkers, non-alcoholic beer intervention showed significant decreases of IL-6r, IL-15, RANTES, and TNF-β levels, as shown in Table 2.

6. Nutrients

It is important to focus on the possible benefits of the intake of specific nutrients to avoid possible deficiencies of these nutrients, which can lead to the development of atherosclerotic disease. We have only included information about fiber, some vitamins, and minerals, but no other nutrients—such as carbohydrates, fats, or proteins—which have also been demonstrated to have a certain effect on the risk of developing atherosclerosis.

6.1. Fiber

A wide number of studies and scientific publications have reported the health benefits of dietary fiber intake decreasing cholesterol concentrations and BP, while a deficiency of fiber intake is associated with CVD development [155].

On the one hand, several meta-analyses have displayed that a higher dietary fiber intake is linked with a lower relative risk of total all-cause mortality among 16–23% [156–158]. On the other hand, several clinical trials have studied the link between diet and inflammation, and more specifically, the impact of dietary fiber. Although, to date, the implicated mechanisms are still unknown, the proposed mechanisms are that dietary fiber decreases the glucose absorption, and down-regulates the expression of oxidative stress related cytokines or the inflammatory response mediated by gut microbiota exposed to fiber [159].

In an observational study of 1958 postmenopausal women (age 50–79 years), dietary fiber consumption was associated with higher levels of inflammatory markers (CRP and IL-6) [160]. Also, in the Women's Health Initiative Observational Study (13,745 US men and women), a higher fiber intake (24.7 g/day) was associated with lower plasma concentrations of IL-6 and TNFR2 compared with the lowest fiber intake group (7.7 g/day) [161]. Similar results were expressed by Qi et al. [162], who observed that concentrations of CRP and TNFR2 were among 8% to 18% lower in the highest quintile of cereal fiber intake compared to the lowest quintile. Similar results were obtained by Estruch et al. [163] for CRP levels and other inflammatory cytokines (IL-6, sICAM-1, sVCAM-1), whose decrease was inversely related with dietary fiber intake, but not significantly. Additionally, cross-sectional data (1088 participants without T2DM at baseline and aged 40 y–60 y) from the Insulin Resistance Atherosclerosis Study [164] showed that whole grain products' intake was inversely related to PAI-1 ($\beta = -0.102$; SEM = 0.038; $p = 0.0077$) and CRP plasma concentrations ($\beta = -0.102$; SEM = 0.048; $p = 0.0340$).

In interventional studies, North et al. [165] studied, in 554 subjects (192 men, 362 women), the associations between dietary fiber and CRP levels, showing significantly lower CRP concentrations (−25–54%) with higher fiber intakes (≥3.3 g/MJ). An interventional cross-over study [166] with a total of 60 participants (50% patients with newly diagnosed T2DM, 50% nondiabetic subjects) received three isoenergetic meals separated by one week intervals: A high-fiber (16.8 g) meal; a high-fat meal; and a low-fiber (4.5 g) meal. Results showed that a high fiber intake showed lower plasma IL-18 concentrations and greater stimulation of adiponectin plasmatic levels. Finally, significant reductions of TNF-α (−3.7 pg/mL; $p < 0.001$), were observed after whole grain product intake in a randomized parallel arm feeding trial in 49 subjects who were overweight or obese, and low fruits, vegetables, and whole grain products' intake [167]. See Table 3.

Table 3. Potential inflammatory effects of different nutrients on CVD.

	Pro- and Anti-Inflammatory Markers and Genes
Fiber	sVCAM-1, sICAM-1, TNF-α, TNFR2, IL-6, IL-18, CRP, PAI-1
Micronutrients	IL-6, CRP, TNF-α, leptin, tHcy

CRP: C-reactive protein; IL: Interleukin; PAI-1: Plasminogen activator inhibitor 1; sICAM-1: Soluble intercellular adhesion molecule 1; sVCAM-1: Soluble vascular cell adhesion molecule; tHcy: Total homocysteine; TNF-α: Tumor necrosis factor; TNFR: Tumor necrosis factor receptor.

6.2. Micronutrients

Nowadays, there are great experimental, epidemiological, and clinical evidence showing how micronutrients' ingestion may protect against CVD [168–170]. Micronutrients exert their protective effect through three possible ways: 1. Reducing endothelial cells damage; 2. improving the production of NO; and 3. inhibiting oxidation of LDL-c [168–170]. Both in adolescence and in adulthood, pro-inflammatory biomarkers have been associated with dietary antioxidants, such as Zn, Se, and vitamin C and E, whose deficiency leads to a higher CVD risk [171–175]. Also, a meta-analysis [176] suggested that Mg supplementation might significantly reduce serum CRP levels (−1.33 mg/L, 95% CI: −2.63, −0.02) after analyzing eight RCTs. Similar results were showed in another meta-analysis [177], where after stratifying by the baseline plasma CRP values of ≤3 and >3 mg/L, found significant reductions of CRP levels (1.12 mg/L, 95% CI: −2.05, −0.18, $p = 0.019$) for the last subgroup. Finally, a recent meta-analysis [178] of seven studies (all RCTs) showed that vitamin D-supplemented groups had lower levels of TNF-α ($p = 0.04$) compared with control groups. No differences between vitamin D and control groups were observed for CRP, IL-10, or IL-6. Relative to vitamin E, a recent meta-analysis [179] suggested that supplementation with vitamin E might reduce serum CRP levels (−0.62 mg/L, 95% CI = −0.92, −0.31; $p < 0.001$) after analyzing 12 trials with 246 participants in the intervention arms and 249 participants in control arms.

De Oliveira Otto et al. [180] investigated the association between dietary micronutrients (heme iron, nonheme iron, zinc (Zn), magnesium (Mg), β-carotene, vitamin C, and vitamin E) with inflammatory markers (CRP, IL-6, total homocysteine (tHcy), fibrinogen, coronary artery calcium, and common and internal carotid artery-IMT) and subclinical atherosclerosis in 5181 participants free of diabetes and CVD from the Multi-Ethnic Study of Atherosclerosis (aged 45 y–84 y). Authors found that Mg and nonheme iron were inversely associated with tHcy concentrations, whilst vitamin C was positively associated with tHcy concentrations. Besides, CRP levels were positively associated with Zn and heme iron, whereas Mg concentrations showed an inverse association with CCA-IMT. Finally, no significant association was observed between dietary intake of carotene or vitamin E and inflammatory markers. Wang et al. [181] also reported that serum vitamin D levels were negatively associated with IL-6 ($r = -0.244$, $p = 0.002$) and hs-CRP ($r = -0.231$, $p = 0.004$) levels after studying 152 acute stroke patients. Furthermore, for vitamin D, several observational studies have reported that reduced vitamin D levels are linked with endothelial dysfunction and higher arterial stiffness [182,183], and this deficiency might be related to foam cell formation and decreased expression of adhesion molecules in endothelial cells [184].

Recently, Tabesh et al. [185] examined the effects of co-supplementation of vitamin D and calcium on inflammatory biomarkers and adipokines in 118 diabetic participants with insufficient vitamin D levels. The placebo-controlled clinical trial after eight weeks with four intervention groups ((1) vitamin D + calcium placebo; (2) calcium + vitamin D placebo; (3) vitamin D + calcium; or (4) vitamin D placebo + calcium placebo) showed that supplementation with calcium and vitamin D decreased leptin (−9 ± 18 ng/mL), IL-6 (−4 ± 1 pg/mL, $p < 0.001$), and TNF-α (−3.4 ± 1.3, $p < 0.05$) concentrations compared with the placebo. Also, Shargorodsky et al. [186] studied the effect of vitamin C (500 mg), vitamin E (200 IU), coenzyme Q10 (60 mg), and selenium (120 μg) on inflammatory markers in the long-term (six months) in participants at high CVD risk. No significant changes were observed for homocysteine, endothelin, aldosterone, and renin in participants who received antioxidant supplementation, while there was a significant decrease in HbA1C and a significant increase in HDL-c. Large and small artery elasticity was also significantly increased after antioxidant supplementation intake. In addition, an RCT conducted by Ellulu et al. [187] in 64 obese patients, who were hypertensive and/or diabetic, reported the potential anti-inflammatory effect of 500 mg of vitamin C, twice daily. Vitamin C might decline hs-CRP ($p = 0.01$), IL-6 ($p = 0.001$), and fasting blood glucose ($p < 0.01$) after eight weeks of treatment. Christen et al. [188] also conducted a randomized, double-blind, placebo-controlled sub-study from the Women's Antioxidant and Folic Acid Cardiovascular Study. They tested a daily combination of folic acid (2.5 mg), vitamin B6 (50 mg), vitamin B_{12} (1 mg), or placebo on 300 participants (half for each group). After 7.3 years, the supplemented group showed significant reductions in homocysteine concentrations (−18%), whereas no changes were observed in CRP, IL-6, sICAM-1, and fibrinogen levels, as shown in Table 3.

7. Bioactive Compounds

Multiple bioactive compounds (omega-3 fatty acids, lycopene, or polyphenols) present in the diet have been associated with beneficial effects on atherosclerosis development. All of them act to reduce levels of LDL-c, improving inflammatory and oxidative stress biomarkers. Next, we analyze those cited above.

7.1. Omega-3 Fatty Acids

PUFAs, as Omega-3 fatty acid (Ω-3 PUFA), α-linolenic acid (ALA), eicosapentaenoic acid (EPA), and docosahexaenoic acid (DHA), have been reported as potential anti-atherogenic agents for the atherosclerotic process [189]. Mechanisms, through which they might reduce CV risk, include improvements in the lipid and lipoprotein profile, oxidation, thrombosis, endothelial function, BP, plaque stability, CV mortality, platelet aggregation, modulating concentration or expression of pro-inflammatory markers (adhesion molecules, cytokines, etc.), and immune cells [190–192].

In a meta-analysis conducted by Wang et al. [193] of 16 randomized placebo-controlled trials in 901 participants, it was reported that Ω-3 PUFA intake (0.45–4.5 g/day, for 56 days) increased FMD by 2.30% (95% CI: 0.89, 3.72%, $p = 0.001$) compared with the placebo group. Additionally, a meta-analysis of 38 RCTs [194] reported reductions by 20–30% in serum triglycerides levels in healthy participants after daily consumption of ≥4 g of EPA and DHA through either supplementation or consumption of enriched foods.

In an observational study [195], 102 Japanese individuals with acute coronary syndrome were analyzed and were stratified in three groups: ≤50, 51–74, and ≥75 years. It was found that low serum DHA concentrations leads to CVD, with DHA useful as a predictive marker of endothelial dysfunction. Similar results were reported by Kelley et al. [196].

On the one hand, Cawood et al. [197] showed that patients in the Ω-3 PUFA group (1.8 g EPA + DHA/day), followed up for 21 days, had a lower number of foam cells ($p = 0.0390$) and T-lymphocytes ($p = 0.0097$), less inflammation ($p = 0.0108$), and improved stability of atheroma plaque ($p = 0.0209$) after analyzing data obtained from a randomized placebo-controlled trial. On the other hand, patients who received Ω-3 PUFAs showed lower expression of mRNA for MMP-7 ($p = 0.0055$), -9 ($p = 0.0048$), -12 ($p = 0.0044$), and for IL-6 ($p = 0.0395$) and sICAM-1 ($p = 0.0142$). Similar results were reported by Thies et al. [198], where after administering dietary fish oil supplements (1.4 g EPA + DHA/day) in patients with advanced atherosclerotic plaque, less inflammation, inhibition of macrophages and lymphocytes infiltration and an increase in plaque stability was observed. Besides, several RCTs have pointed out that Ω-3 PUFAs might modulate the expression of cell adhesion molecules (sICAM-1, sVCAM-1, or sP-selectin) as well as CRP, IL-1β, IL-6, serum amyloid A (SAA), TNF-receptor concentrations, TNF-α, or PAI-1 levels among others [195,199], as shown in Table 4.

Table 4. Potential inflammatory effects of different bioactive compounds on CVD.

	Pro- and Anti-Inflammatory Markers and Genes	Leukocyte Expression
Ω-3 PUFA	sVCAM-1, sICAM-1, sP-selectin, TNF-α, TNFR, IL-1β, IL-6, MMP-7, MMP-9, CRP, PAI-1, SAA	T-lymphocytes
Lycopene	sVCAM-1, IL-6, IL-10, IL-18, MCP-1, tHcy, PAI-1	T-lymphocytes: LFA
Phytosterols	IL-1β, IL-6, CRP	-

CRP: C-reactive protein; IL: Interleukin; LFA: Lymphocyte function-associated antigen 1; MCP-1: Monocyte chemoattractant protein; MMP: Metallopeptidase; PAI-1: Plasminogen activator inhibitor 1; SAA: Serum amyloid A; sICAM-1: Soluble intercellular adhesion molecule 1; sVCAM-1: Soluble vascular cell adhesion molecule; tHcy: Total homocysteine; TNF-α: Tumor necrosis factor; TNFR: Tumor necrosis factor receptor.

7.2. Lycopene

Lycopene is a lipophilic and an unsaturated carotenoid, present in red-colored fruits and vegetables, such as tomatoes, papaya, or watermelons among others. Epidemiological observational and interventional studies [200,201] suggest that lycopene might reduce atherosclerotic risk, particularly in early stages of atherosclerosis, preventing endothelial dysfunction (NO bioavailability and blood flow) and LDL oxidation. Others mechanisms through which lycopene might exert effects is improvement of the metabolic profile (by impairing cholesterol synthesis) and BP, through reductions in arterial stiffness, and modulation of the expression of pro-inflammatory markers and platelet aggregation [202]. Furthermore, dietary lycopene confers CV benefits and significant reduction in CV mortality and major CV events in postmenopausal women free of CVD or cancer [202]. Focusing on the risk of develops atherosclerosis, several studies have pointed out lycopene's antioxidant power as a possible mechanism to explain its health benefits [203].

Additionally, in a recent meta-analysis [204], dietary interventions supplemented with tomatoes significantly decreased LDL-c (−0.22 mmol/L; $p = 0.006$), IL-6 (−0.25; $p = 0.03$) and improved FMD by 2.53% ($p = 0.01$), while lycopene supplementation reduced SBP (−5.66 mmHg; $p = 0.002$). In another study with 40 participants with heart failure [205] (lycopene intervention, 29.4 mg/day of lycopene vs. control group), CRP levels decreased significantly in the intervention group, but only in women ($p = 0.04$).

Dietary data collected from the National Health and Nutrition Examination Survey (NANHES) 2003–2006 [206] showed significant inverse associations with tHcy and CRP for dietary lycopene intake ($p < 0.05$).

Therefore, Valderas-Martinez et al. [201] investigated the postprandial effects of a single dose of raw tomatoes (RT), tomato sauce (TS), and tomato sauce with refined olive oil (TSOO) on CVD. In this randomized, cross-over, controlled feeding trial in 40 subjects free of CVD, authors found that tomato intake significantly decreased some inflammatory biomarkers levels, such as LFA-1, IL-6, IL-18, MCP-1, and VCAM-1, and increased plasma IL-10 levels. In another interventional study [207], with 80 subjects, 40 early atherosclerosis cases, and 40 control subjects, the authors pointed out that serum carotenoids concentrations are linked with the risk of atherosclerosis development. They observed that serum lutein was negatively associated with IL-6 ($p < 0.001$) and directly associated with IFN-γ ($p = 0.002$). Moreover, zeaxanthin was inversely associated with VCAM-1 ($p = 0.001$) and apolipoprotein E ($p = 0.022$) levels, while lycopene was inversely associated with sVCAM-1 ($p = 0.011$) and LDL ($p = 0.046$). However, in a single-blind, randomized controlled intervention trial [208] with healthy volunteers (94 men and 131 women, aged 40 y–65 y), no changes were observed for inflammatory markers (oxLDL, sICAM-1, and IL-6), insulin resistance, and sensitivity markers after 12 weeks of dietary intervention, as shown in Table 4.

7.3. Phytosterols

A large body of scientific evidence has concluded that a daily dose of 2–3 g of plant sterols or phytosterols is associated with an LDL-c reduction of around 6–15% of the total concentration [209,210]. These reductions were also observed in a meta-analysis conducted by Demonty et al. [211], where after administering a daily dose of 2.15 g of phytosterols, LDL-c was reduced by 8.8%. In fact, plant sterols have been proposed as a complement of statins treatment in order to decrease the risk of CVD. However, available data is inconsistent, so more research is needed. In this other meta-analysis [212] of 20 RCTs with 1308 participants, the effect of phytosterols intake on pro-inflammatory markers was evaluated. Significant reductions of CRP levels (−0.10 mg/dL) were observed after plan sterols' intake.

In addition, clinical studies have evaluated the association between phytosterols' consumption and inflammatory markers, such as CRP and cytokines. Although the results showed by Ras et al. [213] are in aggreement with the data reported by Demonty, no changes in CRP levels were observed. In a new study conducted by Ras et al. [209], in 240 hypercholesterolaemic voluntaries who consumed a low-fat spread with added phytosterols (3 g/day) for 12 weeks, no changes in any of the markers evaluated (CRP, SAA, IL-6, IL-8, TNF-α, and sICAM-1) were observed. Devaraj et al. [214] described significant reductions of IL-6 and IL1β levels after the intake of an orange juice-based beverage fortified with sterols (2 g sterols/day). In addition, results of a double-blinded, randomized, crossover trial [215] with 58 hypercholesterolemic participants during 12 weeks of intervention with margarine supplemented with phytosterols (3 g/day) did not show changes in CRP, IL-6, or TNF-α levels, as shown in Table 4.

8. Polyphenols

Polyphenols are the most abundant dietary antioxidants present in most plant origin foods and beverages, which possess a wide range of health effects in the prevention of CVD [216]. The most relevant food sources are fruit and vegetables, red wine, black and green tea, coffee, EVOO, and chocolate, as well as nuts, seeds, herbs, and spices [217].

Numerous scientific reports accumulated in the last years suggest that polyphenols might exert their positive effects by delaying progression of atherosclerosis through several mechanisms: Regulation of signaling and transcription pathways, such as NF-κβ; antioxidant systems; prevention of leukocyte migration and later infiltration inside plaque; reduction of adhesion molecules levels; inhibition of the encoding of pro-inflammatory cytokines; reduction of BP because of the enhanced NO production; and improvements of lipid metabolism, coagulation activity, and endothelial function, among others [218,219].

Several epidemiological studies have reported a negative association between consumption of polyphenols or polyphenols-rich foods and CVD [220–222]. A meta-analysis [223], including 14 prospective cohort studies with 1,279,804 participants and 36,352 CVD cases, showed that moderate consumption of coffee (three to five cups/day) was associated with a lower CVD risk compared with non-consumers. Similar results were found in the meta-analysis conducted by Larsson et al. [224]. Additionally, in another meta-analysis (14 prospective studies, with 513,804 participants and a median of follow-up of 11.5 years), it was found that a daily intake of ≥3 cups of tea was associated with a lower risk of stroke (−13%) and ischemic stroke (−24%) [225].

Focusing on the action of polyphenols on endothelial function, a large number of studies that have tested red wine, grape juice, black tea, soy, or dark cocoa showed an increase in FMD [226]. Besides, a daily consumption of dark chocolate (50 g) was associated with an improvement of FMD by 3.99% in acute and 1.45% in chronic intake and reduced systolic (−5.88 mm Hg) and diastolic BP (−3.30 mm Hg) [227]. Additionally, and contrary to the expected, acute consumption of black tea was associated with an increase of SBP (5.69 mm Hg) and DBP (2.56 mm Hg) while FMD was increased by 3.40%. Contrary to black tea, green tea led to a significant reduction in LDL-c (−0.23 mmol/L) [227]. Furthermore, alcohol and red wine moderate consumption were associated with an increase of FMD after analyzing 801 individual foods with food frequency questionnaires (FFQs) from the 2000 Hoorn Study (women = 399; age 68.5 ± 7.2 years) [228].

So, several meta-analyses of RCTs [229,230] have reported significant reductions of LDL-c, SBP, fasting glucose, BMI, hemoglobin A1c, or TNF-α levels, and significant increments of HDL-c.

Additionally, hs-CRP, IL-1β, and sP-selectin levels were significantly decreased after anthocyanin extract interventions. A larger cross-sectional study (BELSTRESS) with observational data from 1031 healthy men (49 years on average) found that drinking tea might reduce the inflammatory processes underlying CVD, as tea intake leads to lower levels of CRP ($p = 0.004$), SAA ($p = 0.001$), and haptoglobin ($p = 0.02$) [231].

Anti-inflammatory and immune-modulating effects of polyphenols have also been suggested as a potential pathway for polyphenols' health benefits. Several RCTs showed the relationship between cocoa polyphenols and inflammatory biomarkers related to atherosclerosis disease progression. In this sense, Vazquez-Agell et al. [232] suggested that the acute consumption of 40 g of cocoa with water might inhibit the NF-κβ transcription and down-regulate adhesion molecules' production (sICAM-1 and sE-selectin). Moreover, Monagas et al. [233], in a cross-interventional trial, showed that cocoa powder intake led to a reduced expression of adhesion molecules on monocyte surfaces (VLA-4, CD40, and CD36, $p \leq 0.028$; all) and lower serum levels of soluble adhesion molecules (sP-selectin and sICAM-1; both $p = 0.007$) in 42 subjects (mean age 69.7 years) at high risk of CVD and after four weeks of intervention. Additionally, Basu et al. [234] found that green tea consumption, as a beverage or extract, did not alter inflammatory biomarkers (adiponectin, CRP, IL-6, IL-1β, sVCAM-1, sICAM-1, leptin, or leptin:adiponectin ratio) related to MetS after eight weeks of intervention. Only SAA was significantly decreased after green tea beverage and extracts' intake ($p < 0.005$). Also, Zhang et al. [235] reported a significant decrease of CXCL7 by 12.32%, CXCL5 by 9.95%, CXCL8 by 6.07%, CXCL12 by 8.11%, and CCL2 levels by 11.63% after 320 mg intake of purified anthocyanins during 24 weeks. Similar results were observed after administering 320 mg of purified anthocyanins in 150 hypercholesterolemic patients during 24 weeks [236]. In this case, significant reductions of β-thromboglobulin, sP-selectin, and RANTES (regulated on activation, normal T cell expressed and secreted) were observed. Anthocyanins have also been correlated with lower levels of oxidative stress biomarkers. In an interventional study with 42 overweight and smoker participants, significant reductions in oxLDL and 8-iso-prostaglandin F2α were observed after the supplementation with an extract of maqui berry (162 mg anthocyanins) for 40 days [237]. Isoflavones and stilbens (mainly resveratrol) have also shown anti-inflammatory and immunomodulatory effects. Isoflavones improved CRP concentrations of 117 healthy European postmenopausal women after 50 mg/day of isoflavones intake [238]. Other RCT on postmenopausal American women who received daily 25 g of soy protein supplementation, showed reductions of subclinical atherosclerosis progression by 16% as well as reductions of carotid thickness progression by a mean of 68% [239]. Relative to stilbens, Tomé-Carneiro et al. [240] found significant reductions of oxLDL, apolipoprotein B (ApoB), and LDL-c after analyzing the effect of the intake of a grape supplement containing 8 mg resveratrol for six months. Also, a significant reduction of inflammatory markers levels (IL-18, sICAM-1, and sVCAM-1; $p \leq 0.037$; all) related with atheroma plaque development was observed in 44 healthy participants who ingested the resveratrol extract during four weeks [241]. Additionally, a randomized, double-blind, placebo-controlled clinical trial conducted by Seyyedebrahimi et al. [242] with 48 diabetics type 2 participants and supplemented with 800 mg/day of resveratrol for eight weeks showed a reduction in plasmatic levels of protein carbonyl content and ROS in PBMCs. Furthermore, after resveratrol supplement intake, an increase in the expression of Nrf2 and SOD was observed, as shown in Table 5.

Table 5. Potential inflammatory effects of polyphenols on CVD.

	Pro- and Anti-Inflammatory Markers and Genes	Leukocyte Expression	Oxidative Stress Markers
Polyphenols	NF-κβ, sICAM-1, sE- and sP-selectin, IL-1β, IL-18, CRP, SAA, CXCL5, CXCL7, CXCL8, CXCL12, CCL2, TNF-α, β-thromboglobulin, RANTES, ApoB	Monocytes: VLA-4, CD40, CD36	oxLDL, 8-iso-prostaglandin F2α, ROS, SOD, Nrf2

ApoB: Apolipoprotein B; CRP: C-reactive protein; CXCL: Chemokine (C-X-C motif) ligand; IL: Interleukin; NF-κβ: Nuclear factor kappa B; Nrf2: Nuclear factor (erythroid-derived 2)-like 2; oxLDL: Oxidized LDL; RANTES: Regulated on activation, normal T cell expressed and secreted; ROS: Reactive oxygen species; SAA: Serum amyloid A; sICAM-1: Soluble intercellular adhesion molecule 1; SOD: Superoxide dismutase; TNF-α: Tumor necrosis factor.

9. Conclusions

We have shown the intimate relationship between nutrition and CVD. Thus, the challenge is in promoting healthy dietary habits as well as an active lifestyle as early as possible in children and young adults. The evidence favors consumption of healthy dietary patterns, such as the Mediterranean diet or DASH diet, against other unhealthy dietary patterns, such as the Western diet, based on a high consumption of salt, added sugars, and saturated and trans-fats. Despite the fact that strong evidence shows the potential health benefits of a great amount of foods, nutrients, bioactive compounds, and dietary antioxidants, such as polyphenols, may exert on CV risk factors or directly on CVD development, it is necessary to conduct more interventional studies with a higher number of cases and longer follow up. To date, a lot of results obtained have produced few conclusions and sometimes, even contradictions. Therefore, due to a lack of information about possible mechanisms implicated in the cardioprotective effect of diet, foods, nutrients, or bioactive compounds, this needs to be more investigated.

Author Contributions: Conceptualization, R.C. and E.S.; Methodology, R.C. and R.E.; Investigation, R.C. and S.C.-B.; Writing—Original Draft Preparation, R.C. and S.C.-B.; Writing—Review & Editing, R.C. and R.E.; Visualization, E.S.; Supervision, E.S.; project administration, E.S.; funding acquisition, R.E.

Funding: This research was funded by the Instituto de Salud Carlos III, Spain, grant number PIE14/00045; by the Instituto de Salud Carlos III, Spain, grant number PI044504; by Fundació la Marató de TV3, Spain, grant number PI044003; and the Sociedad Española de Medicina Interna (SEMI), spain, grant number DN40585.

Acknowledgments: This research was funded by the Instituto de Salud Carlos III, Spain. CIBER OBN is an initiative of the Instituto de Salud Carlos III, Spain.

Conflicts of Interest: R.E. reports serving on the board of and receiving lecture fees from the Research Foundation on Wine and Nutrition (FIVIN); serving on the boards of the Beer and Health Foundation and the European Foundation for Alcohol Research (ERAB); receiving lecture fees from Cerveceros de España and Sanofi-Aventis; and receiving grant support through his institution from Novartis. The funders had no role in the design of the study; in the collection, analyses, or interpretation of data; in the writing of the manuscript, or in the decision to publish the results. No other potential conflict of interest relevant to this article was reported.

References

1. Benjamin, E.J.; Blaha, M.J.; Chiuve, S.E.; Cushman, M.; Das, S.R.; Deo, R.; de Ferranti, S.D.; Floyd, J.; Fornage, M.; Gillespie, C.; et al. Heart Disease and Stroke Statistics-2017 Update: A Report from the American Heart Association. *Circulation* **2017**, *35*, e146–e603. [CrossRef] [PubMed]
2. Ravera, A.; Carubelli, V.; Sciatti, E.; Bonadei, I.; Gorga, E.; Cani, D.; Vizzardi, E.; Metra, M.; Lombardi, C. Nutrition and cardiovascular disease: Finding the perfect recipe for cardiovascular health. *Nutrients* **2016**, *8*, 363. [CrossRef] [PubMed]
3. Mozaffarian, D.; Benjamin, E.J.; Go, A.S.; Arnett, D.K.; Blaha, M.J.; Cushman, M.; Das, S.R.; de Ferranti, S.; Despres, J.P.; Fullerton, H.J.; et al. Heart disease and stroke statistics-2016 update: A report from the American Heart Association. *Circulation* **2016**, *133*, e38–e360. [CrossRef] [PubMed]
4. Benjamin, E.J.; Virani, S.S.; Callaway, C.W.; Chamberlain, A.M.; Chang, A.R.; Cheng, S.; Chiuve, S.E.; Cushman, M.; Delling, F.N.; Deo, R.; et al. Heart Disease and Stroke Statistics-2018 Update: A report from the American heart association. *Circulation* **2018**, *137*, e67–e492. [CrossRef] [PubMed]
5. Artinian, N.T.; Fletcher, G.F.; Mozaffarian, D.; Kris-Etherton, P.; Van Horn, L.; Lichtenstein, A.H.; Kumanyika, S.; Kraus, W.E.; Fleg, J.L.; Redeker, N.S. Interventions to promote physical activity and dietary lifestyle changes for cardiovascular risk factor reduction in adults: A scientific statement from the American Heart Association. *Circulation* **2010**, *122*, 406–441. [CrossRef] [PubMed]
6. GBD 2013 Mortality and Causes of Death Collaborators. Global, regional, and national age-sex specific all-cause and cause-specific mortality for 240 causes of death, 1990–2013: A systematic analysis for the global burden of disease study 2013. *Lancet* **2015**, *385*, 117–171. [CrossRef]
7. Anand, S.S.; Hawkes, C.; de Souza, R.J.; Mente, A.; Dehghan, M.; Nugent, R.; Zulyniak, M.A.; Weis, T.; Bernstein, A.M.; Krauss, R.M. Food consumption and its impact on cardiovascular disease: Importance of solutions focusedon the globalized food system: A report from the workshop convened by the world heart federation. *J. Am. Coll. Cardiol.* **2015**, *66*, 1590–1614. [CrossRef]
8. Lacroix, S.; Cantin, J.; Nigam, A. Contemporary issues regarding nutrition in cardiovascular rehabilitation. *Ann. Phys. Rehabil. Med.* **2017**, *60*, 36–42. [CrossRef]
9. O'Rourke, K.; Vander Zanden, A.; Shepard, D.; Leach-Kemon, K. For the Institute for Health Metrics and Evaluation. Cardiovascular disease worldwide, 1990–2013. *JAMA* **2015**, *314*, 1905. [CrossRef]
10. Mozaffarian, D.; Ludwig, D.S. Dietary guidelines in the 21st century: A time for food. *JAMA* **2010**, *304*, 681–682. [CrossRef]
11. Libby, P. Interleukin-1 beta as a target for atherosclerosis therapy: Biological basis of CANTOS and beyond. *J. Am. Coll. Cardiol.* **2017**, *70*, 2278–2289. [CrossRef] [PubMed]
12. Libby, P.; Hansson, G.K. Taming immune and inflammatory responses to treat atherosclerosis. *J. Am. Coll. Cardiol.* **2018**, *71*, 173–176. [CrossRef] [PubMed]

13. Micha, R.; Peñalvo, J.L.; Cudhea, F.; Imamura, F.; Rehm, C.D.; Mozaffarian, D. Association between dietary factors and mortality from heart disease, stroke, and type 2 diabetes in the United States. *JAMA* **2017**, *317*, 912–924. [CrossRef] [PubMed]
14. Cardoso Lde, O.; Carvalho, M.S.; Cruz, O.G.; Melere, C.; Luft, V.C.; Molina, M.C.; Faria, C.P.; Benseñor, I.M.; Matos, S.M.; Fonseca, M.D.; et al. Eating patterns in the Brazilian Longitudinal Study of Adult Health (ELSA-Brasil): An exploratory analysis. *Cad. Saude Publica* **2016**, *32*, e00066215. [CrossRef] [PubMed]
15. Lu, H.; Daugherty, A. Recent Highlights of ATVB Atherosclerosis. *Arterioescler. Thromb. Vasc. Biol.* **2015**, *35*, 485–491. [CrossRef] [PubMed]
16. Arbab-Zadeh, A.; Fuster, V. The Myth of the "Vulnerable Plaque": Transitioning from a Focus on Individual Lesions to Atherosclerotic Disease Burden for Coronary Artery Disease Risk Assessment. *J. Am. Coll. Cardiol.* **2015**, *65*, 846–855. [CrossRef] [PubMed]
17. Jolliffe, I.T.; Cadima, J. Principal component analysis: A review and recent developments. *Philos. Trans. A Math. Phys. Eng. Sci.* **2016**, *374*, 20150202. [CrossRef]
18. Ozawa, M.; Shipley, M.; Kivimaki, M.; Singh-Manoux, A.; Brunner, E.J. Dietary pattern, inflammation and cognitive decline: The Whitehall II prospective cohort study. *Clin. Nutr.* **2017**, *36*, 506–512. [CrossRef]
19. Atkins, J.L.; Whincup, P.H.; Morris, R.W.; Lennon, L.T.; Papacosta, O.; Wannamethee, S.G. Dietary patterns and the risk of CVD and all-cause mortality in older British men. *Br. J. Nutr.* **2016**, *116*, 1246–1255. [CrossRef]
20. Viscogliosi, G.; Cipriani, E.; Liguori, M.L.; Marigliano, B.; Saliola, M.; Ettorre, E.; Andreozzi, P. Mediterranean dietary pattern adherence: Associations with prediabetes, metabolic syndrome, and related microinflammation. *Metab. Syndr. Relat. Disord.* **2013**, *11*, 210–216. [CrossRef]
21. Denova-Gutiérrez, E.; Tucker, K.L.; Flores, M.; Barquera, S.; Salmerón, J. Dietary patterns are associated with predicted cardiovascular disease risk in an urban mexican adult population. *J. Nutr.* **2016**, *146*, 90–97. [CrossRef] [PubMed]
22. Okada, E.; Takahashi, K.; Takimoto, H.; Takabayashi, S.; Kishi, T.; Kobayashi, T.; Nakamura, K.; Ukawa, S.; Nakamura, M.; Sasaki, S.; et al. Dietary patterns among Japanese adults: Findings from the National Health and Nutrition Survey, 2012. *Asia Pac. J. Clin. Nutr.* **2018**, *27*, 1120–1130. [CrossRef] [PubMed]
23. Hu, F.B. Dietary patterns analysis: a new direction in nutritional epidemiology. *Curr. Opin. Lipidol.* **2002**, *13*, 3–9. [CrossRef] [PubMed]
24. Wood, A.D.; Strachan, A.A.; Thies, F.; Aucott, L.S.; Reid, D.M.; Hardcastle, A.C.; Mavroeidi, A.; Simpson, W.G.; Duthie, G.G.; Macdonald, H.M. Patterns of dietary intake and serum carotenoid and tocopherol status are associated with biomarkers of chronic low-grade systemic inflammation and cardiovascular risk. *Br. J. Nutr.* **2014**, *112*, 1341–1352. [CrossRef] [PubMed]
25. Sijtsma, F.P.; Meyer, K.A.; Steffen, L.M.; Van Horn, L.; Shikany, J.M.; Odegaard, A.O.; Gross, M.D.; Kromhout, D.; Jacobs, D.R., Jr. Diet quality and markers of endothelial function: The CARDIA study. *Nutr. Metab. Cardiovasc. Dis.* **2014**, *24*, 632–638. [CrossRef] [PubMed]
26. Piccand, E.; Vollenweider, P.; Guessous, I.; Marques-Vidal, P. Association between dietary intake and inflammatory markers: Results from the CoLaus study. *Public Health Nutr.* **2018**, *18*, 1–8. [CrossRef] [PubMed]
27. Slavin, J.L.; Lloyd, B. Health benefits of fruits and vegetables. *Adv. Nutr.* **2012**, *3*, 506–516. [CrossRef]
28. Noad, R.L.; Rooney, C.; McCall, D.; Young, I.S.; McCance, D.; McKinley, M.C.; Woodside, J.V.; McKeown, P.P. Beneficial effect of a polyphenol-rich diet on cardiovascular risk: A randomised control trial. *Heart* **2016**, *102*, 1371–1379. [CrossRef]
29. Quintana Pacheco, D.A.; Sookthai, D.; Wittenbecher, C.; Graf, M.E.; Schübel, R.; Johnson, T.; Katzke, V.; Jakszyn, P.; Kaaks, R.; Kühn, T. Red meat consumption and risk of cardiovascular diseases-is increased iron load a possible link? *Am. J. Clin. Nutr.* **2018**, *107*, 113–119. [CrossRef]
30. Oude Griep, L.M.; Wang, H.; Chan, Q. Empirically-derived dietary patterns, diet quality scores, and markers of inflammation and endothelial dysfunction. *Curr. Nutr. Rep.* **2013**, *2*, 97–104. [CrossRef]
31. Barbaresko, J.; Koch, M.; Schulze, M.B.; Nöthlings, U. Dietary pattern analysis and biomarkers of low-grade inflammation: A systematic literature review. *Nutr. Rev.* **2013**, *71*, 511–527. [CrossRef] [PubMed]
32. Badimon, L.; Chagas, P.; Chiva-Blanch, G. Diet and Cardiovascular Disease: Effects of Foods and Nutrients in Classical and Emerging Cardiovascular Risk Factors. *Curr. Med. Chem.* **2017**. [CrossRef] [PubMed]

33. Ros, E.; Martínez-González, M.A.; Estruch, R.; Salas-Salvadó, J.; Fitó, M.; Martínez, J.A.; Corella, D. Mediterranean diet and cardiovascular health: Teachings of the PREDIMED study. *Adv. Nutr.* **2014**, *5*, 330S–336S. [CrossRef] [PubMed]
34. Ahluwalia, N.; Andreeva, V.A.; Kesse-Guyot, E.; Hercberg, S. Dietary patterns, inflammation and the metabolic syndrome. *Diabetes Metab.* **2013**, *39*, 99–110. [CrossRef] [PubMed]
35. Tapsell, L.C.; Neale, E.P.; Satija, A.; Hu, F.B. Foods, nutrients, and dietary patterns: Interconnections and implications for dietary guidelines. *Adv. Nutr.* **2016**, *7*, 445–454. [CrossRef] [PubMed]
36. Húc, T.; Nowinski, A.; Drapala, A.; Konopelski, P.; Ufnal, M. Indole and indoxyl sulfate, gut bacteria metabolites of tryptophan change arterial blood pressure via peripheral and central mechanisms in rats. *Pharmacol. Res.* **2018**, *130*, 172–179. [CrossRef] [PubMed]
37. Koeth, R.A.; Wang, Z.; Levison, B.S.; Buffa, J.A.; Org, E.; Sheehy, B.T.; Britt, E.B.; Fu, X.; Wu, Y.; Li, L.; et al. Intestinal microbiota metabolism of L-carnitine, a nutrient in red meat, promotes atherosclerosis. *Nat. Med.* **2013**, *19*, 576–585. [CrossRef]
38. Nowinski, A.; Ufnal, M. Trimethylamine N-oxide: A harmful, protective or diagnostic marker in lifestyle diseases? *Nutrition* **2018**, *46*, 7–12. [CrossRef] [PubMed]
39. Mehta, D.; Malik, A.B. Signaling mechanisms regulating endothelial permeability. *Physiol. Rev.* **2016**, *86*, 279–367. [CrossRef] [PubMed]
40. Virmani, R.; Joner, M.; Sakakura, K. Recent highlights of ATVB: Calcification. *Arterioscler. Thromb. Vasc. Biol.* **2014**, *34*, 1329–1332. [CrossRef] [PubMed]
41. Kwon, G.P.; Schroeder, J.L.; Amar, M.J.; Remaley, A.T.; Balaban, R.S. Contribution of macromolecular structure to the retention of low-density lipoprotein at arterial branch points. *Circulation* **2008**, *117*, 2658–2664. [CrossRef] [PubMed]
42. Jennings, L.K. Mechanisms of platelet activation: Need for new strategies to protect against platelet-mediated atherothrombosis. *Thromb. Haemost.* **2009**, *102*, 248–257. [CrossRef] [PubMed]
43. Chistiakov, D.A.; Melnichenko, A.A.; Myasoedova, V.A.; Grenchko, A.V.; Orekhov, A.N. Mechanisms of foam cell formation in atherosclerosis. *J. Mol. Med.* **2017**, *95*, 1153–1165. [CrossRef] [PubMed]
44. Salvayre, R.; Auge, N.; Benoist, H.; Negre-Salvayre, A. Oxidized low-density lipoprotein-induced apoptosis. *Biochim. Biophys. Acta* **2002**, *1585*, 213–221. [CrossRef]
45. Ghosh, S.; Zhao, B.; Bie, J.; Song, J. Macrophage cholesteryl ester mobilization and atherosclerosis. *Vasc. Pharmacol.* **2010**, *52*, 1–10. [CrossRef] [PubMed]
46. Legein, B.; Temmerman, L.; Biessen, E.A.; Lutgens, E. Inflammation and immune system interactions in atherosclerosis. *Cell. Mol. Life Sci.* **2013**, *70*, 3847–3869. [CrossRef]
47. Low Wang, C.C.; Hess, C.N.; Hiatt, W.R.; Goldfine, A.B. Atherosclerotic cardiovascular disease and hearth failure in type 2 diabetes-mechanisms, management, and clinical considerations. *Circulation* **2016**, *133*, 2459–2502. [CrossRef]
48. Harrison, D.; Griendling, K.K.; Landmesser, U.; Hornig, B.; Drexler, H. Role of oxidative stress in atherosclerosis. *Am. J. Cardiol.* **2003**, *91*, 7A–11A. [CrossRef]
49. Brown, D.I.; Griendling, K.K. Regulation of signaling transduction by reactive oxygen species in the cardiovascular system. *Circ. Res.* **2015**, *116*, 531–549. [CrossRef]
50. Littlewood, T.D.; Bennett, M.R. Apoptotic cell death in atherosclerosis. *Curr. Opin. Lipodol.* **2003**, *14*, 469–475. [CrossRef]
51. Li, H.; Horke, S.; Förstermann, U. Vascular oxidative stress, nitric oxide and atherosclerosis. *Atherosclerosis* **2014**, *237*, 208–219. [CrossRef] [PubMed]
52. Reuter, S.; Gupta, S.C.; Chaturvedi, M.M.; Aggarwal, B.B. Oxidative stress, inflammation, and cancer: How are they linked? *Free Radic. Biol. Med.* **2010**, *49*, 1603–1616. [CrossRef] [PubMed]
53. Torres, N.; Guevara-Cruz, M.; Velázquez-Villegas, L.A.; Tovar, A.R. Nutrition and atherosclerosis. *Arch. Med. Res.* **2015**, *46*, 408–426. [CrossRef] [PubMed]
54. Li, H.; Forstermann, U. Uncoupling of endothelial NO synthase in atherosclerosis and vascular disease. *Curr. Opin. Pharmacol.* **2013**, *13*, 161–167. [CrossRef] [PubMed]
55. Badimon, L.; Vilahur, G.; Padro, T. Lipoproteins, platelets and atherothrombosis. *Rev. Esp. Cardiol.* **2009**, *62*, 1161–1178. [CrossRef]
56. Kattoor, A.J.; Pothineni, N.V.K.; Palagiri, D. Oxidative strees in atherosclerosis. *Curr. Atheroscler. Rep.* **2017**, *19*, 42. [CrossRef] [PubMed]

57. Guzik, T.J.; Korbut, R.; Adamek-Guzik, T. Nitric oxide and superoxide in inflammation and immune regulation. *J. Physiol. Pharmacol.* **2003**, *54*, 469–487. [PubMed]
58. Centritto, F.; Iacoviello, L.; di Giuseppe, R.; De Curtis, A.; Costanzo, S.; Zito, F.; Grioni, S.; Sieri, S.; Donati, M.B.; de Gaetano, G.; et al. Dietary patterns, cardiovascular risk factors and C-reactive protein in a healthy Italian population. *Nutr. Metab. Cardiovasc. Dis.* **2009**, *19*, 697–706. [CrossRef]
59. Mozaffarian, D. Dietary and policy priorities for cardiovascular disease, diabetes, and obesity: A comprehensive review. *Circulation* **2016**, *133*, 187–225. [CrossRef]
60. Silveira, B.K.S.; Oliveira, T.M.S.; Andrade, P.A.; Hermsdorff, H.H.M.; Rosa, C.O.B.; Franceschini, S.D.C.C. Dietary pattern and macronutrients profile on the variation of inflammatory biomarkers: Scientific Update. *Cardiol. Res. Pract.* **2018**, *2018*, 4762575. [CrossRef]
61. Esposito, K.; Ciotola, M.; Giugliano, D. Mediterranean diet, endothelial function and vascular inflammatory markers. *Public Health Nutr.* **2006**, *9*, 1073–1076. [CrossRef] [PubMed]
62. Llorente-Cortés, V.; Estruch, R.; Mena, M.P.; Ros, E.; González, M.A.; Fitó, M.; Lamuela-Raventós, R.M.; Badimon, L. Effect of Mediterranean diet on the expression of pro-atherogenic genes in a population at high cardiovascular risk. *Atherosclerosis* **2010**, *208*, 442–450. [CrossRef] [PubMed]
63. Casas, R.; Sacanella, E.; Urpí-Sardà, M.; Chiva-Blanch, G.; Ros, E.; Martínez-González, M.A.; Covas, M.I.; Lamuela-Raventos, R.M.; Salas-Salvadó, J.; Fiol, M.; et al. The effects of the mediterranean diet on biomarkers of vascular wall inflammation and plaque vulnerability in subjects with high risk for cardiovascular disease. A randomized trial. *PLoS ONE* **2014**, *9*, e100084. [CrossRef] [PubMed]
64. Casas, R.; Urpi-Sardà, M.; Sacanella, E.; Arranz, S.; Corella, D.; Castañer, O.; Lamuela-Raventós, R.M.; Salas-Salvadó, J.; Lapetra, J.; Portillo, M.P.; et al. Anti-Inflammatory Effects of the Mediterranean Diet in the Early and Late Stages of Atheroma Plaque Development. *Mediat. Inflamm.* **2017**, *2017*, 3674390. [CrossRef] [PubMed]
65. Kastorini, C.M.; Panagiotakos, D.B.; Chrysohoou, C.; Georgousopoulou, E.; Pitaraki, E.; Puddu, P.E.; Tousoulis, D.; Stefanadis, C.; Pitsavos, C.; ATTICA Study Group. Metabolic syndrome, adherence to the Mediterranean diet and 10-year cardiovascular disease incidence: The ATTICA study. *Atherosclerosis* **2016**, *246*, 87–93. [CrossRef] [PubMed]
66. Cainzos-Achirica, M.; Miedema, M.D.; McEvoy, J.W.; Cushman, M.; Dardari, Z.; Greenland, P.; Nasir, K.; Budoff, M.J.; Al-Mallah, M.H.; Yeboah, J.; et al. The prognostic value of high sensitivity C-reactive protein in a multi-ethnic population after >10 years of follow-up: The Multi-Ethnic Study of Atherosclerosis (MESA). *Int. J. Cardiol.* **2018**, *264*, 158–164. [CrossRef] [PubMed]
67. Lahoz, C.; Castillo, E.; Mostaza, J.M.; de Dios, O.; Salinero-Fort, M.A.; González-Alegre, T.; García-Iglesias, F.; Estirado, E.; Laguna, F.; Sanchez, V.; et al. Relationship of the Adherence to a Mediterranean Diet and Its Main Components with CRP Levels in the Spanish Population. *Nutrients* **2018**, *10*, 379. [CrossRef] [PubMed]
68. Pounis, G.; Costanzo, S.; di Giuseppe, R.; de Lucia, F.; Santimone, I.; Sciarretta, A.; Barisciano, P.; Persichillo, M.; de Curtis, A.; Zito, F.; et al. Consumption of healthy foods at different content of antioxidant vitamins and phytochemicals and metabolic risk factors for cardiovascular disease in men and women of the Moli-sani study. *Eur. J. Clin. Nutr.* **2013**, *67*, 207–213. [CrossRef]
69. Sureda, A.; Bibiloni, M.D.M.; Julibert, A.; Bouzas, C.; Argelich, E.; Llompart, I.; Pons, A.; Tur, J.A. Adherence to the Mediterranean Diet and Inflammatory Markers. *Nutrients* **2018**, *10*, 62. [CrossRef]
70. Estruch, R.; Ros, E.; Salas-Salvadó, J.; Covas, M.I.; Corella, D.; Arós, F.; Gómez-Gracia, E.; Ruiz-Gutiérrez, V.; Fiol, M.; Lapetra, J.; et al. Primary Prevention of Cardiovascular Disease with a Mediterranean Diet Supplemented with Extra-Virgin Olive Oil or Nuts. *N. Engl. J. Med.* **2018**, *378*, e34. [CrossRef]
71. Estruch, R.; Martínez-González, M.A.; Corella, D.; Salas-Salvadó, J.; Ruiz-Gutiérrez, V.; Covas, M.I.; Fiol, M.; Gómez-Gracia, E.; López-Sabater, M.C.; Vinyoles, E.; et al. Effects of a Mediterraneanstyle diet on cardiovascular risk factors: A randomized trial. *Ann. Intern. Med.* **2006**, *145*, 1–11. [CrossRef] [PubMed]
72. Urpi-Sarda, M.; Casas, R.; Chiva-Blanch, G.; Romero-Mamani, E.S.; Valderas-Martínez, P.; Salas-Salvadó, J.; Covas, M.I.; Toledo, E.; Andres-Lacueva, C.; Llorach, R.; et al. The Mediterranean diet pattern and its main components are associated with lower plasma concentrations of tumor necrosis factor receptor 60 in patients at high risk for cardiovascular disease. *J. Nutr.* **2012**, *142*, 1019–1025. [CrossRef] [PubMed]
73. Mena, M.P.; Sacanella, E.; Vazquez-Agell, M.; Morales, M.; Fitó, M.; Escoda, R.; Serrano-Martínez, M.; Salas-Salvadó, J.; Benages, N.; Casas, R.; et al. Inhibition of circulating immune cell activation: A molecular antiinflammatory effect of the Mediterranean diet. *Am. J. Clin. Nutr.* **2009**, *89*, 248–256. [CrossRef] [PubMed]

74. Casas, R.; Sacanella, E.; Urpí-Sardà, M.; Corella, D.; Castañer, O.; Lamuela-Raventos, R.M.; Salas-Salvadó, J.; Martínez-González, M.A.; Ros, E.; Estruch, R. Long-Term Immunomodulatory Effects of a Mediterranean Diet in Adults at High Risk of Cardiovascular Disease in the PREvención con DIeta MEDiterránea (PREDIMED) Randomized Controlled Trial. *J. Nutr.* **2016**, *146*, 1684–1693. [CrossRef] [PubMed]
75. Ambring, A.; Johansson, M.; Axelsen, M.; Gan, L.; Strandvik, B.; Friberg, P. Mediterranean-inspired diet lowers the ratio of serum phospholipid n-6 to n-3 fatty acids, the number of leukocytes and platelets, and vascular endothelial growth factor in healthy subjects. *Am. J. Clin. Nutr.* **2006**, *83*, 575–581. [CrossRef] [PubMed]
76. Esposito, K.; Marfella, R.; Ciotola, M.; Di Palo, C.; Giugliano, F.; Giugliano, G.; D'Armiento, M.; D'Andrea, F.; Giugliano, D. Effect of a Mediterranean-Style Diet on Endothelial Dysfunction and Markers of Vascular Inflammation in the Metabolic Syndrome: A Randomized Trial. *JAMA* **2004**, *292*, 1440. [CrossRef] [PubMed]
77. Azzini, E.; Polito, A.; Fumagalli, A.; Intorre, F.; Venneria, E.; Durazzo, A.; Zaccaria, M.; Ciarapica, D.; Foddai, M.S.; Mauro, B.; et al. Mediterranean diet effect: An Italian picture. *Nutr. J.* **2011**, *10*, 125. [CrossRef] [PubMed]
78. Medina-Remón, A.; Casas, R.; Tressserra-Rimbau, A.; Ros, E.; Martínez-González, M.A.; Fitó, M.; Corella, D.; Salas-Salvadó, J.; Lamuela-Raventos, R.M.; Estruch, R.; et al. Polyphenol intake from a Mediterranean diet decreases inflammatory biomarkers related to atherosclerosis: A substudy of the PREDIMED trial. *Br. J. Clin. Pharmacol.* **2017**, *83*, 114–128. [CrossRef]
79. Arpón, A.; Milagro, F.I.; Razquin, C.; Corella, D.; Estruch, R.; Fitó, M.; Marti, A.; Martínez-González, M.A.; Ros, E.; Salas-Salvadó, J.; et al. Impact of consuming extra-virgin olive oil or nuts within a Mediterranean diet on DNA methylation in peripheral white blood cells within the PREDIMED-Navarra randomized controlled trial: A role for dietary lipids. *Nutrients* **2017**, *10*, 15. [CrossRef]
80. Arpón, A.; Riezu-Boj, J.I.; Milagro, F.I.; Marti, A.; Razquin, C.; Martínez-González, M.A.; Corella, D.; Estruch, R.; Casas, R.; Fitó, M.; et al. Adherence to Mediterranean diet is associated with methylation changes in inflammation-related genes in peripheral blood cells. *J. Physiol. Biochem.* **2016**, *73*, 445–455. [CrossRef]
81. Corella, D.; González, J.I.; Bulló, M.; Carrasco, P.; Portolés, O.; Díez-Espino, J.; Covas, M.I.; Ruíz-Gutierrez, V.; Gómez-Gracia, E.; Arós, F.; et al. Polymorphisms cyclooxygenase-2-765G>C and interleukin-6-174G>C are associated with serum inflammation markers in a high cardiovascular risk population and do not modify the response to a Mediterranean diet supplemented with virgin olive oil or nuts. *J. Nutr.* **2009**, *139*, 128–134. [CrossRef] [PubMed]
82. Corella, D.; Tai, E.S.; Sorlí, J.V.; Chew, S.K.; Coltell, O.; Sotos-Prieto, M.; García-Rios, A.; Estruch, R.; Ordovas, J.M. Association between the APOA2 promoter polymorphism and body weight in Mediterranean and Asian populations: Replication of a gene-saturated fat interaction. *Int. J. Obes.* **2011**, *35*, 666–675. [CrossRef] [PubMed]
83. Corella, D.; Carrasco, P.; Fitó, M.; Martínez-González, M.A.; Salas-Salvadó, J.; Arós, F.; Lapetra, J.; Guillén, M.; Ortega-Azorín, C.; Warnberg, J.; et al. Gene-environment interactions of CETP gene variation in a high cardiovascular risk Mediterranean population. *J. Lipid Res.* **2010**, *51*, 2798–2807. [CrossRef] [PubMed]
84. Corella, D.; Carrasco, P.; Sorlí, J.V.; Estruch, R.; Rico-Sanz, J.; Martínez-González, M.Á.; Salas-Salvadó, J.; Covas, M.I.; Coltell, O.; Arós, F.; et al. Mediterranean diet reduces the adverse effect of the TCF7L2-rs7903146 polymorphism on cardiovascular risk factors and stroke incidence: A randomized controlled trial in a high-cardiovascular-risk population. *Diabetes Care* **2013**, *36*, 3803–3811. [CrossRef] [PubMed]
85. Saneei, P.; Salehi-Abargouei, A.; Esmaillzadeh, A.; Azadbakht, L. Influence of dietary approaches to stop hypertension (DASH) diet on blood pressure: A systematic review and meta-analysis on randomized controlled trials. *Nutr. Metab. Cardiovasc. Dis.* **2014**, *24*, 1253–1261. [CrossRef] [PubMed]
86. Soltani, S.; Shirani, F.; Chitsazi, M.J.; Salehi-Abargouei, A. The effect of dietary approaches to stop hypertension (DASH) diet on weight and body composition in adults: A systematic review and meta-analysis of randomized controlled clinical trials. *Obes. Rev.* **2016**, *17*, 442–454. [CrossRef]
87. Shirani, F.; Salehi-Abargouei, A.; Azadbakht, L. Effects of dietary approaches to stop hypertension (DASH) diet on some risk for developing type 2 diabetes: A systematic review and meta-analysis on controlled clinical trials. *Nutrition* **2013**, *29*, 939–947. [CrossRef]

88. Millen, B.E.; Abrams, S.; Adams-Campbell, L.; Anderson, C.A.; Brenna, J.T.; Campbell, W.W.; Clinton, S.; Hu, F.; Nelson, M.; Neuhouser, M.L.; et al. The 2015 Dietary Guidelines Advisory Committee Scientific Report: Development and Major Conclusions. *Adv. Nutr.* **2016**, *7*, 438–444. [CrossRef]
89. Soltani, S.; Chitsazi, M.J.; Salehi-Abargouei, A. The effect of dietary approaches to stop hypertension (DASH) on serum inflammatory markers: A systematic review and meta-analysis of randomized trials. *Clin. Nutr.* **2018**, *37*, 542–550. [CrossRef]
90. Woo, J.; Yu, B.W.M.; Chan, R.S.M.; Leung, J. Influence of Dietary Patterns and Inflammatory Markers on Atherosclerosis Using Ankle Brachial Index as a Surrogate. *J. Nutr. Health Aging* **2018**, *22*, 619–626. [CrossRef]
91. Rifai, L.; Pisano, C.; Hayden, J.; Sulo, S.; Silver, M.A. Impact of the DASH diet on endothelial function, exercise capacity, and quality of life in patientswith heart failure. *Bayl. Univ. Med. Cent. Proc.* **2015**, *28*, 151–156. [CrossRef]
92. d'El-Rei, J.; Cunha, A.R.; Trindade, M.; Neves, M.F. Beneficial Effects of Dietary Nitrate on Endothelial Function and Blood Pressure Levels. *Int. J. Hypertens.* **2016**, *2016*, 6791519. [CrossRef]
93. Marques, F.Z.; Mackay, C.R.; Kaye, D.M. Beyond gut feelings: How the gut microbiota regulates blood pressure. *Nat. Rev. Cardiol.* **2018**, *15*, 20–32. [CrossRef] [PubMed]
94. Derkach, A.; Sampson, J.; Joseph, J.; Playdon, M.C.; Stolzenberg-Solomon, R.Z. Effects of dietary sodium on metabolites: The Dietary Approaches to Stop Hypertension (DASH)-Sodium Feeding Study. *Am. J. Clin. Nutr.* **2017**, *106*, 1131–1141. [CrossRef] [PubMed]
95. Mertens, E.; Markey, O.; Geleijnse, J.M.; Lovegrove, J.A.; Givens, D.I. Adherence to a healthy diet in relation to cardiovascular incidence and risk markers: Evidence from the Caerphilly Prospective Study. *Eur. J. Nutr.* **2018**, *57*, 1245–1258. [CrossRef] [PubMed]
96. Jones, N.R.V.; Forouhi, N.G.; Khaw, K.T.; Wareham, N.J.; Monsivais, P. Accordance to the Dietary Approaches to Stop Hypertension diet pattern and cardiovascular disease in a British, population-based cohort. *Eur. J. Epidemiol.* **2018**, *33*, 235–244. [CrossRef]
97. Saglimbene, V.M.; Wong, G.; Craig, J.C.; Ruospo, M.; Palmer, S.C.; Campbell, K.; Garcia-Larsen, V.; Natale, P.; Teixeira-Pinto, A.; Carrero, J.J.; et al. The Association of Mediterranean and DASH Diets with Mortality in Adults on Hemodialysis: The DIET-HD Multinational Cohort Study. *J. Am. Soc. Nephrol.* **2018**, *29*, 1741–1751. [CrossRef]
98. Neelakantan, N.; Koh, W.P.; Yuan, J.M.; van Dam, R.M. Diet-quality indexes are associated with a lower risk of cardiovascular, respiratory, and all-cause mortality among Chinese adults. *J. Nutr.* **2018**, *148*, 1323–1332. [CrossRef]
99. Rai, S.K.; Fung, T.T.; Lu, N.; Keller, S.F.; Curhan, G.C.; Choi, H.K. The Dietary Approaches to Stop Hypertension (DASH) diet, Western diet, and risk of gout in men: Prospective cohort study. *BMJ* **2017**, *357*, j1794. [CrossRef]
100. Neale, E.P.; Batterham, M.J.; Tapsell, L.C. Consumption of a healthy dietary pattern results in significant reductions in C-reactive protein levels in adults: A meta-analysis. *Nutr. Res.* **2016**, *36*, 391–401. [CrossRef]
101. Eichelmann, F.; Schwingshackl, L.; Fedirko, V.; Aleksandrova, K. Effect of plantbased diets on obesity-related inflammatory profiles: A systematic review and meta-analysis of intervention trials. *Obes. Rev.* **2016**, *17*, 1067–1079. [CrossRef] [PubMed]
102. Phillips, C.M.; Harrington, J.M.; Perry, I.J. Relationship between dietary quality, determined by DASH score, and cardiometabolic health biomarkers: A cross-sectional analysis in adults. *Clin. Nutr.* **2018**, *S0261–S5614*, in press. [CrossRef] [PubMed]
103. Asemi, Z.; Samimi, M.; Tabassi, Z.; Sabihi, S.S.; Esmaillzadeh, A. A randomized controlled clinical trial investigating the effect of DASH diet on insulin resistance, inflammation, and oxidative stress in gestational diabetes. *Nutrition* **2013**, *29*, 619–624. [CrossRef] [PubMed]
104. Kawamura, A.; Kajiya, K.; Kishi, H.; Inagaki, J.; Mitarai, M.; Oda, H.; Umemoto, S.; Kobayashi, S. Effects of the DASH-JUMP dietary intervention in Japanese participants with high-normal blood pressure and stage 1 hypertension: An open-label single-arm trial. *Hypertens. Res.* **2016**, *39*, 777–785. [CrossRef] [PubMed]
105. Saneei, P.; Hashemipour, M.; Kelishadi, R.; Esmaillzadeh, A. The Dietary Approaches to Stop Hypertension (DASH) diet affects inflammation in childhood metabolic syndrome: A randomized cross-over clinical trial. *Ann. Nutr. Metab.* **2014**, *64*, 20–27. [CrossRef] [PubMed]
106. Dauchet, L.; Amouyel, P.; Hercberg, S.; Dallongeville, J. Fruit and vegetable consumption and risk of coronary heart disease: A metaanalysis of cohort studies. *J. Nutr.* **2006**, *136*, 2588–2593. [CrossRef]

107. Dauchet, L.; Amouyel, P.; Dallongeville, J. Fruits, vegetables and coronary heart disease. *Nat. Rev. Cardiol.* **2009**, *6*, 599–608. [CrossRef] [PubMed]
108. Hosseini, B.; Berthon, B.S.; Saedisomeolia, A.; Starkey, M.R.; Collison, A.; Wark, P.A.B.; Wood, L.G. Effects of fruit and vegetable consumption on inflammatory biomarkers and immune cell populations: A systematic literature review and meta-analysis. *Am. J. Clin. Nutr.* **2018**, *108*, 136–155. [CrossRef]
109. Graham, I.; Atar, D.; Borch-Johnsen, K.; Boysen, G.; Burell, G.; Cifkova, R.; Dallongeville, J.; De Backer, G.; Ebrahim, S.; Gjelsvik, B.; et al. European guidelines on cardiovascular disease prevention in clinical practice: Executive summary: Fourth Joint Task Force of the European Society of Cardiology and Other Societies on Cardiovascular Disease Prevention in Clinical Practice (Constituted by representatives of nine societies and by invited experts). *Eur. Heart J.* **2007**, *28*, 2375–2414. [CrossRef]
110. American Heart Association Nutrition Committee; Lichtenstein, A.H.; Appel, L.J.; Brands, M.; Carnethon, M.; Daniels, S.; Franch, H.A.; Franklin, B.; Kris-Etherton, P.; Harris, W.S.; et al. Diet and lifestyle recommendations revision 2006: A scientific statement from the American Heart Association Nutrition Committee. *Circulation* **2006**, *114*, 82–96. [CrossRef]
111. Corley, J.; Kyle, J.A.; Starr, J.M.; McNeill, G.; Deary, I.J. Dietary factors and biomarkers of systemic inflammation in older people: The Lothian Birth Cohort 1936. *Br. J. Nutr.* **2015**, *114*, 1088–1098. [CrossRef] [PubMed]
112. Holt, E.M.; Steffen, L.M.; Moran, A.; Basu, S.; Steinberger, J.; Ross, J.A.; Hong, C.P.; Sinaiko, A.R. Fruit and vegetable consumption and its relation to markers of inflammation and oxidative stressin adolescents. *J. Am. Diet Assoc.* **2009**, *109*, 414–421. [CrossRef] [PubMed]
113. Arouca, A.; Michels, N.; Moreno, L.A.; González-Gil, E.M.; Marcos, A.; Gómez, S.; Díaz, L.E.; Widhalm, K.; Molnár, D.; Manios, Y.; et al. Associations between a Mediterranean diet pattern and inflammatory biomarkers in Europeanadolescents. *Eur. J. Nutr.* **2018**, *57*, 1747–1760. [CrossRef] [PubMed]
114. Jiang, Y.; Wu, S.H.; Shu, X.O.; Xiang, Y.B.; Ji, B.T.; Milne, G.L.; Cai, Q.; Zhang, X.; Gao, Y.T.; Zheng, W.; et al. Cruciferous vegetable intake is inversely correlated with circulating levels of proinflammatory markers in women. *J. Acad. Nutr. Diet.* **2014**, *114*, 700–708. [CrossRef] [PubMed]
115. Wongwarawipat, T.; Papageorgiou, N.; Bertsias, D.; Siasos, G.; Tousoulis, D. Olive Oil-related Anti-inflammatory effects on atherosclerosis: Potential clinical implications. *Endocr. Metab. Immune Disord. Drug Targets* **2018**, *18*, 51–62. [CrossRef] [PubMed]
116. Schwingshackl, L.; Christoph, M.; Hoffmann, G. Effects of olive oil on markers of inflammation and endothelial function-A systematic review and meta-analysis. *Nutrients* **2015**, *7*, 7651–7675. [CrossRef] [PubMed]
117. Casas, R.; Estruch, R.; Sacanella, E. The protective effects of extra virgin olive oil on immune-mediated inflammatory responses. *Endocr. Metab. Immune Disord. Drug Targets* **2018**, *18*, 23–35. [CrossRef]
118. Schwingshackl, L.; Hoffmann, G. Monounsaturated fatty acids, olive oil and health status: A systematic review and meta-analysis of cohort studies. *Lipids Health Dis.* **2014**, *13*, 154. [CrossRef]
119. Fitó, M.; Estruch, R.; Salas-Salvadõ, J.; Martínez-Gonzalez, M.A.; Arós, F.; Vila, J.; Corella, D.; Díaz, O.; Sáez, G.; De La Torre, R.; et al. Effect of the mediterranean diet on heart failure biomarkers: A randomized sample from the PREDIMED trial. *Eur. J. Heart Fail.* **2014**, *16*, 543–550. [CrossRef]
120. Sala-Vila, A.; Romero-Mamani, E.S.; Gilabert, R.; Núñez, I.; De La Torre, R.; Corella, D.; Ruiz-Gutiérrez, V.; López-Sabater, M.C.; Pintó, X.; Rekondo, J.; et al. Changes in ultrasound-assessed carotid intima-media thickness and plaque with a mediterranean diet: A substudy of the PREDIMED trial. *Arterioscler. Thromb. Vasc. Biol.* **2014**, *34*, 439–445. [CrossRef]
121. Murie-Fernandez, M.; Irimia, P.; Toledo, E.; Martínez-Vila, E.; Buil-Cosiales, P.; Serrano-Martínez, M.; Ruiz-Gutiérrez, V.; Ros, E.; Estruch, R.; Martínez-González, M.A. Carotid intima-media thickness changes with mediterranean diet: A randomized trial (PREDIMED-Navarra). *Atherosclerosis* **2011**, *219*, 158–162. [CrossRef] [PubMed]
122. Moreno-Luna, R.; Muñoz-Hernandez, R.; Miranda, M.L.; Costa, A.F.; Jimenez-Jimenez, L.; Vallejo-Vaz, A.J.; Muriana, F.J.G.; Villar, J.; Stiefel, P. Olive oil polyphenols decrease blood pressure and improve endothelial function in young women with mild hypertension. *Am. J. Hypertens.* **2012**, *25*, 1299–1304. [CrossRef] [PubMed]

123. Storniolo, C.E.; Casillas, R.; Bulló, M.; Castañer, O.; Ros, E.; Sáez, G.T.; Toledo, E.; Estruch, R.; Ruiz-Gutiérrez, V.; Fitó, M.; et al. A mediterranean diet supplemented with extra virgin olive oil or nuts improves endothelial markers involved in blood pressure control in hypertensive women. *Eur. J. Nutr.* **2017**, *56*, 89–97. [CrossRef] [PubMed]
124. Murphy, K.J. A Mediterranean diet lowers blood pressure and improves endothelial function: Results from the medley randomized intervention trial. *Am. J. Clin. Nutr.* **2017**, *105*, 1305–1313. [CrossRef]
125. Camargo, A.; Delgado-Lista, J.; Garcia-Rios, A.; Cruz-Teno, C.; Yubero-Serrano, E.M.; Perez-Martinez, P.; Gutierrez-Mariscal, F.M.; Lora-Aguilar, P.; Rodriguez-Cantalejo, F.; Fuentes-Jimenez, F.; et al. Expression of proinflammatory; proatherogenic genes is reduced by the Mediterranean diet in elderly people. *Br. J. Nutr.* **2012**, *108*, 500–508. [CrossRef] [PubMed]
126. Widmer, R.J.; Freund, M.; Flamme, J.; Sexton, J.; Lennon, R.; Romani, A.; Mulinacci, N.; Vinceri, F.; Lerman, L.O.; Lerman, A. Beneficial effects of polyphenol-rich olive oil in patients with early atherosclerosis. *Eur. J. Nutr.* **2013**, *52*, 1223–1231. [CrossRef] [PubMed]
127. Oliveras-Lópeza, M.J.; Bernáab, G.; Jurado-Ruizab, E.; de la Serrana, L.G.; Martínab, F. Consumption of extra-virgin olive oil rich in phenolic compounds has beneficial antioxidant effects in healthy human adults. *J. Funct. Foods* **2014**, *10*, 475–484. [CrossRef]
128. Visioli, F.; Caruso, D.; Grande, S.; Bosisio, R.; Villa, M.; Galli, G.; Sirtori, C.; Galli, C. Virgin Olive Oil Study (VOLOS): Vasoprotective potential of extra virgin olive oil in mildly dyslipidemic patients. *Eur. J. Nutr.* **2005**, *44*, 121–127. [CrossRef]
129. Guasch-Ferré, M.; Liu, X.; Malik, V.S.; Sun, Q.; Willett, W.C.; Manson, J.E.; Rexrode, K.M.; Li, Y.; Hu, F.B.; Bhupathiraju, S.N. Nut Consumption and Risk of Cardiovascular Disease. *J. Am. Coll. Cardiol.* **2017**, *70*, 2519–2532. [CrossRef]
130. Aune, D.; Keum, N.; Giovannucci, E.; Fadnes, L.T.; Boffetta, P.; Greenwood, D.C.; Tonstad, S.; Vatten, L.J.; Riboli, E.; Norat, T. Nut consumption and risk of cardiovascular disease, total cancer, all-cause and cause-specific mortality: A systematic review and dose-response meta-analysis of prospective studies. *BMC Med.* **2016**, *14*, 207. [CrossRef]
131. Mente, A.; de Koning, L.; Shannon, H.S.; Anand, S.S. A systematic review of the evidence supporting a causal link between dietary factors and coronary heart disease. *Arch. Intern. Med.* **2009**, *169*, 659–669. [CrossRef] [PubMed]
132. Mozaffarian, D.; Hao, T.; Rimm, E.B.; Willett, W.C.; Hu, F.B. Changes in diet and lifestyle and long-term weight gain in women and men. *N. Engl. J. Med.* **2011**, *364*, 2392–2404. [CrossRef] [PubMed]
133. Del Gobbo, L.C.; Falk, M.C.; Feldman, R.; Lewis, K.; Mozaffarian, D. Effects of tree nuts on blood lipids, apolipoproteins, and blood pressure: Systematic review, meta-analysis, and dose-response of 61 controlled intervention trials. *Am. J. Clin. Nutr.* **2015**, *102*, 1347–1356. [CrossRef] [PubMed]
134. Banel, D.K.; Hu, F.B. Effects of walnut consumption on blood lipids and other cardiovascular risk factors: A meta-analysis and systematic review. *Am. J. Clin. Nutr.* **2009**, *90*, 56–63. [CrossRef] [PubMed]
135. Zhou, D.; Yu, H.; He, F.; Reilly, K.H.; Zhang, J.; Li, S.; Zhang, T.; Wang, B.; Ding, Y.; Xi, B. Nut consumption in relation to cardiovascular disease risk and type 2 diabetes: A systematic review and meta-analysis of prospective studies. *Am. J. Clin. Nutr.* **2014**, *100*, 270–277. [CrossRef] [PubMed]
136. Guo, K.; Zhou, Z.; Jiang, Y.; Li, W.; Li, Y. Meta-analysis of prospective studies on the effects of nut consumption on hypertension and type 2 diabetes mellitus. *J. Diabetes* **2015**, *7*, 202–212. [CrossRef] [PubMed]
137. Yu, Z.; Malik, V.S.; Keum, N.; Hu, F.B.; Giovannucci, E.L.; Stampfer, M.J.; Willett, W.C.; Fuchs, C.S.; Bao, Y. Associations between nut consumption and inflammatory biomarkers. *Am. J. Clin. Nutr.* **2016**, *104*, 722–728. [CrossRef] [PubMed]
138. Neale, E.P.; Tapsell, L.C.; Guan, V.; Batterham, M.J. The effect of nut consumption on markers of inflammation and endothelial function: A systematic review and meta-analysis of randomised controlled trials. *BMJ Open* **2017**, *7*, e016863. [CrossRef] [PubMed]
139. Xiao, Y.; Xia, J.; Ke, Y.; Cheng, J.; Yuan, J.; Wu, S.; Lv, Z.; Huang, S.; Kim, J.H.; Wong, S.Y.; et al. Effects of nut consumption on selected inflammatory markers: A systematic review and meta-analysis of randomized controlled trials. *Nutrition* **2018**, *54*, 129–143. [CrossRef] [PubMed]
140. Casas-Agustench, P.; López-Uriarte, P.; Bulló, M.; Ros, E.; Cabré-Vila, J.J.; Salas-Salvadó, J. Effects of one serving of mixed nuts on serum lipids, insulin resistance and inflammatory markersin patients with the metabolic syndrome. *Nutr. Metab. Cardiovasc. Dis.* **2011**, *21*, 126–135. [CrossRef] [PubMed]

141. Hernández-Alonso, P.; Salas-Salvadó, J.; Baldrich-Mora, M.; Juanola-Falgarona, M.; Bulló, M. Beneficial effect of pistachio consumption on glucose metabolism, insulin resistance, inflammation, and related metabolic risk markers: A randomized clinical trial. *Diabetes Care* **2014**, *37*, 3098–3105. [CrossRef] [PubMed]
142. Liu, J.F.; Liu, Y.H.; Chen, C.M.; Chang, W.H.; Chen, C.Y. The effect of almonds on inflammation and oxidative stress in Chinese patients with type 2 diabetes mellitus: A randomized crossover controlled feeding trial. *Eur. J. Nutr.* **2013**, *52*, 927–935. [CrossRef] [PubMed]
143. Chiva-Blanch, G.; Arranz, S.; Lamuela-Raventos, R.M.; Estruch, R. Effects of wine, alcohol and polyphenols on cardiovascular disease risk factors: Evidences from human studies. *Alcohol Alcohol.* **2013**, *48*, 270–277. [CrossRef] [PubMed]
144. Haseeb, S.; Alexander, B.; Baranchuk, A. Wine and cardiovascular health: A comprehensive review. *Circulation* **2017**, *136*, 1434–1448. [CrossRef] [PubMed]
145. Costanzo, S.; Di Castelnuovo, A.; Donati, M.B.; Iacoviello, L.; de Gaetano, G. Wine, beer or spirit drinking in relation to fatal and non-fatal cardiovascular events: A meta-analysis. *Eur. J. Epidemiol.* **2011**, *26*, 833–850. [CrossRef] [PubMed]
146. Haseeb, S.; Alexander, B.; Santi, R.L.; Liprandi, A.S.; Baranchuk, A. What's in wine? A clinician's perspective. *Trends Cardiovasc. Med.* **2018**, *S1050–S1738*. in press. [CrossRef] [PubMed]
147. Janssen, I.; Landay, A.L.; Ruppert, K.; Powell, L.H. Moderate wine consumption is associated with lower hemostatic and inflammatory risk factorsover 8 years: The study of women's health across the nation (SWAN). *Nutr. Aging* **2014**, *2*, 91–99. [CrossRef]
148. Estruch, R.; Sacanella, E.; Badia, E.; Antúnez, E.; Nicolás, J.M.; Fernández-Solá, J.; Rotilio, D.; de Gaetano, G.; Rubin, E.; Urbano-Márquez, A. Different effects of red wine and gin consumption on inflammatory biomarkers of atherosclerosis: A prospective randomized crossover trial. Effects of wine on inflammatory markers. *Atherosclerosis* **2004**, *175*, 117–123. [CrossRef]
149. Chiva-Blanch, G.; Urpi-Sarda, M.; Llorach, R.; Rotches-Ribalta, M.; Guillén, M.; Casas, R.; Arranz, S.; Valderas-Martinez, P.; Portoles, O.; Corella, D.; et al. Differential effects of polyphenols and alcohol of red wine on the expression of adhesion molecules and inflammatory cytokines related to atherosclerosis: A randomized clinical trial. *Am. J. Clin. Nutr.* **2012**, *95*, 326–334. [CrossRef]
150. Roth, I.; Casas, R.; Medina-Remón, A.; Lamuela-Raventós, R.M.; Estruch, R. Consumption of aged white wine modulates cardiovascular risk factors via circulating endothelial progenitor cells and inflammatory biomarkers. *Clin. Nutr.* **2018**, *S0261–S5614*, 30219-X. [CrossRef]
151. Estruch, R.; Sacanella, E.; Mota, F.; Chiva-Blanch, G.; Antúnez, E.; Casals, E.; Deulofeu, R.; Rotilio, D.; Andres-Lacueva, C.; Lamuela-Raventos, R.M.; et al. Moderate consumption of red wine, but not gin, decreases erythrocyte superoxide dismutase activity: A randomized cross-over trial. *Nutr. Metab. Cardiovasc. Dis.* **2011**, *21*, 46–53. [CrossRef] [PubMed]
152. De Gaetano, G.; Constanzo, S.; Castelnuovo, A.D.; Badimon, L.; Bejko, D.; Alkerwi, A.; Chiva-Blanch, G.; Estruch, R.; La Vecchia, C.; Panico, S. Effects of moderate beer consumption on health and disease: A consensus document. *Nutr. Metab. Cardiovasc. Dis.* **2016**, *26*, 443–467. [CrossRef] [PubMed]
153. Wannamethee, S.G.; Shaper, A.G. Type of alcoholic drink and risk of major coronary heart disease event and all-cause mortality. *Am. J. Public Health* **1999**, *89*, 685–690. [CrossRef] [PubMed]
154. Chiva-Blanch, G.; Magraner, E.; Condines, X.; Valderas-Martínez, P.; Roth, I.; Arranz, S.; Casas, R.; Navarro, M.; Hervas, A.; Sisó, A.; et al. Effects of alcohol and polyphenols from beer on atherosclerotic biomarkers in high cardiovascular risk men: A randomized feeding trial. *Nutr. Metab. Cardiovasc. Dis.* **2015**, *24*, 36–45. [CrossRef] [PubMed]
155. Sánchez-Muniz, F.J. Dietary fibre and cardiovascular health. *Nutr. Hosp.* **2012**, *27*, 31–45. [CrossRef] [PubMed]
156. Huang, T.; Xu, M.; Lee, A.; Cho, S.; Qi, L. Consumption of whole grains and cereal fiber and total and cause-specific mortality: Prospective analysis of 367,442 individuals. *BMC Med.* **2015**, *13*, 59. [CrossRef]
157. Yang, Y.; Zhao, L.G.; Wu, Q.J.; Ma, X.; Xiang, Y.B. Association between dietary fiber and lower risk of all-cause mortality: A meta-analysis of cohort studies. *Am. J. Epidemiol.* **2015**, *181*, 83–91. [CrossRef]
158. McRae, M.P. Dietary fiber is beneficial for the prevention of cardiovascular disease: An umbrella review of Meta-analyses. *J. Chiropr. Med.* **2017**, *16*, 289–299. [CrossRef]

159. Effects of Dietary Fiber Intake on Cardiovascular Risk Factors. Available online: https://www.intechopen.com/books/recent-advances-in-cardiovascular-risk-factors/effects-of-dietary-fiber-intake-on-cardiovascular-risk-factors (accessed on 26 October 2018).
160. Ma, Y.; Hébert, J.R.; Li, W.; Bertone-Johnson, E.R.; Olendzki, B.; Pagoto, S.L.; Tinker, L.; Rosal, M.C.; Ockene, I.S.; Ockene, J.K.; et al. Association between dietary fiber and markers of systemic inflammation in the Women's Health Initiative Observational Study. *Nutrition* **2008**, *24*, 941–949. [CrossRef]
161. Nguyen, N.T.; Magno, C.P.; Lane, K.T.; Hinojosa, M.W.; Lane, J.S. Definition of Metabolic Syndrome-Scott M. Grundy. Association of Hypertension, Diabetes, Dyslipidemia, and Metabolic Syndrome with Obesity: Findings from the National Health and Nutrition Examination Survey, 1999 to 2004. *J. Am. Coll. Surg.* **2008**, *207*, 928–934. [CrossRef]
162. Qi, L.; van Dam, R.M.; Liu, S.; Franz, M.; Mantzoros, C.; Hu, F.B. Whole-grain, bran, and cereal fiber intakes and markers of systemic inflammation in diabetic women. *Diabetes Care* **2006**, *29*, 207–211. [CrossRef]
163. Estruch, R.; Martínez-González, M.A.; Corella, D.; Basora-Gallisá, J.; Ruiz-Gutiérrez, V.; Covas, M.I.; Fiol, M.; Gómez-Gracia, E.; López-Sabater, M.C.; Escoda, R.; et al. Effects of dietary fibre intake on risk factors for cardiovascular disease in subjects at high risk. *J. Epidemiol. Community Health* **2009**, *63*, 582–588. [CrossRef]
164. Masters, R.C.; Liese, A.D.; Haffner, S.M.; Wagenknecht, L.E.; Hanley, A.J. Whole and refined grain intakes are related to inflammatory protein concentrations in human plasma. *J. Nutr.* **2010**, *140*, 587–594. [CrossRef] [PubMed]
165. North, C.J.; Venter, C.S.; Jerling, J.C. The effects of dietary fibre on C-reactive protein, an inflammation marker predicting cardiovascular disease. *Eur. J. Clin. Nutr.* **2009**, *63*, 921–933. [CrossRef] [PubMed]
166. Esposito, K.; Nappo, F.; Giugliano, F.; Di Palo, C.; Ciotola, M.; Barbieri, M.; Paolisso, G.; Giugliano, D. Meal modulation of circulating interleukin 18 and adiponectin concentrations in healthy subjects and in patients with type 2 diabetes mellitus. *Am. J. Clin. Nutr.* **2003**, *78*, 1135–1140. [CrossRef]
167. Kopf, J.C.; Suhr, M.J.; Clarke, J.; Eyun, S.I.; Riethoven, J.M.; Ramer-Tait, A.E.; Rose, D.J. Role of whole grains versus fruits and vegetables in reducing subclinical inflammation and promoting gastrointestinal health in individuals affected by overweight and obesity: A randomized controlled trial. *Nutr. J.* **2018**, *17*, 72. [CrossRef] [PubMed]
168. Toole, J.F.; Malinow, M.R.; Chambless, L.E.; Spence, J.D.; Pettigrew, L.C.; Howard, V.J.; Sides, E.G.; Wang, C.H.; Stampfer, M. Lowering homocysteine in patients with ischemic strole to prevent recurrent stroke, myocardial infarction and death: The Vitamin Intervention for Stroke Prevention (VISP) randomized controlled-trial. *JAMA* **2004**, *291*, 565–575. [CrossRef]
169. Spence, J.D.; Bang, H.; Chambles, L.E.; Stampfer, M.J. Vitamin Intervention for Stroke Prevention trial: An efficacy analysis. *Stroke* **2005**, *36*, 2404–2409. [CrossRef] [PubMed]
170. Spence, J.D. Homocysteine: Call off the funeral. *Stroke* **2006**, *37*, 282–283. [CrossRef] [PubMed]
171. Root, M.M.; McGinn, M.C.; Nieman, D.C.; Henson, D.A.; Heinz, S.A.; Shanely, R.A.; Knab, A.M.; Jin, F. Combined fruit and vegetable intake is correlated with improved inflammatory and oxidant status from a cross-sectional study in a community setting. *Nutrients* **2012**, *4*, 29–41. [CrossRef] [PubMed]
172. Helmersson, J.; Arnlöv, J.; Larsson, A.; Basu, S. Low dietary intake of alpha-carotene, beta-tocopherol and ascorbic acid is associated with increased inflammatory and oxidative stress status in a Swedish cohort. *Br. J. Nutr.* **2009**, *101*, 1775–1782. [CrossRef] [PubMed]
173. Wang, L.; Gaziano, J.M.; Norkus, E.P.; Buring, J.E.; Sesso, H.D. Associations of plasma carotenoids with risk factors and biomarkers related to cardiovascular disease in middle-aged and older women. *Am. J. Clin. Nutr.* **2008**, *88*, 747–754. [CrossRef] [PubMed]
174. Lee, H.; Lee, I.S.; Choue, R. Obesity, inflammation and diet. *Pediatr. Gastroenterol. Hepatol. Nutr.* **2013**, *16*, 143–152. [CrossRef] [PubMed]
175. Gammoh, N.Z.; Rink, L. Zinc in Infection and Inflammation. *Nutrients* **2017**, *9*, 624. [CrossRef] [PubMed]
176. Mazidi, M.; Rezaie, P.; Banach, M. Effect of magnesium supplements on serum C-reactive protein: A systematic review and meta-analysis. *Arch. Med. Sci.* **2018**, *14*, 707–716. [CrossRef] [PubMed]
177. Simental-Mendia, L.E.; Sahebkar, A.; Rodriguez-Moran, M.; Zambrano-Galvan, G.; Guerrero-Romero, F. Effect of magnesium supplementation on plasma C-reactive protein concentrations: A systematic review and Meta-analysis of randomized controlled trials. *Curr. Pharm. Des.* **2017**, *23*, 4678–4686. [CrossRef] [PubMed]

178. Rodriguez, A.J.; Mousa, A.; Ebeling, P.R.; Scott, D.; de Courten, B. Effects of vitamin D supplementation on inflammatory markers in heart failure: A systematic review and meta-analysis of randomized controlled trials. *Sci. Rep.* **2018**, *8*, 1169. [CrossRef]
179. Saboori, S.; Shab-Bidar, S.; Speakman, J.R.; Rad, E.Y.; Djafarian, K. Effect of vitamin E supplementation on serum C-reactive protein level: A meta-analysis of randomized controlled trials. *Eur. J. Clin. Nutr.* **2015**, *69*, 867–873. [CrossRef]
180. De Oliveira Otto, M.C.; Alonso, A.; Lee, D.H.; Delclos, G.L.; Jenny, N.S.; Jiang, R.; Lima, J.A.; Symanski, E.; Jacobs, D.R., Jr.; Nettleton, J.A. Dietary micronutrient intakes are associated with markers of inflammation but not with markers of subclinical atherosclerosis. *J. Nutr.* **2011**, *141*, 1508–1515. [CrossRef]
181. Wang, Q.; Zhu, Z.; Liu, Y.; Tu, X.; He, J. Relationship between serum vitamin D levels and inflammatory markers in acute stroke patients. *Brain Behav.* **2018**, *8*, e00885. [CrossRef]
182. Pilz, S.; Tomaschitz, A.; März, W.; Drechsler, C.; Ritz, E.; Zittermann, A.; Cavalier, E.; Pieber, T.R.; Lappe, J.M.; Grant, W.B.; et al. Vitamin D, cardiovascular disease and mortality. *Clin. Endocrinol.* **2011**, *75*, 575–584. [CrossRef] [PubMed]
183. Brewer, L.C.; Michos, E.D.; Reis, J.P. Vitamin D in atherosclerosis, vascular disease, and endothelial function. *Curr. Drug Targets* **2011**, *12*, 54–60. [CrossRef] [PubMed]
184. Oh, J.; Weng, S.; Felton, S.K.; Bhandare, S.; Riek, A.; Butler, B.; Proctor, B.M.; Petty, M.; Chen, Z.; Schechtman, K.B.; et al. 1,25(OH)2 vitamin d inhibits foam cell formation and suppresses macrophage cholesterol uptakein patients with type 2 diabetes mellitus. *Circulation* **2009**, *120*, 687–698. [CrossRef] [PubMed]
185. Tabesh, M.; Azadbakht, L.; Faghihimani, E.; Tabesh, M.; Esmaillzadeh, A. Calcium-Vitamin D cosupplementation influences circulating inflammatory biomarkers and adipocytokines in vitamin D-insufficient diabetics: A randomized controlled clinical trial. *J. Clin. Endocrinol. Metab.* **2014**, *99*, E2485–E2493. [CrossRef] [PubMed]
186. Shargorodsky, M.; Debby, O.; Matas, Z.; Zimlichman, R. Effect of long-term treatment with antioxidants (vitamin C, vitamin E, coenzyme Q10 and selenium) on arterial compliance, humoral factors and inflammatory markers in patients with multiple cardiovascular risk factors. *Nutr. Metab.* **2010**, *7*, 55. [CrossRef]
187. Ellulu, M.S.; Rahmat, A.; Patimah, I.; Khaza'ai, H.; Abed, Y. Effect of vitamin C on inflammation and metabolic markers in hypertensive and/or diabetic obese adults: A randomized controlled trial. *Drug Des. Dev. Ther.* **2015**, *9*, 3405–3412. [CrossRef]
188. Christen, W.G.; Cook, N.R.; Van Denburgh, M.; Zaharris, E.; Albert, C.M.; Manson, J.E. Effect of Combined Treatment with folic acid, Vitamin B6, and Vitamin B12 on plasmabiomarkers of inflammation and endothelial dysfunction in women. *J. Am. Heart Assoc.* **2018**, *7*, e008517. [CrossRef]
189. Massaro, M.; Scoditti, E.; Carluccio, M.A.; De Caterina, R. Nutraceuticals and prevention of atherosclerosis: Focus on omega-3 polyunsaturated fatty acids and Mediterranean diet polyphenols. *Cardiovasc. Ther.* **2010**, *28*, e13–e19. [CrossRef]
190. Hamer, M.; Steptoe, A. Influence of specific nutrients on progression of atherosclerosis, vascular function, haemostasis and inflammation in coronary heart disease patients: A systematic review. *Br. J. Nutr.* **2006**, *95*, 849–859. [CrossRef]
191. Burke, M.F.; Burke, F.M.; Soffer, D.E. Review of cardiometabolic effects of prescription Omega-3 fatty acids. *Curr. Atheroscler. Rep.* **2017**, *19*, 6. [CrossRef]
192. Calder, P.C. The role of marine omega-3 (n-3) fatty acids in inflammatory processes, atherosclerosis and plaque stability. *Mol. Nutr. Food Res.* **2012**, *56*, 1073–1080. [CrossRef] [PubMed]
193. Wang, Q.; Liang, X.; Wang, L.; Lu, X.; Huang, J.; Cao, J.; Li, H.; Gu, D. Effect of omega-3 fatty acids supplementation on endothelial function: A meta-analysis of randomized controlled trials. *Atherosclerosis* **2012**, *221*, 536–543. [CrossRef] [PubMed]
194. Leslie, M.A.; Cohen, D.J.A.; Liddle, D.M.; Robinson, L.E.; Ma, D.W.L. A review of the effect of omega-3 polyunsaturated fatty acids on blood triacylglycerol levels in normolipidemic and borderline hyperlipidemic individuals. *Lipids Health Dis.* **2015**, *14*, 53. [CrossRef] [PubMed]
195. Yagi, S.; Aihara, K.I.; Fukuda, D.; Takashima, A.; Hara, T.; Hotchi, J.; Ise, T.; Yamaguchi, K.; Tobiume, T.; Iwase, T.; et al. Effects of Docosahexaenoic Acid on the Endothelial Function in Patients with Coronary Artery Disease. *J. Atheroscler. Thromb.* **2015**, *22*, 447–454. [CrossRef] [PubMed]

196. Kelley, D.S.; Adkins, Y. Similarities and differences between the effects of EPA and DHA on markers of atherosclerosis in human subjects. *Proc. Nutr. Soc.* **2012**, *71*, 322–331. [CrossRef] [PubMed]
197. Cawood, A.L.; Ding, R.; Napper, F.L.; Young, R.H.; Williams, J.A.; Ward, M.J.; Gudmundsen, O.; Vige, R.; Payne, S.P.; Ye, S.; et al. Eicosapentaenoic acid (EPA) from highly concentrated n-3 fatty acid ethyl esters is incorporated into advanced atherosclerotic plaques and higher plaque EPA is associated with decreasedplaque inflammation and increased stability. *Atherosclerosis* **2010**, *212*, 252–259. [CrossRef] [PubMed]
198. Thies, F.; Garry, J.M.C.; Yaqoob, P.; Rerkasem, K.; Williams, J.; Shearman, C.P.; Gallagher, P.J.; Calder, P.C.; Grimble, R.F. Association of n-3 polyunsaturated fatty acids with stability of atherosclerotic plaques: A randomised controlled trial. *Lancet* **2003**, *361*, 477–485. [CrossRef]
199. Robinson, J.G.; Stone, N.J. Antiatherosclerotic and antithrombotic effects of omega-3 fatty acids. *Am. J. Cardiol.* **2006**, *98*, 39i–49i. [CrossRef]
200. Kaliora, A.C.; Dedoussis, G.V. Natural antioxidant compounds in risk factors for CVD. *Pharmacol. Res.* **2007**, *56*, 99–109. [CrossRef]
201. Valderas-Martinez, P.; Chiva-Blanch, G.; Casas, R.; Arranz, S.; Martínez-Huélamo, M.; Urpi-Sarda, M.; Torrado, X.; Corella, D.; Lamuela-Raventós, R.M.; Estruch, R. Tomato sauce enriched with olive oil exerts greater effects on cardiovascular disease risk factors than raw tomato and tomato sauce: A randomized trial. *Nutrients* **2016**, *8*, 170. [CrossRef]
202. Mozos, I.; Stoian, D.; Caraba, A.; Malainer, C.; Horbańczuk, J.O.; Atanasov, A.G. Lycopene and Vascular Health. *Front. Pharmacol.* **2018**, *9*, 521. [CrossRef]
203. Costa-Rodrigues, J.; Pinho, O.; Monteiro, P.R.R. Can lycopene be considered an effective protection against cardiovascular disease? *Food Chem.* **2018**, *245*, 1148–1153. [CrossRef] [PubMed]
204. Cheng, H.M.; Koutsidis, G.; Lodge, J.K.; Ashor, A.; Siervo, M.; Lara, J. Tomato and lycopene supplementation and cardiovascular risk factors: A systematic review and meta-analysis. *Atherosclerosis* **2017**, *257*, 100–108. [CrossRef] [PubMed]
205. Biddle, M.J.; Lennie, T.R.; Bricker, G.V.; Kopec, R.E.; Schwartz, S.J.; Moser, D.K. Lycopene dietary intervention: A pilot study in patients with heart failure. *Cardiovasc. Nurs.* **2016**, *30*, 205–212. [CrossRef] [PubMed]
206. Wang, Y.; Chung, S.J.; McCullough, M.L.; Song, W.O.; Fernandez, M.L.; Koo, S.I.; Chun, O.K. Dietary carotenoids are associated with cardiovascular disease risk biomarkers mediated by serum carotenoid concentrations. *J. Nutr.* **2014**, *144*, 1067–1074. [CrossRef] [PubMed]
207. Xu, X.R.; Zou, Z.Y.; Huang, Y.M.; Xiao, X.; Ma, L.; Lin, X.M. Serum carotenoids in relation to risk factors for development of atherosclerosis. *Clin. Biochem.* **2012**, *45*, 1357–1361. [CrossRef]
208. Thies, F.; Masson, L.F.; Rudd, A.; Vaughan, N.; Tsang, C.; Brittenden, J.; Simpson, W.G.; Duthie, S.; Horgan, G.W.; Duthie, G. Effect of a tomato-rich diet on markers of cardiovascular disease risk in moderately overweight, disease-free, middle-aged adults: A randomized controlled trial. *Am. J. Clin. Nutr.* **2012**, *95*, 1013–1022. [CrossRef]
209. Ras, R.T.; Geleijnse, J.M.; Trautwein, E.A. LDL-cholesterol-lowering effect of plant sterols and stanols across different dose ranges: A meta-analysis of randomised controlled studies. *Br. J. Nutr.* **2014**, *112*, 214–219. [CrossRef]
210. Cabra, C.E.; Simas-TorresKlein, M.R. Phytosterols in the Treatment of Hypercholesterolemia and Prevention of Cardiovascular Diseases. *Arq. Brasil. Cardiol.* **2017**, *109*, 475–482. [CrossRef]
211. Demonty, I.; Ras, R.T.; van der Knaap, H.C.; Duchateau, G.S.; Meijer, L.; Zock, P.L.; Geleijnse, J.M.; Trautwein, E.A. Continuous dose-response relationship of the LDL-cholesterol-lowering effect of phytosterol intake. *J. Nutr.* **2009**, *139*, 271–284. [CrossRef]
212. Rocha, V.Z.; Ras, R.T.; Gagliardi, A.C.; Mangili, L.C.; Trautwein, E.A.; Santos, R.D. Effects of phytosterols on markers of inflammation: A systematic review and meta-analysis. *Atherosclerosis* **2016**, *48*, 76–83. [CrossRef] [PubMed]
213. Ras, R.T.; Fuchs, D.; Koppenol, W.P.; Schalkwijk, C.G.; Otten-Hofman, A.; Garczarek, U.; Greyling, A.; Wagner, F.; Trautwein, E.A. Effect of a plant sterol-enriched spread on biomarkers of endothelial dysfunction and low grade inflammation in hypercholesterolaemic subjects. *J. Nutr. Sci.* **2016**, *5*, e44. [CrossRef] [PubMed]
214. Devaraj, S.; Jialal, I.; Rockwood, J.; Zak, D. Effect of orange juice and beverage with phytosterols on cytokines and PAI1 activity. *Clin. Nutr.* **2011**, *30*, 668–671. [CrossRef] [PubMed]

215. Heggen, E.; Kirkhus, B.; Pedersen, J.I.; Tonstad, S. Effects of margarine enriched with plant sterol esters from rapeseed and tall oils on markers of endothelial function, inflammation and hemostasis. *Scand. J. Clin. Lab. Investig.* **2015**, *75*, 189–192. [CrossRef] [PubMed]
216. Mozaffarian, D.; Wu, J.H.Y. Flavonoids, Dairy Foods and Cardiovascular and Metabolic Health: A review of emerging biologic pathways. *Circ. Res.* **2018**, *122*, 369–384. [CrossRef] [PubMed]
217. Tressera-Rimbau, A.; Arranz, S.; Eder, M.; Vallverdú-Queralt, A. Dietary Polyphenols in the Prevention of Stroke. *Oxid. Med. Cell. Longev.* **2017**, *2017*, 7467962. [CrossRef] [PubMed]
218. González-Gallego, J.; García-Mediavilla, M.V.; Sánchez-Campos, S.; Tuñón, M.J. Fruit polyphenols immunity and inflammation. *Br. J. Nutr.* **2010**, *104*, S15–S27. [CrossRef]
219. Bahramsoltani, R.; Ebrahimi, F.; Farzaei, M.H.; Baratpourmoghaddam, A.; Ahmadi, P.; Rostamiasrabadi, P.; Rasouli Amirabadi, A.H.; Rahimi, R. Dietary polyphenols for atherosclerosis: A comprehensive review and future perspectives. *Crit. Rev. Food Sci. Nutr.* **2017**, 1–19. [CrossRef]
220. Witkowska, A.M.; Waśkiewicz, A.; Zujko, M.E.; Szcześniewska, D.; Pająk, A.; Stepaniak, U.; Drygas, W. Dietary polyphenol intake but not the dietary total antioxidant capacity is inversely related to cardiovascular disease in postmenopausal polish women: Results of WOBASZ and WOBASZ II Studies. *Oxid. Med. Cell. Longev.* **2017**, *2017*, 5982809. [CrossRef]
221. Chiva-Blanch, G.; Badimon, L. Effects of polyphenol intake on metabolic syndrome: Current Evidences from Human Trials. *Oxid. Med. Cell. Longev.* **2017**, *2017*, 5812401. [CrossRef]
222. Miranda, A.M.; Steluti, J.; Fisberg, R.M.; Marchioni, D.M. Association between Coffee Consumption and Its Polyphenols with Cardiovascular Risk Factors: A Population-Based Study. *Nutrients* **2017**, *9*, 276. [CrossRef]
223. Ding, M.; Bhupathiraju, S.N.; Satija, A.; van Dam, R.M.; Hu, F.B. Long-term coffee consumption and risk of cardiovascular disease: A systematic review and a dose-response meta-analysis of prospective cohort studies. *Circulation* **2014**, *129*, 643–659. [CrossRef] [PubMed]
224. Larsson, S.C.; Orsini, N. Coffee consumption and risk of stroke: A dose-response meta-analysis of prospective studies. *Am. J. Epidemiol.* **2011**, *174*, 993–1001. [CrossRef] [PubMed]
225. Shen, L.; Song, L.G.; Ma, H.; Jin, C.N.; Wang, J.A.; Xiang, M.X. Tea consumption and risk of stroke: A dose-response meta-analysis of prospective studies. *J. Zhejiang Univ. Sci. B* **2012**, *13*, 652–662. [CrossRef] [PubMed]
226. Manach, C.; Mazur, A.; Scalbert, A. Polyphenols and Prevention of cardiovascular diseases. *Curr. Opin. Lipidol.* **2005**, *16*, 77–84. [CrossRef] [PubMed]
227. Hooper, L.; Kroon, P.A.; Rimm, E.B.; Cohn, J.S.; Harvey, I.; Le Cornu, K.A.; Ryder, J.J.; Hall, W.L.; Cassidy, A. Flavonoids, flavonoid-rich foods, and cardiovascular risk: A meta-analysis of randomized controlled trials. *Am. J. Clin. Nutr.* **2008**, *88*, 38–50. [CrossRef]
228. Van Bussel, B.C.T.; Henry, R.M.A.; Schalkwijk, C.G.; Dekker, J.M.; Nijpels, G.; Feskens, E.J.M.; Stehouwer, C.D.A. Alcohol and red wine consumption but not fruit vegetables fish or dairy products are associated with less endothelial dysfunction and less low-grade inflammation: The Hoorn Study. *Eur. J. Nutr.* **2018**, *57*, 1409–1419. [CrossRef] [PubMed]
229. Huang, H.; Chen, G.; Liao, D.; Zhu, Y.; Xue, X. Effects of Berries Consumption on Cardiovascular Risk Factors: A Meta-analysis with Trial Sequential Analysis of Randomized Controlled Trials. *Sci. Rep.* **2016**, *6*, 23625. [CrossRef]
230. Luís, Â.; Domingues, F.; Pereira, L. Association between berries intake and cardiovascular diseases risk factors: A systematic review with meta-analysis and trial sequential analysis of randomized controlled trials. *Food Funct.* **2018**, *9*, 740–757. [CrossRef]
231. De Bacquer, D.; Clays, E.; Delanghe, J.; De Backer, G. Epidemiological evidence for an association between habitual tea consumption and markers of chronic inflammation. *Atherosclerosis* **2006**, *189*, 428–435. [CrossRef]
232. Vazquez-Agell, M.; Urpi-Sarda, M.; Sacanella, E.; Camino-Lopez, S.; Chiva-Blanch, G.; Llorente-Cortes, V.; Tobias, E.; Roura, E.; Andres-Lacueva, C.; Lamuela-Raventos, R.M.; et al. Cocoa consumption reduces NF-kB activation in peripheral blood mononuclear cells in humans. *Nutr. Metab. Cardiovasc. Dis.* **2013**, *23*, 257–263. [CrossRef] [PubMed]
233. Monagas, M.; Khan, N.; Andres-Lacueva, C.; Casas, R.; Urpí-Sardà, M.; Llorach, R.; Lamuela-Raventós, R.M.; Estruch, R. Effect of cocoa powder on the modulation of inflammatory biomarkers in patients at high risk of cardiovascular disease. *Am. J. Clin. Nutr.* **2009**, *90*, 1144–1150. [CrossRef] [PubMed]

234. Basu, A.; Du, M.; Sanchez, K.; Leyva, M.J.; Betts, N.M.; Blevins, S.; Wu, M.; Aston, C.E.; Lyons, T.J. Green tea minimally affects biomarkers of inflammation in obese subjects with metabolicsyndrome. *Nutrition* **2011**, *27*, 206–213. [CrossRef] [PubMed]
235. Zhang, X.; Zhu, Y.; Song, F.; Yao, Y.; Ya, F.; Li, D.; Ling, W.; Yang, Y. Effects of purified anthocyanin supplementation on platelet chemokines in hypocholesterolemic individuals: A randomized controlled trial. *Nutr. Metab.* **2016**, *13*, 86. [CrossRef] [PubMed]
236. Song, F.; Zhu, Y.; Shi, Z.; Tian, J.; Deng, X.; Ren, J.; Andrews, M.C.; Ni, H.; Ling, W.; Yang, Y. Plant food anthocyanins inhibit platelet granule secretion in hypercholesterolaemia: Involving the signaling pathway of PI3K–Akt. *Thromb. Haemost.* **2014**, *112*, 981–991. [CrossRef] [PubMed]
237. Davinelli, S.; Bertoglio, J.C.; Zarrelli, A.; Pina, R.; Scapagnini, G. A Randomized clinical trial evaluating the efficacy of an anthocyanin-maqui berry extract (Delphinol®) on oxidative stress biomarkers. *J. Am. Coll. Nutr.* **2015**, *34*, 28–33. [CrossRef] [PubMed]
238. Hall, W.L.; Vafeiadou, K.; Hallund, J.; Bügel, S.; Koebnick, C.; Reimann, M.; Ferrari, M.; Branca, F.; Talbot, D.; Dadd, T.; et al. Soy-isoflavone-enriched foods and inflammatory biomarkers of cardiovascular disease risk in postmenopausal women: Interactions with genotype and equol production. *Am. J. Clin. Nutr.* **2005**, *82*, 1260–1268. [CrossRef] [PubMed]
239. Hodis, H.N.; Mack, W.J.; Kono, N.; Azen, S.P.; Shoupe, D.; Hwang-Levine, J.; Petitti, D.; Whitfield-Maxwell, L.; Yan, M.; Franke, A.A.; et al. Isoflavone soy protein supplementation and atherosclerosis progression in healthy postmenopausal women: A randomized controlled trial. *Stroke* **2011**, *42*, 3168–3175. [CrossRef] [PubMed]
240. Tomé-Carneiro, J.; Gonzálvez, M.; Larrosa, M.; García-Almagro, F.J.; Avilés-Plaza, F.; Parra, S.; Yáñez-Gascón, M.J.; Ruiz-Ros, J.A.; García-Conesa, M.T.; Tomás-Barberán, F.A.; et al. Consumption of a grape extract supplement containing resveratrol decreases oxidized LDL and ApoB in patients undergoing primary prevention of cardiovascular disease: A triple-blind, 6-month follow-up, placebo-controlled, randomized trial. *Mol. Nutr. Food Res.* **2012**, *56*, 810–821. [CrossRef] [PubMed]
241. Agarwal, B.; Campen, M.J.; Channell, M.M.; Wherry, S.J.; Varamini, B.; Davis, J.G.; Baur, J.A.; Smoliga, J.M. Resveratrol for primary prevention of atherosclerosis: Clinical trial evidence for improved gene expression in vascular endothelium. *Int. J. Cardiol.* **2013**, *166*, 246–248. [CrossRef] [PubMed]
242. Seyyedebrahimi, S.; Khodabandehloo, H.; Nasli Esfahani, E.; Meshkani, R. The effects of resveratrol on markers of oxidative stress in patients with type 2 diabetes: A randomized, double-blind, placebo-controlled clinical trial. *Acta Diabetol.* **2018**, *55*, 341–353. [CrossRef] [PubMed]

International Journal of
Molecular Sciences

Review

Marine Omega-3 (N-3) Fatty Acids for Cardiovascular Health: An Update for 2020

Jacqueline K. Innes [1] and Philip C. Calder [1,2,*]

1 School of Human Development and Health, Faculty of Medicine, University of Southampton, Southampton SO16 6YD, UK; innesjackie@gmail.com
2 National Institute for Health Research Southampton Biomedical Research Centre, University Hospital Southampton NHS Foundation Trust and University of Southampton, Southampton SO16 6YD, UK
* Correspondence: pcc@soton.ac.uk; Tel.: +44-23281-205250

Received: 25 January 2020; Accepted: 14 February 2020; Published: 18 February 2020

Abstract: The omega-3 (n-3) fatty acids, eicosapentaenoic acid (EPA) and docosahexaenoic acid (DHA), are found in seafood (especially fatty fish), supplements and concentrated pharmaceutical preparations. Long-term prospective cohort studies consistently demonstrate an association between higher intakes of fish, fatty fish and marine n-3 fatty acids (EPA + DHA) or higher levels of EPA and DHA in the body and lower risk of developing cardiovascular disease (CVD), especially coronary heart disease (CHD) and myocardial infarction (MI), and cardiovascular mortality in the general population. This cardioprotective effect of EPA and DHA is most likely due to the beneficial modulation of a number of known risk factors for CVD, such as blood lipids, blood pressure, heart rate and heart rate variability, platelet aggregation, endothelial function, and inflammation. Evidence for primary prevention of CVD through randomised controlled trials (RCTs) is relatively weak. In high-risk patients, especially in the secondary prevention setting (e.g., post-MI), a number of large RCTs support the use of EPA + DHA (or EPA alone) as confirmed through a recent meta-analysis. This review presents some of the key studies that have investigated EPA and DHA in the primary and secondary prevention of CVD, describes potential mechanisms for their cardioprotective effect, and evaluates the more recently published RCTs in the context of existing scientific literature.

Keywords: eicosapentaenoic acid; docosahexaenoic acid; omega-3 polyunsaturated fatty acids; cardiovascular disease; coronary heart disease

---

1. Marine Omega-3 Fatty Acids: Sources and Intakes

Omega-3 (n-3) fatty acids are a family of polyunsaturated fatty acids. They are characterised by, and named according to, the presence of the closest double bond to the methyl end of the hydrocarbon (acyl) chain being on carbon number three, if the methyl carbon is counted as number one. The most functionally important n-3 fatty acids appear to be eicosapentaenoic acid (EPA; 20:5n-3) and docosahexaenoic acid (DHA; 22:6n-3) [1]; however, roles for docosapentaenoic acid (22:5n-3) have also emerged now [2]. The best dietary source of EPA and DHA (and also docosapentaenoic acid) is seafood, especially fatty fish (also called 'oily fish'). The blubber and tissues of sea mammals, such as whales and seals, also contain EPA and DHA in significant amounts. Various supplements, including fish oils, cod liver oil, krill oil and some algal oils, contain EPA and DHA. Finally, concentrated pharmaceutical-grade preparations of EPA and DHA, or EPA alone, are available. Typical values for the EPA and DHA content of selected fish, n-3 fatty acid supplements and pharmaceutical preparations are shown in Table 1. EPA and DHA are often referred to as marine n-3 fatty acids because of their association with seafood.

Table 1. Content of EPA and DHA in fatty fish, lean fish, supplements and pharmaceuticals.

Fish Type	Typical EPA + DHA per Adult Serving	Comment
Fatty (e.g., salmon, trout, mackerel, sardines and herring)	1–3.5 g	Usually more EPA than DHA; content depends on the type of fish, season, water temperature, diet, stage of life cycle, wild or farmed and method of cooking
Lean (e.g., cod, plaice, haddock and sea bass)	0.1–0.3 g	Usually more EPA than DHA
Supplement Type	**Typical EPA + DHA Content per g of oil**	
Cod liver oil	200 mg	Usually more EPA than DHA
Standard "fish oil"	300 mg	Usually more EPA than DHA
Fish oil concentrate	450–600 mg	Usually more EPA than DHA
Tuna oil	460 mg	More DHA than EPA
Krill oil	205 mg	Usually more EPA than DHA; some in phospholipid form
Algal oil	400 mg	Mainly DHA
Flaxseed oil	0 mg	Contains α-linolenic acid, but not EPA or DHA
Pharmaceuticals	**Typical EPA + DHA Content per g of oil**	
Omacor/Lovaza	460 mg EPA + 380 mg DHA	In ethyl ester form
Omtryg	465 mg EPA + 375 mg DHA	In ethyl ester form
Epanova	550 mg EPA + 200 mg DHA	In free fatty acid form
Vascepa/icosapent ethyl	900 mg EPA	In ethyl ester form

Abbreviations: DHA, docosahexaenoic acid; EPA, eicosapentaenoic acid.

The various n-3 fatty acids are related metabolically to one another, and the pathway of conversion of plant-derived n-3 fatty acids (e.g., α-linolenic acid (ALA; 18:3n-3)) to EPA and then to DHA is shown in Figure 1. Studies in humans have identified that there is a fairly low rate of conversion of ALA along this pathway, especially all the way to DHA [3,4]. It is now recognised that this conversion is influenced by several factors, including the stage of life course, age, sex, various hormones, genetics and diseases [4].

The intake of EPA and DHA from diet is strongly influenced by fish consumption because fish, in general, and fatty fish, in particular, are the richest dietary source of these fatty acids. The intake of fish and fatty fish is high in some countries, such as Japan, but it is low in many Western countries, including the USA and the United Kingdom. As a result, the intake of EPA + DHA among adults varies among different populations and is low in most Western countries; it is generally considered that in non-fatty fish eaters, the intake of EPA + DHA is <0.2 g/day [5,6]. This is lower than the recommended intake for the general population [7–9]. Nevertheless, despite these low intakes, it is evident that the recommendations (typically 0.2–0.5 g/day depending upon the authority making the recommendation) can be met by including fatty fish in the diet on a regular basis or, if that is not possible, by using supplements that contain EPA and DHA (Table 1).

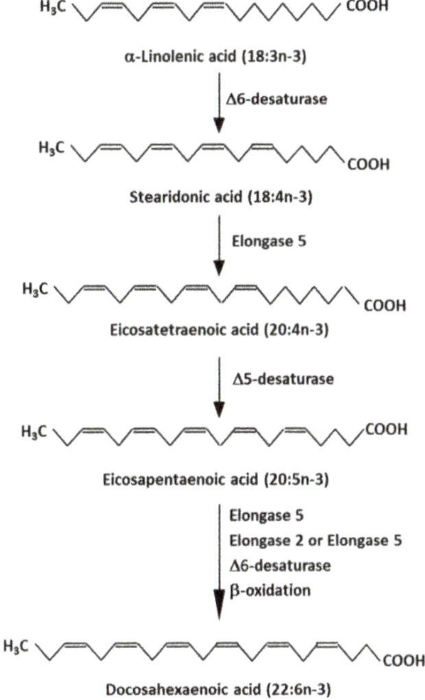

Figure 1. Metabolic pathway of conversion of the plant essential n-3 fatty acid, α-linolenic acid (18:3n-3), to eicosapentaenoic acid (20:5n-3) and docosahexaenoic acid (22:6n-3).

2. Strong Evidence for a Protective Effect of EPA and DHA Towards Cardiovascular Disease Emerges from Ecological, Case Control and Cohort Studies

The potential for EPA and DHA to have a role in reducing the risk of cardiovascular disease (CVD) was first identified by studies in the Greenland Inuit, where the low rate of mortality from myocardial infarction (MI) and ischaemic heart disease [10,11] was linked to the very high dietary intake of EPA and DHA [12]. These observations were replicated in other native Arctic populations [13] and the Japanese population [14]. Subsequently, substantial evidence accumulated from epidemiological and case-control studies in Western populations indicating that consumption of fish, fatty fish, or EPA and

DHA is associated with reduced risk of mortality from CVD, especially coronary heart disease (CHD) (reviewed in [15]). For example, in the Nurse's Health Study, there was an inverse dose-dependent association of risk for developing CHD, having a non-fatal MI or dying from CHD across quintiles of intake of EPA + DHA [16]. The risk for all three outcomes was about 50% in the group with the highest intake compared with the group with the lowest intake of EPA+DHA. The intake of EPA and DHA is highly correlated with their concentrations in blood lipids and red blood cells [17]. A number of studies have associated the concentrations of EPA + DHA (often expressed as a proportion of total fatty acids) in blood plasma or serum, plasma lipid fractions, whole blood, red blood cells and adipose tissue with lower cardiovascular morbidity and mortality (reviewed in [15]). For example, in the Physician's Health Study, there was an inverse dose-dependent association of risk for sudden death across quartiles of whole blood EPA + DHA, with an 80% lower risk in those with the highest whole blood EPA+DHA concentration compared to those with the lowest whole blood EPA + DHA concentration [18]. More recently, the largest prospective cohort study conducted to date included ~420,000 participants from the National Institutes of Health AARP Diet and Health Study with a 16-year follow-up and reported a significant inverse association between fish and EPA + DHA intake and various mortality outcomes [19]. Comparing the highest with lowest quintiles of fish intake, both men and women had 10% lower CVD mortality. EPA + DHA intake was associated with 15% and 18% lower CVD mortality in men and women, respectively, across extreme quintiles.

Cohort studies associating the intake of fish or marine n-3 fatty acids with cardiovascular or coronary outcomes have been subject to a number of meta-analyses. These include a 2012 aggregation of seven prospective cohort studies, including 176,441 participants, which investigated the association between dietary fish, EPA+DHA intake or plasma EPA + DHA concentrations and heart failure [20]. The investigators found a 15% risk reduction of heart failure associated with the highest versus lowest fish intake and a 14% lower risk of heart failure for those with the highest intake compared to those with the lowest dietary intake or plasma concentrations of EPA + DHA. A comprehensive meta-analysis, published in 2014, investigated the association between dietary intakes or blood levels of different classes of fatty acids (including n-3 fatty acids) and combined coronary disease outcomes [21]. The aggregation of data from 16 studies involving over 422,000 individuals showed a risk reduction of 13% for those in the top tertile of dietary EPA + DHA intake compared with those in the lower tertile of intake. Furthermore, the aggregation of data from 13 studies involving over 20,000 individuals showed risk reductions of 22%, 21% and 25% for those in the top tertile of circulating EPA, DHA and EPA + DHA, respectively, compared with those in the lower tertile. Alexander et al. [22] brought together data from prospective cohort studies examining the association of dietary EPA and DHA with risk of various coronary outcomes. The aggregation of data from 17 studies showed an 18% risk reduction for any CHD event for those with higher dietary intake of EPA + DHA compared to those with lower intake. There were also significant reductions of 23%, 19% and 47% in the risk for fatal coronary events, coronary death and sudden cardiac death, respectively.

The association between EPA or DHA concentration in a body compartment, such as plasma, serum, red blood cells or adipose tissue, and risk of future CHD in adults who were healthy at study entry was investigated by pooling data from 19 studies involving over 45,000 individuals [23]. EPA and DHA were each independently associated with a lower risk of fatal CHD, with a 10% lower risk for each one standard deviation increase in content [23]. The omega-3 index is the red blood cell content of EPA+DHA expressed as a proportion of total fatty acids [24]. Omega-3 index is a marker of both long-term dietary intake of these fatty acids and their tissue levels and is suggested to be a marker of CHD risk [24]. Harris et al. [25] used data from 10 cohort studies and identified a 15% reduction in risk of fatal CHD for each one standard deviation increase in omega-3 index.

3. Mechanisms by which EPA and DHA Reduce the Risk of Cardiovascular Disease

Prospective cohort studies have the advantage of a very long follow-up time to observe health outcomes in what starts as a generally healthy study population, something which is typically not possible in randomised control trials (RCTs). There are well-recognised limitations of such cohort studies, including the lack of ability to show causation. Despite this significant limitation, the considerable number of large prospective cohort studies conducted to date that have consistently shown an inverse association between dietary, blood or tissue EPA and DHA and incidence of mortality from CVD provide important evidence for the key role of marine n-3 fatty acids in the prevention of CVD. As such, there has been much interest in the mechanisms by which n-3 fatty acids, specifically EPA and DHA, achieve their cardioprotective action, with much attention being focused on the potential modulation of key cardiovascular risk factors. These risk factors include high blood pressure, high serum triglycerides, low high-density lipoprotein (HDL)-cholesterol, elevated post-prandial lipaemia, endothelial dysfunction, cardiac arrhythmia, heart rate and heart rate variability and a tendency towards thrombosis and inflammation. Large numbers of studies, including many RCTs, in humans have investigated the effect of the combination of EPA and DHA on these risk factors and many of these studies have been included in a number of meta-analyses performed in recent years (Table 2). These meta-analyses demonstrate that EPA and DHA lower triglycerides [26], lower the blood pressure (both systolic and diastolic) [26,27], reduce the heart rate and increase heart rate variability [26,28–30] and reduce platelet aggregation [31], whilst appearing to increase both low density lipoprotein (LDL)- and HDL-cholesterol [26]. Regarding vascular endothelial function, EPA and DHA have been demonstrated to improve flow-mediated dilatation [32,33] and arterial compliance [34]. Concerning the effect of EPA and DHA on inflammation, several meta-analyses have reported that they lower blood concentrations of the acute phase protein, C-reactive protein (CRP), and the pro-inflammatory cytokines, tumour necrosis factor (TNF)-α and interleukin (IL)-6 [26,35], although the effect may be dependent on the health status of the individual. Furthermore, EPA and DHA have been reported to decrease the plasma or serum concentrations of pro-inflammatory eicosanoids like thromboxane B_2 and leukotriene B_4 [36].

Whilst the majority of the scientific evidence base to-date has focused on the administration of EPA and DHA in combination (as occurs naturally in fish and most supplements), there has been much interest in the potential for EPA and DHA to have independent roles in cardiovascular risk reduction. A recent systematic review of the scientific literature concluded that EPA and DHA appear to have differential effects on a number of cardiometabolic outcomes [37]. For example, regarding modulation of blood lipids, whilst both EPA and DHA lowered blood triglycerides, there was evidence for a slightly larger triglyceride-lowering effect for DHA [38,39]. Whilst neither EPA nor DHA affected total cholesterol concentrations to a significant degree, there was an independent effect on other blood lipid parameters, with EPA lowering the HDL3-cholesterol subfraction and DHA increasing the more cardioprotective HDL2-cholesterol [40,41]. DHA also increased LDL-cholesterol more than EPA, an effect observed more in men than in women, and increased LDL particle size, an effect which was not observed with EPA [39–41]. From the more limited trial data, DHA appears to be more effective than EPA at lowering blood pressure and heart rate in normotensive individuals, whilst neither EPA nor DHA had any effect in hypertensive diabetic patients [40–43]. DHA also appeared to increase vasodilatory effects and reduce constrictor effects in the vasculature [44]. Both EPA and DHA were equally effective at increasing systemic arterial compliance [45]. In terms of platelet function, only EPA decreased platelet count and volume [46], whilst only DHA decreased collagen-stimulated platelet aggregation and platelet-derived thromboxane B_2 [47]. Interestingly, neither EPA nor DHA had any effect on fibrinolytic function [47]. Furthermore, from the limited comparative studies available, DHA seemed to be more effective than EPA at lowering a wide range of pro-inflammatory biomarkers in subjects with subclinical inflammation [39,48]. Both EPA and DHA, however, were effective at reducing biomarkers of oxidative stress (F2 isoprostanes) [49–51]. Thus, whilst there are relatively few trials that have been conducted to date that directly compare EPA and DHA, the limited data suggest that EPA and DHA have different effects with regard to cardiovascular risk factors. More research, however, is necessary to be more certain about this.

Table 2. Selected meta-analyses of the effect of marine n-3 fatty acids on cardiovascular risk factors.

Study	Cardiovascular Risk Factors Assessed	Study Design	Form and Dosage of n-3 Fatty Acids	Duration of n-3 Fatty Acid Treatment	Pooled Effects of n-3 Fatty Acids Versus Placebo
AbuMweis et al., 2018 [26]	Blood lipids, heart rate, blood pressure, inflammatory markers, platelet function and flow-mediated dilatation	Meta-analysis of 171 RCTs (up to Feb 2013) in participants in various states of health (note: the number of studies used for the analysis of different outcomes varied from 110 for triglycerides and HDL-cholesterol to 9 for flow-mediated dilatation)	Oral marine n-3 fatty acid supplements providing 0.18–15 g/d EPA+DHA	4–240 weeks	Significant dose-dependent decrease in triglycerides (MD = −0.368 mmol/L; 95% CI: −0.427–−0.309) Significant decrease in systolic blood pressure (MD = −2.195 mmHg; 95% CI: −3.171–−1.217) Significant decrease in diastolic blood pressure (MD = −1.37 mmHg; 95% CI: −2.415–−0.325) Significant decrease in heart rate (MD = −1.37 bpm; 95% CI: −2.41–−0.325) Significant decrease in CRP (MD = −0.343 mg/L; 95% CI: −0.454–−0.232) Significant increase in LDL-cholesterol (MD = 0.150 mmol/L; 95% CI: 0.058–0.243) and HDL-cholesterol (MD = 0.039 mmol/L; 95% CI: 0.024–0.054) No significant effect on total cholesterol, TNF-α, fibrinogen, platelet count, soluble intercellular adhesion molecule 1, soluble vascular cell adhesion molecule 1 or flow-mediated dilatation
Gao et al., 2013 [31]	Platelet aggregation	Meta-analysis of 15 RCTs (up to Jul 2011) including 742 participants in various states of health	Oral marine n-3 fatty acid supplements providing 0.84–6.8 g/d EPA+DHA	2–16 weeks	Significant decrease in adenosine diphosphate-induced platelet aggregation (SMD = −1.23; 95% CI: −2.24–−0.23; p = 0.02) Significant decrease in platelet aggregation units (SMD = −6.78; 95% CI: −12.58–−0.98; p = 0.02) Non-significant trend towards decreased collagen-induced and arachidonic acid-induced platelet aggregation Greater effect observed in non-healthy participants
Hidayat et al., 2017 [30]	Heart rate	Meta-analysis of 51 RCTs (up to May 2017) including ~3000 participants in various states of health	Oral marine n-3 fatty acid supplements providing 0.5–15.0 g/d EPA+DHA	2–52 weeks	Significant decrease in heart rate (WMD = −2.23 bpm; 95% CI: −3.07–−1.40); observed to be due to DHA, not EPA
Jiang et al., 2016 [36]	Pro-inflammatory eicosanoids	Meta-analysis of 18 RCTS (up to November 2015) including 826 subjects in various states of health	Oral marine n-3 fatty acid supplements providing 0.18–4.05 g/d EPA+DHA or EPA alone	4–24 weeks	Significant decrease in serum/plasma thromboxane B_2 in participants with high risk of CVD (SMD = −1.26; 95% CI: −1.65–−0.86) Significant decrease in neutrophil leukotriene B_4 in unhealthy subjects (SMD = −0.59; 95% CI: −1.02–−0.16)
Li et al., 2014 [35]	Pro-inflammatory cytokines	Meta-analysis of 68 RCTs (up to 2013) including 4601 participants in various states of health	Oral marine n-3 fatty acid supplements or dietary intake providing 0.3–6.6 g/d EPA+DHA	4–12 months	Participants with chronic disease: Significant decrease in CRP (WMD = −0.20 mg/L; 95% CI: −0.28–−0.12) and IL-6 (WMD = −0.22 pg/mL; 95%CI: −0.38–−0.06) No significant effect on TNF-α Healthy participants: Significant decrease in CRP (WMD = −0.18 mg/L; 95% CI: −0.28–−0.08) and TNF-α (WMD = −0.12 pg/mL; 95% CI: −0.16–−0.07) No significant effect on IL-6

Table 2. *Cont.*

Study	Cardiovascular Risk Factors Assessed	Study Design	Form and Dosage of n-3 Fatty Acids	Duration of n-3 Fatty Acid Treatment	Pooled Effects of n-3 Fatty Acids Versus Placebo
Miller et al., 2014 [27]	Blood pressure	Meta-analysis of 70 RCTs (up to February 2013) in normotensive and hypertensive subjects	Oral marine n-3 fatty acids from seafood, fortified foods, fish oil, algal oil and purified ethyl esters; mean EPA+DHA dose: 3.8 g/d	>3 weeks (mean study duration: 69 days)	Significant decrease in systolic blood pressure (WMD = −1.52 mmHg; 95% CI: −2.25–−0.79) Significant decrease in diastolic blood pressure (WMD = −0.99 mmHg; 95% CI: −1.54–−0.44) Significant decrease in systolic blood pressure (WMD = −4.51 mmHg; 95% CI: −6.12–−2.83) and diastolic blood pressure (WMD = −3.05 mmHg; 95% CI: −4.35–−1.74) in hypertensive individuals
Mozaffarian et al., 2005 [28]	Heart rate	Meta-analysis of 30 RCTs (up to January 2005) including 1678 healthy participants	Oral marine n-3 fatty acid supplements; median EPA+DHA intake: 3.5 g/d	>2 weeks (median study duration: 8 weeks)	Significant decrease in heart rate (WMD = −1.6 bpm; 95% CI: 0.6–2.5) In those with baseline heart rate ≥ 69 bpm, heart rate decreased by 2.5 bpm (95% CI: 1.4–3.5)
Pase et al., 2011 [34]	Arterial stiffness	Meta-analysis of 10 RCTs (up to September 2010) including 550 participants in various states of health	Oral marine n-3 fatty acid supplements providing 0.64–3 g/d EPA+DHA	6–105 weeks	Significant improvement in pulse wave velocity (SMD = 0.33; 95% CI: 0.12–0.56) Significant improvement in arterial compliance (SMD = 0.48; 95% CI: 0.24–0.72)
Wang et al., 2012 [33]	Vascular endothelial function	Meta-analysis of 16 RCTs (up to August 2011) including 901 participants in various states of health	Oral marine n-3 fatty acid supplements and dietary intake providing 0.45–4.7 g/d EPA+DHA	2 weeks to 12 months (median: 56 days)	Significant increase in flow-mediated dilatation (WMD = 2.3%; 95% CI: 0.89–3.72) No significant change in endothelium-independent vasodilation
Xin et al., 2012 [32]	Vascular endothelial function	Meta-analysis of 16 RCTs (up to February 2012) including 1385 participants in various states of health	Oral marine n-3 fatty acid supplements providing 0.45–4.53 g/d EPA+DHA	2-52 weeks	Significant increase in flow-mediated dilatation (WMD = 1.49%; 95% CI: 0.48–2.5)
Xin et al., 2013 [29]	Heart rate variability	Meta-analysis of 15 RCTS including 692 participants in various states of health	Oral marine n-3 fatty acid supplements providing 0.64–5.9 g/d EPA+DHA	6-24 weeks	Significant increase in high frequency power value of heart rate variability (SMD = 0.30). A sensitivity analysis demonstrated a significant reduction in low frequency power/high frequency power ratio with >1 g/d EPA+DHA

Abbreviations: CI, confidence interval; CVD, cardiovascular disease; CRP, C-reactive protein; DHA, docosahexaenoic acid; EPA, eicosapentaenoic acid; HDL, high-density lipoprotein; IL-6, interleukin-6; LDL, low-density lipoprotein; MD, mean difference; RCT, randomized controlled trial; SMD, standard mean difference; TNF-α, tumour necrosis factor-α; WMD, weighted mean difference.

4. RCTs of Primary Prevention of Cardiovascular Disease with Marine n-3 Fatty Acids

When compared to the large number of observational studies investigating the association between marine n-3 fatty acid exposure and cardiovascular outcomes that have been conducted to date (Section 2), only a few RCTs of sufficient size or duration have investigated the cardioprotective effects of marine n-3 fatty acids in generally healthy populations. The open-label, Japan EPA Lipid Intervention Study (JELIS) directly investigated the use of 1.8 g/d EPA (as an ethyl ester) plus statin versus statin alone in 18,645 hypercholesterolaemic participants [52]. A number of the participants were on existing cardiovascular medication (in addition to statin) and the study included hypercholesterolemic, but otherwise, healthy subjects as well as those with pre-existing CHD, with all patients being followed up for ~5 years. The primary outcome was any major coronary event, including sudden cardiac death, fatal and non-fatal MI and other non-fatal events, including unstable angina pectoris, angioplasty, stenting, and coronary artery bypass grafting. The addition of EPA to statin had no effect over statin alone on the primary outcome in the primary prevention arm of the trial. Two large primary prevention RCTs were published in late 2018 [53,54]. The ASCEND (A Study of Cardiovascular Events iN Diabetes) trial randomised 15,480 people with diabetics and no evidence of CVD to receive either marine n-3 fatty acids (840 mg/d EPA + DHA) or olive oil placebo [53]. The primary outcome was the first serious vascular event and after a mean follow-up of 7.4 years, there was no difference in the primary outcome between the two groups. In exploratory analyses, there were significantly fewer deaths from vascular events in the marine n-3 fatty acid arm (rate ratio: 0.81; 95% CI: 0.67–0.99), as well as a trend towards reduced risk of death from CHD (rate ratio: 0.79; 95% CI: 0.61–1.02). The Vitamin D and Omega-3 Trial (VITAL) trial randomised 25,871 healthy participants aged over 50 years (men) and 55 years (women) to receive marine n-3 fatty acids (840 mg/d EPA+DHA) and/or vitamin D (2000 IU/d) or placebo [54]. After a median follow-up of 5.3 years, there was no difference in the primary outcome of major cardiovascular events (a composite of MI, stroke or death from cardiovascular causes) in those participants supplemented with marine n-3 fatty acids versus placebo. An analysis of the individual components of the composite showed a significant reduction in the n-3 fatty acid arm in MI (hazard ratio: 0.72; 95% CI: 0.59–0.90) and CHD (hazard ratio: 0.83; 95% CI: 0.71–0.97). Correspondingly, there was also a reduced risk of death from these two non-prespecified outcomes (for MI—hazard ratio: 0.50, 95% CI: 0.26–0.97; for CHD—hazard ratio: 0.76, 95% CI: 0.49–1.16). Thus, whilst RCT evidence in primary prevention is less clear than that from the prospective cohort studies, there is now some indication of benefit from marine n-3 fatty acids towards cardiovascular health, especially CHD, from recent large and long RCTs, such as ASCEND and VITAL.

5. RCTs of Secondary Prevention of Cardiovascular Disease with Marine n-3 Fatty Acids

A number of large, randomized, controlled, secondary prevention trials or trials in high-risk patients have been conducted to investigate the effect of EPA and DHA in patients with established CVD. These trials generate a changing picture with time.

5.1. *Secondary Prevention Trials and Meta-Analyses Published Prior to 2010*

Several large, secondary prevention trials of marine n-3 fatty acids were conducted prior to 2010. The Diet and Reinfarction Trial (DART) included 2,033 recent (mean: 41 days) MI survivors, who were given dietary advice concerning fat, fish and fibre intake and followed up for 2 years [55]. Those patients advised to eat at least two portions of fatty fish per week (or to take fish oil supplements) had a 29% reduction in total mortality as well as a reduced risk of death from ischaemic heart disease at 2 years compared to those patients given other advice. The landmark Gruppo Italiano per lo Studio della Sopravvivenza nell'Infarto miocardico (GISSI)–Prevenzione trial investigated the effect of supplementation with 840 mg/d EPA+DHA in 11,324 recent (≤3 months) MI survivors versus vitamin E supplementation, supplementation with both EPA + DHA and vitamin E, and placebo control [56]. After 3.5 years, patients who received EPA + DHA had a 20% reduction in total mortality, 30% reduction in cardiovascular death and 45% decrease in sudden death compared to those that did not receive EPA + DHA. There was no benefit reported on non-fatal MI or stroke. The beneficial effect of EPA + DHA on total mortality and sudden cardiac death was observed after 3 months and 4 months of supplementation, respectively, and raised interest in the potential anti-arrhythmic action of EPA and DHA [57]. The GISSI investigators undertook a separate RCT in 6,975 patients with chronic heart failure (GISSI-HF) to investigate the effect of 840 mg/d EPA+DHA versus placebo over a period of ~4 years in this patient population [58]. The investigators reported a small (9%), but significant, reduction in all-cause mortality and a small (8%) reduction in combined all-cause mortality or admission to hospital for cardiovascular reasons in this high-risk population following supplementation with marine n-3 fatty acids. In line with the increased prescription of statins to prevent all-cause mortality and cardiovascular events at this time, JELIS directly investigated the use of 1.8 g/d EPA (as an ethyl ester) plus statin versus statin alone in 18,645 hypercholesterolaemic participants who were followed up for ~5 years [52]. As mentioned above, the addition of EPA to statin had no effect over statin alone in the primary prevention arm of JELIS, but in the secondary prevention arm, EPA caused a 19% decrease in non-fatal coronary events compared with the statin alone group [52]. Unlike GISSI, JELIS found no beneficial effect on cardiovascular mortality. A subsequent analysis found an inverse association between plasma EPA levels and risk of major coronary events; participants with the highest levels of plasma EPA (≥150 μg/mL) were 20% less likely to experience a major coronary event [59].

Meta-analyses of RCTs conducted prior to 2010 confirmed the reductions in mortality seen in individual trials of marine n-3 fatty acids ([60–63]; Table 3). For example, a 2002 meta-analysis of 11 RCTs in 15,806 patients with CHD found 20% lower risk of non-fatal MI, 30% lower risk of fatal MI, 30% lower risk for sudden death and 20% lower risk for all-cause mortality in patients who received marine n-3 fatty acids versus control [60]. Another meta-analysis from 2002 of 14 RCTs and 20,260 participants, both with and without CVD, found a 23% reduction in overall mortality and 32% reduction in cardiac mortality for those patients who were given marine n-3 fatty acids versus controls [61]. A further meta-analysis in 2009 focused on 8 RCTs and 20,997 patients with CHD found a 57% reduction in sudden death in patients with prior MI who were taking marine n-3 fatty acids compared with placebo [62]. In 2009, a meta-analysis of 11 RCTs representing 39,044 patients with all stages of CVD, including both low- and high-risk patients, found a 13% reduction in cardiovascular death and in sudden death and an 8% reduction in all-cause mortality in high-risk patients who were taking marine n-3 fatty acids compared to controls [63]. The investigators also found an 8% reduction in non-fatal cardiovascular events in moderate-risk patients.

5.2. RCTs with Marine n-3 Fatty Acids in High-Risk Patients Published in the Period 2010–2013

Three RCTs published in 2010 failed to replicate the findings of earlier trials [64–66]. The OMEGA study investigated the effect of supplementation with 840 mg/d EPA + DHA for 1 year in 3,851 early post-MI patients, with the primary endpoint of sudden cardiac death [64]. Marine n-3 fatty acids did not decrease the rate of sudden cardiac death, total mortality, major adverse cerebrovascular and cardiovascular events or revascularisation compared to control. The Supplémentation en Folates et Omega-3 (SU.FOL.OM3) trial investigated the effect of supplementation of B vitamins and/or marine n-3 fatty acids (600 mg/d EPA + DHA) in 2,501 patients with documented MI, unstable angina or ischaemic stroke for ~5 years [65]; the primary outcomes were cardiovascular death, stroke and non-fatal MI. Marine n-3 fatty acids did not have any effect on these outcomes. The Alpha Omega study included 4,837 post-MI patients who were given margarine fortified with low doses of EPA + DHA (400 mg/d), ALA (2 g/d), EPA + DHA + ALA or placebo, and followed up for 40 months [66]. There was no reduction in cardiovascular events in any group. On further analysis, those patients with diabetes in the EPA + DHA as well as ALA-fortified margarine groups, however, experienced a reduction in fatal CHD and arrhythmia-related events [66].

The Outcome Reduction with an Initial Glargine Intervention (ORIGIN) trial, published in 2012, assessed the effect of supplementation with 840 mg/d EPA + DHA in 12,536 dysglycaemic patients at high risk of cardiovascular events, together with insulin glargine or standard care for a median of 6 years [67]. No effect of marine n-3 fatty acids was reported for total mortality, death from cardiovascular causes or arrhythmia compared to placebo. The Risk and Prevention Study, published in 2013, investigated the effect of supplementation with 840 mg/d marine n-3 fatty acids in 12,513 patients with multiple cardiovascular risk factors or atherosclerotic vascular disease (but no MI) for a median of 5 years, and reported no effect on hospitalisation or death from cardiovascular causes compared to placebo [68]. In a prespecified subgroup analysis, compared with placebo, marine n-3 fatty acids resulted in an 18% lower incidence of the revised primary endpoint among women. Most of the secondary endpoints (e.g., fatal or non-fatal MI or stroke, fatal or non-fatal coronary event, and sudden death) did not differ between groups; however, admissions for heart failure were significantly lower in the marine n-3 fatty acid group.

Whilst these RCTs failed to show any benefit of marine n-3 fatty acids, they did have acknowledged limitations [69,70], which are worthy of mention. OMEGA was underpowered to detect any benefit on its primary outcome, sudden cardiac death, as power calculations were based on earlier RCTs in patients without more recent treatments and hence with higher background rates of sudden cardiac death compared to those seen in OMEGA. Furthermore, reported fish consumption was relatively high among the enrolled patients and increased during the trial, thereby increasing dietary EPA + DHA intake and introducing a confounding factor. The trial also had a relatively short follow-up period of just one year. Regarding SU.FOL.OM3, the length of time between the occurrence of primary cardiovascular event and the commencement of supplementation (median of 101 days) was longer than that of GISSI-Prevenzione (≤3 months) and the dose of EPA + DHA used was lower than that of GISSI-Prevenzione (840 mg/d EPA + DHA) and JELIS (1.8 g/d EPA). There were also fewer than expected background cardiac deaths, perhaps also due to more effective pharmacological management of cardiovascular risk factors. With respect to Alpha Omega, despite the long follow-up time, the dose of EPA+DHA used was modest and enrolment to the trial occurred on average at 4 years post-MI, despite the existence of data suggesting that early intervention with EPA + DHA is required to see a beneficial effect post-MI. ORIGIN included a large number of participants taking effective cardiovascular risk-reducing medications compared to earlier trials, and this may have made the cardioprotective effect of EPA + DHA harder to detect. The Risk and Prevention Study set out to examine the effect of marine n-3 fatty acids on the composite primary outcome of death, non-fatal MI and non-fatal stroke; however, the event rate at one year was lower than anticipated and the primary outcome was revised to hospitalisation or death from cardiovascular causes over the follow-up period.

Unsurprisingly, several meta-analyses conducted over the last ten years reflect the inclusion of these null RCTs and report more mixed conclusions than earlier meta-analyses ([21,22,71–79]; Table 3). A meta-analysis from 2013 (representing 11 RCTs and 15,348 patients with CVD) reported a 32% reduction in cardiac death, 33% reduction in sudden death and 25% reduction in myocardial infarction, with no effect on all-cause mortality or stroke [75]. This meta-analysis included only those RCTs with EPA + DHA dosages of at least 1 g/d and with a duration of at least 1 year. Another meta-analysis published in 2014, which included a broader range of marine n-3 fatty acid dosages and durations and focused only on patients with ischaemic CHD, reported that whilst there were no effects on cardiovascular events, there was a 12% reduction in death from cardiac causes, 14% reduction in sudden cardiac death and 8% reduction in all-cause mortality in those taking EPA + DHA versus placebo [76]. Published in 2017, a meta-analysis focusing solely on cardiac death as the outcome examined 14 RCTs (involving 71,899 subjects in both mixed and secondary prevention settings) with marine n-3 fatty acid supplementation for a duration of at least 6 months [77]. The study found an 8% lower risk of cardiac death in the primary analysis and a 13%–29% lower risk of cardiac death in those participants with intakes of EPA + DHA of at least 1g/d and for certain subgroups, such as subjects with high plasma triglycerides or cholesterol, and where <40% subjects were taking statins.

Table 3. Meta-analyses published up to 2018 of RCTs investigating the effect of marine n-3 fatty acids on cardiovascular outcomes.

Study	Study Design	Form & Dosage of Marine-3 Fatty Acids	Duration of Treatment with Marine n-3 Fatty Acids	Pooled Effects of Marine n-3 Fatty Acids Versus Placebo
Bucher et al., 2002 [60]	11 RCTs (up to August 1999) representing 15,806 patients with CHD	Dietary (2 RCTs) and supplemental (9 RCTs) marine n-3 fatty acids with a dose range of 0.3–6.0 g/d EPA and 0.6–3.7 g/d DHA	6–46 months (mean: 20 months)	30% reduction in fatal MI 30% reduction in sudden death 20% reduction in overall mortality
Studer et al., 2005 [61]	14 RCTs (up to June 2003) representing 20,260 participants in primary and secondary prevention settings	Supplemental marine n-3 fatty acids; dose range not given	Mean: 1.9 ± 1.2 years	23% reduction in overall mortality 32% reduction in cardiovascular mortality
Zhao et al., 2009 [62]	8 RCTs (up to June 2008) representing 20,997 patients with CHD	Dietary (3 RCTs) and supplemental (5 RCTs) marine n-3 fatty acids with a dose range of 0.3–4.1 g/d EPA and 0.4–2.8 g/d DHA	9–108 months (mean: 33 months)	57% reduction in sudden death in patients with prior MI 39% increased risk of sudden death in patients with angina 29% reduction in cardiac death (NS) 23% reduction in all-cause mortality (NS)
Marik and Varon, 2009 [63]	11 RCTs (up to December 2008) representing 39,044 patients with all stages of CVD including high-risk and low-risk subjects	Supplemental marine n-3 fatty acids with a dose range of 0.7–4.8 g/d EPA + DHA (mean: 1.8 ± 1.2 g/d)	1–4.6 years (mean: 2.2 ± 1.2 years)	13% reduction in cardiovascular death in high-risk patients 13% reduction in sudden cardiac death in high-risk patients 8% reduction in all-cause mortality in high-risk patients 8% reduction in non-fatal cardiovascular events in moderate-risk patients.
Kotwal et al., 2012 [71]	20 RCTs (up to March 2011) representing 62,851 patients in primary and secondary prevention settings	Diet (3 RCTs) and supplemental (17 RCTs) marine n-3 fatty acids with a dose range of 0.8–3.4 g/d EPA + DHA	6 months–6 years	14% reduction in vascular death No effect on cardiovascular events, total mortality, coronary events, arrhythmia or cerebrovascular events
Kwak et al., 2012 [72]	14 RCTs (up to April 2011) representing 20,485 patients with CVD	Supplemental marine n-3 fatty acids with a dose range of 0.4–4.8 g/d EPA + DHA (mean: 1.7 g/d EPA + DHA)	1–4.7 years (mean: 2 years)	9% reduction in cardiovascular death No effect on cardiovascular events, all-cause mortality, sudden cardiac death, MI, congestive heart failure or stroke
Trikalinos et al., 2012 [73]	18 RCTs (up to May 2011) representing 51,264 patients	Supplemental marine n-3 fatty acids with a dose range of 0.27–6.0 g/d EPA + DHA	1–5 years	11% reduction in cardiovascular mortality
Rizos et al., 2012 [74]	20 RCTs (up to August 2012) representing 68,680 patients in primary and secondary prevention settings	Diet (2 RCTs) and supplemental (18 RCTs) marine n-3 fatty acids with a dose range of 0.53–1.80 g/d EPA + DHA (median EPA + DHA dose: 1 g/d)	1–6.2 years (median: 2 years)	No effect on all-cause mortality, cardiac death, sudden death, MI or stroke

Table 3. *Cont.*

Study	Study Design	Form & Dosage of Marine-3 Fatty Acids	Duration of Treatment with Marine n-3 Fatty Acids	Pooled Effects of Marine n-3 Fatty Acids Versus Placebo
Casula et al., 2013 [75]	11 RCTs (up to March 2013) representing 15,348 patients with CVD	Supplemental marine n-3 fatty acids with a dose range of 1–6 g/d EPA + DHA	≥ 1 year (duration ranged from 1–3.5 years)	32% reduction in cardiac death 33% reduction in sudden death 25% reduction in MI 11% reduction in all-cause mortality (NS) No effect on stroke
Wen et al., 2014 [76]	14 RCTs (up to May 2013) representing 32,656 patients with CHD	Supplemental marine n-3 fatty acids with a dose range of 0.4–6.9 g/d EPA + DHA	< 3 months to 4.6 years	12% reduction in death from cardiac causes 14% reduction in sudden cardiac death 8% reduction in all-cause mortality 7% reduction in cardiovascular events (NS)
Chowdhury et al., 2014 [21]	17 RCTs (up to June 2013) representing 76,580 participants	Supplemental marine n-3 fatty acids with a dose range of 0.3 g/d EPA to 6 g/d EPA + DHA.	0.1–8 years	7% reduction in coronary outcomes (NS)
Alexander et al., 2017 [22]	18 RCTs (up to November 2015)	Supplemental marine n-3 fatty acids with a dose range of 0.4–5.0 g/d EPA + DHA	0.5–7 years	14%–16% reduction in CHD in high-risk subgroups i.e., those with elevated triglycerides and LDL-cholesterol
Maki et al., 2017 [77]	14 RCTs (up to December 2016) representing 71,899 patients in a mixed/secondary prevention setting	Supplemental marine n-3 fatty acids with a dose range of 0.27–5.0 g/d EPA + DHA	≥ 6 months (range 0.5–6.2 years)	8% reduction in cardiac death ~13%–29% reduction in cardiac death in the subgroup with high-risk individuals (secondary prevention, high triglycerides, high LDL-cholesterol and <40% statin use) and with EPA+DHA > 1 g/d
Aung et al., 2018 [78]	10 RCTs representing 77,917 high-risk patients (prior CHD or stroke)	Supplemental marine n-3 fatty acids with a dose range of 0.2–1.8 g/d EPA and 0–1.7 g/d DHA	1–6.2 years (mean: 4.4 years)	7% reduction in CHD death (NS) No effect on non-fatal MI, CHD events or major vascular events
Abdelhamid et al., 2018 [79]	79 RCTs (up to April 2017) representing 112,059 participants in primary and secondary prevention settings	Dietary or supplemental marine n-3 fatty acids with a dose range from 0.5 g/d to ~5 g/d EPA + DHA	1–7 years	7% reduction in CHD events No effect on all-cause mortality, cardiovascular mortality, cardiovascular events, CHD mortality, stroke or arrhythmia.

Abbreviations: CHD, coronary heart disease; DHA, docosahexaenoic acid; EPA, eicosapentaenoic acid; LDL, low-density lipoprotein; MI, myocardial infarction; NS, not significant; RCT, randomised controlled trial.

5.3. An Important RCT and a New Meta-Analysis were Published in 2019

The Reduction of Cardiovascular Events with Icosapent Ethyl–Intervention Trial (REDUCE-IT), published in early 2019, included 8,179 participants (29% in a primary prevention cohort with diabetes plus another cardiovascular risk factor and 71% in a secondary prevention cohort) supplemented daily with 4 g of icosapent ethyl (a highly purified EPA ethyl ester) or mineral oil placebo and followed up for a median of 4.9 years [80]. All patients were being treated with statins and had triglyceride concentrations of 135–499 mg/dL (1.52–5.63 mmol/L). The primary outcome was a composite of cardiovascular death, non-fatal MI, non-fatal stroke, coronary revascularisation or unstable angina. The patients who received icosapent ethyl had a statistically significant reduction in the primary outcome compared to placebo (hazard ratio: 0.75; 95% CI: 0.68–0.83; $p < 0.001$), the pre-specified secondary outcome (composite of cardiovascular death, non-fatal MI or non-fatal stroke) (hazard ratio: 0.80; 95% CI: 0.66–0.98; $p = 0.03$) and a whole range of other pre-specified outcomes (Figure 2). This effect was greater in the secondary prevention cohort than in the primary prevention cohort [80]. Interestingly, as with JELIS, REDUCE-IT used EPA only, but at a much higher dose (4 g daily versus 1.8 g daily) and included participants with high triglyceride levels and on statin medication. Accordingly, the participants in the icosapent ethyl arm had a significant reduction in triglycerides (a decrease of 0.5 mmol/L; $p < 0.001$) and LDL-cholesterol (a decrease of 0.13 mmol/L; $p < 0.001$) relative to placebo at 1 year. Of note, the improvement in the primary outcome in the icosapent ethyl arm of the trial was not dependent on the baseline triglyceride level or the degree of subsequent triglyceride lowering, suggesting that the reduction in cardiovascular risk in this population may be via a different mechanism independent of (or in addition to) triglyceride lowering. REDUCE-IT is of key interest as it demonstrates that even in at-risk populations that are well managed with modern pharmacological treatments, a suitably high dosage of EPA (i.e., 4 g daily) can provide an additional benefit in reducing cardiovascular-related events and mortality.

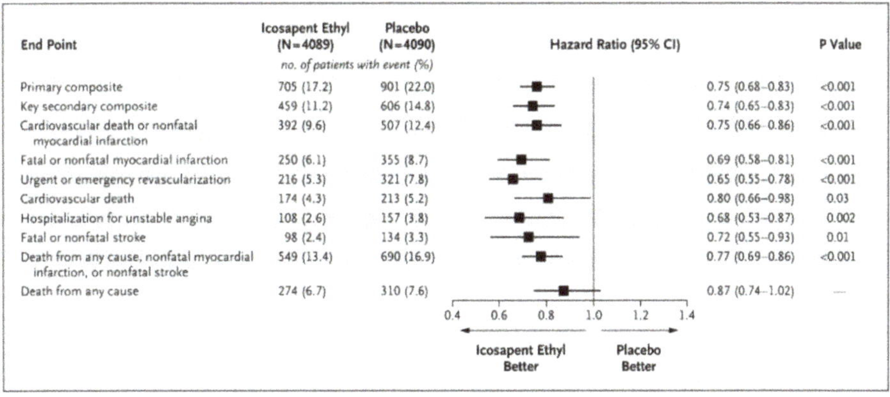

Figure 2. Effect of icosapent ethyl compared with placebo on different endpoints in REDUCE-IT [80]. "From New England Journal of Medicine, Bhatt D.L., Steg P.G., Miller M., Brinton E.A., Jacobson T.A., Ketchum S.B., Doyle R.T. Jr., Juliano R.A., Jiao L., Granowitz C., Tardif J.C., Ballantyne C.M.; REDUCE-IT Investigators, Cardiovascular Risk Reduction with Icosapent Ethyl for Hypertriglyceridemia, Volume No. 380, Page 11–22, Copyright © (2020) Massachusetts Medical Society. Reprinted with permission from Massachusetts Medical Society."

A recent meta-analysis, published in 2019, included data from 13 RCTs, including GISSI-Prevenzione, JELIS, GISSI-HF, SU.FOL.OM3, Alpha Omega, OMEGA, ORIGIN, VITAL, ASCEND and REDUCE-IT [81]. Trials had to have a sample size of at least 500 patients and a follow-up for at least one year to be included. The sample sizes ranged from 563 [82] to 25,871 [54], the mean duration of follow-up ranged from 1.0 [64] to 7.4 [53] years and the mean dose of EPA+DHA ranged from

0.37 [66] to 4.0 [80] g/d. The total sample size of the aggregated trials was 127,477, while the mean duration of follow-up was 5 years. The outcomes of interest included MI, CHD death, total CHD, total stroke, CVD death, total CVD and major vascular events. In the analysis excluding REDUCE-IT, marine n-3 fatty acid supplementation was associated with a significantly lower risk of MI, CHD death, CVD death and total CVD (Table 4). Inverse associations for all outcomes were strengthened after including REDUCE-IT (Table 4). Statistically significant linear dose–response relationships were found for several outcomes: for example, every 1 g/d EPA + DHA corresponded to 9% and 7% lower risk of MI and total CHD, respectively. The meta-analysis concluded that "marine [n-3 fatty acid] supplementation lowers risk for MI, CHD death, total CHD, CVD death, and total CVD, even after exclusion of REDUCE-IT. Risk reductions appeared to be linearly related to marine [n-3 fatty acid] dose."

Table 4. Summary of findings from the meta-analysis published by Hu et al. [81] in 2019.

Outcome	Number of Studies	Finding for Marine n-3 Fatty Acids Versus Placebo Rate Ratio [95% Confidence Interval] (p)	Finding if Data from REDUCE-IT Removed Rate Ratio [95% Confidence Interval] (p)
Myocardial infarction	13	0.88 [0.83, 0.94] (<0.001)	0.92 [0.86, 0.99] (0.020)
CHD death	12	0.92 [0.86, 0.98] (0.014)	Outcome not reported in REDUCE-IT
Total CHD	13	0.93 [0.89, 0.96] (<0.001)	0.95 [0.91, 0.99] (0.008)
Total stroke	13	1.02 [0.95, 1.10] (0.569)	1.05 [0.98, 1.14] (0.183)
CVD Death	12	0.92 [0.88, 0.97] (0.003)	0.93 [0.88, 0.99] (0.013)
Total CVD	13	0.95 [0.82, 0.98] (<0.001)	0.97 [0.94, 0.99] (0.015)
Major vascular events	13	0.95 [0.93, 0.98] (<0.001)	0.97 [0.94, 1.00] 90.058)

Abbreviations: CHD, coronary heart disease; CVD, cardiovascular disease.

6. Trusted Authority Views on Marine n-3 Fatty Acids and Cardiovascular Disease

In 2010, the European Food Safety Authority (EFSA) concluded that EPA and DHA help to maintain normal cardiac function, normal blood pressure and normal blood concentrations of triglycerides in the general population [9]. In terms of the intake of EPA and DHA required to bring about these health claims, EFSA recommend an intake of 250 mg/day of EPA+DHA to maintain normal cardiac function, 3 g/d to maintain normal blood pressure and 2 g/d to maintain normal blood concentrations of triglycerides, as part of a healthy diet [9]. The American Heart Association (AHA) also supports the use of marine n-3 fatty acids. In 2018, the AHA published guidance in support of the dietary intake of fish in primary prevention [83]; specifically, the AHA advisory recommends 1 to 2 seafood meals per week to reduce the risk of congestive heart failure, CHD, ischemic stroke and sudden cardiac death. In recognition of the triglyceride-lowering effect of EPA and DHA, the AHA recently updated its earlier recommendation for the use of 2 to 4 g/d EPA + DHA for this indication: "we conclude that prescription n-3 fatty acids, whether EPA + DHA or EPA-only, at a dose of 4 g/d, are clinically useful for reducing triglycerides, after any underlying causes are addressed and diet and lifestyle strategies are implemented, either as monotherapy or as an adjunct to other triglyceride-lowering therapies" [84]. On the basis of the positive outcomes of REDUCE-IT, the European Society of Cardiology and the European Atherosclerosis Society issued an update to the "Clinical Practice Guidelines for the Management of Dyslipidaemias" specifically recommending that "in high-risk patients with [triglyceride] levels between 1.5 and 5.6 mmol/L (135–499 mg/dL) despite statin treatment, n-3 polyunsaturated fatty acids (icosapent ethyl 2 × 2 g/day) should be considered with a statin" [85]. Furthermore, the American Diabetes Association makes a recommendation "that icosapent ethyl be considered for patients with diabetes and atherosclerotic cardiovascular disease or other cardiac risk factors on a statin with controlled low-density cholesterol, but with elevated triglycerides (135–499 mg/dL) to reduce cardiovascular risk" [86]. Finally, the National Lipid Association's position is "that for patients aged ≥45 years with clinical [atherosclerotic cardiovascular disease] ASCVD, or aged ≥50 years with diabetes mellitus requiring medication plus ≥1 additional risk factor, with fasting triglycerides 135 to 499 mg/dL on high-intensity or maximally tolerated statin therapy (±ezetimibe), treatment with icosapent ethyl is recommended for ASCVD risk reduction" [87]. In 2017, the AHA reinforced its earlier support for EPA+DHA in people with CVD [88] and extended [89] that "the recommendation for patients with prevalent CHD such as a recent MI remains essentially unchanged: Treatment with n-3 fatty acid supplements is reasonable for these patients. Even a potential modest reduction in CHD mortality (10%) in this clinical population would justify treatment with a relatively safe therapy. We now recommend treatment for patients with prevalent heart failure without preserved left ventricular function to reduce mortality and hospitalizations (9%) on the basis of a single, large RCT. Although we do not recommend treatment for patients with diabetes mellitus and prediabetes to prevent CHD, there was a lack of consensus on the recommendation for patients at high CVD risk. Because there are no reported RCTs related to the primary prevention of CHD, heart failure, and atrial fibrillation, we were not able to make recommendations for these indications." In contrast to these well thought out and carefully worded statements by the AHA, in 2014, the National Institute for Health and Clinical Excellence in the UK recommended not to use marine n-3 fatty acids for the primary prevention of CVD, for the secondary prevention of CVD or in people with diabetes [90].

7. Summary, Discussion and Conclusions

There is a large body of evidence gathered from long-term prospective cohort studies that consistently demonstrates an association between higher intakes of fish, fatty fish and marine n-3 fatty acids (EPA+DHA) or higher levels of EPA and DHA in the body and lower risk of developing CVD, especially CHD, having an MI and cardiovascular mortality in the general population. This cardioprotective effect of EPA and DHA is plausible considering the robust identification of mechanisms that show that EPA and DHA beneficially modulate a number of known risk factors for CVD, such as blood lipids, blood pressure, heart rate and heart rate variability, platelet aggregation, endothelial function and inflammation. Despite this, evidence for primary prevention of CVD through RCTs is relatively weak. In high-risk patients, especially in the secondary prevention setting (e.g., post-MI), a number of large RCTs support the use of EPA + DHA (or EPA alone) as confirmed through a very recent meta-analysis [81]. Surveying these trials serves to highlight a number of factors which may have contributed to the positive outcomes reported and why other trials have had less conclusive or even null findings. Such factors include a sufficiently high dose of EPA alone or of a combination of EPA + DHA; sufficient duration of supplementation; supplementation in post-MI patients to begin relatively soon post-MI; and the adequate powering of studies to detect an effect on the primary outcome which may have a low rate of occurrence. Given these considerations, it is unsurprising that studies of short duration or using low doses of EPA + DHA conducted in the setting of multiple pharmaceutical (and other) approaches to controlling risk have failed to demonstrate the effects of EPA + DHA. As discussed elsewhere [91,92], there are also other factors to consider when evaluating and interpreting the literature in this field. The first is that the marine n-3 fatty acid naive condition is unlikely to occur, such that any placebo-controlled trial of EPA and DHA is conducted against a certain background intake of those fatty acids in all participants, although that background intake may be very low [5,6]. Nevertheless, background intakes of EPA and DHA can be highly variable both within and between individuals, with significant changes occurring simply by eating a single meal of fatty fish. Finally, the bioavailability of EPA and DHA can vary [93] (a) among individuals due to physiological differences, (b) according to the intake of meals in relation to supplements and (c) perhaps according to the time of day, thus influencing how much of these bioactive fatty acids is actually available to exert their effects. Taken together, these factors highlight the challenges to conducting robust trials in humans with supplemental n-3 fatty acids, and they are likely to contribute to the variable outcomes from such trials, especially when low doses of EPA and DHA are used.

Recent AHA advisories support the use of marine n-3 fatty acids in the treatment of hypertriglyceridemia [84] and in a range of patients with CVD [89] and the use of fish for primary prevention of CVD [83]. Both EPA and DHA beneficially modify a range of risk factors, although DHA may be more effective [37]. Nevertheless, the highly successful REDUCE-IT used pure EPA, although at a high dose [80]. Differentiating the dose-dependent effects of EPA and DHA on cardiovascular risk factors and on clinical outcomes is going to be important. Furthermore, robust primary prevention trials are still needed. In the meantime, recent trials reviewed herein and the most recent meta-analysis support the use of marine n-3 fatty acids to reduce cardiovascular mortality.

Author Contributions: Conceptualization, P.C.C.; writing—original draft preparation, J.K.I.; writing—review and editing, J.K.I. and P.C.C. All authors have read and agreed to the published version of the manuscript.

Funding: This research received no external funding.

Conflicts of Interest: J.K.I. has no conflicts of interest to declare. P.C.C. has research funding from BASF AS; acts as an advisor/consultant to BASF AS, DSM, Cargill, Smartfish and Pfizer (now GSK) Consumer Healthcare; and has received reimbursement for travel and/or speaking from Fresenius Kabi, Pfizer (now GSK) Consumer Healthcare and Smartfish.

Abbreviations

AHA	American Heart Association
ALA	alpha-linolenic acid
CHD	coronary heart disease
CI	confidence interval
CRP	C-reactive protein
CVD	cardiovascular disease
DHA	docosahexaenoic acid
DPA	docosapentaenoic acid
EFSA	European Food Safety Authority
EPA	eicosapentaenoic acid
HDL	high-density lipoprotein
IL	interleukin
LDL	low-density lipoprotein
MD	mean difference
MI	myocardial infarction
NS	not significant
RCT	randomised controlled trial
SMD	standard mean difference
TNF	tumour necrosis factor
WMD	weighted mean difference

References

1. Calder, P.C. Very long-chain n-3 fatty acids and human health: Fact, fiction and the future. *Proc. Nutr. Soc.* **2018**, *77*, 52–72. [CrossRef]
2. Kaur, G.; Cameron-Smith, D.; Garg, M.; Sinclair, A.J. Docosapentaenoic acid (22:5n-3): A review of its biological effects. *Prog. Lipid Res.* **2011**, *50*, 28–34. [CrossRef]
3. Arterburn, L.M.; Hall, E.B.; Oken, H. Distribution, interconversion, and dose response of n-3 fatty acids in humans. *Am. J. Clin. Nutr.* **2006**, *83*, 1467S. [CrossRef]
4. Baker, E.J.; Miles, E.A.; Burdge, G.C.; Yaqoob, P.; Calder, P.C. Metabolism and functional effects of plant-derived omega-3 fatty acids in humans. *Prog. Lipid Res.* **2016**, *64*, 30–56. [CrossRef]
5. Meyer, B.J.; Mann, N.J.; Lewis, J.L.; Milligan, G.C.; Sinclair, A.J.; Howe, P.R. Dietary intakes and food sources of omega-6 and omega-3 polyunsaturated fatty acids. *Lipids* **2003**, *38*, 391–398. [CrossRef]
6. Howe, P.; Meyer, B.; Record, S.; Baghurst, K. Dietary intake of long-chain omega-3 polyunsaturated fatty acids: Contribution of meat sources. *Nutrition* **2006**, *22*, 47–53. [CrossRef]

7. Scientific Advisory Committee on Nutrition/Committee on Toxicity. *Advice on Fish Consumption: Benefits and Risks*; TSO: London, UK, 2004.
8. Food and Agricultural Organisation of the United Nations. *Fat and Fatty Acids in Human Nutrition: Report of an Expert Consultation*; Food and Agricultural Organisation of the United Nations: Rome, Italy, 2010.
9. European Food Safety Authority. Panel on Dietetic Products, Nutrition and Allergies (NDA); Scientific Opinion on the substantiation of health claims related to eicosapentaenoic acid (EPA), docosahexaenoic acid (DHA), docosapentaenoic acid (DPA) and maintenance. *EFSA J.* **2010**, *8*, 1796. [CrossRef]
10. Kromann, N.; Green, A. Epidemiological studies in the Upernavik district, Greenland. *J. Intern. Med.* **1980**, *208*, 401–406. [CrossRef]
11. Bjerregaard, P.; Dyerberg, J. Mortality from ischaemic heart disease and cerebrovascular disease in Greenland. *Int. J. Epidemiol.* **1988**, *17*, 514–519. [CrossRef]
12. Bang, H.; Dyerberg, J.; Hjørne, N. The composition of food consumed by Greenland Eskimos. *J. Intern. Med.* **1976**, *200*, 69–73. [CrossRef]
13. Newman, W.; Middaugh, J.; Propst, M.; Rogers, D. Atherosclerosis in Alaska Natives and non-Natives. *Lancet* **1993**, *341*, 1056–1057. [CrossRef]
14. Yano, K.; MacLean, C.J.; Reed, D.M.; Shimizu, Y.; Sasaki, H.; Kodama, K.; Kato, H.; Kagan, A. A comparison of the 12-year mortality and predictive factors of coronary heart disease among Japanese men in Japan and Hawaii. *Am. J. Epidemiol.* **1988**, *127*, 476–487. [CrossRef]
15. Calder, P.C. n-3 Fatty acids and cardiovascular disease: Evidence explained and mechanisms explored. *Clin. Sci.* **2004**, *107*, 1–11. [CrossRef]
16. Hu, F.; Bronner, L.; Willett, W.; Stampfer, M.; Rexrode, K.M.; Albert, C.; Hunter, D.; Manson, J. Fish and omega-3 fatty acid intake and risk of coronary heart disease in women. *JAMA* **2002**, *287*, 1815–1821. [CrossRef]
17. Browning, L.M.; Walker, C.G.; Mander, A.P.; West, A.L.; Madden, J.; Gambell, J.M.; Young, S.; Wang, L.; Jebb, S.A.; Calder, P.C. Incorporation of eicosapentaenoic and docosahexaenoic acids into lipid pools when given as supplements providing doses equivalent to typical intakes of oily fish. *Am. J. Clin. Nutr.* **2012**, *96*, 748–758. [CrossRef]
18. Albert, C.; Campos, H.; Stampfer, M.; Ridker, P.; Manson, J.; Willett, W.; Ma, J. Blood levels of long-chain n–3 fatty acids and the risk of sudden death. *N. Engl. J. Med.* **2002**, *346*, 1113–1118. [CrossRef]
19. Zhang, Y.; Zhuang, P.; He, W.; Chen, J.N.; Wang, W.Q.; Freedman, N.D.; Abnet, C.C.; Wang, J.B.; Jiao, J.J. Association of fish and long-chain omega-3 fatty acids intakes with total and cause-specific mortality: Prospective analysis of 421 309 individuals. *J. Intern. Med.* **2018**, *284*, 399–417. [CrossRef]
20. Djoussé, L.; Akinkuolie, A.; Wu, J.; Ding, E.; Gaziano, J. Fish consumption, omega-3 fatty acids and risk of heart failure: A meta-analysis. *Clin. Nutr.* **2012**, *31*, 846–853. [CrossRef]
21. Chowdhury, R.; Warnakula, S.; Kunutsor, S.; Crowe, F.; Ward, H.; Johnson, L.; Franco, O.; Butterworth, A.; Forouhi, N.; Thompson, S.; et al. Association of dietary, circulating, and supplement fatty acids with coronary risk: A systematic review and meta-analysis. *Ann. Intern. Med.* **2014**, *160*, 398–406. [CrossRef]
22. Alexander, D.D.; Miller, P.E.; Van Elswyk, M.E.; Kuratko, C.N.; Bylsma, L.C. A meta-analysis of randomized controlled trials and prospective cohort studies of eicosapentaenoic and docosahexaenoic long-chain omega-3 fatty acids and coronary heart disease risk. *Mayo Clin. Proc.* **2017**, *92*, 15–29. [CrossRef]
23. Del Gobbo, L.C.; Imamura, F.; Aslibekyan, S.; Marklund, M.; Virtanen, J.K.; Wennberg, M.; Yakoob, M.Y.; Chiuve, S.E.; Dela Cruz, L.; Frazier-Wood, A.C.; et al. ω-3 Polyunsaturated fatty acid biomarkers and coronary heart disease: Pooling project of 19 cohort studies. *JAMA Intern. Med.* **2016**, *176*, 1155–1166. [CrossRef] [PubMed]
24. Harris, W.S.; Von Schacky, C. The Omega-3 Index: A new risk factor for death from coronary heart disease? *Prev. Med.* **2004**, *39*, 212–220. [CrossRef] [PubMed]
25. Harris, W.S.; Del Gobbo, L.; Tintle, N.L. The omega-3 index and relative risk for coronary heart disease mortality: Estimation from 10 cohort studies. *Atherosclerosis* **2017**, *262*, 51–54. [CrossRef] [PubMed]
26. AbuMweis, S.; Jew, S.; Tayyem, R.; Agraib, L. Eicosapentaenoic acid and docosahexaenoic acid containing supplements modulate risk factors for cardiovascular disease: A meta-analysis of randomised placebo-control human clinical trials. *J. Hum. Nutr. Diet.* **2018**, *31*, 67–84. [CrossRef]

27. Miller, P.E.; Van Elswyk, M.; Alexander, D.D. Long-chain omega-3 fatty acids eicosapentaenoic acid and docosahexaenoic acid and blood pressure: A meta-analysis of randomized controlled trials. *Am. J. Hypertens.* **2014**, *27*, 885–896. [CrossRef]
28. Mozaffarian, D.; Geelen, A.; Brouwer, I.A.; Geleijnse, J.M.; Zock, P.L.; Katan, M.B. Effect of fish oil on heart rate in humans: A meta-analysis of randomized controlled trials. *Circulation* **2005**, *112*, 1945–1952. [CrossRef]
29. Xin, W.; Wei, W.; Li, X. Short-term effects of fish-oil supplementation on heart rate variability in humans: A meta-analysis of randomized controlled trials. *Am. J. Clin. Nutr.* **2013**, *97*, 926–935. [CrossRef]
30. Hidayat, K.; Yang, J.; Zhang, Z.; Chen, G.-C.; Qin, L.-Q.; Eggersdorfer, M.; Zhang, W. Effect of omega-3 long-chain polyunsaturated fatty acid supplementation on heart rate: A meta-analysis of randomized controlled trials. *Eur. J. Clin. Nutr.* **2017**. [CrossRef]
31. Gao, L.; Cao, J.; Mao, Q.X.; Lu, X.X.; Zhou, X.L.; Fan, L. Influence of omega-3 polyunsaturated fatty acid-supplementation on platelet aggregation in humans: A meta-analysis of randomized controlled trials. *Atherosclerosis* **2013**, *226*, 328–334. [CrossRef]
32. Xin, W.; Wei, W.; Li, X. Effect of fish oil supplementation on fasting vascular endothelial function in humans: A meta-analysis of randomized controlled trials. *PLoS ONE* **2012**, *7*, e46028. [CrossRef]
33. Wang, Q.; Liang, X.; Wang, L.; Lu, X.; Huang, J.; Cao, J.; Li, H.; Gu, D. Effect of omega-3 fatty acids supplementation on endothelial function: A meta-analysis of randomized controlled trials. *Atherosclerosis* **2012**, *221*, 536–543. [CrossRef]
34. Pase, M.P.; Grima, N.A.; Sarris, J. Do long-chain n-3 fatty acids reduce arterial stiffness? A meta-analysis of randomised controlled trials. *Br. J. Nutr.* **2011**, *106*, 974–980. [CrossRef]
35. Li, K.; Huang, T.; Zheng, J.; Wu, K.; Li, D. Effect of marine-derived n-3 polyunsaturated fatty acids on C-reactive protein, interleukin-6 and tumor necrosis factor alpha: A meta-analysis. *PLoS ONE* **2014**, *9*, e88103. [CrossRef]
36. Jiang, J.; Li, K.; Wang, F.; Yang, B.; Fu, Y.; Zheng, J.; Li, D. Effect of marine-derived n-3 polyunsaturated fatty acids on major eicosanoids: A systematic review and meta-analysis from 18 randomized controlled trials. *PLoS ONE* **2016**, *11*, e0147351. [CrossRef] [PubMed]
37. Innes, J.K.; Calder, P.C. The differential effects of eicosapentaenoic acid and docosahexaenoic acid on cardiometabolic risk factors: A systematic review. *Int. J. Mol. Sci.* **2018**, *19*, 532. [CrossRef] [PubMed]
38. Grimsgaard, S.; Bonna, K.H.; Hansen, J.-B.; Nordøy, A. Highly purified eicosapentaenoic acid and docosahexaenoic acid in humans have similar triacylglycerol-lowering effects but divergent effects on serum fatty acids. *Am. J. Clin. Nutr.* **1997**, *66*, 649–659. [CrossRef] [PubMed]
39. Allaire, J.; Couture, P.; Leclerc, M.; Charest, A.; Marin, J.; Lépine, M.-C.; Talbot, D.; Tchernof, A.; Lamarche, B. A randomized, crossover, head-to-head comparison of eicosapentaenoic acid and docosahexaenoic acid supplementation to reduce inflammation markers in men and women: The Comparing EPA to DHA (ComparED) Study. *Am. J. Clin. Nutr.* **2016**, *104*, 280–287. [CrossRef]
40. Woodman, R.J.; Mori, T.A.; Burke, V.; Puddey, I.B.; Watts, G.F.; Beilin, L.J. Effects of purified eicosapentaenoic and docosahexaenoic acids on glycemic control, blood pressure, and serum lipids in type 2 diabetic patients with treated hypertension. *Am. J. Clin. Nutr.* **2002**, *76*, 1007–1015. [CrossRef]
41. Mori, T.A.; Burke, V.; Puddey, I.B.; Watts, G.F.; O'Neal, D.N.; Best, J.D.; Beilin, L.J. Purified eicosapentaenoic and docosahexaenoic acids have differential effects on serum lipids and lipoproteins, LDL particle size, glucose, and insulin in mildly hyperlipidemic men. *Am. J. Clin. Nutr.* **2000**, *71*, 1085–1094. [CrossRef]
42. Grimsgaard, S.; Bonaa, K.; Hansen, J.; Myhre, E. Effects of highly purified eicosapentaenoic acid and docosahexaenoic acid on hemodynamics in humans. *Am. J. Clin. Nutr.* **1998**, *68*, 52–59. [CrossRef]
43. Mori, T.; Bao, D.; Burke, V.; Puddey, I.; Beilin, L. Docosahexaenoic acid but not eicosapentaenoic acid lowers ambulatory blood pressure and heart rate in humans. *Hypertension* **1999**, *34*, 253–260. [CrossRef] [PubMed]
44. Mori, T.A.; Watts, G.F.; Burke, V.; Hilme, E.; Puddey, I.B.; Beilin, L.J. Differential effects of eicosapentaenoic acid and docosahexaenoic acid on vascular reactivity of the forearm microcirculation in hyperlipidaemic, overweight men. *Circulation* **2000**, *102*, 1264–1269. [CrossRef] [PubMed]
45. Nestel, P.; Shige, H.; Pomeroy, S.; Cehun, M.; Abbey, M.; Raederstorff, D. The n-3 fatty acids eicosapentaenoic acid and docosahexaenoic acid increase systemic arterial compliance in humans. *Am. J. Clin. Nutr.* **2002**, *76*, 326–330. [CrossRef] [PubMed]
46. Park, Y.; Harris, W. EPA, but not DHA, decreases mean platelet volume in normal subjects. *Lipids* **2002**, *37*, 941–946. [CrossRef] [PubMed]

47. Woodman, R.J.; Mori, T.A.; Burke, V.; Puddey, I.B.; Barden, A.; Watts, G.F.; Beilin, L.J. Effects of purified eicosapentaenoic acid and docosahexaenoic acid on platelet, fibrinolytic and vascular function in hypertensive type 2 diabetic patients. *Atherosclerosis* **2003**, *166*, 85–93. [CrossRef]
48. Vors, C.; Allaire, J.; Marin, J.; Lepine, M.C.; Charest, A.; Tchernof, A.; Couture, P.; Lamarche, B. Inflammatory gene expression in whole blood cells after EPA vs. DHA supplementation: Results from the ComparED study. *Atherosclerosis* **2017**, *257*, 116–122. [CrossRef] [PubMed]
49. Mori, T.A.; Puddey, I.B.; Burke, V.; Croft, K.D.; Dunstan, D.W.; Rivera, J.H.; Beilin, L.J. Effect of omega 3 fatty acids on oxidative stress in humans: GC-MS measurement of urinary F-2-isoprostane excretion. *Redox Rep.* **2000**, *5*, 45–46. [CrossRef]
50. Mas, E.; Woodman, R.J.; Burke, V.; Puddey, I.B.; Beilin, L.J.; Durand, T.; Mori, T.A. The omega-3 fatty acids EPA and DHA decrease plasma F2-isoprostanes: Results from two placebo-controlled interventions. *Free Radic. Res.* **2010**, *44*, 983–990. [CrossRef]
51. Mori, T.A.; Woodman, R.J.; Burke, V.; Puddey, I.B.; Croft, K.D.; Beilin, L.J. Effect of eicosapentaenoic acid and docosahexaenoic acid on oxidative stress and inflammatory markers in treated-hypertensive type 2 diabetic subjects. *Free Radic. Biol. Med.* **2003**, *35*, 772–781. [CrossRef]
52. Yokoyama, M.; Origasa, H.; Matsuzaki, M.; Matsuzawa, Y.; Saito, Y.; Ishikawa, Y.; Oikawa, S.; Sasaki, J.; Hishida, H.; Itakura, H.; et al. Effects of eicosapentaenoic acid on major coronary events in hypercholesterolaemic patients (JELIS): A randomised open- label, blinded endpoint analysis. *Lancet* **2007**, *369*, 1090–1098. [CrossRef]
53. Bowman, L.; Mafham, M.; Wallendszus, K.; Stevens, W.; Buck, G.; Barton, J.; Murphy, K.; Aung, T.; Haynes, R.; Cox, J.; et al. Effects of n-3 fatty acid supplements in diabetes mellitus. *N. Engl. J. Med.* **2018**, *379*, 1540–1550. [CrossRef] [PubMed]
54. Manson, J.A.E.; Cook, N.R.; Lee, I.M.; Christen, W.; Bassuk, S.S.; Mora, S.; Gibson, H.; Albert, C.M.; Gordon, D.; Copeland, T.; et al. Marine n-3 fatty acids and prevention of cardiovascular disease and cancer. *N. Engl. J. Med.* **2019**, *380*, 23–32. [CrossRef] [PubMed]
55. Burr, M.L.; Gilbert, J.F.; Holliday, R.M.; Elwood, P.C.; Fehily, A.M.; Rogers, S.; Sweetnam, P.M.; Deadman, N.M. Effects of changes in fat, fish, and fibre intakes on death and myocardial reinfarction: Diet and reinfarction trial (DART). *Lancet* **1989**, *344*, 757–761. [CrossRef]
56. The GISSI-Prevenzione Investigators. Dietary supplementation with n-3 polyunsaturated fatty acids and vitamin E after myocardial infarction: Results of the GISSI-Prevenzione trial. *Lancet* **1999**, *354*, 447–455. [CrossRef]
57. Marchioli, R.; Barzi, F.; Bomba, E.; Chieffo, C.; Di Gregorio, D.; Di Mascio, R.; Franzosi, M.G.; Geraci, E.; Levantesi, G.; Maggioni, A.; et al. Early protection against sudden death by n-3 polyunsaturated fatty acids after myocardial infarction: Time-course analysis of the results of the Gruppo Italiano per lo Studio della Sopravvivenza nell'Infarto Miocardico (GISSI)-Prevenzione. *Circulation* **2002**, *105*, 1897–1903. [CrossRef]
58. Romagna, E. The GISSI-HF Investigators Effect of n-3 polyunsaturated fatty acids in patients with chronic heart failure (the GISSI-HF trial): A randomised, double-blind, placebo-controlled trial. *Lancet* **2008**, *372*, 1223–1230. [CrossRef]
59. Itakura, H.; Yokoyama, M.; Matsuzaki, M.; Saito, Y.; Origasa, H.; Ishikawa, Y.; Oikawa, S.; Sasaki, J.; Hishida, H.; Kita, T.; et al. Relationships between plasma fatty acid composition and coronary artery disease. *J. Atheroscler. Thromb.* **2011**, *18*, 99–107. [CrossRef]
60. Bucher, H.C.; Hengstler, P.; Schindler, C.; Meier, G. N-3 polyunsaturated fatty acids in coronary heart disease: A meta-analysis of randomized controlled trials. *Am. J. Med.* **2002**, *112*, 298–304. [CrossRef]
61. Studer, M.; Briel, M.; Leimenstoll, B.; Glass, T.R.; Bucher, H.C. Effect of different antilipidemic agents and diets on mortality: A systematic review. *Arch. Intern. Med.* **2005**, *165*, 725–730. [CrossRef]
62. Zhao, Y.-T.; Chen, Q.; Sun, Y.-X.; Li, X.-B.; Zhang, P.; Xu, Y.; Guo, J.-H. Prevention of sudden cardiac death with omega-3 fatty acids in patients with coronary heart disease: A meta-analysis of randomized controlled trials. *Ann. Med.* **2009**, *41*, 301–310. [CrossRef]
63. Marik, P.E.; Varon, J. Omega-3 dietary supplements and the risk of cardiovascular events: A systematic review. *Clin. Cardiol.* **2009**, *32*, 365–372. [CrossRef] [PubMed]
64. Rauch, B.; Schiele, R.; Schneider, S.; Diller, F.; Victor, N.; Gohlke, H.; Gottwik, M.; Steinbeck, G.; Del Castillo, U.; Sack, R.; et al. OMEGA, a randomized, placebo-controlled trial to test the effect of highly purified omega-3

fatty acids on top of modern guideline-adjusted therapy after myocardial infarction. *Circulation* **2010**, *122*, 2152–2159. [CrossRef] [PubMed]
65. Galan, P.; Kesse-Guyot, E.; Czernichow, S.; Briancon, S.; Blacher, J.; Hercberg, S. Effects of B vitamins and omega 3 fatty acids on cardiovascular diseases: A randomised placebo controlled trial. *BMJ* **2010**, *341*, c6273. [CrossRef] [PubMed]
66. Kromhout, D.; Giltay, E.J.; Geleijnse, J.M. n–3 Fatty acids and cardiovascular events after myocardial infarction. *N. Engl. J. Med.* **2010**, *363*, 2015–2026. [CrossRef]
67. Bosch, J.; Gerstein, H.; Dagenais, G.; Diaz, R.; Dyal, L.; Jung, H.; Maggiono, A.; Probstfield, J.; Ramachandran, A.; Riddle, M.; et al. n–3 Fatty acids and cardiovascular outcomes in patients with dysglycemia (ORIGIN). *N. Engl. J. Med.* **2012**, *367*, 309–318. [CrossRef]
68. Roncaglioni, M.; Tombesi, M.; Avanzini, F.; Barlera, S.; Caimi, V.; Longoni, P.; Marzona, I.; Milani, V.; Silletta, M.; Tognoni, G.; et al. n–3 Fatty acids in patients with multiple cardiovascular risk factors (Risk and Prevention Study). *N. Engl. J. Med.* **2013**, *368*, 1800–1808. [CrossRef]
69. Calder, P.C.; Yaqoob, P. Marine omega-3 fatty acids and coronary heart disease. *Curr. Opin. Cardiol.* **2012**, *27*, 412–419. [CrossRef]
70. Calder, P.C. Limited impact of omega-3 fatty acids in patients with multiple cardiovascular risk factors. *Evid. Based Med.* **2014**, *19*, 18. [CrossRef]
71. Kotwal, S.; Jun, M.; Sullivan, D.; Perkovic, V.; Neal, B. Omega 3 fatty acids and cardiovascular outcomes: Systematic review and meta-analysis. *Circ. Cardiovasc. Qual. Outcomes* **2012**, *5*, 808–818. [CrossRef]
72. Kwak, S.M.; Myung, S.K.; Lee, Y.J.; Seo, H.G. Efficacy of omega-3 fatty acid supplements (eicosapentaenoic acid and docosahexaenoic acid) in the secondary prevention of cardiovascular disease: A meta-analysis of randomized, double-blind, placebo-controlled trials. *Arch. Intern. Med.* **2012**, *172*, 686–694. [CrossRef]
73. Trikalinos, T.; Lee, J.; Moorthy, D.; Yu, W.; Lau, J.; Lichtenstein, A.; Chung, M. Effects of eicosapentanoic acid and docosahexanoic acid on mortality across diverse settings: Systematic review and meta-analysis of randomized trials and prospective cohorts. *Tech. Rev.* **2012**, *17*, 4.
74. Rizos, E.; Ntzani, E.; Bika, E.; Kostapanos, M.; Elisaf, M. Association between omega-3 fatty acid supplementation and risk of major cardiovascular disease events: A systematic review and meta-analysis. *JAMA* **2012**, *308*, 1024–1033. [CrossRef] [PubMed]
75. Casula, M.; Soranna, D.; Catapano, A.L.; Corrao, G. Long-term effect of high dose omega-3 fatty acid supplementation for secondary prevention of cardiovascular outcomes: A meta-analysis of randomized, double blind, placebo controlled trials. *Atheroscler. Suppl.* **2013**, *14*, 243–251. [CrossRef]
76. Wen, Y.T.; Dai, J.H.; Gao, Q. Effects of omega-3 fatty acid on major cardiovascular events and mortality in patients with coronary heart disease: A meta-analysis of randomized controlled trials. *Nutr. Metab. Cardiovasc. Dis.* **2014**, *24*, 470–475. [CrossRef]
77. Maki, K.C.; Palacios, O.M.; Bell, M.; Toth, P.P. Use of supplemental long-chain omega-3 fatty acids and risk for cardiac death: An updated meta-analysis and review of research gaps. *J. Clin. Lipidol.* **2017**, *11*, 1152–1160.e2. [CrossRef]
78. Aung, T.; Halsey, J.; Kromhout, D.; Gerstein, H.C.; Marchioli, R.; Tavazzi, L.; Geleijnse, J.M.; Rauch, B.; Ness, A.; Galan, P.; et al. Associations of omega-3 fatty acid supplement use with cardiovascular disease risks. *JAMA Cardiol.* **2018**, *3*, 225–234. [CrossRef]
79. Abdelhamid, A.; Brown, T.; Brainard, J.; Biswas, P.; Thorpe, G.; Moore, H.; Deane, K.; AlAbdulghafoor, F.; Summerbell, C.; Worthington, H.; et al. Omega-3 fatty acids for the primary and secondary prevention of cardiovascular disease. *Cochrane Database Syst. Rev.* **2018**, CD003177. [CrossRef]
80. Bhatt, D.L.; Steg, P.G.; Miller, M.; Brinton, E.A.; Jacobson, T.A.; Ketchum, S.B.; Doyle, R.T.; Juliano, R.A.; Jiao, L.; Granowitz, C.; et al. Cardiovascular risk reduction with icosapent ethyl for hypertriglyceridemia. *N. Engl. J. Med.* **2019**, *380*, 11–22. [CrossRef]
81. Hu, Y.; Hu, F.B.; Manson, J.E. Marine omega-3 supplementation and cardiovascular disease: An updated meta-analysis of 13 randomized controlled trials involving 127,477 participants. *J. Am. Heart Assoc.* **2019**, *8*, e013543. [CrossRef]
82. Einvik, G.; Ole Klemsdal, T.; Sandvik, L.; Hjerkinn, E.M. A randomized clinical trial on n-3 polyunsaturated fatty acids supplementation and all-cause mortality in elderly men at high cardiovascular risk. *Eur. J. Cardiovasc. Prev. Rehabil.* **2010**, *17*, 588–592. [CrossRef]

83. Rimm, E.B.; Appel, L.J.; Chiuve, S.E.; Djoussé, L.; Engler, M.B.; Kris-Etherton, P.M.; Mozaffarian, D.; Siscovick, D.S.; Lichtenstein, A.H. Seafood long-chain n-3 polyunsaturated fatty acids and cardiovascular disease: A science advisory from the American Heart Association. *Circulation* **2018**, *138*, 35–47. [CrossRef] [PubMed]
84. Skulas-Ray, A.C.; Wilson, P.W.F.; Harris, W.S.; Brinton, E.A.; Kris-Etherton, P.M.; Richter, C.K.; Jacobson, T.A.; Engler, M.B.; Miller, M.; Robinson, J.G.; et al. Omega-3 fatty acids for the management of hypertriglyceridemia: A science advisory from the American Heart Association. *Circulation* **2019**, *140*, e673–e691. [CrossRef] [PubMed]
85. Mach, F.; Baigent, C.; Carapano, A.L.; Koskinas, K.C.; Casula, M.; Badimon, L.; Chapman, M.J.; de Backer, G.G.; Delgado, V.; Ference, B.A.; et al. 2019 ESC/EAS guidelines for the management of dyslipidaemias: Lipid modification to reduce cardiovascular risk. *Eur. Heart J.* **2020**, *41*, 111–188. [CrossRef] [PubMed]
86. The Living Standards of Medical Care in Diabetes—Updates to the standards of medical care in diabetes. Available online: https://care.diabetesjournals.org/living-standards#March%2027,%202019 (accessed on 7 February 2020).
87. Orringer, C.E.; Jacobson, T.A.; Maki, K.C. National Lipid Association Scientific Statement on the use of icosapent ethyl in statin-treated patients with elevated triglycerides and high or very-high ASCVD risk. *J. Clin. Lipidol.* **2019**, *13*, 860–872. [CrossRef]
88. Kris-Etherton, P.M.; Harris, W.S.; Appel, L.J. Fish consumption, fish oil, omega-3 fatty acids, and cardiovascular disease. *Circulation* **2002**, *106*, 2747–2757. [CrossRef]
89. Siscovick, D.S.; Barringer, T.A.; Fretts, A.M.; Wu, J.H.Y.; Lichtenstein, A.H.; Costello, R.B.; Kris-Etherton, P.M.; Jacobson, T.A.; Engler, M.B.; Alger, H.M.; et al. Omega-3 polyunsaturated fatty acid (fish oil) supplementation and the prevention of clinical cardiovascular disease: A Science Advisory from the American Heart Association. *Circulation* **2017**, *135*, e867–e884. [CrossRef]
90. NICE Cardiovascular disease: Risk assessment and reduction, including lipid modification (clinical guideline CG181). Available online: https://www.nice.org.uk/guidance/cg181 (accessed on 14 July 2014).
91. Von Schacky, C. Omega-3 fatty acids in cardiovascular disease—An uphill battle. *Prostagland. Leukot. Essent Fatty Acids* **2015**, *92*, 41–47. [CrossRef]
92. Rice, H.B.; Bernasconi, A.; Maki, K.C.; Harris, W.S.; von Schacky, C.; Calder, P.C. Conducting omega-3 clinical trials with cardiovascular outcomes: Proceedings of a Workshop held at ISSFAL 2014. *Prostagland. Leukot. Essent. Fatty Acids* **2016**, *107*, 30–42. [CrossRef]
93. Schuchardt, J.P.; Hahn, A. Bioavailability of long-chain omega-3 fatty acids. *Prostagland. Leukot. Essent. Fatty Acids* **2013**, *89*, 1–9. [CrossRef]

International Journal of
Molecular Sciences

Review

Fish, Fish Oils and Cardioprotection: Promise or Fish Tale?

Akshay Goel [1], Naga Venkata Pothineni [1], Mayank Singhal [2], Hakan Paydak [1], Tom Saldeen [1] and Jawahar L. Mehta [1,*]

1 Division of Cardiology, University of Arkansas for Medical Sciences and Central Arkansas Veterans Healthcare System, Little Rock, AR 72205, USA; AGoel@uams.edu (A.G.); NVPothineni@uams.edu (N.V.P.); HPaydak@uams.edu (H.P.); tom.saldeen@uppsalapharma.se (T.S.)
2 Cape Fear Valley Hospital, Fayetteville, NC 28304, USA; drmayanksinghal@gmail.com
* Correspondence: MehtaJL@uams.edu; Tel.: +1-501-296-1426; Fax: +1-501-686-6180

Received: 8 November 2018; Accepted: 20 November 2018; Published: 22 November 2018

Abstract: Fish and commercially available fish oil preparations are rich sources of long-chain omega-3 polyunsaturated fatty acids. Eicosapentaenoic acid (EPA) and docosahexaenoic acid (DHA) are the most important fatty acids in fish oil. Following dietary intake, these fatty acids get incorporated into the cell membrane phospholipids throughout the body, especially in the heart and brain. They play an important role in early brain development during infancy, and have also been shown to be of benefit in dementia, depression, and other neuropsychiatric disorders. Early epidemiologic studies show an inverse relationship between fish consumption and the risk of coronary heart disease. This led to the identification of the cardioprotective role of these marine-derived fatty acids. Many experimental studies and some clinical trials have documented the benefits of fish oil supplementation in decreasing the incidence and progression of atherosclerosis, myocardial infarction, heart failure, arrhythmias, and stroke. Possible mechanisms include reduction in triglycerides, alteration in membrane fluidity, modulation of cardiac ion channels, and anti-inflammatory, anti-thrombotic, and anti-arrhythmic effects. Fish oil supplements are generally safe, and the risk of toxicity with methylmercury, an environmental toxin found in fish, is minimal. Current guidelines recommend the consumption of either one to two servings of oily fish per week or daily fish oil supplements (around 1 g of omega-3 polyunsaturated fatty acids per day) in adults. However, recent large-scale studies have failed to demonstrate any benefit of fish oil supplements on cardiovascular outcomes and mortality. Here, we review the different trials that evaluated the role of fish oil in cardiovascular diseases.

Keywords: fish oil; omega-3 fatty acids; eicosapentaenoic acid (EPA); docosahexaenoic acid (DHA); cardiovascular disease

---

1. Introduction

Cardiovascular disease (CVD) is a leading cause of death in the United States [1]. However, despite extensive advances in our knowledge of nutritional options in the prevention and therapy of CVD, the benefits of dietary fish and fish oil supplementation on CVD remains debatable. Fish oil is a rich source of long-chain omega-3 (ω-3) polyunsaturated fatty acids (PUFAs)—eicosapentaenoic acid (EPA), and docosahexaenoic acid (DHA). The possible cardioprotective role of fish consumption was first identified in the early studies of Greenland's Inuit population who were found to have a low incidence of myocardial infarction (MI) compared to their Danish counterparts. The benefit was attributed to the high fish consumption by the Inuits [2,3]. This was followed by nearly four decades of research, including animal studies, epidemiological studies, randomized controlled trials (RCTs), meta-analyses of epidemiological cohort studies, and trial meta-analyses. Many of these studies demonstrated the cardioprotective effects of fish consumption and fish oil supplementation [4–10].

The rising amount of evidence led to recommendations regarding consumption of seafood/fish and/or their dietary supplementation with ω-3 PUFA from the Food and Drug Administration (FDA) [11]. In 2012, a prescription fish oil formulation was approved by the US FDA. Fish oil continues to be the most popular natural supplement in the US, used by nearly 18.8 million adults [12]. However, several recent large-scale studies have failed to demonstrate any significant benefits of fish oil supplements on cardiovascular outcomes and mortality [13–17]. In this review, we discuss the structure and metabolism of marine-derived ω-3 PUFAs, their proposed benefits and molecular mechanisms of action, and the evidence regarding their role in CVD. We conclude by providing possible reasons for the conflicting evidence and recommendations regarding the dietary intake of fish and fish oil supplements.

2. Structure and Metabolism of PUFAs

Fatty acids are long-chain hydrocarbons with a carboxylic acid group at one end (alpha terminal) and methyl group at the other end (omega terminal). They can be classified based on the number of double bonds in their side chains—saturated fatty acids (no double bond), monounsaturated fatty acids or MUFAs (single double bond), and polyunsaturated fatty acids or PUFAs (two or more double bonds). PUFAs can be classified further by the length of the carbon chain and the position of the first double bond from the methyl terminal into omega-6 (ω-6 or n-6) or omega-3 (ω-3 or n-3). For example, linoleic acid (LA) or 9,12-octadecadienoic acid ($C_{18:2}$) has 18 carbon atoms with 2 double bonds. Since the first double bond from the methyl terminal is at the sixth position, it is an ω-6 PUFA. Similarly, alpha-linolenic acid (ALA) is 9,12,15-octadecatrienoic acid ($C_{18:3}$), meaning it has 18 carbon atoms with 3 double bonds. However, in this case, the first double bond is at the third position from the methyl terminal, and hence it is an ω-3 PUFA. Both LA and ALA are considered essential fatty acids since they cannot be synthesized by humans and must be ingested via their diet.

The essential fatty acids LA and ALA are then metabolized to other fatty acids through desaturase and elongase enzymes. LA (ω-6) is metabolized to arachidonic acid (5,8,11,14-eicosatetraenoic acid, $C_{20:4}$, ω-6). Similarly, ALA (ω-3) is converted to EPA (5,8,11,14,17-eicosapentaenoic acid, $C_{20:5}$, ω-3) and DHA (4,7,10,13,16,19-docosahexaenoic acid, $C_{22:6}$, ω-3). Thus, EPA and DHA are traditionally considered non-essential since, technically speaking, they can be synthesized from ALA. However, this pathway is slow and inefficient [18]. Therefore, for all practical purposes, the dietary intake of EPA and DHA is "essential" and crucial to obtain health benefits.

After absorption from the intestine as chylomicrons, fatty acids are transported to the liver and other tissues. PUFAs are subsequently incorporated into the phospholipid bilayer of plasma membranes, and affect membrane fluidity and signaling. ω-6 and ω-3 PUFAs have opposite effects in the body. Diets rich in ω-6 are precursors of eicosanoids associated with inflammation, vasoconstriction, and platelet aggregation [19]. Acute self-limited inflammation is a protective response to infection and injury. However, excessive inappropriate inflammation has been linked to atherosclerosis and cancer. On the other hand, ω-3 PUFAs are precursors of anti-inflammatory molecules and provide benefits against chronic inflammatory conditions, like diabetes, ischemic heart disease, and cancer [20]. These molecular mechanisms are discussed in further detail in a later section of this review.

The structure and metabolism of major PUFAs is depicted in Figure 1.

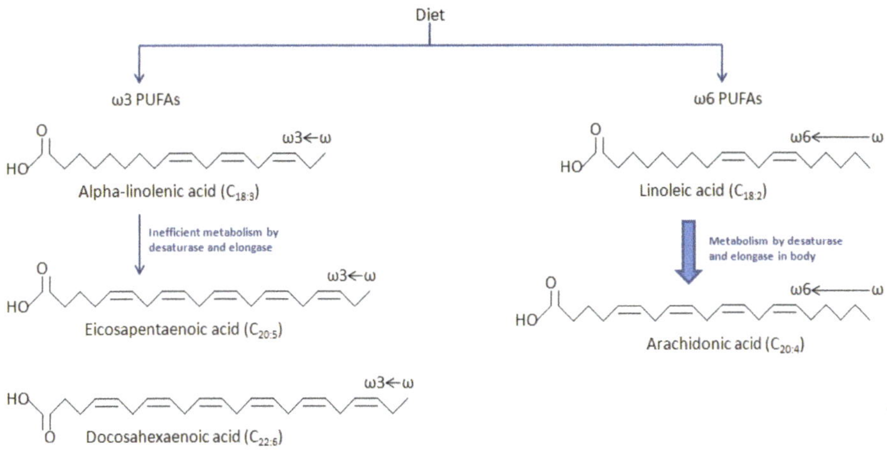

Figure 1. Structure and metabolism of major polyunsaturated fatty acids (PUFAs).

3. Dietary Sources of Major PUFAs

Most vegetable oils and crop seeds, like corn, sunflower, soybean, and canola oils, are a rich source of ω-6 LA with lesser amounts of ω-3 ALA. On the other hand, flax, walnuts, and chia seeds are a rich source of ALA, as are some green leafy vegetables.

Fatty fish and other seafood are the most important dietary sources of EPA and DHA. Wild (marine) fish feed on phyto- and zoo-planktons, and are therefore a richer source of EPA and DHA than cultivated (farmed) fish which feed on cereals and vegetable oils [20]. Cod liver and algal oil have been proposed to be non-traditional marine sources of EPA and DHA. Finally, marine-derived ω-3 fortified food products, like cereals, pastas, dairy products, eggs, meat, salad dressings, and oils are now available. They are a potential option for vegetarians and those who dislike seafood [18]. The principal dietary sources of PUFAs are shown in Table 1 [21,22].

Table 1. Dietary sources of major PUFAs.

PUFA	Dietary Source
Linoleic acid (ω-6)	Corn, safflower, soybean, sunflower oils
Alpha-linolenic acid (plant-derived ω-3)	Flaxseed oil, canola (rapeseed) oil, walnuts, seeds of chia, perilla, green leafy vegetables
Eicosapentaenoic and Docosahexaenoic acids (marine-derived ω-3)	Fish, fish oil, other seafood, beef, lamb, ω-3 fortified foods

4. Proposed Cardioprotective Benefits of ω-3 PUFAs and Their Molecular Mechanisms of Action

Fish and fish oil consumption has been shown to decrease cardiovascular events and mortality in many secondary prevention trials that included high-risk patients with recent myocardial infarction or heart failure. Several molecular mechanisms for this cardioprotective effect of ω-3 PUFAs have been proposed.

4.1. Anti-Inflammatory Effect

ω-3 PUFAs have pleiotropic anti-inflammatory effects. They readily replace arachidonic acid in cell membranes. This results in decreased production of ω-6 arachidonic acid-derived inflammatory mediators, like prostaglandin (PG) E_2, thromboxane (TX) A_2, and leukotrienes (LTs) A_4, B_4, C_4, D_4, and E_4 [23,24]. ω-3 PUFAs generate less potent inflammatory substrates, like 3-series PGs and

TXs and 5-series LTs [24]. 3-series PGs have less potent biological effects (compared to 2-series PGs) and TX A_3 lacks pro-platelet aggregatory properties. 5-series LTs are relatively less effective as pro-inflammatory molecules. EPA and DHA are metabolized to anti-inflammatory mediators, like resolvins, protectins, and the G protein-coupled receptor 120 [24,25]. Additionally, cell membranes with higher ω-3: ω-6 PUFA ratio have higher fluidity. Marine-derived ω-3 PUFAs have been shown to reduce the circulating levels of pro-inflammatory cytokines like interleukin (IL)-1, IL-6, and the tumor necrosis factor (TNF)-α [26]. ω-3 PUFAs also regulate intracellular signaling pathways to inactivate nuclear transcriptional factors. They decrease expression of inflammatory genes via the downregulation of nuclear factor (NF)-κB. The inhibition of NF-κB is mediated by the activation of peroxisome proliferator-activated receptors (PPAR) [27].

4.2. *Improved Endothelial Function*

Endothelial dysfunction from loss of endothelial-derived nitric oxide (NO) synthesis results in predisposition to accelerated atherosclerosis and adverse vascular events. Marine-derived ω-3 PUFAs cause translocation and activation of endothelial NO synthase (eNOS) into the cytosol, resulting in vasodilation and improved endothelial function [25,28]. Recently, EPA was shown to prevent saturated fatty acid-induced vascular endothelial dysfunction through regulation of long-chain acyl-coA synthetase expression [29]. Endothelial function is additionally improved by reduced expression of endothelial vascular cell adhesion molecules, resulting in attenuated leukocyte adhesion to endothelium [30,31].

4.3. *Atherosclerotic Plaque Stabilization*

EPA and DHA inhibit the proliferation and migration of smooth muscle cells (SMCs), a central step in atherosclerotic plaque formation and progression. Vasa vasorum is a network of microvessels extending up to the plaque base, which is vital for plaque progression. ω-3 PUFAs interfere with the neovascularization of vasa vasorum, thereby suppressing plaque development [25].

Apart from reducing plaque progression, ω-3 PUFAs also contribute to plaque stability. Plaque vulnerability predisposes people to plaque erosion or rupture, which causes acute coronary syndrome. High tissue levels of EPA and DHA decrease macrophage infiltration and the release of matrix metalloproteinases (MMPs), resulting in greater plaque stability [23]. The addition of EPA to statin therapy has been shown to reduce the lipid core in coronary plaques [32]. Plaques in patients receiving fish oil are more likely to be fibrous cap atheromas, with fewer macrophages and lesser inflammation, and are therefore more likely to be stable [33].

4.4. *Effect on Lipid Metabolism*

ω-3 PUFAs modulate the activity of genes that control lipid homeostasis. Large doses of fish oil interfere with the synthesis of very low-density lipoprotein (VLDL) via inhibition of sterol receptor element-binding protein-1c. This results in a marked lowering of serum TG levels [23]. Although ω-3 fatty acids do not affect the serum levels of total cholesterol and low-density lipoprotein (LDL), fish oil has been shown to reduce remnant lipoproteins and post-prandial lipemia after fatty meals [30,31]. Remnant lipoprotein levels and post-prandial lipemia are involved in the pathogenesis of sudden cardiac death (SCD) [23]. Marine-derived ω-3 fatty acids also cause a favorable change in high-density lipoprotein (HDL) by increasing the large, cholesterol-rich HDL2 fraction and lowering the small, TG-rich HDL3 fraction [30].

4.5. *Anti-Thrombotic Effect*

EPA inhibits the synthesis of platelet TXA_2 which causes platelet aggregation and vasoconstriction. Both EPA and DHA antagonize the TXA_2 and PGH_2 receptors in human platelets [34]. There are reports that ω-3 PUFA reduce fibrinogen levels and increase tissue plasminogen-activator concentrations [30,31,35].

4.6. Anti-Arrhythmic Effect

ω-3 PUFAs modulate the activity of multiple ion channels and stabilize the cardiomyocyte membrane, thereby preventing tachyarrhythmias and SCD [5]. EPA and DHA inhibit the voltage-gated sodium channels in cardiac myocytes, increase the voltage threshold for depolarization, and prolong the refractory period. They also modulate certain calcium channels, decrease free cytosolic calcium, and reduce membrane excitability further [24]. Part of the anti-arrhythmic action of these fatty acids is also due to their autonomic effects, like increased vagal tone [36]. In addition, low serum levels of EPA and DHA have been found to increase the risk of cardiogenic syncope in patients with Brugada syndrome [37].

4.7. Cardiac Remodeling

The OMEGA-REMODEL trial reported the beneficial effects of high-dose ω-3 PUFA therapy on cardiac remodeling in patients with MI. Cardiac magnetic resonance confirmed reduced ventricular remodeling and myocardial fibrosis after PUFA supplementation [38]. Similar findings of the preferential effects of ω-3 PUFAs on cardiac remodeling and heart failure were also seen in the Gruppo Italiano per lo Studio della Streptochinasi nell'Infarto Miocardico-Heart Failure (GISSI-HF) trial [10].

Cardiac myofibroblasts, activated by inflammatory signals or pressure overload, cause cardiac remodeling. Recently, a novel EPA metabolite called 18-hydroxy eicosapentaenoic acid (18-HEPE) has been identified, which prevents cardiac remodeling under pressure overload [39]. Higher dietary intake of EPA ethyl ester increases the plasma concentration of 18-HEPE, and may have beneficial long-term effects by preventing cardiac fibrosis [24].

Moreover, EPA and DHA improve cardiac mitochondrial function by increasing the efficiency of adenosine triphosphate (ATP) production [40].

4.8. Improved Exercise Tolerance

Decreased exercise capacity is a known risk factor for CVD. Recently, ω-3 PUFAs have been shown to improve exercise tolerance. This may be due to favorable effects on erythrocyte rheology and skeletal muscle function [25,41]. Improved exercise capacity may contribute to a lower risk of adverse cardiac events.

4.9. Improved Cognitive Function

Poor cognitive function is a risk factor for cardiovascular events [42]. Serum EPA levels have been shown to be independently associated with cognitive function in CVD patients [43]. Thus, improved dietary ω-3 PUFA intake might conceivably improve cognition and decrease the risk of cardiovascular events [25].

Figure 2 summarizes the various molecular mechanisms by which ω-3 PUFAs, like EPA and DHA, have been proposed to exert their cardioprotective effects.

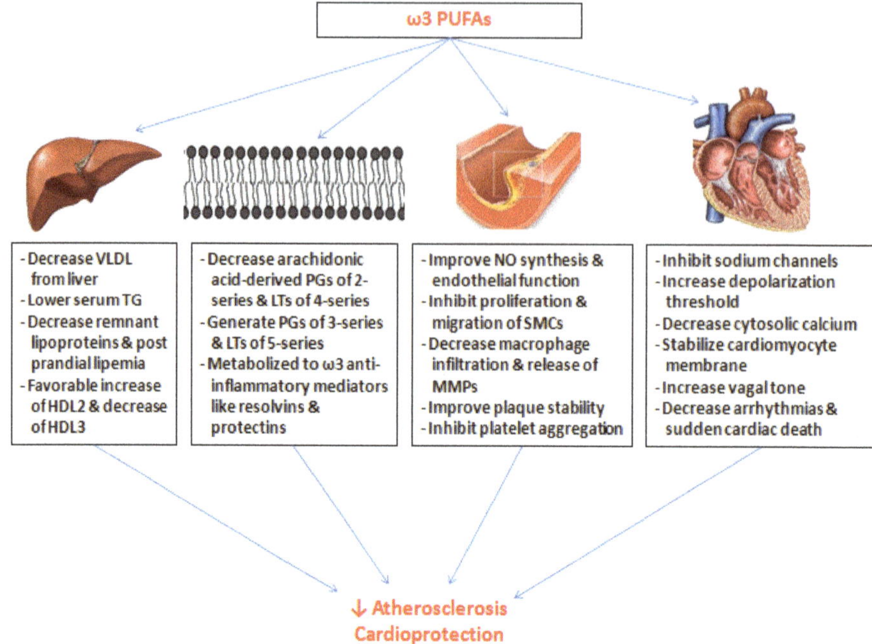

Figure 2. Pleiotropic cardioprotective effects of ω-3 PUFAs.

5. Evidence from Trials and Meta-Analyses

The Diet and Reinfarction Trial (DART) was the first RCT to assess the benefits of dietary fish and fish oil in the secondary prevention of MI [4]. A total of 2033 patients with recent MI were followed for two years. Consumption of at least two servings (200–400 g) of fish per week was associated with a reduction in all-cause mortality. Similar beneficial effects were also noticed in participants who took fish oil capsules instead of fish.

The GISSI-Prevenzione trial included 11,324 patients with previous MI (within three months) [6]. The intervention group received one fish oil capsule per day, in addition to standard care. The intervention group was noted to have a 41% reduction in all-cause mortality and 53% reduction in SCD as early as four months into the study. The difference in all-cause mortality remained significant after a follow-up of 3.5 years, and was proposed to be primarily due to the anti-arrhythmic effects of EPA and DHA.

The DART-2 was a secondary prevention study conducted in men with stable angina (and not previous MI, unlike the first DART study) [44]. Patients who were advised to consume two portions of oily fish every week or take three fish oil capsules daily had a higher risk of cardiac mortality and SCD. Subgroup analysis revealed that this difference was driven by the fish oil capsule group. The study was largely criticized for its poor design and lack of blinding.

The Japan EPA Lipid Intervention Study (JELIS), a combined primary (14,981 patients) and secondary intervention (3664 patients) trial, included a total of 18,645 patients [9]. The EPA (1.8 g daily) plus statin group showed a reduction in major coronary events compared to the statin-alone group. Subgroup analysis revealed that the significant reduction in coronary events was mainly seen in patients with a history of coronary artery disease, and there was no benefit of EPA in primary prevention. However, the results of this study may have been diluted, as the consumption of fish in the Japanese population is high at baseline.

The GISSI-HF trial was designed to study the effects of daily EPA and DHA supplementation in patients with HF [10]. The ω-3 PUFA group had a reduction in all-cause mortality and cardiovascular hospitalizations compared to the control group.

The Alpha Omega trial was a secondary prevention study using 4837 patients with prior MI [14]. Low-dose EPA and DHA in margarine were given daily to the intervention group, while the control group received only plain margarine. There was no difference in cardiovascular events between the two groups. The lack of benefit from EPA and DHA in the Alpha Omega study [14], compared to the GISSI Prevenzione [6] and JELIS [9] trials (all secondary prevention trials in prior MI patients), may have been, in part, due to the lower treatment dose of the intervention arm.

The OMEGA study included German patients with MI in the two weeks prior to enrolment [13]. The treatment group received EPA and DHA daily, versus the control group, which received olive oil. After a follow-up of 1 year, no significant difference was found in the rates of adverse cardiovascular events, SCD, and all-cause mortality between the two groups. However, the study was concluded to be underpowered due to the lower-than-expected event rates and overestimation of the effect of ω-3 PUFAs.

SU.FOL.OM3 was a randomized trial evaluating the effects of B-vitamin and ω-3 PUFA supplementation in 2501 French patients with a recent acute coronary or cerebral ischemic event [15]. ω-3 PUFA intake did not significantly affect the major cardiovascular event rate. Like the OMEGA study, the event rate was lower than anticipated and so the trial was underpowered.

The Outcome Reduction with an Initial Glargine Intervention (ORIGIN) trial was the first study to investigate the effects of ω-3 PUFA supplementation on cardiovascular events in patients with pre-diabetes and diabetes [16]. Analysis of 12,536 patients over 6.2 years did not reveal any benefits of ω-3 PUFAs in reducing cardiovascular death, compared to the olive oil given to the control group. These results were in contrast to the findings of the GISSI Prevenzione [6] and GISSI-HF [10] trials. Inter-trial differences in patient baseline characteristics and concomitant therapies could be a possible explanation.

The Risk and Prevention study investigated the efficacy of ω-3 PUFAs in Italian patients with high risk of CVD but without previous MI [17]. After a follow-up of five years, there was no difference in the primary endpoint of cardiovascular death and hospitalization between the ω-3 PUFA treatment group and olive oil control group. However, subgroup analysis revealed that the ω-3 PUFA group had fewer heart failure hospitalizations. Additionally, women in the treatment group had lower rates of primary endpoint than the control group. The results of this study may not be generalizable due to the Mediterranean dietary habits in the Italian population.

The PREDIMED trial in 7447 patients reported that a Mediterranean diet, with 50 g of extra-virgin olive oil daily, significantly reduced cardiovascular events and death at a follow-up period of 4.8 years [45]. It remains unknown whether the benefit was due to the Mediterranean diet alone, olive oil alone, or whether it was a combined effect. The OMEGA [13], ORIGIN [16], and Risk and Prevention [17] trials did not detect a significant difference between the overall outcomes in the ω-3 PUFA group and olive oil control group. However, the results of the PREDIMED trial [45] indicate that olive oil itself may have some cardioprotective benefits and is likely not an ideal control.

The Age-Related Eye Disease Study 2 (AREDS2) trial included 4203 patients at high risk of CVD and either intermediate or advanced macular degeneration [46]. Patients were randomly grouped to the ω-3 PUFA group (650 mg EPA plus 350 mg DHA), macular xanthophyll group (10 mg lutein plus 2 mg zeaxanthin), combination therapy, or matching placebos. Long-chain ω-3 PUFAs or macular xanthophylls did not reduce the risk of CVD events.

Rizos et al. performed a meta-analysis of 20 RCTs including 68,680 patients to study the role of ω-3 PUFA supplementation on major cardiovascular outcomes [47]. Overall, ω-3 PUFAs were not associated with cardiovascular benefits.

Studies have also evaluated the role of ω-3 PUFAs in the secondary prevention of atrial fibrillation [48,49]. The FORWARD trial included 586 participants with previous atrial fibrillation who were randomized to receive either ω-3 PUFA for 1 g per day, or a placebo for one year [48]. PUFA supplementation did not reduce recurrent atrial fibrillation. Mariani et al. performed a meta-analysis of 16 trials covering 4677 patients and concluded that ω-3 PUFAs have no effect in preventing recurrent or post-operative atrial fibrillation [49].

The results of the ASCEND trial were recently published [50]. It was a randomized, placebo-controlled, blinded trial in 15,480 patients followed for 7.4 years. The study aimed to assess the efficacy and safety of taking 100 mg of aspirin daily in preventing cardiovascular events and cancer in diabetic patients without known CVD. This study also investigated whether daily ω-3 PUFA supplementation decreased cardiovascular events in this population. Aspirin use was noted to prevent cardiovascular events in patients, but also caused major bleeding events. Compared to the placebo group that received olive oil capsules, 1 g of ω-3 PUFA supplementation daily failed to decrease the risk of serious vascular events in diabetics without known CVD.

The Reduction of Cardiovascular Events with EPA-Intervention Trial (REDUCE-IT) was recently concluded [51]. The study involved 8179 high-risk patients with hypertriglyceridemia on statin therapy who were randomized to receive either 4 g of ethyl EPA daily, or a placebo. After a median follow-up of 4.9 years, there was an approximately 25% reduction in the risk of major adverse cardiovascular events in the treatment group.

The Vitamin D and Omega-3 Trial (VITAL) results were recently announced [52]. The study included 25,871 participants with an objective to assess the effect of daily vitamin D_3 (2000 IU) and fish oil supplement (1 g) in the primary prevention of cancer and CVD. After a follow-up of over five years, ω-3 PUFA supplementation did not result in a lower incidence of cardiovascular events or cancer, compared to the placebo.

The STRENGTH study is an ongoing RCT that will enroll approximately 13,000 patients with hypertriglyceridemia, low HDL, and a high risk for CVD [53]. Patients are being randomized to receive either statin with corn oil or statin with prescription ω-3 carboxylic acids. The study is anticipated to be completed in 2019.

Results of various published trials and meta-analyses discussed above are presented in Table 2.

Table 2. Trials and meta-analyses of ω-3 PUFAs in cardiovascular disease (CVD).

Study	Study Design	Number of Patients	Intervention	Follow-Up	Outcome
DART [4] 1989	Secondary prevention RCT	2033	200–400 g fish per week	2 years	29% reduction in mortality
GISSI Prevenzione [6] 1999	Secondary prevention RCT	11,324	882 mg EPA and DHA daily	3.5 years	15–20% reduction in mortality and CV events
DART-2 [44] 2003	Secondary prevention RCT	3114	2 fish servings per week or 3 fish oil capsules daily	3-9 years	Higher cardiac mortality and SCD
JELIS [9] 2007	Primary and secondary prevention RCT	18,645	1.8 g EPA daily	4.6 years	19% reduction in coronary events in CAD patients, no benefit in primary prevention
GISSI-HF [10] 2008	Secondary prevention RCT	6975	840 mg EPA and DHA daily	3.9 years	9% reduction in mortality and 8% reduction in hospitalizations
Alpha Omega [14] 2010	Secondary prevention RCT	4837	226 mg EPA and 150 mg DHA daily	3.4 years	No benefit
OMEGA [13] 2010	Secondary prevention RCT	3851	460 mg EPA and 380 mg DHA daily (vs olive oil control)	1 year	No benefit
SU.FOL.OM3 [15] 2010	Secondary prevention RCT	2501	600 mg EPA and DHA daily	4.7 years	No benefit

Table 2. *Cont.*

Study	Study Design	Number of Patients	Intervention	Follow-Up	Outcome
ORIGIN [16] 2012	Secondary prevention RCT	12,536	465 mg EPA and 375 mg DHA daily (vs olive oil control)	6.2 years	No benefit
Rizos et al. [47] 2012	Meta-analysis of 20 RCTs	68,680	1000 mg EPA and DHA daily (median)	-	No benefit
Risk and Prevention [17] 2013	Primary prevention RCT	12,513	850 mg of EPA and DHA daily (vs olive oil control)	5 years	No benefit
AREDS2 [46] 2014	Primary prevention RCT	4203	650 mg EPA plus 350 mg DHA daily	4.8 years	No benefit
ASCEND [50] 2018	Primary prevention RCT	15,480	1 g ω-3 PUFA daily (vs olive oil control)	7.4 years	No benefit
REDUCE-IT [51] 2018	Primary prevention RCT	8179	4 g ethyl EPA daily	4.9 years	25% reduction in major CV events
VITAL [52] 2018	Primary prevention RCT	25,871	1 g ω-3 PUFA daily	5.3 years	No benefit

6. Available ω-3 PUFA Formulations

ω-3 PUFA ethyl esters was the first formulation approved by the FDA in the year 2004. It was marketed under the trade name Omacor® by Reliant Pharmaceuticals and approved for use in patients with serum triglyceride levels greater than 500 mg/dL. It was later renamed to Lovaza® (GlaxoSmithKline). Ethyl EPA or icosapent ethyl, marketed under the name Vascepa® by Amarin Pharmaceuticals, was approved in 2012. It differed from the earlier formulation in that it contained only ethyl esters of EPA without any DHA [54].

In 2014, ω-3 carboxylic acids, marketed under the name Epanova® by AstraZeneca, was approved for hypertriglyceridemia greater than 500 mg/dL. This formulation consists of free fatty acids instead of the prodrug, and therefore does not require pancreatic lipase for conversion to active form. Thus, it can be taken independent of meals with good bioavailability [54]. The ECLIPSE [55] and ECLIPSE II [56] studies compared the pharmacokinetics of ethyl esters and carboxylic acids. Both studies reported much higher bioavailability with the carboxylic acid formulations.

Despite newer options, ethyl esters continue to be the most commonly prescribed due to its generic options.

Differences between various ω-3 prescription products are shown in Table 3 [22].

Table 3. ω-3 PUFA formulations.

	Ethyl Esters of EPA and DHA	Ethyl Esters of EPA Only	Free Fatty Acids of EPA and DHA
Brand name	Lovaza® (GlaxoSmithKline)	Vascepa® (Amarin Pharmaceuticals)	Epanova® (AstraZeneca)
Approval date	2004	2012	2014
EPA/DHA (g per capsule)	EPA 0.465 g DHA0.375 g	EPA 1 g	EPA 0.550 g DHA 0.200 g
Dosing	2 g (2 capsules) twice daily or 4 g (4 capsules)once daily WITH MEALS	2 g (2 capsules) twice daily WITH MEALS	2 g (2 capsules)or 4 g (4 capsules) once daily WITH OR WITHOUT MEALS

7. Side-Effects and Safety Concerns

Fish oil supplements are generally well tolerated. The most common side effects are gastrointestinal, like nausea, eructation, and diarrhea. Certain large fish (like king mackerel, shark, swordfish) have a higher chance of being contaminated with methyl mercury and therefore should be avoided by pregnant/breastfeeding women and children. There have been some concerns about an increased risk of minor bleeding with ω-3 fatty acid supplementation, especially in patients taking aspirin and statins. However, no major bleeding events have been reported in trials to date.

8. Conclusions

Data from early epidemiologic and observational studies have shown there to be cardioprotective benefits from fish and fish oil consumption. However, most primary prevention trials and recent secondary prevention trials have failed to replicate similar results. Several possibilities could explain the difference in results—the high efficacy of modern-day pharmacotherapy (like statins) and revascularization that attenuates the benefit of ω-3 fatty acids, a lower dose of EPA and DHA supplementation in trials than what was needed, insufficient length of follow-up to see benefits, an improved diet with a higher consumption of fish and other seafood which may account for the decreased magnitude of benefits from fish oil capsules over time, improper study design with use of olive oil as a control (olive oil itself has cardioprotective properties making it an unideal control), and fewer-than-anticipated events leading to underpowered studies. Some trials with a large sample size and strong study design are ongoing, and may shed useful light on the subject.

Based on the current evidence, individuals are advised to consume a healthy diet with two servings of fatty fish every week. Such a food-based approach also supplies several other beneficial nutrients apart from ω-3 PUFAs. For those who cannot consume fish, fish oil supplements containing EPA and DHA have a good safety profile and may be reasonable options, especially in patients with pre-existing CVD, heart failure, and hypertriglyceridemia.

Author Contributions: All authors contributed to literature review and writing of this manuscript.

Funding: This research received no external funding.

Conflicts of Interest: The authors declare no conflicts of interest.

References

1. Bauer, U.E.; Briss, P.A.; Goodman, R.A.; Bowman, B.A. Prevention of chronic disease in the 21st century: Elimination of the leading preventable causes of premature death and disability in the USA. *Lancet* **2014**, *384*, 45–52. [CrossRef]
2. Bang, H.O.; Dyerberg, J.; Nielsen, A. Plasma lipid and lipoprotein pattern in Greenlandic West-coast Eskimos. *Lancet* **1971**, *297*, 1143–1146. [CrossRef]
3. Bang, H.O.; Dyerberg, J.; Sinclair, H.M. The composition of the Eskimo food in north western Greenland. *Am. J. Clin. Nutr.* **1980**, *33*, 2657–2661. [CrossRef] [PubMed]
4. Burr, M.L.; Gilbert, J.F.; Holliday, R.A.; Elwood, P.C.; Fehily, A.M.; Rogers, S.; Sweetnam, P.M.; Deadman, N.M. Effects of changes in fat, fish, and fibre intakes on death and myocardial reinfarction: Diet and reinfarction trial (DART). *Lancet* **1989**, *334*, 757–761. [CrossRef]
5. Yang, B.; Saldeen, T.G.; Nichols, W.W.; Mehta, J.L. Dietary fish oil supplementation attenuates myocardial dysfunction and injury caused by global ischemia and reperfusion in isolated rat hearts. *J. Nutr.* **1993**, *123*, 2067–2074. [PubMed]
6. GISSI-Prevenzione Investigators. Dietary supplementation with n-3 polyunsaturated fatty acids and vitamin E after myocardial infarction: Results of the GISSI-Prevenzione trial. *Lancet* **1999**, *354*, 447–455. [CrossRef]
7. Whelton, S.P.; He, J.; Whelton, P.K.; Muntner, P. Meta-analysis of observational studies on fish intake and coronary heart disease. *Am. J. Cardiol.* **2004**, *93*, 1119–1123. [CrossRef] [PubMed]
8. He, K.; Song, Y.; Daviglus, M.L.; Liu, K.; Van Horn, L.; Dyer, A.R.; Greenland, P. Accumulated evidence on fish consumption and coronary heart disease mortality: A Meta-analysis of cohort studies. *Circulation* **2004**, *109*, 2705–2711. [CrossRef] [PubMed]
9. Yokoyama, M.; Origasa, H.; Matsuzaki, M.; Matsuzawa, Y.; Saito, Y.; Ishikawa, Y.; Oikawa, S.; Sasaki, J.; Hishida, H.; Itakura, H.; et al. Effects of eicosapentaenoic acid on major coronary events in hypercholesterolaemic patients (JELIS): A randomised open-label, blinded endpoint analysis. *Lancet* **2007**, *369*, 1090–1098. [CrossRef]
10. Tavazzi, L.; Maggioni, A.P.; Marchioli, R.; Barlera, S.; Franzosi, M.G.; Latini, R.; Lucci, D.; Nicolosi, G.L.; Porcu, M.; Tognoni, G. Effect of n-3 polyunsaturated fatty acids in patients with chronic heart failure (the GISSI-HF trial): A randomised, double-blind, placebo-controlled trial. *Lancet* **2008**, *372*, 1223–1230. [PubMed]

11. Kris-Etherton, P.M.; Harris, W.S.; Appel, L.J. Fish consumption, fish oil, omega-3 fatty acids, and cardiovascular disease. *Circulation* **2002**, *106*, 2747–2757. [CrossRef] [PubMed]
12. Clarke, T.C.; Black, L.I.; Stussman, B.J.; Barnes, P.M.; Nahin, R.L. *Trends in the Use of Complementary Health Approaches among Adults: United States, 2002–2012*; National Health Statistics Reports No. 79; National Center for Health Statistics: Hyattsville, MD, USA, 2015; pp. 1–16.
13. Rauch, B.; Schiele, R.; Schneider, S.; Diller, F.; Victor, N.; Gohlke, H.; Gottwik, M.; Steinbeck, G.; Del Castillo, U.; Sack, R.; et al. OMEGA, a randomized, placebo-controlled trial to test the effect of highly purified omega-3 fatty acids on top of modern guideline-adjusted therapy after myocardial infarction. *Circulation* **2010**, *122*, 2152–2159. [CrossRef] [PubMed]
14. Kromhout, D.; Giltay, E.J.; Geleijnse, J.M. n–3 Fatty acids and cardiovascular events after myocardial infarction. *N. Engl. J. Med.* **2010**, *363*, 2015–2026. [CrossRef] [PubMed]
15. Galan, P.; Kesse-Guyot, E.; Czernichow, S.; Briancon, S.; Blacher, J.; Hercberg, S. Effects of B vitamins and omega 3 fatty acids on cardiovascular diseases: A randomised placebo controlled trial. *BMJ* **2010**, *341*, c6273. [CrossRef] [PubMed]
16. ORIGIN Trial Investigators. n–3 Fatty acids and cardiovascular outcomes in patients with dysglycemia. *N. Engl. J. Med.* **2012**, *367*, 309–318. [CrossRef] [PubMed]
17. Risk and Prevention Study Collaborative Group. N–3 fatty acids in patients with multiple cardiovascular risk factors. *N. Engl. J. Med.* **2013**, *368*, 1800–1808. [CrossRef] [PubMed]
18. Bowen, K.J.; Harris, W.S.; Kris-Etherton, P.M. Omega-3 fatty acids and cardiovascular disease: Are there benefits? *Curr. Treat. Opt. Cardiovasc. Med.* **2016**, *18*, 69. [CrossRef] [PubMed]
19. Dennis, E.A.; Norris, P.C. Eicosanoid storm in infection and inflammation. *Nat. Rev. Immunol.* **2015**, *15*, 511. [CrossRef] [PubMed]
20. Saini, R.K.; Keum, Y.S. Omega-3 and omega-6 polyunsaturated fatty acids: Dietary sources, metabolism, and significance—A review. *Life Sci.* **2018**, *203*, 255–267. [CrossRef] [PubMed]
21. Dietary Guidelines Advisory Committee. *Scientific Report of the 2015 Dietary Guidelines Advisory Committee*; US Department of Health and Human Services: Washington, DC, USA, 2015.
22. Burke, M.F.; Burke, F.M.; Soffer, D.E. Review of cardiometabolic effects of prescription omega-3 fatty acids. *Curr. Atheroscler. Rep.* **2017**, *19*, 60. [CrossRef] [PubMed]
23. Kromhout, D.; Yasuda, S.; Geleijnse, J.M.; Shimokawa, H. Fish oil and omega-3 fatty acids in cardiovascular disease: Do they really work? *Eur. Heart J.* **2011**, *33*, 436–443. [CrossRef] [PubMed]
24. Endo, J.; Arita, M. Cardioprotective mechanism of omega-3 polyunsaturated fatty acids. *J. Cardiol.* **2016**, *67*, 22–27. [CrossRef] [PubMed]
25. Yagi, S.; Fukuda, D.; Aihara, K.I.; Akaike, M.; Shimabukuro, M.; Sata, M. N-3 polyunsaturated fatty acids: Promising nutrients for preventing cardiovascular disease. *J. Atheroscler. Thromb.* **2017**, *24*, 999–1010. [CrossRef] [PubMed]
26. Calder, P.C. The role of marine omega-3 (n-fatty acids in inflammatory processes, atherosclerosis and plaque stability. *Mol. Nutr. Food Res.* **2012**, *56*, 1073–1080. [CrossRef] [PubMed]
27. De Winther, M.P.; Kanters, E.; Kraal, G.; Hofker, M.H. Nuclear factor κB signaling in atherogenesis. *Arter. Thromb. Vasc. Biol.* **2005**, *25*, 904–914. [CrossRef] [PubMed]
28. Omura, M.; Kobayashi, S.; Mizukami, Y.; Mogami, K.; Todoroki-Ikeda, N.; Miyake, T.; Matsuzaki, M. Eicosapentaenoic acid (EPA) induces Ca2+-independent activation and translocation of endothelial nitric oxide synthase and endothelium-dependent vasorelaxation. *FEBS Lett.* **2001**, *487*, 361–366. [CrossRef]
29. Ishida, T.; Naoe, S.; Nakakuki, M.; Kawano, H.; Imada, K. Eicosapentaenoic acid prevents saturated fatty acid-induced vascular endothelial dysfunction: Involvement of long-chain acyl-CoA synthetase. *J. Atheroscler. Thromb.* **2015**, *22*, 1172–1185. [CrossRef] [PubMed]
30. Jain, A.P.; Aggarwal, K.K.; Zhang, P.Y. Omega-3 fatty acids and cardiovascular disease. *Eur. Rev. Med. Pharmacol. Sci.* **2015**, *19*, 441–445. [PubMed]
31. Haglund, O.; Mehta, J.L.; Saldeen, T. Effects of fish oil on some parameters of fibrinolysis and lipoprotein(a) in healthy subjects. *Am. J. Cardiol.* **1994**, *74*, 189–192. [CrossRef]
32. Niki, T.; Wakatsuki, T.; Yamaguchi, K.; Taketani, Y.; Oeduka, H.; Kusunose, K.; Ise, T.; Iwase, T.; Yamada, H.; Soeki, T.; et al. Effects of the addition of eicosapentaenoic acid to strong statin therapy on inflammatory cytokines and coronary plaque components assessed by integrated backscatter intravascular ultrasound. *Circ. J.* **2016**, *80*, 450–460. [CrossRef] [PubMed]

33. Thies, F.; Garry, J.M.; Yaqoob, P.; Rerkasem, K.; Williams, J.; Shearman, C.P.; Gallagher, P.J.; Calder, P.C.; Grimble, R.F. Association of n-3 polyunsaturated fatty acids with stability of atherosclerotic plaques: A randomised controlled trial. *Lancet* **2003**, *361*, 477–485. [CrossRef]
34. Swann, P.G.; Venton, D.L.; Le Breton, G.C. Eicosapentaenoic acid and docosahexaenoic acid are antagonists at the thromboxane A2/prostaglandin H2 receptor in human platelets. *FEBS Lett.* **1989**, *243*, 244–246. [CrossRef]
35. Atakisi, O.; Atakisi, E.; Ozcan, A.; Karapehlivan, M.; Kart, A. Protective effect of omega-3 fatty acids on diethylnitrosamine toxicity in rats. *Eur. Rev. Med. Pharmacol. Sci.* **2013**, *17*, 467–471. [PubMed]
36. Christensen, J.H.; Schmidt, E.B. Autonomic nervous system, heart rate variability and n-3 fatty acids. *J. Cardiovasc. Med.* **2007**, *8*, 19–22. [CrossRef] [PubMed]
37. Yagi, S.; Soeki, T.; Aihara, K.I.; Fukuda, D.; Ise, T.; Kadota, M.; Bando, S.; Matsuura, T.; Tobiume, T.; Yamaguchi, K.; et al. Low serum levels of eicosapentaenoic acid and docosahexaenoic acid are risk factors for cardiogenic syncope in patients with Brugada syndrome. *Int. Heart J.* **2017**, *58*, 720–723. [CrossRef] [PubMed]
38. Heydari, B.; Abdullah, S.; Pottala, J.V.; Shah, R.; Abbasi, S.; Mandry, D.; Francis, S.A.; Lumish, H.; Ghoshhajra, B.B.; Hoffmann, U.; et al. Effect of Omega-3 Acid Ethyl Esters on Left Ventricular Remodeling After Acute Myocardial Infarction Clinical Perspective: The OMEGA-REMODEL Randomized Clinical Trial. *Circulation* **2016**, *134*, 378–391. [CrossRef] [PubMed]
39. Endo, J.; Sano, M.; Isobe, Y.; Fukuda, K.; Kang, J.X.; Arai, H.; Arita, M. 18-HEPE, an n-3 fatty acid metabolite released by macrophages, prevents pressure overload–induced maladaptive cardiac remodeling. *J. Exp. Med.* **2014**, *211*, 1673–1687. [CrossRef] [PubMed]
40. Duda, M.K.; O'shea, K.M.; Stanley, W.C. ω-3 polyunsaturated fatty acid supplementation for the treatment of heart failure: Mechanisms and clinical potential. *Cardiovasc. Res.* **2009**, *84*, 33–41. [CrossRef] [PubMed]
41. Da Boit, M.; Hunter, A.M.; Gray, S.R. Fit with good fat? The role of n-3 polyunsaturated fatty acids on exercise performance. *Metabolism* **2017**, *66*, 45–54. [CrossRef] [PubMed]
42. Yagi, S.; Akaike, M.; Ise, T.; Ueda, Y.; Iwase, T.; Sata, M. Renin–angiotensin–aldosterone system has a pivotal role in cognitive impairment. *Hypertens. Res.* **2013**, *36*, 753. [CrossRef] [PubMed]
43. Yagi, S.; Hara, T.; Ueno, R.; Aihara, K.I.; Fukuda, D.; Takashima, A.; Hotchi, J.; Ise, T.; Yamaguchi, K.; Tobiume, T.; et al. Serum concentration of eicosapentaenoic acid is associated with cognitive function in patients with coronary artery disease. *Nutr. J.* **2014**, *13*, 112. [CrossRef] [PubMed]
44. Burr, M.L.; Ashfield-Watt, P.A.; Dunstan, F.D.; Fehily, A.M.; Breay, P.; Ashton, T.; Zotos, P.C.; Haboubi, N.A.; Elwood, P.C. Lack of benefit of dietary advice to men with angina: Results of a controlled trial. *Eur. J. Clin. Nutr.* **2003**, *57*, 193. [CrossRef] [PubMed]
45. Estruch, R.; Ros, E.; Salas-Salvadó, J.; Covas, M.I.; Corella, D.; Arós, F.; Gómez-Gracia, E.; Ruiz-Gutiérrez, V.; Fiol, M.; Lapetra, J.; et al. Primary prevention of cardiovascular disease with a Mediterranean diet. *N. Engl. J. Med.* **2013**, *368*, 1279–1290. [CrossRef] [PubMed]
46. Bonds, D.E.; Harrington, M.; Worrall, B.B.; Bertoni, A.G.; Eaton, C.B.; Hsia, J.; Robinson, J.; Clemons, T.E.; Fine, L.J.; Chew, E.Y. Effect of long-chain ω-3 fatty acids and lutein+ zeaxanthin supplements on cardiovascular outcomes: Results of the Age-Related Eye Disease Study 2 (AREDSrandomized clinical trial. *JAMA Intern. Med.* **2014**, *174*, 763–771. [CrossRef] [PubMed]
47. Rizos, E.C.; Ntzani, E.E.; Bika, E.; Kostapanos, M.S.; Elisaf, M.S. Association between omega-3 fatty acid supplementation and risk of major cardiovascular disease events: A systematic review and meta-analysis. *JAMA* **2012**, *308*, 1024–1033. [CrossRef] [PubMed]
48. Macchia, A.; Grancelli, H.; Varini, S.; Nul, D.; Laffaye, N.; Mariani, J.; Ferrante, D.; Badra, R.; Figal, J.; Ramos, S.; et al. Omega-3 fatty acids for the prevention of recurrent symptomatic atrial fibrillation: Results of the FORWARD (Randomized Trial to Assess Efficacy of PUFA for the Maintenance of Sinus Rhythm in Persistent Atrial Fibrillation) trial. *J. Am. Coll. Cardiol.* **2013**, *61*, 463–468. [CrossRef] [PubMed]
49. Mariani, J.; Doval, H.C.; Nul, D.; Varini, S.; Grancelli, H.; Ferrante, D.; Tognoni, G.; Macchia, A. N-3 polyunsaturated fatty acids to prevent atrial fibrillation: Updated systematic review and meta-analysis of randomized controlled trials. *J. Am. Heart Assoc.* **2013**, *2*, e005033. [CrossRef] [PubMed]
50. The ASCEND Study Collaborative Group. Effects of n-3 fatty acid supplements in diabetes mellitus. *N. Engl. J. Med.* **2018**, *379*, 1540–1550. [CrossRef] [PubMed]

51. A Study of AMR101 to Evaluate Its Ability to Reduce Cardiovascular Events in High Risk Patients with Hypertriglyceridemia and on Statin. 2016. Available online: https://clinicaltrials.gov/ct2/show/NCT01492361 (accessed on 13 October 2018).
52. Manson, J.E.; Cook, N.R.; Lee, I.M.; Christen, W.; Bassuk, S.S.; Mora, S.; Gibson, H.; Albert, C.M.; Gordon, D.; Copeland, T.; et al. Marine n− 3 fatty acids and prevention of cardiovascular disease and cancer. *N. Engl. J. Med.* **2018**. [CrossRef] [PubMed]
53. US National Institutes of Health. *Outcomes Study to Assess STatin Residual Risk Reduction With EpaNova in HiGh CV Risk PatienTs With Hypertriglyceridemia (STRENGTH)*; US National Institutes of Health: Bethesda, MD, USA, 2015.
54. Benes, L.B.; Bassi, N.S.; Kalot, M.A.; Davidson, M.H. Evolution of Omega-3 Fatty Acid Therapy and Current and Future Role in the Management of Dyslipidemia. *Cardiol. Clin.* **2018**, *36*, 277–285. [CrossRef] [PubMed]
55. Davidson, M.H.; Johnson, J.; Rooney, M.W.; Kyle, M.L.; Kling, D.F. A novel omega-3 free fatty acid formulation has dramatically improved bioavailability during a low-fat diet compared with omega-3-acid ethyl esters: The ECLIPSE (Epanova® compared to Lovaza® in a pharmacokinetic single-dose evaluation) study. *J. Clin. Lipidol.* **2012**, *6*, 573–584. [CrossRef] [PubMed]
56. Offman, E.; Marenco, T.; Ferber, S.; Johnson, J.; Kling, D.; Curcio, D.; Davidson, M. Steady-state bioavailability of prescription omega-3 on a low-fat diet is significantly improved with a free fatty acid formulation compared with an ethyl ester formulation: The ECLIPSE II study. *Vasc. Health Risk Manag.* **2013**, *9*, 563. [CrossRef] [PubMed]

International Journal of
Molecular Sciences

Review

Micronutrient Depletion in Heart Failure: Common, Clinically Relevant and Treatable

Natasa Cvetinovic [1], Goran Loncar [2,3], Andjelka M. Isakovic [3], Stephan von Haehling [4,5], Wolfram Doehner [6], Mitja Lainscak [7,8,*] and Jerneja Farkas [9,10,*]

1 Department of Cardiology, University Clinical Hospital Center "Dr. Dragisa Misovic—Dedinje", 11000 Belgrade, Serbia; ncvetinovic@gmail.com
2 Institute for Cardiovascular Diseases Dedinje, 11000 Belgrade, Serbia; loncar_goran@yahoo.com
3 School of Medicine, University of Belgrade, 11000 Belgrade, Serbia; andjelka.isakovic@gmail.com
4 Department of Cardiology and Pneumology, University of Goettingen Medical Center, DE-37075 Goettingen, Germany; stephan.von.haehling@med.uni-goettingen.de
5 German Center for Cardiovascular Research (DZHK), partner site Goettingen, DE-37099 Goettingen, Germany
6 Berlin Institute of Health Center for Regenerative Therapies (BCRT) and Department of Cardiology (Virchow Klinikum), German Centre for Cardiovascular Research (DZHK) partner site Berlin, Charité Universitätsmedizin Berlin, DE-13353 Berlin, Germany; wolfram.doehner@charite.de
7 Division of Cardiology, General Hospital Murska Sobota, SI-9000 Murska Sobota, Slovenia
8 Faculty of Medicine, University of Ljubljana, SI-1000 Ljubljana, Slovenia
9 Research Unit, General Hospital Murska Sobota, SI-9000 Murska Sobota, Slovenia
10 National Institute of Public Health, SI-1000 Ljubljana, Slovenia
* Correspondence: mitja.lainscak@guest.arnes.si (M.L.); jerneja.farkas@sb-ms.si (J.F.)

Received: 14 October 2019; Accepted: 4 November 2019; Published: 11 November 2019

Abstract: Heart failure (HF) is a chronic condition with many imbalances, including nutritional issues. Next to sarcopenia and cachexia which are clinically evident, micronutrient deficiency is also present in HF. It is involved in HF pathophysiology and has prognostic implications. In general, most widely known micronutrients are depleted in HF, which is associated with symptoms and adverse outcomes. Nutritional intake is important but is not the only factor reducing the micronutrient availability for bodily processes, because absorption, distribution, and patient comorbidity may play a major role. In this context, interventional studies with parenteral micronutrient supplementation provide evidence that normalization of micronutrients is associated with improvement in physical performance and quality of life. Outcome studies are underway and should be reported in the following years.

Keywords: heart failure; micronutrients; iron; vitamins; trace elements

---

1. Introduction

Heart failure (HF) is a complex disease with many potential underlying causes, which affects the function of other tissues and is often followed by comorbidities [1], especially in the elderly [2]. Pathogenesis of HF has been elucidated and includes many mechanisms, such as increased hemodynamic overload, ischemic-related dysfunction, ventricular remodeling, excessive neurohormonal stimulation, abnormal cardiomyocyte calcium cycling, excessive or inadequate accumulation of extracellular matrix, accelerated apoptosis, and genetic mutations [1]. Although no single unifying pathogenetic theory can explain HF completely, there is evidence suggesting that declining bioenergy plays a major role [3]. Fatty acids and carbohydrates are the main energy source for cardiomyocytes. The conversion of these macronutrients to biological energy requires micronutrients such as coenzyme Q10, thiamine, riboflavin, etc., which are essential cofactors for energy production, energy transfer, and maintenance of the physiological heart function [4]. Moreover, patients with HF have oxidative stress, inappropriate food intake (Table 1), altered metabolism and intestinal function, and pro-inflammatory status, which leads to a deficiency of micronutrients (e.g., iron, selenium, and zinc) and consequently affects prognosis. This review aims to underline the role of micronutrients in the pathophysiology of HF, their prognostic implications, and the effects of supplementation. We reviewed micronutrients that were most represented in the literature and could potentially have the most beneficial effects when supplemented in HF patients, namely B complex vitamins (B1, B2, B6, B12), vitamin C, trace elements selenium, zinc, and iron, coenzyme Q10, with a special attention on vitamin D due to a long history of its research and supplementation, and the abundance of available data.

Table 1. Recommended micronutrient dietary allowance for the general population.

Micronutrient	RDA	Reference
Vitamin D	15–20 μg/day	[5,6]
Thiamin (B1)	1.1–1.2 mg/day	[7]
Riboflavin (B2)	0.9–1.3 mg/day	[8]
Pyridoxine (B6)	1.6–2.0 mg/day	[9]
Cobalamin (B12)	2.0–2.4 μg/day	[10]
Vitamin C	75–90 mg/day	[11]
Coenzyme Q10 *	n.a.	
Selenium	55 μg/day	[10]
Zinc	8–11 mg/day	[11]
Iron	8–18 mg/day	[11]

RDA: recommended dietary allowance. n.a.: not assessed. *: no RDA has been established.

2. Vitamin D

Although many diseases (e.g., cancer, autoimmune disorders, infertility and pregnancy complications, insulin resistance, and type 2 diabetes mellitus) are associated with vitamin D levels, its metabolism, and vitamin D-related genes [12,13], the association between vitamin D and cardiovascular disease is still controversial. Vitamin D regulates numerous processes involved in the pathogenesis of HF, such as cell proliferation and differentiation, apoptosis, oxidative stress, inflammation, endothelial function, vascular calcification, and activation of the renin-angiotensin system [14]. Observational studies have reported that subjects with lower vitamin D levels are under higher risk of developing HF [15,16]. Vitamin D deficiency (Table 2) is one of the most common types of hypovitaminosis worldwide, with a prevalence of almost 50% among the elderly [17], and is the most common vitamin deficiency in HF [18]. It is more common among patients with HF compared with healthy control individuals, independently of age [19]. While 90–99% of elderly HF patients are affected by vitamin D deficiency [17,20,21], severe deficiency is reported in 17–68% of HF patients, dependent on age, gender, functional status, HF severity, and left ventricle ejection fraction (LVEF) [17,20,22]. Pathophysiology of vitamin D deficiency includes reduced food intake and intestinal absorption of nutrients, and limited exposure to solar ultraviolet light (Table 1). Accordingly, both aging and HF predispose patients to hypovitaminosis, and they may have an additive effect on vitamin D level in elderly patients. In light of the above, it is reasonable to evaluate vitamin D status in HF patients separately according to age.

Table 2. Serum normal concentrations and definition of deficiency of micronutrients.

Micronutrient	Reference Range	Insufficiency	Deficiency	Reference
Vitamin D	>30 ng/mL >75 nmol/L	21–29 ng/mL 52.5–72.5 nmol/L	<20 ng/mL <50 nmol/L	[23]
Thiamin (B1) *	25–75 ng/mL 75–195 nmol/L			[24]
Riboflavin (B2) [a]			EGRAC ≥ 1.3	[25]
Pyridoxine (B6) [b]	5–50 ng/mL 23–223 nmol/L	20–30 nmol/L	<20 nmol/L	[26–28]
Cobalamin (B12)	180–950 pg/mL 125–701 pmol/L		<200 pg/mL <148 pmol/L	[29,30]
Vitamin C	>50 μmol/L	10–50 μmol/L	<10 μmol/L	[31]
Coenzyme Q10	0.5–1.7 μmol/L			[32]
Selenium	70–150 ng/mL 0.9–1.8 μmol/L			[33]
Zinc	0.7–1.6 μg/mL 11–25 μmol/L			[34]
Iron	45–160 μg/mL 8–23 μmol/L		<45 μg/mL <8 μmol/L	[35]

*: exact range depends on the laboratory; different cutoff values are used. [a]: expressed as erythrocyte glutathione reductase activation coefficient (EGRAC). [b]: measured as pyridoxal phosphate.

Vitamin D is associated with functional status, illness severity, and prognosis in HF [36]. Vitamin D level positively correlates with cardiopulmonary stress test performance [22] and six-minute walk test distance [37], and negatively with NYHA class [17,20]. In a study of severe HF patients, vitamin D concentration was significantly lower in hospitalized subjects requiring intravenous inotropic agents or left ventricular assist devices, compared with those treated as outpatients [22]. Vitamin D also correlates with NT-proANP [19], NT-proBNP [17,38] and left ventricle ejection fraction (LVEF) [20], which are predictors of prognosis in HF. Furthermore, vitamin D levels independently predicted all-cause mortality in an HF study with an 18-month follow-up [38] and strongly predicted mortality in a study with a longer follow-up period of 4 years [36]. Additionally, severe vitamin D deficiency was found as an independent predictor of death due to HF in a study of 3299 enrolled subjects with suspect coronary artery disease [17], and lower levels of vitamin D independently predicts mortality in end-stage HF patients [39].

Although it has been strongly confirmed that vitamin D plays an important role in the development and prognosis of HF, evidence about the effect of vitamin D supplementation in HF patients is inconsistent. It has been documented that vitamin D supplementation (800 IU/day) might protect people >60 years from HF [40]. A recent meta-analysis, which included 81 interventional studies with vitamin D supplementation in HF patients, has confirmed a positive effect on cardiovascular risk factors such as high blood pressure, dyslipidemia, and inflammation [41]. Thus, results suggest that the required dose of vitamin D for improvements in risk factors is ≥4000 IU (100 g)/day, which is above the tolerable upper intake level for adults and can be associated with toxic side effects [42,43]. Supplementation with lower doses (≤2000 IU/day) in HF patients significantly increased vitamin D levels but had no benefits on LVEF, functional capacity, or quality of life [44,45]. The EVITA, a randomized clinical trial (RCT) that compared vitamin D 4000 IU/day to a placebo in patients with advanced HF and vitamin D deficiency (vitamin D ≤ 75 nmol/L), found no difference in mortality and hospitalization rate. Furthermore, supplementation was associated with a greater need for mechanical circulatory support implant (especially in patients with initial vitamin D concentrations ≥ 30 nmol/L) and higher incidence of hypercalcemia [46]. Accordingly, vitamin D supplementation in HF can be considered but requires caution, especially in patients with no evidence of significant deficiency.

3. Other Fat-Soluble Vitamins

Although experimental data have provided evidence that vitamin A regulates the cardiac renin-angiotensin system, and that vitamin A receptors impact the development of diabetes-induced cardiac remodeling and HF in patients with diabetes mellitus [47], the association between vitamin A and HF has not been confirmed in observational studies. Only one study has evaluated vitamin A status in HF, and it found no significant difference between HF patients and sex-matched healthy controls [48]. Dietary vitamin A intake showed no association with HF mortality in the Japan Collaborative Cohort Study, which enrolled 58,696 subjects [49].

Observational studies have found no significant differences in vitamin E status between HF patients and healthy controls [48,50]. The same result was found in a rat model of HF [51]. However, in the Japan Collaborative Cohort Study, high dietary intake of vitamin E was associated with a reduced risk of HF mortality in women but not in men [49]. Although vitamin E supplementation (400 mg/day) reduces oxidative stress in HF patients [52], it has no impact on symptoms, physical functioning, or mortality [53–56]. Moreover, there are indications that vitamin E treatment may contribute to HF development [54,55].

It is known that vitamin K status is associated with reduced coronary artery calcification and cardiovascular mortality risk. Vitamin K levels, intake, and supplementation have not been evaluated in HF patients, although epidemiological studies suggest a protective role of vitamin K in cardiovascular diseases [57].

4. B Vitamins

Thiamine (vitamin B1) is a cofactor in the metabolism of carbohydrates and amino acids and is also essential for aerobic metabolism and adenosine triphosphate (ATP) production. Thiamine deficiency is associated with many cardiovascular diseases, including HF, chronic vascular inflammation, myocardial infarction, and conduction defects [58]. Furthermore, the deficiency can cause HF by depriving the heart of ATP [59]. On the other hand, HF induces thiamin deficiency by increasing urinary excretion due to therapy with loop diuretics, and poor absorption of thiamine due to cardiac cachexia and splanchnic congestion [60,61]. The addition of spironoloctone to loop diuretics has beneficial effects on thiamine levels in HF patients [62].

The prevalence of thiamine deficiency (Table 2) in HF patients is significantly higher compared with non-HF subjects and ranges from 3% in ambulatory settings to 91% in hospital settings [63]. The first observational study that evaluated the effects of thiamine supplementation in HF enrolled only six patients who were treated by intravenous thiamine (100 mg/day) for 7 days [64]. The results confirmed significant improvement in thiamine levels as assessed by LVEF and NYHA, which encouraged further research, but findings were controversial. An RCT using seven days of intravenous thiamine versus a placebo in a double-blind manner, followed by six weeks of oral thiamine (200 mg/day) in all patients, indicated significant improvement of LVEF in the thiamine group after seven days (LVEF increase was 4%), and impressive improvement of 22% in all patients at the end of treatment [65]. An RCT with oral thiamine (300 mg/day), had a small sample size (18 patients) and a short follow-up period (28 days), but confirmed LVEF improvement after supplementation [66]. However, these results were not confirmed in an RCT that enrolled 52 patients. The main difference between these two RCTs with similar designs was the use of diuretic therapy, which significantly influenced thiamine level. In the trial with no LVEF improvement, the proportion of patients who were on furosemide was low (<20%) and spironolactone was prescribed to most patients (>80%) in the thiamine group [62]. Oral supplementation may be less effective than intravenous, due to the impaired enteral absorption in HF. Limitations of all performed studies, and potential cause of inconclusive results, included small sample sizes, short follow-up periods, and subjective measures used as endpoints. Accordingly, further research is needed to elucidate the effects of thiamine supplementation on mortality, hospitalization rate, functional status, and quality of life in HF.

Riboflavin (vitamin B2) is an essential cofactor in cellular energy production, and its deficiency may contribute to the depletion of energy reserves observed in the failing heart. It is water-soluble, has limited tissue storage, and depends on intake (Table 1) and renal excretion. Prevalence of riboflavin deficiency (Table 2) is significantly higher (27%) in HF patients than in healthy controls (2.2%) [67,68]. A study on animal models of HF suggests that riboflavin supplementation significantly improves left ventricular systolic and diastolic function [69]. Further research on HF and riboflavin deficiency in humans is required.

Pyridoxine (vitamin B6) plays an important role in intermediary metabolism, as a cofactor mainly in the metabolism of amino acids, but also of carbohydrates and lipids, as well as in the biosynthesis of neurotransmitters. Pyridoxine deficiency (Table 2) prevalence is also higher in HF patients (38%) than in healthy controls [68]. There are no investigations of isolated pyridoxine supplementation in HF patients.

Cobalamin (vitamin B12) deficiency has been evaluated in the setting of HF, but results are controversial. The largest study evaluating vitamin B12 in HF, with almost 1000 patients, provided unexpected results: vitamin B12 had a weak negative correlation with LVEF and a significant positive correlation with NYHA class [70]. A possible reason for these results is the unclear selection of HF patients: a large number of patients had no systolic dysfunction (mean LVEF were 58% and 65% for men and women, respectively) and no clinical manifestations of HF (about 35% patients were NYHA class I). An observational study found that vitamin B12 deficiency (serum level <200 pg/mL) is relatively rare in HF (5%), and that vitamin B12 level is not related to prognosis [71]. However, the study was not limited to patients with reduced LVEF, and 13% of enrolled patients had LVEF >45%. Although parenteral replacement therapy should be started soon after the vitamin B12 deficiency has been established [72], an interventional study with vitamin B12 supplementation, in addition to folate and vitamin B6, in elderly HF patients (mean age 81 years) suggested no association with NT-proBNP levels [73]. In the future, vitamin B12 should be evaluated in the setting of HF with reduced ejection fraction.

5. Vitamin C

Three observational studies [48,50,74] showed significantly lower vitamin C levels among HF patients (mean 39.7 µmol/L, 61 µmol/L, 22 µmol/L, respectively) than in healthy controls, although in all of them, concentrations were above the cutoff for deficiency (Table 2) [75]. In a more recent study, patients with HF had lower vitamin C levels (43.3 µmol/L in NYHA II and 46.8 µmol/L in NYHA III/IV) than the control group had (57.2 µmol/L), but the difference was not significant [67]. Two studies, which determined vitamin C deficiency by a 3-day food diary, found a similar prevalence of the dietary deficiency (39% and 40%, respectively) among HF patients [76,77]. Furthermore, dietary deficiency predicted shorter cardiac event-free survival in both of them. Although some interventional studies have shown that vitamin C supplementation enhances the contractile response to dobutamine [78] and improves endothelial function [79] in HF, there have been no RCTs evaluating vitamin C supplementation in HF. For further research, we propose RCTs that evaluate the effects of vitamin C supplementation on biomarker status, echocardiography parameters, functional status, quality of life, and long-term outcomes in HF patients with confirmed vitamin C deficiency.

6. Coenzyme Q10

Coenzyme Q10 (CoQ10) is essential for ATP synthesis, as well as a powerful lipid-soluble antioxidant [80]. Myocardial CoQ10 deficiency has been observed in patients with HF, and it correlates with the severity of symptoms and the LVEF [81]. CoQ10 supplementation also significantly increases its concentration in the blood [82]. Lower levels of CoQ10 (Table 2) are observed in elderly patients and those with more advanced HF [83]. Since 1976, when the first interventional study with CoQ10 administration in HF suggested its therapeutic potential [84], many RCTs have been performed that confirm its benefit in the treatment of HF [85]. CoQ10 significantly improves cardiac output, cardiac index, stroke volume, and LVEF [86,87]. The administration of CoQ10 in patients with end-stage HF awaiting cardiac transplantation significantly improved functional status, clinical symptoms, and quality of life, but there were no changes in echo parameters [88]. An RCT performed 25 years ago that enrolled 641 HF patients (NYHA class III and IV) demonstrated that the addition of CoQ10 to conventional therapy (2 mg/kg daily for 1 year) significantly reduces hospitalization due to HF worsening and the incidence of serious complications [89]. The most recent Q-SYMBIO RCT (CoQ10 100 mg 3 times/day for 2 years vs. placebo) confirmed that CoQ10 as an adjunctive treatment of HF is safe, improves symptoms and functional status, and reduces major adverse cardiovascular events, such as death from HF, sudden cardiac death, and hospitalization due to HF worsening [90]. According to all of the above, we suggest that physicians should consider CoQ10 supplementation in advanced HF, especially if the deficiency is confirmed.

7. Selenium

Selenium is an essential nutrient and one of the most important antioxidants in the body. It is found within the body in various selenoproteins, such as glutathione peroxidase (GPx), thioredoxin reductase and selenoprotein P [91]. Moreover, selenium has an important role in converting thyroxin into the biologically active triiodothyronine. This may be an additional mechanism by which low concentrations of selenium compromise cardiovascular conditions [92]. The dietary intake of selenium (Table 1) differs throughout the world. In Europe, due to poor selenium content in the soil, the estimated mean intake of selenium (40 µg/day) [93] is significantly lower than the proposed dose for a normal-weighted Caucasian (75 µg/day), which is needed for optimal function of selenoprotein P [94] and cancer protection [33]. Furthermore, it is well known that selenium deficiency may be a cause of reversible HF, a condition known as Keshan Disease [95]. Also, it has been suggested that patients with HF tend to have lower circulating levels of selenium (Table 2) than individuals free from the condition [43]. A case-control-pair longitudinal study with 11,000 enrolled participants found that subjects with low selenium levels have a higher risk for myocardial infarction and cardiovascular mortality [96], but RCTs with selenium supplementation have found no evidence of cardiovascular protection [97].

The metabolism of selenium and CoQ10 are strongly associated with each other, and a deficiency of one can reduce levels or function of the other [98–100]. Accordingly, they are often used and evaluated together. An RCT of an elderly population (mean age 78) who were supplemented with 200 µg/day of organic selenium and 200 mg/day of CoQ10 versus a placebo, found that long-term supplementation reduces NT-proBNP levels and cardiovascular mortality in those with mild to moderate impaired cardiac function [101]. Although an RCT with multiple micronutrient supplementation that included selenium, suggested a beneficial effect on LVEF and quality of life in 30 HF patients (LVEF ≤ 35%, HF due to ischemic heart disease) [102], data about its effects on HF when used alone are insufficient. A recent RCT that compared selenium 200 µg/day with a placebo indicated that supplementation for 12 weeks to HF patients has beneficial effects on insulin metabolism and few markers of cardio-metabolic risk [103].

8. Zinc

Zinc, which has antioxidant properties [104], is an essential element for humans and is required for enzyme function, multiple signaling pathways and transcription factors, and is also the second most abundant trace metal in humans [105]. Although zinc is widely distributed in human tissue and has numerous functions, it has an important role in controlling cardiac contractility in cardyomyocites [106].

One of the oldest studies on zinc levels in HF patients evaluated zinc status in 20 younger HF patients (34–64 years), and indicated significantly lower levels in subjects with dilated cardiomyopathy than in healthy controls (74.5 µg/dL or 11.4 µmol/L vs. 93.1 µg/dL or 14.2 µmol/L, respectively) [107]. This finding has been supported by studies on patients with LVEF <40% and NYHA class II-IV [48,108–110], and finally confirmed by a recent meta-analysis published in 2018 [111]. Moreover, subgroup analysis found that patients with idiopathic dilated cardiomyopathy had lower zinc levels than control subjects, except for patients with ischemic cardiomyopathy. A recent study enrolled 968 hospitalized patients with decompensated HF who were divided into 3 groups based on serum zinc levels, and found the highest cardiac and all-cause mortality in the third tertile (<62 μg/dL). Serum zinc level was a predictor of cardiac and all-cause mortality, independently of age, gender, comorbidities, medications, other micronutrient levels, B-type natriuretic peptide, and LVEF [112].

Although the role of zinc in cardiovascular medicine has been well represented in molecular research and observational studies in the past few years, there have been no interventional studies or RCTs evaluating zinc supplementation in HF. Two previous studies that included zinc in multiple micronutrient supplements have suggested an association with improvement of cardiac function and quality of life [103,113].

According to the above, further research into zinc supplementation in HF is needed. We propose a study population of elderly patients with advanced HF due to non-ischemic cardiomyopathy and confirmed zinc deficiency, who could have the greatest benefit of the supplementation.

9. Iron

Iron is an essential microelement, required for transport, storage and usage of oxygen in humans. In HF, iron deficiency (Table 2) is one of the most common comorbidities, affecting 37–61% of patients [114–116]. The deficiency, even before the onset of anemia, can be severe among patients with CHF, aggravating symptoms, quality of life, functional status, and clinical outcomes, and is associated with an increased risk of mortality [114,117–120]. The 2016 European Society of Cardiology (ESC) guidelines for the diagnosis and treatment of acute and chronic HF recommend that all patients with HF should be tested for iron deficiency [121]. Moreover, intravenous iron, ferric carboxymaltose, is specifically recommended to be considered for the treatment of iron deficiency in HF, in order to alleviate symptoms and improve exercise capacity and quality of life [122,123]. Oral iron supplementation has no benefit in the setting of HF [124,125] and is not recommended by ESC guidelines. The impaired enteral absorption and other conditions characterized by immune activation are considered as causes for the ineffectiveness of oral iron administration. Findings of the IRONOUT HF RCT (oral iron therapy vs. placebo) [124] are in contrast to results from RCTs of intravenous iron repletion (FAIR-HF and CONFIRM-HF) [122,123], although patient populations were similar. Moreover, in the IRONOUT HF oral therapy produced improvement in iron stores, though the improvement was minimal and clinically not significant.

Iron deficiency is also a frequent co-morbidity in HF with preserved EF (HFpEF), which associated with reduced exercise capacity and quality of life, and its prevalence increases with increasing severity of diastolic dysfunction [126]. Beneficial effects of intravenous iron therapy in an animal model of HFpEF [127] encourage further research, which should elucidate the effects of iron therapy in HF with preserved LVEF.

10. Conclusions and Future Perspectives

Although micronutrient deficiencies are frequent among HF patients, their role in pathogenesis and treatment has not yet been completely elucidated. Besides iron supplementation, which is recommended by ESC, CoQ10 supplementation has the strongest evidence of benefit in HF. The available data suggest consideration of vitamin D supplementation in cases of confirmed deficiency in patients with HF. Some micronutrients (selenium, zinc, vitamin B2, and vitamin B6) showed improvement of various clinical parameters in HF, but were investigated only in a multiple supplementation setting, while the effects of vitamins B1 and B12 were inconsistent due to unclear selection of patients and poorly defined endpoints. The influence of vitamin C supplementation in HF is yet to be determined. Thus, interventional studies with strict selection criteria and clear endpoints for each micronutrient are still needed to gain better insight into their role in HF. Furthermore, the question remains as to whether the deficiency cutoff values drawn from the general population apply to HF patients; this should be addressed in future research. Supplementation trials should take into account reduced food intake and poor intestinal absorption of nutrients in these patients, and should investigate not only dosage with beneficial effects, but also the routes of administration (i.e., oral or intravenous). In addition, the available data on micronutrient supplementation in patients with HFpEF are scarce and should be thoroughly researched in the future.

Funding: The authors acknowledge the projects (Epidemiology, patophysiology and clinical significance of iron deficiency in chronic cardiopulmonary disease; ID J3-9284 and Burden of cachexia and sarcopenia in patients with chronic diseases: epidemiology, pathophysiology and outcomes; ID J3-9292) were financially supported by the Slovenian Research Agency.

Conflicts of Interest: The autohors declare no conflict of interest.

References

1. Braunwald, E. Heart Failure. *JACC Hear. Fail.* **2013**, *1*, 1–20. [CrossRef] [PubMed]
2. Cvetinovic, N.; Loncar, G.; Farkas, J. Heart failure management in the elderly—A public health challenge. *Wien. Klin. Wochenschr.* **2016**, *128*, 466–473. [CrossRef] [PubMed]
3. Wong, A.-P.; Niedzwiecki, A.; Rath, M. Myocardial energetics and the role of micronutrients in heart failure: A critical review. *Am. J. Cardiovasc. Dis.* **2016**, *15*, 81–92.
4. Witte, K.K.; Clark, A.L. Micronutrients and their supplementation in chronic cardiac failure. An update beyond theoretical perspectives. *Heart Fail. Rev.* **2006**, *11*, 65–74. [CrossRef] [PubMed]
5. Ross, A.C.; Taylor, C.L.; Yaktine, A.L.; del Valle, H.B. *Dietary Reference Values for Vitamin D*; National Academies Press: Washington, DC, USA, 2016.
6. Ross, A.C.; Taylor, C.L.; Yaktine, A.L.; del Valle, H.B. *Dietary Reference Intakes for Calcium and Vitamin, D*; National Academies Press: Washington, DC, USA, 2011.
7. Ross, A.C.; Taylor, C.L.; Yaktine, A.L.; del Valle, H.B. *Dietary Reference Intakes for Thiamin, Riboflavin, Niacin, Vitamin B6, Folate, Vitamin B12, Pantothenic Acid, Biotin, and Choline*; National Academies Press: Washington, DC, USA, 1998.
8. Ross, A.C.; Taylor, C.L.; Yaktine, A.L.; del Valle, H.B. *Dietary Reference Intakes*; National Academies Press: Washington, DC, USA, 2003.
9. Said, H.M. Water-Soluble Vitamins. *World Rev. Nutr. Diet.* **2014**, *111*, 30–37. [PubMed]
10. Ross, A.C.; Taylor, C.L.; Yaktine, A.L.; del Valle, H.B. *Dietary Reference Intakes for Vitamin C, Vitamin E, Selenium, and Carotenoids*; National Academies Press: Washington, DC, USA, 2000.
11. Ross, A.C.; Taylor, C.L.; Yaktine, A.L.; del Valle, H.B. *Dietary Reference Intakes for Vitamin A, Vitamin K, Arsenic, Boron, Chromium, Copper, Iodine, Iron, Manganese, Molybdenum, Nickel, Silicon, Vanadium, and Zinc*; National Academies Press: Washington, DC, USA, 2001.
12. Wimalawansa, S.J.; Razzaque, M.S.; Al-Daghri, N.M. Calcium and vitamin D in human health: Hype or real? *J. Steroid Biochem. Mol. Biol.* **2018**, *180*, 4–14. [CrossRef] [PubMed]
13. Zeljic, K.; Kandolf-Sekulovic, L.; Supic, G.; Pejovic, J.; Novakovic, M.; Mijuskovic, Z. Melanoma risk is associated with vitamin D receptor gene polymorphisms. *Melanoma Res.* **2014**, *24*, 273–279. [CrossRef] [PubMed]
14. Parker, J.; Hashmi, O.; Dutton, D.; Mavrodaris, A.; Stranges, S.; Kandala, N.B.; Clarke, A.; Franco, O.H. Levels of vitamin D and cardiometabolic disorders: Systematic review and meta-analysis. *Maturitas* **2010**, *65*, 225–236. [CrossRef] [PubMed]
15. Kim, D.H.; Sabour, S.; Sagar, U.N.; Adams, S.; Whellan, D.J. Prevalence of Hypovitaminosis D in Cardiovascular Diseases (from the National Health and Nutrition Examination Survey 2001 to 2004). *Am. J. Cardiol.* **2008**, *102*, 1540–1544. [CrossRef] [PubMed]
16. Anderson, J.L.; May, H.T.; Horne, B.D.; Bair, T.L.; Hall, N.L.; Carlquist, J.F.; Lappé, D.L.; Muhlestein, J.B. Relation of vitamin D deficiency to cardiovascular risk factors, disease status, and incident events in a general healthcare population. *Am. J. Cardiol.* **2010**, *106*, 963–968. [CrossRef] [PubMed]
17. Wellnitz, B.; Seelhorst, U.; Dimai, H.P.; Dobnig, H.; Pilz, S.; Maäz, W.; Fahrleitner-Pammer, A.; Boehm, B.O. Association of vitamin D deficiency with heart failure and sudden cardiac death in a large cross-sectional study of patients referred for coronary angiography. *J. Clin. Endocrinol. Metab.* **2008**, *93*, 3927–3935.
18. Song, E.K.; Kang, S.M. Micronutrient Deficiency Independently Predicts Adverse Health Outcomes in Patients with Heart Failure. *J. Cardiovasc. Nurs.* **2017**, *32*, 47–53. [CrossRef] [PubMed]

19. Zittermann, A.; Schulze Schleithoff, S.; Tenderich, G.; Berthold, H.K.; Körfer, R.; Stehle, P. Low vitamin D status: A contributing factor in the pathogenesis of congestive heart failure? *J. Am. Coll. Cardiol.* **2003**, *41*, 105–112. [CrossRef]
20. Ameri, P.; Ronco, D.; Casu, M.; Denegri, A.; Bovio, M.; Menoni, S.; Ferone, D.; Murialdo, G. High prevalence of vitamin D deficiency and its association with left ventricular dilation: An echocardiography study in elderly patients with chronic heart failure. *Nutr. Metab. Cardiovasc. Dis.* **2010**, *20*, 633–640. [CrossRef] [PubMed]
21. Loncar, G.; Bozic, B.; Cvetinovic, N.; Dungen, H.D.; Lainscak, M.; von Haehling, S.; Doehner, W.; Radojicic, Z.; Putnikovic, B.; Trippel, T.; et al. Secondary hyperparathyroidism prevalence and prognostic role in elderly males with heart failure. *J. Endocrinol. Investig.* **2017**, *40*, 297–304. [CrossRef] [PubMed]
22. Shane, E.; Mancini, D.; Aaronson, K.; Silverberg, S.J.; Seibel, M.J.; Addesso, V.; McMahon, D.J. Bone mass, vitamin D deficiency, and hyperparathyroidism in congestive heart failure. *Am. J. Med.* **1997**, *103*, 197–207. [CrossRef]
23. Holick, M.F. The vitamin D deficiency pandemic: Approaches for diagnosis, treatment and prevention. *Rev. Endocr. Metab. Disord.* **2017**, *18*, 153–165. [CrossRef] [PubMed]
24. Whitfield, K.C.; Bourassa, M.W.; Adamolekun, B.; Bergeron, G.; Bettendorff, L.; Brown, K.H.; Cox, L.; Fattal-Valevski, A.; Fischer, P.R.; Frank, E.L.; et al. Thiamine deficiency disorders: Diagnosis, prevalence, and a roadmap for global control programs. *Ann. N. Y. Acad. Sci.* **2018**, *1430*, 3–43. [CrossRef] [PubMed]
25. Hill, M.H.E.; Bradley, A.; Mushtaq, S.; Williams, E.A.; Powers, H.J. Effects of methodological variation on assessment of riboflavin status using the erythrocyte glutathione reductase activation coefficient assay. *Br. J. Nutr.* **2009**, *102*, 273. [CrossRef] [PubMed]
26. McPherson, R.A.; Pincus, M.R. *Henry's Clinical Diagnosis and Management by Laboratory Methods*; Elsevier: Amsterdam, Switzerland, 2016.
27. Leklem, J.E. Vitamin B-6: A Status Report. *J. Nutr.* **1990**, *120*, 1503–1507. [CrossRef] [PubMed]
28. Panton, K.K.; Farup, P.G.; Sagen, E.; Sirum, U.F.; Åsberg, A. Vitamin B6 in plasma—Sample stability and the reference limits. *Scand. J. Clin. Lab. Investig.* **2013**, *73*, 476–479. [CrossRef] [PubMed]
29. Hunt, A.; Harrington, D.; Robinson, S. Vitamin B12 deficiency. *BMJ* **2014**, *349*, g5226. [CrossRef] [PubMed]
30. Klee, G.G. Cobalamin and folate evaluation: Measurement of methylmalonic acid and homocysteine vs. vitamin B(12) and folate. *Clin. Chem.* **2000**, *46*, 1277–1283. [PubMed]
31. Scientific Opinion on Dietary Reference Values for vitamin C. *EFSA J.* **2013**, *11*, 3418. [CrossRef]
32. Molyneux, S.L.; Young, J.M.; Florkowski, C.M.; Lever, M.; George, P.M. Coenzyme Q10: Is there a clinical role and a case for measurement? *Clin. Biochem. Rev.* **2008**, *29*, 71–82. [PubMed]
33. Thomson, C.D. Assessment of requirements for selenium and adequacy of selenium status: A review. *Eur. J. Clin. Nutr.* **2004**, *58*, 391–402. [CrossRef] [PubMed]
34. Livingstone, C. Zinc: Physiology, deficiency, and parenteral nutrition. *Nutr. Clin. Pract.* **2015**, *30*, 371–382. [CrossRef] [PubMed]
35. Gomella, L.G.; Haist, S.A. *Clinician's Pocket Reference*; University of Kentucky, College of Medicine, McGraw-Hill Companies, Inc: New York, NY, USA, 2007.
36. Gruson, D.; Ferracin, B.; Ahn, S.A.; Zierold, C.; Blocki, F.; Hawkins, D.M.; Bonelli, F.; Rousseau, M.F. 1,25-Dihydroxyvitamin D to PTH(1-84) ratios strongly predict cardiovascular death in heart failure. *PLoS ONE* **2015**, *10*, e0135427. [CrossRef] [PubMed]
37. Boxer, R.S.; Dauser, D.A.; Walsh, S.J.; Hager, W.D.; Kenny, A.M. The association between vitamin D and inflammation with the 6-minute walk and frailty in patients with heart failure. *J. Am. Geriatr. Soc.* **2008**, *56*, 454–461. [CrossRef] [PubMed]
38. Liu, L.C.Y.; Voors, A.A.; Van Veldhuisen, D.J.; Van Der Veer, E.; Belonje, A.M.; Szymanski, M.K.; Silljé, H.H.; Van Gilst, W.H.; Jaarsma, T.; De Boer, R.A. Vitamin D status and outcomes in heart failure patients. *Eur. J. Heart Fail.* **2011**, *13*, 619–625. [CrossRef] [PubMed]
39. Zittermann, A.; Schleithoff, S.S.; Götting, C.; Dronow, O.; Fuchs, U.; Kühn, J.; Kleesiek, K.; Tenderich, G.; Koerfer, R. Poor outcome in end-stage heart failure patients with low circulating calcitriol levels. *Eur. J. Heart Fail.* **2008**, *10*, 321–327. [CrossRef] [PubMed]
40. Ford, J.A.; MacLennan, G.S.; Avenell, A.; Bolland, M.; Grey, A.; Witham, M. Cardiovascular disease and vitamin D supplementation: Trial analysis, systematic review, and meta-analysis. *Am. J. Clin. Nutr.* **2014**, *100*, 746–755. [CrossRef] [PubMed]

41. Mirhosseini, N.; Rainsbury, J.; Kimball, S.M. Vitamin D Supplementation, Serum 25(OH)D Concentrations and Cardiovascular Disease Risk Factors: A Systematic Review and Meta-Analysis. *Front. Cardiovasc. Med.* **2018**, *5*, 87. [CrossRef] [PubMed]
42. Ross, A.C.; Manson, J.E.; Abrams, S.A.; Aloia, J.F.; Brannon, P.M.; Clinton, S.K.; Durazo-Arvizu, R.A.; Gallagher, J.C.; Gallo, R.L.; Jones, G.; et al. The 2011 Dietary Reference Intakes for Calcium and Vitamin D: What Dietetics Practitioners Need to Know. *J. Am. Diet. Assoc.* **2011**, *111*, 524–527. [CrossRef] [PubMed]
43. McKeag, N.A.; McKinley, M.C.; Woodside, J.V.; Harbinson, M.T.; McKeown, P.P. The Role of Micronutrients in Heart Failure. *J. Acad. Nutr. Diet.* **2012**, *112*, 870–886. [CrossRef] [PubMed]
44. Schleithoff, S.S.; Zittermann, A.; Tenderich, G.; Berthold, H.K.; Stehle, P.; Koerfer, R. Vitamin D supplementation improves cytokine profiles in patients with congestive heart failure: A double-blind, randomized, placebo-controlled trial. *Am. J. Clin. Nutr.* **2006**, *83*, 754–759. [CrossRef] [PubMed]
45. Witham, M.D.; Crighton, L.J.; Gillespie, N.D.; Struthers, A.D.; McMurdo, M.E.T. The effects of vitamin D supplementation on physical function and quality of life in older patients with heart failure a randomized controlled trial. *Circ. Heart Fail.* **2010**, *3*, 195–201. [CrossRef] [PubMed]
46. Zittermann, A.; Ernst, J.B.; Prokop, S.; Fuchs, U.; Dreier, J.; Kuhn, J.; Knabbe, C.; Birschmann, I.; Schulz, U.; Berthold, H.K.; et al. Effect of Vitamin D on all-causemortality in heart failure (EVITA): A 3-year randomized clinical trial with 4000 IU Vitamin D daily. *Eur. Heart J.* **2017**, *38*, 2279–2286. [CrossRef] [PubMed]
47. Pan, J.; Guleria, R.; Zhu, S.; Baker, K. Molecular Mechanisms of Retinoid Receptors in Diabetes-Induced Cardiac Remodeling. *J. Clin. Med.* **2014**, *4*, 566–594. [CrossRef] [PubMed]
48. De Lorgeril, M.; Salen, P.; Accominotti, M.; Cadau, M.; Steghens, J.-P.; Boucher, F.; De Leiris, J. Dietary and blood ANTIOXIDANTS in patients with chronic heart failure. Insights into the potential importance of selenium in heart failure. *Eur. J. Heart Fail.* **2001**, *3*, 661–669. [CrossRef]
49. Eshak, E.S.; Iso, H.; Yamagishi, K.; Cui, R.; Tamakoshi, A. Dietary intakes of fat soluble vitamins as predictors of mortality from heart failure in a large prospective cohort study. *Nutrition* **2018**, *47*, 50–55. [CrossRef] [PubMed]
50. Keith, M.; Geranmayegan, A.; Sole, M.J.; Kurian, R.; Robinson, A.; Omran, A.S.; Jeejeebhoy, K.N. Increased oxidative stress in patients with congestive heart failure. *J. Am. Coll. Cardiol.* **1998**, *31*, 1352–1356. [CrossRef]
51. Marcocci, L.; Suzuki, Y.J. Metabolomics studies to assess biological functions of vitamin E nicotinate. *Antioxidants* **2019**, *8*, 127. [CrossRef] [PubMed]
52. Ghatak, A.; Brar, M.J.S.; Agarwal, A.; Goel, N.; Rastogi, A.K.; Vaish, A.K.; Sircar, A.R.; Chandra, M. Oxy free radical system in heart failure and therapeutic role of oral vitamin E. *Int. J. Cardiol.* **1996**, *57*, 119–127. [CrossRef]
53. Yusuf, S.; Dagenais, G.; Pogue, J.; Bosch, J.; Sleight, P. Vitamin E supplementation and cardiovascular events in high-risk patients. The Heart Outcomes Prevention Evaluation Study Investigators. *N. Engl. J. Med.* **2000**, *342*, 154–160. [PubMed]
54. Lonn, E. Effects of long-term vitamin E supplementation on cardiovascular events and cancer: A randomized controlled trial. *J. Am. Med. Assoc.* **2005**, *293*, 1338–1347.
55. Marchioli, R.; Levantesi, G.; Macchia, A.; Marfisi, R.M.; Nicolosi, G.L.; Tavazzi, L.; Tognoni, G.; Valagussa, F. Vitamin E increases the risk of developing heart failure after myocardial infarction: Results from the GISSI-Prevenzione trial. *J. Cardiovasc. Med.* **2006**, *7*, 347–350. [CrossRef] [PubMed]
56. E Keith, M.; Jeejeebhoy, K.N.; Langer, A.; Kurian, R.; Barr, A.; O'Kelly, B.; Sole, M.J. A controlled clinical trial of vitamin E supplementation in patients with congestive heart failure. *Am. J. Clin. Nutr.* **2001**, *73*, 219–224. [CrossRef] [PubMed]
57. Tsugawa, N. Cardiovascular diseases and fat soluble vitamins: Vitamin D and vitamin K. *J. Nutr. Sci. Vitam. (Tokyo)* **2015**, *61*, S170–S172. [CrossRef] [PubMed]
58. Eshak, E.S.; Arafa, A.E. Thiamine deficiency and cardiovascular disorders. *Nutr. Metab. Cardiovasc. Dis.* **2018**, *28*, 965–972. [CrossRef] [PubMed]
59. Attaluri, P.; Castillo, A.; Edriss, H.; Nugent, K. Thiamine Deficiency: An Important Consideration in Critically Ill Patients. *Am. J. Med. Sci.* **2018**, *356*, 382–390. [CrossRef] [PubMed]
60. Zenuk, C.; Healey, J.; Donnelly, J.; Vaillancourt, R.; Almalki, Y.; Smith, S. Thiamine deficiency in congestive heart failure patients receiving long term furosemide therapy. *Can J. Clin. Pharmacol.* **2003**, *10*, 184–188. [PubMed]

61. Kattoor, A.J.; Goel, A.; Mehta, J.L. Thiamine Therapy for Heart Failure: A Promise or Fiction? *Cardiovasc. Drugs Ther.* **2018**, *32*, 313–317. [CrossRef] [PubMed]
62. Rocha, R.M.; Silva, G.V.; de Albuquerque, D.C.; Tura, B.R.; Albanesi Filho, F.M. Influence of spironolactone therapy on thiamine blood levels in patients with heart failure. *Arq. Bras. Cardiol.* **2008**, *90*, 324–328. [PubMed]
63. Jain, A.; Mehta, R.; Al-Ani, M.; Hill, J.A.; Winchester, D.E. Determining the Role of Thiamine Deficiency in Systolic Heart Failure: A Meta-Analysis and Systematic Review. *J. Card. Fail.* **2015**, *21*, 1000–1007. [CrossRef] [PubMed]
64. Seligmann, H.; Halkin, H.; Rauchfleisch, S.; Kaufmann, N.; Tal, R.; Motro, M.; Vered, Z.; Ezra, D. Thiamine deficiency in patients with congestive heart failure receiving long-term furosemide therapy: A pilot study. *Am. J. Med.* **1991**, *91*, 151–155. [CrossRef]
65. Shimon, H.; Almog, S.; Vered, Z.; Seligmann, H.; Shefi, M.; Peleg, E.; Rosenthal, T.; Motro, M.; Halkin, H.; Ezra, D. Improved left ventricular function after thiamine supplementation in patients with congestive heart failure receiving long-term furosemide therapy. *Am. J. Med.* **1995**, *98*, 485–490. [CrossRef]
66. Schoenenberger, A.W.; Schoenenberger-Berzins, R.; Der Maur, C.A.; Suter, P.M.; Vergopoulos, A.; Erne, P. Thiamine supplementation in symptomatic chronic heart failure: A randomized, double-blind, placebo-controlled, cross-over pilot study. *Clin. Res. Cardiol.* **2012**, *101*, 159–164. [CrossRef] [PubMed]
67. Hughes, C.M.; Woodside, J.V.; McGartland, C.; Roberts, M.J.; Nicholls, D.P.; McKeown, P.P. Nutritional intake and oxidative stress in chronic heart failure. *Nutr. Metab. Cardiovasc. Dis.* **2012**, *22*, 376–382. [CrossRef] [PubMed]
68. Keith, M.E.; Walsh, N.A.; Darling, P.B.; Hanninen, S.A.; Thirugnanam, S.; Leong-Poi, H.; Barr, A.; Sole, M.J. B-Vitamin Deficiency in Hospitalized Patients with Heart Failure. *J. Am. Diet. Assoc.* **2009**, *109*, 1406–1410. [CrossRef] [PubMed]
69. Wang, G.; Li, W.; Lu, X.; Zhao, X. Riboflavin Alleviates Cardiac Failure in Type I Diabetic Cardiomyopathy. *Heart Int.* **2011**, *6*, e21. [CrossRef] [PubMed]
70. Herrmann, M.; Müller, S.; Kindermann, I.; Günther, L.; König, J.; Böhm, M.; Herrmann, W. Plasma B vitamins and their relation to the severity of chronic heart failure. *Am. J. Clin. Nutr.* **2007**, *85*, 117–123. [CrossRef] [PubMed]
71. Van Der Wal, H.H.; Comin-Colet, J.; Klip, I.T.; Enjuanes, C.; Beverborg, N.G.; Voors, A.A.; Banasiak, W.; van Veldhuisen, D.J.; Bruguera, J.; Ponikowski, P.; et al. Vitamin B12 and folate deficiency in chronic heart failure. *Heart* **2015**, *101*, 302–310. [CrossRef] [PubMed]
72. Briani, C.; Torre, C.D.; Citton, V.; Manara, R.; Pompanin, S.; Binotto, G.; Adami, F. Cobalamin deficiency: Clinical picture and radiological findings. *Nutrients* **2013**, *5*, 4521–4539. [CrossRef] [PubMed]
73. Andersson, S.E.; Edvinsson, M.L.; Edvinsson, L. Reduction of homocysteine in elderly with heart failure improved vascular function and blood pressure control but did not affect inflammatory activity. *Basic Clin. Pharmacol. Toxicol.* **2005**, *97*, 306–310. [CrossRef] [PubMed]
74. Demirbag, R.; Yilmaz, R.; Erel, O.; Gultekin, U.; Asci, D.E.Z. The relationship between potency of oxidative stress and severity of dilated cardiomyopathy. *Can. J. Cardiol.* **2005**, *21*, 851–855. [PubMed]
75. Jacob, R. Assessment of human vitamin C status. *J. Nutr.* **1990**, *11*, 1480–1485. [CrossRef] [PubMed]
76. Song, E.K.; Kang, S.M. Vitamin C Deficiency, High-Sensitivity C-Reactive Protein, and Cardiac Event-Free Survival in Patients with Heart Failure. *J. Cardiovasc. Nurs.* **2018**, *33*, 6–12. [CrossRef] [PubMed]
77. Wu, J.-R.; Song, E.K.; Moser, D.K.; Lennie, T.A. Dietary Vitamin C Deficiency Is Associated With Health-Related Quality of Life and Cardiac Event-free Survival in Adults With Heart Failure. *J. Cardiovasc. Nurs.* **2018**, *34*, 29–35. [CrossRef] [PubMed]
78. Shinke, T.; Shite, J.; Takaoka, H.; Hata, K.; Inoue, N.; Yoshikawa, R.; Matsumoto, H.; Masai, H.; Watanabe, S.; Ozawa, T.; et al. Vitamin C restores the contractile response to dobutamine and improves myocardial efficiency in patients with heart failure after anterior myocardial infarction. *Am. Heart J.* **2007**, *154*, e1–e8. [CrossRef] [PubMed]
79. Hornig, B.; Arakawa, N.; Kohler, C.; Drexler, H. Vitamin C improves endothelial function of conduit arteries in patients with chronic heart failure. *Circulation* **1998**, *97*, 363–368. [CrossRef] [PubMed]
80. Bentinger, M.; Brismar, K.; Dallner, G. The antioxidant role of coenzyme Q. *Mitochondrion* **2007**, *7*, S41–S50. [CrossRef] [PubMed]
81. Folkers, K.; Vadhanavikit, S.; Mortensent, S.A. Biochemical rationale and myocardial tissue data on the effective therapy of cardiomyopathy with coenzyme Q10. *Med. Sci.* **1985**, *82*, 901–904. [CrossRef] [PubMed]

82. Madmani, M.E.; Yusuf Solaiman, A.; Tamr Agha, K.; Madmani, Y.; Shahrour, Y.; Essali, A.; Kadro, W. Coenzyme Q10 for heart failure. *Cochrane Database Syst. Rev.* **2014**, *2*, CD008684. [CrossRef] [PubMed]
83. McMurray, J.J.V.; Dunselman, P.; Wedel, H.; Cleland, J.G.F.; Lindberg, M.; Hjalmarson, A.; Kjekshus, J.; Waagstein, F.; Apetrei, E.; Barrios, V.; et al. Coenzyme Q10, rosuvastatin, and clinical outcomes in heart failure: A pre-specified substudy of CORONA (Controlled Rosuvastatin Multinational Study in Heart Failure). *J. Am. Coll. Cardiol.* **2010**, *56*, 1196–1204. [CrossRef] [PubMed]
84. Ishiyama, T.; Morita, Y.; Toyama, S.; Yamagami, T.; Tsukamoto, N. A clinical study of the effect of coenzyme Q on congestive heart failure. *Jpn. Heart J.* **1976**, *17*, 32–42. [CrossRef] [PubMed]
85. Jafari, M.; Mousavi, S.M.; Asgharzadeh, A.; Yazdani, N. Coenzyme Q10 in the treatment of heart failure: A systematic review of systematic reviews. *Indian Heart J.* **2018**, *70*, S111–S117. [CrossRef] [PubMed]
86. Sander, S.; Coleman, C.I.; Patel, A.A.; Kluger, J.; Michael White, C. The Impact of Coenzyme Q10 on Systolic Function in Patients With Chronic Heart Failure. *J. Card. Fail.* **2006**, *12*, 464–472. [CrossRef] [PubMed]
87. Rosenfeldt, F.; Hilton, D.; Pepe, S.; Krum, H. Systematic review of effect of coenzyme Q10 in physical exercise, hypertension and heart failure. *Biofactors* **2003**, *18*, 91–100. [CrossRef] [PubMed]
88. Sinatra, S.T.; Berman, M. Coenzyme Q10 in patients with end-stage heart failure awaiting cardiac transplantation: A randomized, placebo-controlled study. Clin. Cardiol. 2004, 27, A26–A30. Coenzyme Q10 in patients with end-stage heart failure awaiting cardiac transplantation: A randomized, placebo-controlled study. *Clin. Cardiol.* **2004**, *27*, 295–299.
89. Morisco, C.; Trimarco, B.; Condorelli, M. Effect of coenzyme Q10therapy in patients with congestive heart failure: A long-term multicenter randomized study. *Clin. Investig.* **1993**, *71*, S134–S136. [CrossRef] [PubMed]
90. Mortensen, S.A.; Rosenfeldt, F.; Kumar, A.; Dolliner, P.; Filipiak, K.J.; Pella, D.; Alehagen, U.; Steurer, G.; Littarru, G.P. Q-SYMBIO Study Investigators. The effect of coenzyme Q10on morbidity and mortality in chronic heart failure: Results from Q-SYMBIO: A randomized double-blind trial. *JACC Hear Fail.* **2014**, *2*, 641–649. [CrossRef] [PubMed]
91. Fairweather-Tait, S.J.; Bao, Y.; Broadley, M.R.; Collings, R.; Ford, D.; Hesketh, J.E.; Hurst, R. Selenium in Human Health and Disease. *Antioxid. Redox Signal.* **2011**, *14*, 1337–1383. [CrossRef] [PubMed]
92. de Lorgeril, M.; Salen, P. Selenium and antioxidant defenses as major mediators in the development of chronic heart failure. *Heart Fail. Rev.* **2006**, *11*, 13–17. [CrossRef] [PubMed]
93. Rayman, M.P. Selenium and human health. *Lancet* **2012**, *379*, 1256–1268. [CrossRef]
94. Xia, Y.; E Hill, K.; Li, P.; Xu, J.; Zhou, D.; Motley, A.K.; Wang, L.; Byrne, D.W.; Burk, R.F. Optimization of selenoprotein P and other plasma selenium biomarkers for the assessment of the selenium nutritional requirement: A placebo-controlled, double-blind study of selenomethionine supplementation in selenium-deficient Chinese subjects. *Am. J. Clin. Nutr.* **2010**, *92*, 525–531. [CrossRef] [PubMed]
95. Ge, K.; Yang, G. The epidemiology of selenium deficiency in the etiological study of endemic diseases in China. *Am. J. Clin. Nutr.* **1993**, *57*, 259S–263S. [CrossRef] [PubMed]
96. Salonen, J.T.; Alfthan, G.; Huttunen, J.K.; Pikkarainen, J.; Puska, P. Association between cardiovascular death and myocardial infarction and serum selenium in a matched-pair longitudinal study. *Lancet* **1982**, *2*, 175–179. [CrossRef]
97. Navas-Acien, A.; Bleys, J.; Guallar, E. Selenium intake and cardiovascular risk: What is new? *Curr. Opin. Lipidol.* **2008**, *19*, 43–49. [CrossRef] [PubMed]
98. Nordman, T.; Xia, L.; Björkhem-Bergman, L.; Damdimopoulos, A.; Nalvarte, I.; Arnér, E.S.J.; Spyrou, G.; Eriksson, L.C.; Björnstedt, M.; Olsson, J.M. Regeneration of the antioxidant ubiquinol by lipoamide dehydrogenase, thioredoxin reductase and glutathione reductase. *Biofactors* **2003**, *18*, 45–50. [CrossRef] [PubMed]
99. Moosmann, B.; Behl, C. Selenoproteins, cholesterol-lowering drugs, and the consequences: Revisiting of the mevalonate pathway. *Trends Cardiovasc. Med.* **2004**, *14*, 273–281. [CrossRef] [PubMed]
100. Alehagen, U.; Aaseth, J. Selenium and coenzyme Q10 interrelationship in cardiovascular diseases-A clinician's point of view. *J. Trace Elem. Med. Biol.* **2015**, *31*, 157–162. [CrossRef] [PubMed]
101. Johansson, P.; Dahlström, Ö.; Dahlström, U.; Alehagen, U. Effect of selenium and Q10 on the cardiac biomarker NT-proBNP. *Scand Cardiovasc. J.* **2013**, *47*, 281–288. [CrossRef] [PubMed]
102. Witte, K.K.; Nikitin, N.P.; Parker, A.C.; Von Haehling, S.; Volk, H.-D.; Anker, S.D.; Clark, A.L.; Cleland, J.G. The effect of micronutrient supplementation on quality-of-life and left ventricular function in elderly patients with chronic heart failure. *Eur. Heart J.* **2005**, *26*, 2238–2244. [CrossRef] [PubMed]

103. Raygan, F.; Behnejad, M.; Ostadmohammadi, V.; Bahmani, F.; Mansournia, M.A.; Karamali, F.; Asemi, Z. Selenium supplementation lowers insulin resistance and markers of cardio-metabolic risk in patients with congestive heart failure: A randomised, double-blind, placebo-controlled trial. *Br. J. Nutr.* **2018**, *120*, 33–40. [CrossRef] [PubMed]
104. Vasto, S.; Mocchegiani, E.; Candore, G.; Listì, F.; Colonna-Romano, G.; Lio, D.; Malavolta, M.; Giacconi, R.; Cipriano, C.; Caruso, C. Inflammation, genes and zinc in ageing and age-related diseases. *Biogerontology* **2006**, *7*, 315–327. [CrossRef] [PubMed]
105. Cherasse, Y.; Urade, Y. Dietary zinc acts as a sleep modulator. *Int. J. Mol. Sci.* **2017**, *18*, 2334. [CrossRef] [PubMed]
106. Turan, B.; Tuncay, E. Impact of labile zinc on heart function: From physiology to pathophysiology. *Int. J. Mol. Sci.* **2017**, *18*, 2395. [CrossRef] [PubMed]
107. Oster, O. Trace element concentrations (Cu, Zn, Fe) in sera from patients with dilated cardiomyopathy. *Clin. Chim. Acta* **1993**, *214*, 209–218. [CrossRef]
108. Ripa, S.; Ripa, R.; Giustiniani, S. Are failured cardiomyopathies a zinc-deficit related disease? A study on Zn and Cu in patients with chronic failured dilated and hypertrophic cardiomyopathies. *Minerva Med.* **1998**, *89*, 397–403. [PubMed]
109. Topuzoglu, G.; Erbay, A.R.; Karul, A.B.; Yensel, N. Concentations of Copper, Zinc, and Magnesium in Sera from Patients with Idiopathic Dilated Cardiomyopathy. *Biol. Trace Elem. Res.* **2003**, *95*, 11–17. [CrossRef]
110. Koşar, F.; Sahin, I.; Taşkapan, C.; Küçükbay, Z.; Güllü, H.; Taşkapan, H.; Cehreli, S. Trace element status (Se, Zn, Cu) in heart failure. *Anadolu Kardiyol. Derg.* **2006**, *6*, 216–220. [PubMed]
111. Yu, X.; Huang, L.; Zhao, J.; Wang, Z.; Yao, W.; Wu, X.; Huang, J.; Bian, B. The Relationship between Serum Zinc Level and Heart Failure: A Meta-Analysis. *Biomed Res Int.* **2018**, *2018*, 1–9. [CrossRef] [PubMed]
112. Yoshihisa, A.; Abe, S.; Kiko, T.; Kimishima, Y.; Sato, Y.; Watanabe, S.; Kanno, Y.; Miyata-Tatsumi, M.; Misaka, T.; Sato, T.; et al. Association of Serum Zinc Level With Prognosis in Patients With Heart Failure. *J. Card Fail.* **2018**, *24*, 375–383. [CrossRef] [PubMed]
113. Jeejeebhoy, F.; Keith, M.; Freeman, M.; Barr, A.; McCall, M.; Kurian, R.; Mazer, D.; Errett, L. Nutritional supplementation with MyoVive repletes essential cardiac myocyte nutrients and reduces left ventricular size in patients with left ventricular dysfunction. *Am. Heart J.* **2002**, *143*, 1092–1100. [CrossRef] [PubMed]
114. Jankowska, E.A.; Rozentryt, P.; Witkowska, A.; Nowak, J.; Hartmann, O.; Ponikowska, B.; Borodulin-Nadzieja, L.; Banasiak, W.; Polonski, L.; Filippatos, G.; et al. Iron deficiency: An ominous sign in patients with systolic chronic heart failure. *Eur. Heart J.* **2010**, *31*, 1872–1880. [CrossRef] [PubMed]
115. Peyrin-Biroulet, L.; Williet, N.; Cacoub, P. Guidelines on the diagnosis and treatment of iron deficiency across indications: A systematic review. *Am. J. Clin. Nutr.* **2015**, *102*, 1585–1594. [CrossRef] [PubMed]
116. Klip, I.T.; Comín-Colet, J.; Voors, A.A.; Ponikowski, P.; Enjuanes, C.; Banasiak, W.; Lok, D.J.; Rosentryt, P.; Torrens, A.; Polonski, L.; et al. Iron deficiency in chronic heart failure: An international pooled analysis. *Am. Heart J.* **2013**, *165*, 575–582.e3. [CrossRef] [PubMed]
117. Jankowska, E.A.; Von Haehling, S.; Anker, S.D.; MacDougall, I.C.; Ponikowski, P. Iron deficiency and heart failure: Diagnostic dilemmas and therapeutic perspectives. *Eur. Heart J.* **2013**, *34*, 816–829. [CrossRef] [PubMed]
118. Anker, S.D.; Kirwan, B.A.; van Veldhuisen, D.J.; Filippatos, G.; Comin-Colet, J.; Ruschitzka, F.; Lüscher, T.F.; Arutyunov, G.P.; Motr, M.; Mori, C.; et al. Effects of ferric carboxymaltose on hospitalisations and mortality rates in iron-deficient heart failure patients: An individual patient data meta-analysis. *Eur. J. Heart Fail.* **2018**, *20*, 125–133. [CrossRef] [PubMed]
119. Enjuanes, C.; Bruguera, J.; Grau, M.; Cladellas, M.; Gonzalez, G.; Meroño, O.; Moliner-Borja, P.; Verdú, J.M.; Farré, N.; Comín-Colet, J. Iron Status in Chronic Heart Failure: Impact on Symptoms, Functional Class and Submaximal Exercise Capacity. *Rev. Esp. Cardiol. (Engl. Ed)* **2016**, *69*, 247–255. [CrossRef] [PubMed]
120. Lam, C.S.P.; Doehner, W.; Comin-Colet, J.; Lam, C.S. Iron deficiency in chronic heart failure: Case-based practical guidance. *ESC Heart Fail.* **2018**, *5*, 764–771. [CrossRef] [PubMed]
121. Ponikowski, P.; Voors, A.A.; Anker, S.D.; Bueno, H.; Cleland, J.G.F.; Coats, A.J.S.; Falk, V.; González-Juanatey, J.R.; Harjola, V.-P.; Jankowska, E.A.; et al. 2016 ESC Guidelines for the diagnosis and treatment of acute and chronic heart failure. *Eur. Heart J.* **2016**, *37*, 2129–2200. [CrossRef] [PubMed]

122. Anker, S.D.; Colet, J.C.; Filippatos, G.; Willenheimer, R.; Dickstein, K.; Drexler, H.; Lüscher, T.F.; Bart, B.; Banasiak, W.; Niegowska, J.; et al. Ferric Carboxymaltose in Patients with Heart Failure and Iron Deficiency. *N. Engl. J. Med.* **2009**, *361*, 2436–2448. [CrossRef] [PubMed]
123. Ponikowski, P.; Van Veldhuisen, D.J.; Comin-Colet, J.; Ertl, G.; Komajda, M.; Mareev, V.; McDonagh, T.; Parkhomenko, A.; Tavazzi, L.; Levesque, V.; et al. Beneficial effects of long-term intravenous iron therapy with ferric carboxymaltose in patients with symptomatic heart failure and iron deficiency. *Eur. Heart J.* **2015**, *36*, 657–668. [CrossRef] [PubMed]
124. Lewis, G.D.; Malhotra, R.; Hernandez, A.F.; McNulty, S.E.; Smith, A.; Felker, G.M.; Tang, W.H.W.; LaRue, S.J.; Redfield, M.M.; Semigran, M.J.; et al. Effect of oral iron repletion on exercise capacity in patients with heart failure with reduced ejection fraction and iron deficiency the IRONOUT HF randomized clinical trial. *JAMA* **2017**, *317*, 1958–1966. [CrossRef] [PubMed]
125. Beck-Da-Silva, L.; Piardi, D.; Söder, S.; Rohde, L.E.; Pereira-Barretto, A.C.; De Albuquerque, D.; Bocchi, E.; Vilas-Boas, F.; Moura, L.Z.; Montera, M.W.; et al. IRON-HF study: A randomized trial to assess the effects of iron in heart failure patients with anemia. *Int. J. Cardiol.* **2013**, *168*, 3439–3442. [CrossRef] [PubMed]
126. Bekfani, T.; Pellicori, P.; Morris, D.; Ebner, N.; Valentova, M.; Sandek, A.; Doehner, W.; Cleland, J.G.; Lainscak, M.; Schulze, P.C.; et al. Iron deficiency in patients with heart failure with preserved ejection fraction and its association with reduced exercise capacity, muscle strength and quality of life. *Clin. Res. Cardiol.* **2019**, *108*, 203–211. [CrossRef] [PubMed]
127. Paterek, A.; Kępska, M.; Sochanowicz, B.; Chajduk, E.; Kołodziejczyk, J.; Polkowska-Motrenko, H.; Kruszewski, M.; Leszek, P.; Mackiewicz, U.; Mączewski, M. Beneficial effects of intravenous iron therapy in a rat model of heart failure with preserved systemic iron status but depleted intracellular cardiac stores. *Sci. Rep.* **2018**, *8*, 15758. Available online: http://www.nature.com/articles/s41598-018-33277-2 (accessed on 4 November 2018). [CrossRef] [PubMed]

International Journal of
Molecular Sciences

Article

Correlation of Vitamin D with Inflammatory Cytokines, Atherosclerotic Parameters, and Lifestyle Factors in the Setting of Heart Failure: A 12-Month Follow-Up Study

Daniel N. Roffe-Vazquez [1], Anna S. Huerta-Delgado [1], Elena C. Castillo [2], José R. Villarreal-Calderón [1], Adrian M. Gonzalez-Gil [1], Cecilio Enriquez [2], Gerardo Garcia-Rivas [2,3,*] and Leticia Elizondo-Montemayor [1,3,*]

1 Tecnologico de Monterrey, Center for Research in Clinical Nutrition, Escuela de Medicina, Monterrey 64710, N.L., Mexico; daniel.roffe@hotmail.com (D.N.R.-V.); anna.sofy@hotmail.com (A.S.H.-D.); joser.villarreal@udem.edu (J.R.V.-C.); amgonzalezgil@hotmail.com (A.M.G.-G.)
2 Tecnologico de Monterrey, Centro de Investigacion Biomedica, Hospital Zambrano Hellion, San Pedro Garza-Garcia 66278, N.L., Mexico; ecgonzalez@tec.mx (E.C.C.); dr.cecilioenriquez@medicos.tecsalud.mx (C.E.)
3 Tecnologico de Monterrey, Cardiovascular and Metabolomics Research Group, Escuela de Medicina, San Pedro Garza-Garcia 66278, N.L., Mexico
* Correspondence: gdejesus@tec.mx (G.G.-R.); lelizond@tec.mx (L.E.-M.)

Received: 15 October 2019; Accepted: 13 November 2019; Published: 19 November 2019

Abstract: Vitamin D deficiency is highly prevalent worldwide. It has been associated with heart failure (HF) given its immunoregulatory functions. In-vitro and animal models have shown protective roles through mechanisms involving procollagen-1, JNK2, calcineurin/NFAT, NF-κB, MAPK, Th1, Th2, Th17, cytokines, cholesterol-efflux, oxLDL, and GLUT4, among others. A 12-month follow-up in HF patients showed a high prevalence of vitamin D deficiency, with no seasonal variation (64.7–82.4%). A positive correlation between serum 25(OH)D concentration and dietary intake of vitamin D-rich foods was found. A significant inverse correlation with IL-1β ($R = -0.78$), TNF-α ($R = -0.53$), IL-6 ($R = -0.42$), IL-8 ($R = -0.41$), IL-17A ($R = -0.31$), LDL-cholesterol ($R = -0.51$), Apo-B ($R = -0.57$), total-cholesterol ($R = -0.48$), and triglycerides ($R = -0.32$) was shown. Cluster analysis demonstrated that patients from cluster three, with the lowest 25(OH)D levels, presented the lowermost vitamin D intake, IL-10 (1.0 ± 0.9 pg/mL), and IL-12p70 (0.5 ± 0.4 pg/mL), but the highest TNF-α (9.1 ± 3.5 pg/mL), IL-8 (55.6 ± 117.1 pg/mL), IL-17A (3.5 ± 2.0 pg/mL), total-cholesterol (193.9 ± 61.4 mg/dL), LDL-cholesterol (127.7 ± 58.2 mg/dL), and Apo-B (101.4 ± 33.4 mg/dL) levels, compared with patients from cluster one. Although the role of vitamin D in the pathogenesis of HF in humans is still uncertain, we applied the molecular mechanisms of in-vitro and animal models to explain our findings. Vitamin D deficiency might contribute to inflammation, remodeling, fibrosis, and atherosclerosis in patients with HF.

Keywords: vitamin D; heart failure; inflammation; seasonal variation; lifestyle; cytokines; lipids; mechanisms; immunoregulatory

1. Introduction

It has long been recognized that vitamin D plays a major role in bone and mineral metabolism, but until recent years, it has also been implicated in immunity and cardiovascular health. The first rate-limiting step in the endogenous synthesis of vitamin D is the skin's sun exposure, as ultraviolet B (UVB) radiation is necessary to convert 7-dehydrocholesterol into vitamin D3 (cholecalciferol), with estimates of cutaneous synthesis providing between 80% and 100% of the vitamin D requirements of the body [1]. In the liver, 25-hydroxylase (CYP2R1) converts vitamin D3 into 25-hydroxyvitamin D [25(OH)D], which is then transformed in the kidney by means of 1-alpha-hydroxylase (CYP27B1), into 1,25-dihydroxyvitamin D [1,25(OH)2D3], also known as calcitriol, the hormone's biologically active form [2]. The enzyme 1-alpha-hydroxylase is activated by the parathyroid hormone (PTH) and inhibited by fibroblast growth factor-23 (FGF-23). On the other hand, 25-hydroxyvitamin D3 24-hydroxylase (CYP24A1) is involved in calcitriol's catabolism. This enzyme limits the generation of calcitriol by converting it into hydroxylated derivatives (which will later be excreted), or by converting 25(OH)D3 into 24,25(OH)2D3 or 23,25(OH)2D3, further decreasing its availability for 1-hydroxylation. CYP24A1 is inhibited by PTH and induced by FGF-23 [3]. By feedback mechanisms, both 1-alpha-hydroxylase and 25-hydroxyvitamin D3 24-hydroxylase regulate 1,25(OH)2D3's bioavailability as a means to protect against hypercalcemia [3,4].

Several risk factors have been described to predispose to vitamin D deficiency. Living in higher latitudes (above 35 degrees) is a well-known risk factor [5]; however, vitamin D deficiency has been shown to be highly prevalent even in populations living near the equator [6]. Other risk factors for vitamin D deficiency include skin phototype, sunscreen usage, indoor working environments, minimal outdoor activity, obesity, high body mass index (BMI), and waist circumference (WC), as well as older age, which is accompanied by a 75% decrease in vitamin D synthesis capacity [1,6]. Among these factors, the combination of low UVB availability and/or underexposure, in addition to insufficient dietary intake, have been described as the most important [1].

Notably, nutrition surveillance data from various countries have indicated that vitamin D ingestion is lower than the recommended minimum amount [1,7]. There is also evidence that suggests that the recommended intake established by guidelines (600 IU) is actually insufficient and should be increased to at least 800 IU [5].

Vitamin D also plays a role in immune regulation, as immune cells express 1-alpha-hydroxylase to regulate their own local concentration of calcitriol. This effect has been related to the milieu of cytokines [8], given that adequate levels of serum 25(OH)D have been associated with increased levels of the anti-inflammatory cytokines interleukins (IL) 4 and 10, and to lower levels of the proinflammatory cytokines IL-1 and IL-6 [9]. On the contrary, vitamin D deficiency has been associated with chronic inflammatory states and a proinflammatory cytokine profile. Vitamin D deficiency has also been linked with augmented collagen synthesis, oxidative stress, and fibrosis, which are potential mechanisms underlying cardiovascular diseases, such as heart failure [10].

Heart failure (HF) is one of the most common chronic medical conditions worldwide. It is estimated that as of 2019, more than 6 million Americans are living with this disease [11–13]. HF with reduced ejection fraction (HFrEF) is defined by a left ventricular ejection fraction (LVEF) of $< 40\%$ [14]. Most of the risk factors for HFrEF in Western populations have been described in association with ischemic heart disease [15]. It has been proposed that vitamin D might play a role in disease pathogenesis, as serum 25(OH)D levels have been demonstrated to be lower in patients with HF compared with control subjects [16]. Specifically, prospective studies have shown that the risk of developing HF is increased in patients with vitamin D deficiency [17]. One of the mechanistic links regarding this association relies on a proinflammatory cytokine state, including both innate and adaptive immunity [18–20].

Regarding innate immunity, vitamin D deficiency has been shown to increase the expression of nuclear-factor kappa B (NF-κB), leading to a higher secretion and release of the main proinflammatory cytokines, such as IL-6 and monocyte chemoattractant protein-1 (MCP-1) [10]. Cardiomyocytes themselves are activated by hemodynamic stress and are able to secrete several inflammatory cytokines as well [21]. On the other hand, by activation of mitogen-activated protein kinase (MAPK) phosphatase-1 and subsequent inhibition of p38 MAPK, vitamin D has been demonstrated to switch the cytokine profile into an anti-inflammatory state, with inhibition of the production of tumor necrosis factor-alpha (TNF-α) and IL-6 [22]. In the case of adaptive immunity, vitamin D has been found to inhibit the CD4+ T-cell course towards the Th17 lineage [9]. Additionally, vitamin D is able to hinder IL-17 production of those CD4+ T-cells that have been already committed towards the Th17 lineage [23]. Not only does vitamin D exert its effects through immune regulation, but it is also able to control extracellular matrix metabolism. Vitamin D deficiency has been related to increased expression of matrix metalloproteinase (MMP) 2 and 9 [10], as well as to augmented tissue macrophage infiltration [24]. These mechanisms may be associated with some of the structural abnormalities seen in HF, such as ventricular remodeling, tissue fibrosis, and a systemic proinflammatory state. Other mechanisms through which vitamin D deficiency predisposes to HF include: overactivation of the renin-angiotensin-aldosterone system (RAAS); dysfunction of the intracellular calcium handling by the cardiomyocyte; overexpression of procollagen-1, which leads to increased fibrosis; diminished protein kinase A (PKA) levels that result in impaired contractility; and activation of calcineurin signaling, which promotes cardiac hypertrophy [25–28].

Given the evidence provided relating the mechanisms through which vitamin D deficiency is associated with HF, it is important to study the seasonal variation of vitamin D deficiency and the cytokine profile, as well as the lifestyle factors that exert an influence in patients with HF. Thus, the aim of this study is to describe the seasonal variation of vitamin D deficiency and its association with 13 inflammatory cytokines, biochemical, and lifestyle factors during a 12-month follow-up in a cohort of patients with HFrEF, as well as to describe the molecular inflammatory and atherosclerotic mechanisms related to vitamin D deficiency. An original figure summarizing these mechanisms is also shown.

2. Results

2.1. Demographic, Anthropometric, and Lifestyle Parameters

The mean and standard deviation of serum 25(OH)D concentrations, demographic data, lifestyle, and anthropometric characteristics are presented in Table 1. The mean age of the cohort (n = 17) was 64.2 years. Most of the patients in the study were males (82.4%). In order to assess the level of skin pigmentation in our patients, the Fitzpatrick's classification of phototypes was clinically determined. Most of the patients belonged to phototype IV (58.8%), followed by V (29.4%), and lastly by phototype III (11.8%). All of the patients were of Hispanic ethnicity. Heart failure severity was clinically determined by the New York Heart Association (NYHA) classification of functional capacity [29]. All of the patients in our cohort had an NYHA class III status, which describes subjects that are asymptomatic at rest but develop fatigue, dyspnea, or palpitations when performing less than ordinary physical activity. Mean BMI was 28.5 kg/m^2, body fat percentage (BF%) was 30.5%, WC was 99.1 cm, and fat mass was 23.4 kg.

Table 1. Serum 25(OH)D levels, demographic, lifestyle, and anthropometric parameters in patients with heart failure.

Parameter	Mean (Standard Deviation)
Age (years)	64.2 (± 8.9)
Gender	Male 82.4%/Female 17.6%
Phototype	III 11.8%/IV 58.8%/V 29.4%
Smoking	Yes 11.8%/No 88.2%
Height (m)	1.6 (± 0.1)
Weight (kg)	76.8 (± 11.7)
Body fat (%)	30.5 (± 6.3)
Fat mass (kg)	23.4 (± 5.6)
Waist circumference (cm)	99.1 (± 9)
BMI (kg/m^2)	28.5 (± 2.8)
Vitamin D intake (IU/day)	224 (± 113.1)
Sun exposure (min/week)	307.3 (± 444.2)
25(OH)D (ng/mL)	24.3 (± 7.7)

Data are expressed as mean (± standard deviation) unless specified otherwise as a percentage (%). BMI = body mass index; 25(OH)D = 25-hydroxyvitamin D.

Mean vitamin D intake per day was calculated to be 224 ± 113.1 IU; all of the patients were found to have an overall consumption throughout the year of less than 400 IU/day. Sun exposure had a large variability among subjects; therefore, this data was further analyzed by grouping patients into those who reported a null amount of sun exposure (negative) versus those who had any time of sun exposure (positive) per day (Figure 1). Since individual patients reported different amounts of sun exposure throughout the seasons, the data was analyzed considering four observations per patient, each corresponding to a season of the year, with a total of 68 observations. From these, 51 of the observations were categorized as positive and 17 as negative. A vitamin D sufficient status was statistically more prevalent in patients with sun exposure (31.4%) compared to those with no exposure (5.9%) ($p < 0.044$).

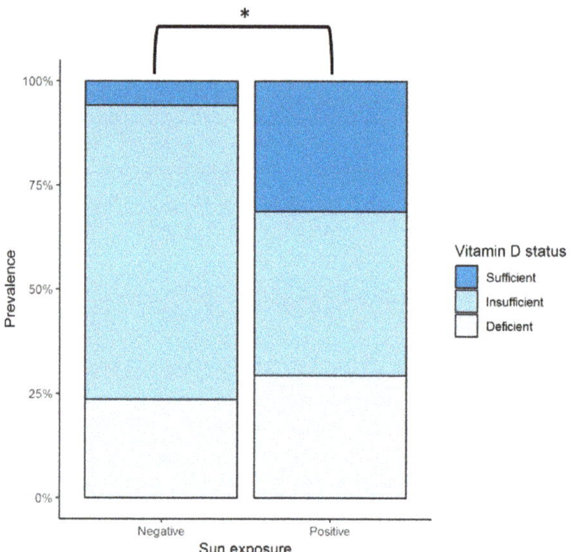

Figure 1. Sun exposure and vitamin D status in patients with heart failure during the 12-month follow up. Negative = observations with null sun exposure; positive = observations with any amount of sun exposure; * represents a statistically significant difference in the prevalence of vitamin D sufficiency in patients with some amount of sun exposure (positive), 31.4%, versus those with null exposure (negative), 5.9% ($p < 0.044$). Sufficient vitamin D status is defined as serum 25(OH)D levels > 30 ng/mL, insufficient as serum 25(OH)D levels > 20 and < 30 ng/mL, and deficient status as serum 25(OH)D levels < 20 ng/mL. [30]. 25(OH)D = 25-hydroxyvitamin D.

2.2. Cluster Analysis

Principal component analysis (PCA) was performed with numeric demographic, anthropometric, and lifestyle variables, as well as with serum 25(OH)D concentration, in order to better understand the nature of the observations in our cohort. The 68 observations (four observations per patient, each one according to the corresponding season) were classified into three clusters by partitioning around medoids (PAM). Graphical representation of the grouping of each observation is shown in Figure 2a, while the contribution of each variable to the total variance of PCA is seen in Figure 2b.

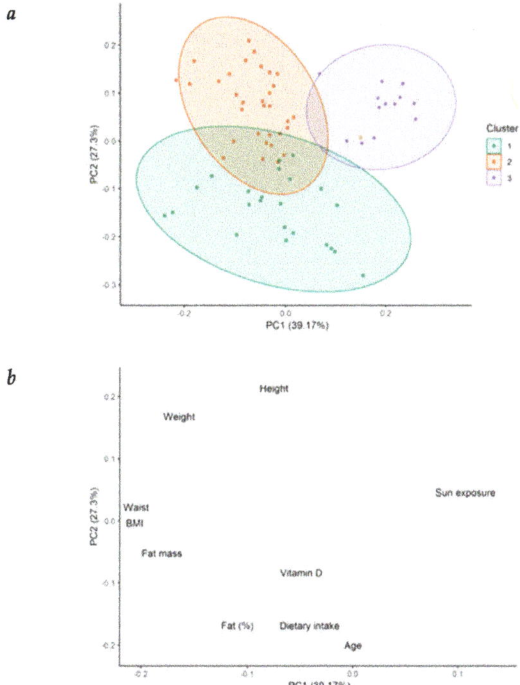

Figure 2. Cluster partitioning in patients with heart failure. (**a**) Cluster partitioning and its graphical representation of the principal component analysis (PCA) in patients with heart failure (HF); (**b**) graphical representation of the contribution of each individual variable to the variance of the PCA. BMI = body mass index.

2.2.1. Cluster Analysis for Lifestyle and Anthropometric Parameters

Differences in data of patients grouped into three clusters is summarized in Table 2. Patients from cluster 1 were characterized by the highest levels of serum 25(OH)D concentration (28.3 ± 8.0 ng/mL). These patients presented the highest vitamin D dietary intake (322.9 ± 103.4 IU/day), in spite of being older (72.2 ± 7.5 years), as well as the highest BF% (35.8% ± 4.9%), compared with patients from clusters 3 and 2. However, the total fat mass in kg and the BMI were significantly different from those shown by patients from cluster 3, but not from those of cluster 2. In contrast, patients from cluster 3 presented the lowest serum 25(OH)D levels (19 ± 4.5 ng/mL). These patients showed the highest amount of sun exposure (1178.5 ± 925.6 min/week) compared with patients from the other two clusters, but lower vitamin D ingestion (174 ± 72.9 IU/day) compared with patients from cluster 1. They also displayed lower BMI (24.6 ± 1.2 kg/m^2), BF% (22.7% ± 3.3%), and fat mass in kg (15.0 ± 2.6 kg) than patients from cluster 1. Finally, patients from cluster 2 showed the highest body weight (85.4 ± 8.0 kg) and WC (103.9 ± 7.1 cm); lower vitamin D intake (173.1 ± 83.9 IU/day), but no difference in sun exposure compared with patients from cluster 1. No significant difference among the clusters was demonstrated for LVEF (%): cluster 1 (28.8% ± 7.1%), cluster 2 (30.5% ± 13.0%), cluster 3 (27.3% ± 4.3%) (Figure A1-Appendix A).

2.2.2. Biochemical and Metabolic Parameters among the Clusters of Patients with HF

Using the same cluster classification, Table 3 shows the differences among clusters for biochemical and metabolic parameters. Patients from cluster 3 were found to present with the highest levels of total cholesterol (TC) (193.9 ± 61.4 mg/dL), low-density lipoprotein (LDL) cholesterol (127.7 ± 58.2 mg/dL), non-high-density lipoprotein (non-HDL) cholesterol (152.3 ± 58.2 mg/dL), Apo-B (101.4 ± 33.4 mg/dL), Apo-B/Apo-A ratio (0.7 ± 0.2), and TC/HDL ratio (4.7 ± 1.4). All of these metabolic parameters were significantly different compared with patients from cluster 1. These patients also presented the lowest insulin levels (8.6 ± 4.5 µU/mL) and Homeostatic Model Assessment (HOMA) index (2.7 ± 1.6), both of which were significantly different regarding cluster 2. Figure 3 shows the PCA, including the biochemical and metabolic variables shown in Table 3.

Table 2. 25(OH)D levels, lifestyle, anthropometrical, and clinical data for patients with heart failure grouped into three clusters according to the PCA analysis.

Parameter	Cluster 1	Cluster 2	Cluster 3
Age (years)	72.2 (± 7.5) [b, c]	60.4 (± 6.6) [a]	59.6 (± 6.6) [a]
Height (m)	1.56 (± 0.06) [b, c]	1.69 (± 0.06) [a, c]	1.64 (± 0.06) [a, b]
Weight (kg)	70.7 (± 9.8) [b]	85.4 (± 8.0) [a, c]	66.3 (± 5.4) [b]
Body fat (%)	35.8 (± 4.9) [b, c]	29.8 (± 4.3) [a, c]	22.7 (± 3.3) [a, b]
Fat mass (kg)	25.4 (± 5.6) [c]	25.3 (± 2.7) [c]	15 (± 2.6) [a, b]
Waist circumference (cm)	98.9 (± 7.7) [b, c]	103.9 (± 7.1) [a, c]	87.7 (± 4.2) [a, b]
BMI (kg/m^2)	28.9 (± 2.8) [c]	29.7 (± 1.7) [c]	24.6 (± 1.2) [a, b]
Vitamin D intake (IU/day)	322.9 (± 103.4) [b, c]	173.1 (± 83.9) [a]	174 (± 72.9) [a]
Sun exposure (min/week)	202.2 (± 218.9) [c]	171.7 (± 234.2) [c]	1178.5 (± 925.6) [a, b]
Vitamin D (ng/mL)	28.3 (± 8.0) [b, c]	23.6 (± 7.2) [a]	19 (± 4.5) [a]
LVEF (%)	28.8 (± 7.1)	30.5 (± 13.0)	27.3 (± 4.3)

Tukey honest significant difference (HSD); data are expressed as mean (± standard deviation). BMI = body mass index; LVEF = left ventricular ejection fraction. [a] = statistical difference when compared vs. cluster 1 ($p < 0.05$); [b] = statistical difference when compared vs. cluster 2 ($p < 0.05$); [c] = statistical difference when compared vs. cluster 3 ($p < 0.05$). Cluster 1 includes 23 observations from 7 different patients, cluster 2 includes 32 observations from 10 different patients, and cluster 3 includes 13 observations from 4 different patients.

Table 3. Biochemical and metabolic parameters in the three clusters of patients with heart failure.

Analyte (Units)	Cluster 1	Cluster 2	Cluster 3
BNP (pg/mL)	125.5 (± 104.1)	138.5 (± 158.9)	183.1 (± 142.3)
hsCRP (mg/dL)	0.5 (± 1.1)	0.5 (± 0.6)	0.1 (± 0.1)
Calcium (mg/dL)	9.4 (± 0.3) [b, c]	9.2 (± 0.2) [a]	9 (± 0.3) [a]
PTH (pg/mL)	80.6 (± 54.0)	61.8 (± 23.4)	58.1 (± 15.2)
Total cholesterol (mg/dL)	138.1 (± 50.2) [c]	171.5 (± 38.3)	193.9 (± 61.4) [a]
LDL (mg/dL)	63.5 (± 35.9) [c]	93 (± 32.5)	127.7 (± 58.2) [a]
HDL (mg/dL)	38.3 (± 10.8)	42.1 (± 8.5)	41.6 (± 6.8)
non-HDL (mg/dL)	99.8 (± 44.8) [c]	128.1 (± 31.0)	152.3 (± 58.2) [a]
VLDL (mg/dL)	30 (± 15.3)	36.4 (± 16.1)	39.8 (± 42.6)
Triglycerides (mg/dL)	149.8 (± 76.7)	181.9 (± 80.7)	199 (± 212.8)
Ratio cholesterol/HDL	3.5 (± 0.6) [c]	4.1 (± 0.7)	4.7 (± 1.4) [a]
Apo A (mg/dL)	139.1 (± 16.6)	141.9 (± 20.2)	134.4 (± 11.6)
Apo B (mg/dL)	70.8 (± 26.1) [b, c]	97.7 (± 18.4) [a]	101.4 (± 33.4) [a]
Ratio Apo B/Apo A	0.5 (± 0.2) [b, c]	0.7 (± 0.1) [a]	0.7 (± 0.2) [a]
Glucose (mg/dL)	132.4 (± 50.2)	137.5 (± 54.9)	134 (± 56.4)
Insulin (µU/mL)	16.3 (± 11.3)	20.8 (± 14.7) [c]	8.6 (± 4.5) [b]
HOMA index	5.2 (± 4.0)	6.3 (± 3.9) [c]	2.7 (± 1.6) [b]

Tukey honest significant difference (HSD); data are expressed as mean (± standard deviation). BNP = brain natriuretic peptide; hsCRP = high sensitivity C-reactive protein; PTH = parathyroid hormone; LDL = low-density lipoprotein; HDL = high-density lipoprotein; non-HDL = non-high-density lipoprotein; VLDL = very low-density lipoprotein; Apo = apolipoprotein; HOMA = Homeostatic Model Assessment. [a] = statistical difference when compared vs. cluster 1 ($p < 0.05$); [b] = statistical difference when compared vs. cluster 2 ($p < 0.05$); [c] = statistical difference when compared vs. cluster 3 ($p < 0.05$).

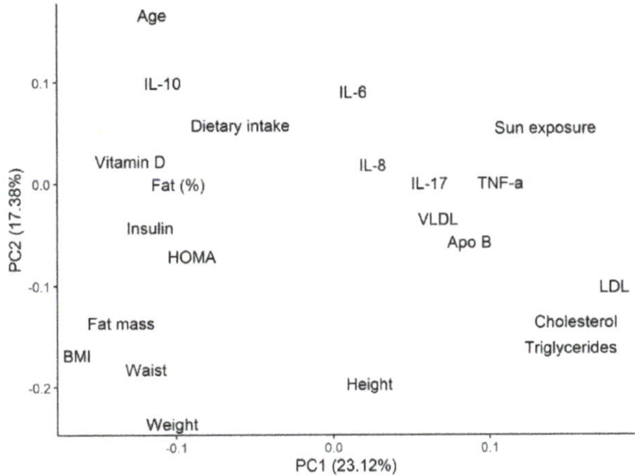

Figure 3. PCA analysis for the metabolic and biochemical variables described in the cluster's analysis. PCA = principal component analysis; IL= interleukin; HOMA = Homeostatic Model Assessment; BMI = body mass index; TNF = tumor necrosis factor; VLDL = very low-density lipoprotein; Apo = apolipoprotein; LDL = low-density lipoprotein.

2.2.3. Cytokine Differences among the Clusters of Patients with Heart Failure

Cytokine profile analysis, according to the patient's cluster grouping, is shown in Table 4. Patients from cluster 1 showed the lowest levels of TNF-α (3.3 ± 4.1 pg/mL) and IL-17A (2.4 ± 2.3 pg/mL), as well as the highest levels of IL-10 (2.7 ± 2.0 pg/mL), IL-6 (6.2 ± 6.9 pg/mL), IL-18 (471.2 ± 400.6 pg/mL), IL-12p70 (4.5 ± 6.0 pg/mL), and interferon (IFN)-α2 (74.5 ± 99.7 pg/mL). Among these cytokines, TNF-α, IL-12p70, IFN-α2, and IL-10 were significantly different from cluster 3. Opposite to these results, patients from cluster 3 showed the highest levels of TNF-α (9.1 ± 3.5 pg/mL), IL-8 (55.6 ± 117.1 pg/mL), and IL-17A (3.5 ± 2.0 pg/mL), as well as the lowest levels of IL-10 (1.0 ± 0.9 pg/mL), IL-12p70 (0.5 ± 0.4 pg/mL), and IFN-α2 (1.1 ± 0.4 pg/mL). From these, TNF-α, IL-12p70, IL-10, and IFN-α2 levels were significantly different regarding cluster 1. Patients from cluster 2 were shown to have significant lower levels of IL-10 (1.1 ± 0.8 pg/mL), IL-12p70 (0.6 ± 0.6 pg/mL), and IFNα-2 (4.3 ± 6.5 pg/mL) compared with those from cluster 1. Figure 3 shows the PCA, including the cytokine variables shown in Table 4.

Table 4. Differences in cytokine concentration for the three clusters of patients with heart failure.

Cytokine (pg/mL)	Cluster 1	Cluster 2	Cluster 3
IFN-α2	74.5 (± 99.7) [b, c]	4.3 (± 6.5) [a]	1.1 (± 0.4) [a]
IFN-Ɣ	65.7 (± 238.1)	48.6 (± 196.2)	8.2 (± 14.7)
TNF-α	3.3 (± 4.1) [c]	5.3 (± 5.8)	9.1 (± 3.5) [a]
MCP-1	543.6 (± 374.9)	490.9 (± 439.4)	252 (± 321.6)
IL-6	6.2 (± 6.9)	3.3 (± 2.1)	4.4 (± 2.6)
IL-8	38.2 (± 82.4)	18.8 (± 34.4)	55.6 (± 117.1)
IL-10	2.7 (± 2.0) [b, c]	1.1 (± 0.8) [a]	1 (± 0.9) [a]
IL-12p70	4.5 (± 6.0) [b, c]	0.6 (± 0.6) [a]	0.5 (± 0.4) [a]
IL-17A	2.4 (± 2.3)	2.8 (± 2.8)	3.5 (± 2.0)
IL-18	471.2 (± 400.6)	284 (± 217.6)	221.9 (± 127.8)
IL-23	36.7 (± 76.8) [b]	3.6 (± 2.8) [a]	3 (± 2.2)
IL-33	0.9 (± 0.6)	1 (± 0.9)	2.1 (± 2.7)

Tukey honest significant difference (HSD); data are expressed as mean (± standard deviation). IFN = interferon; TNF = tumor necrosis factor; MCP = monocyte chemoattractant protein; IL = interleukin. [a] = statistical difference when compared vs. cluster 1 ($p < 0.05$); [b] = statistical difference when compared vs. cluster 2 ($p < 0.05$); [c] = statistical difference when compared vs. cluster 3 ($p < 0.05$).

2.3. Correlation between 25(OH)D Levels and Ventricular Function, Biochemical, Lifestyle, and Anthropometric Parameters in Patients with Heart Failure

A significant positive correlation was found between 25(OH)D serum levels and age ($R = 0.386$, $p < 0.01$), BMI ($R = 0.265$, $p < 0.05$), dietary intake of vitamin D-rich foods ($R = 0.276$, $p < 0.05$), and calcium ($R = 0.354$, $p < 0.05$) (Tables 5 and 6). No association was observed between sun exposure and 25(OH)D concentration when the total data was analyzed; however, as was shown in Figure 1, a difference was demonstrated when the observations were subgrouped into null sun exposure or any time of sun exposure. There was no correlation between 25(OH)D levels and ventricular function parameters (LVEF and brain natriuretic peptide [BNP]) (Table 6 and Figure A2-Appendix A).

Table 5. Association of serum 25(OH)D levels with ventricular function, lifestyle, and anthropometric parameters in patients with heart failure for the 12-month follow-up.

Variable	Spearman's Rho
Age (years)	0.386 **
Weight (kg)	0.031
Body fat (%)	0.160
Fat mass (kg)	0.150
Waist circumference (cm)	0.150
BMI (kg/m^2)	0.265 *
Vitamin D intake (IU/day)	0.276 *
Sun exposure (min/week)	−0.075

Spearman's rank correlation test; * = $p < 0.05$; ** = $p < 0.01$. Values represent the four measurements corresponding to each season of the year per patient (only correlations for serum 25(OH)D levels are shown). LVEF = left ventricular ejection fraction; BMI = body mass index.

Table 6. Association of serum 25(OH)D levels with biochemical and metabolic parameters in patients with heart failure for the 12-month follow-up.

Variable	Spearman's Rho
LVEF (%)	0.251
BNP (pg/mL)	0.221
hsCRP (mg/dL)	0.133
Calcium (mg/dL)	0.354 *
PTH (pg/mL)	0.283
Total cholesterol (mg/dL)	−0.479 **
LDL (mg/dL)	−0.507 ***
HDL (mg/dL)	−0.196
non-HDL (mg/dL)	−0.481 **
VLDL (mg/dL)	−0.317 *
Triglycerides (mg/dL)	−0.317 *
Ratio cholesterol/HDL	−0.423 **
Apo A (mg/dL)	−0.299
Apo B (mg/dL)	−0.566 ***
Ratio Apo B/Apo A	−0.496 ***
Glucose (mg/dL)	−0.073
Insulin (μU/mL)	−0.017
HOMA index	−0.077

Spearman's rank correlation test; * = $p < 0.05$; ** = $p < 0.01$; *** = $p < 0.001$. Values represent the four measurements corresponding to each season of the year per patient (only correlations for serum 25(OH)D levels are shown). LVEF = left ventricular ejection fraction; BNP = brain natriuretic peptide; hsCRP = high sensitivity C-reactive protein; PTH = parathyroid hormone; LDL = low-density lipoprotein; HDL = high-density lipoprotein; VLDL = very low-density lipoprotein; Apo = apolipoprotein; HOMA = Homeostatic Model Assessment.

2.4. Lipid Parameters Are Negatively Correlated with 25(OH)D Levels in Patients with Heart Failure

A significant inverse correlation was found between serum 25(OH)D concentration and LDL cholesterol ($R = -0.507$, $p < 0.001$), Apo-B ($R = -0.566$, $p < 0.001$), Apo-B/Apo-A ratio ($R = -0.496$,

$p < 0.001$), total cholesterol ($R = -0.479$, $p < 0.01$), non-HDL cholesterol ($R = -0.481$, $p < 0.01$), and triglycerides ($R = -0.317$, $p < 0.05$) (Table 6). No correlation with HDL cholesterol or Apo-A was found. A decision-tree analysis was created in order to identify which variables might have the greatest prediction power to determine 25(OH)D status. Despite multiple variables being associated with serum 25(OH)D levels in the univariate correlation analysis, after multivariate analysis, Apo-B was the only variable that remained with a significant correlation. An Apo-B level > 76 mg/dL was shown to significantly predict a higher risk of being vitamin D insufficient or deficient ($p = 0.025$) (Figure 4). No seasonal variation was found for any of the lipid parameters in the HFrEF patients, nor was there a correlation between serum 25(OH)D levels and glycemic parameters such as fasting glucose, insulin levels, or HOMA index, or any seasonal variation either.

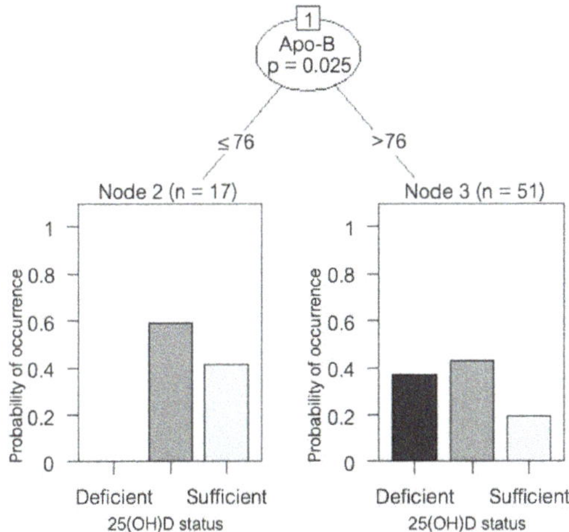

Figure 4. Decision tree predicting vitamin D status in patients with heart failure. Variable name and p-value is shown inside the circle; the corresponding lines show cut-off values. Bar graphs at the bottom show 25(OH)D status distribution amongst patients. Apo = apolipoprotein; sufficiency = 25(OH)D ≥ 30 ng/mL, insufficiency = 25(OH)D ≥ 20 and < 30 ng/mL, and deficiency = 25(OH)D < 20 ng/mL [30]. 25(OH)D = 25-hydroxyvitamin D.

2.5. *Inflammatory Cytokines Are Inversely Correlated with 25(OH)D Levels in Patients with Heart Failure*

Table 7 shows a strong inverse correlation between serum 25(OH)D levels and the proinflammatory cytokines IL-1β ($R = -0.779$, $p < 0.001$), TNF-α ($R = -0.530$, $p < 0.001$), IL-6 ($R = -0.418$, $p < 0.01$), and IL-8 ($R = -0.414$, p < 0.01), and a weaker negative correlation with IL-17A ($R = -0.309$, $p < 0.05$), IL-18 ($R = -0.349$, $p < 0.05$), and IL-33 ($R = -0.357$, $p < 0.05$) as well. There was no seasonal variation amongst the cytokine milieu. Furthermore, there was no significant correlation between BMI and cytokine levels.

2.6. *Vitamin D Insufficiency/Deficiency Was Highly Prevalent Throughout All Year*

Most of the patients with HF in our cohort were vitamin D insufficient or deficient throughout all of the year (Figure 5a). The highest prevalence of vitamin D insufficiency/deficiency was found to be in winter and spring (82.4%), followed by that in autumn (70.6%), and remaining high during the summer (64.7%). An ANOVA analysis was performed to evaluate whether serum 25(OH)D concentrations varied throughout the year. As observed in Figure 5b and Table 8, 25(OH)D levels were found to be lowest in winter (23.6 ± 7.8 ng/mL), followed by spring (24.0 ± 7.8 ng/mL) and autumn (24.4 ± 8.9 ng/mL), while the highest concentrations were found in summer (25.2 ± 6.8 ng/mL). Although no significant seasonal variation was observed, there was a tendency towards a higher prevalence of vitamin D deficiency during winter.

Table 7. Association of serum 25(OH)D concentration and cytokine levels in patients with heart failure for the 12-month follow-up.

Variable	Spearman's Rho
IL-1β	−0.779 ***
IFN-α	0.464 ***
IFN-ɣ	−0.046
TNF-α	−0.530 ***
MCP-1	0.257
IL-6	−0.418 **
IL-8	−0.414 **
IL-10	0.001
IL-12p70	0.188
IL-17A	−0.309 *
IL-18	−0.349 *
IL-23	0.114
IL-33	−0.357 *

Spearman's rank correlation test; * = $p < 0.05$; ** = $p < 0.01$; *** = $p = 0.001$. Values include the four measurements corresponding to each season of the year. IL = interleukin; IFN = interferon; TNF = tumor necrosis factor; MCP = monocyte chemoattractant protein.

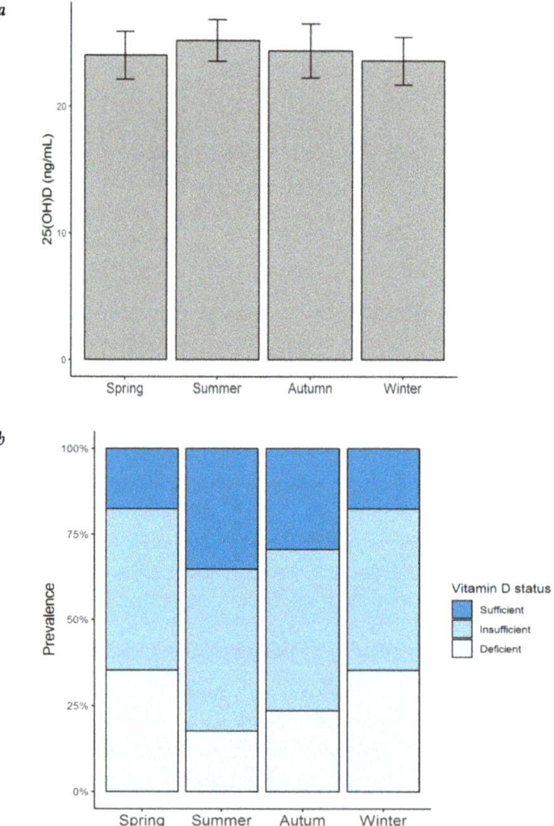

Figure 5. Seasonal variation of serum 25(OH)D concentration and prevalence of vitamin D status in patients with heart failure. (**a**) Seasonal variation of serum 25(OH)D concentration in patients with HF; bar height = mean; error bars = ± standard error; (**b**) prevalence of vitamin D status by season in the same patients. 25(OH)D = 25-hydroxyvitamin D.

Table 8. Seasonal variation of serum 25(OH)D levels and prevalence of deficiency, insufficiency, and sufficiency in patients with heart failure.

Season	25(OH)D (ng/mL)	Vitamin D Deficiency *n* (%)	Vitamin D Insufficiency *n* (%)	Vitamin D Sufficiency *n* (%)
Spring	24 (± 7.8)	6 (35.3)	8 (47.1)	3 (17.6)
Summer	25.2 (± 6.8)	3 (17.6)	8 (47.1)	6 (35.3)
Autumn	24.4 (± 8.9)	4 (23.5)	8 (47.1)	5 (29.4)
Winter	23.6 (± 7.8)	6 (35.3)	8 (47.1)	3 (17.6)

Data are expressed as mean (± standard deviation) for serum 25(OH)D levels, and as absolute value (*n*) and percentage (%) of the population for each season. 25(OH)D = 25-hydroxyvitamin D.

3. Discussion

3.1. Vitamin D Deficiency in the Setting of Patients with HFrEF

One of the core aspects of recent molecular and clinical research into HFrEF has been to better understand the role of vitamin D and the vitamin D receptor in the initiation and progression of HF. There are several mechanisms underlying this association. Studies in animals and humans have shown that serum 25(OH)D levels are inversely correlated with the RAAS activity, possibly because active vitamin D inhibits renin biosynthesis [25]. Vitamin D deficiency may, therefore, contribute to a hyper-activated RAAS axis, which then promotes the deleterious hemodynamic consequences of salt and water retention, vasoconstriction, and ventricular remodeling. Another mechanism associating vitamin D deficiency and ventricular dysfunction is intracellular calcium handling by the cardiomyocyte. In a mouse model of cardiac dysfunction, mice lacking calcitriol have been found to exhibit abnormal calcium handling, impaired ventricular functioning, and adverse cardiac remodeling and fibrosis [26]. Furthermore, activation of the Vitamin D receptor has been shown to regulate the expression of procollagen 1, which in turn may regulate profibrotic activity in the myocardium [27]. In another animal model, the deletion of the vitamin D receptor gene was also found to activate the calcineurin/nuclear factor of activated T-cells (NFAT) intracellular signaling pathway, a potent pro-hypertrophic signal [28]. Moreover, myocardial contractility may also be enhanced in the presence of sufficient vitamin D levels, which correlates with increased intracellular concentrations of cyclic adenosine monophosphate (cAMP), PKA, and PKC [27]. Overall, these findings demonstrate several mechanistic associations in which vitamin D deficiency may participate in HF progression, hemodynamic abnormalities, and structural dysfunction. Our results show that most of the patients with HFrEF in our cohort have low serum 25(OH)D levels and are in line with other studies. A significant high prevalence of vitamin D deficiency in patients with HF, when comparing low and very low vs. normal serum 25(OH)D levels (HR = 1.33 and 2.19, respectively), has been found [31]. A high prevalence of vitamin D deficiency among patients with HF and an independent association of low 25(OH)D levels with hospitalization and mortality rates have also been described [32]. On the other hand, in a different type of study to evaluate whether inadequate serum levels of 25(OH)D predict the prevalence of chronic conditions such as cardiovascular disease, type 2 diabetes, and obesity, among others, no association with any chronic condition, including heart failure, was found; this study did not report prevalence of vitamin D deficiency [33]. Both cellular and hemodynamic mechanisms may be implicated in this association, which not only explains an increased prevalence of vitamin D deficiency among patients with HF, but may also recognize a connection between vitamin D deficiency and higher risk of adverse outcomes in patients with HF. In our cohort of patients with HF, vitamin D deficiency may thus contribute to ventricular dysfunction, cardiac remodeling, hypertrophy, fibrosis, and disease progression by the molecular and cellular signaling mechanisms described above.

3.2. Correlation of Atherosclerotic and Metabolic Parameters with Vitamin D Deficiency in Patients with HFrEF

Diverse types of dyslipidemias have been recognized as risk factors for the development of HF [34], and in conjunction with vitamin D deficiency, they may explain some of the mechanisms for disease progression. Atherosclerosis plays a predominant role in the pathophysiology of ischemic heart disease, which by itself, is the leading cause of HF. LDL-cholesterol particles turn into oxidized LDL cholesterol (oxLDL) by reactive oxygen species. Macrophages, which express high-capacity scavenger receptors that are not down-regulated in the presence of high oxLDL concentration, are modified into foam cells [35]. Clusters of foam cells accumulated in the sub-endothelium, up-regulate the expression of NF-κB, which in turn triggers the membrane expression of vascular cell adhesion protein 1 (VCAM-1) and the endothelial adhesion molecule E-selectin. Another mechanism that increases NF-κB transcription in the endothelial cells is the increased proinflammatory cytokines IL-1, IL-8, and MCP-1 that characterize the inflammatory state in HF [36]. Adhesion molecules VCAM-1 and E-selectin contribute to increased cellular inflammation in the endothelial surface, as macrophages and T-cells are attracted to these proteins [35,36]. This propagates the inflammatory cycle, eventually leading to disturbed blood flow in the vessel's lumen and progression of atherosclerosis.

Regarding these mechanisms, vitamin D may have lipid-regulating effects. Therefore, adequate levels of 25(OH)D might act as a protective factor in coronary disease and HF, but in the case of vitamin D deficiency, there may be deleterious consequences. Figure 6 shows the possible immunomodulatory role of vitamin D regarding inflammation in the context of HF progression. In an animal model of hypercholesterolemia, vitamin D was shown to promote cholesterol efflux from macrophage-derived foam cells by augmenting the expression of ATP-binding membrane cassette transporter types A1 and G1 (ABCA1/G1) [37]. Hepatocyte uptake of cholesterol was also shown to be increased in the presence of higher levels of 25(OH)D concentrations. In addition, vitamin D has also been described to downregulate the c-Jun N-terminal protein kinase 2 (JNK2) cellular pathway in macrophages, which decreases oxLDL uptake and subsequently inhibits transformation into a foam cell phenotype [38]. Given the high prevalence of vitamin D deficiency throughout the four seasons of the year in our cohort of patients with HF, these protective effects of vitamin D may likely be absent. Furthermore, our results show an inverse correlation between 25(OH)D levels and total cholesterol, LDL cholesterol, non-HDL cholesterol, triglycerides, Apo-B, and Apo-B/Apo-A ratio. These correlations may also reflect the mechanisms described above, pointing towards the lack of the possible protective role of vitamin D in the setting of dyslipidemias. Our results are consistent with the prospective study, *The Atherosclerosis Risk in Communities*, in which lower 25(OH)D concentrations were shown to be associated with higher LDL cholesterol and total cholesterol, as well as an overall greater risk of dyslipidemias [39]. Similarly, Forrest et al. found that vitamin D deficiency was significantly more common in individuals with high LDL cholesterol [40].

Dysregulation of glucose metabolism has been linked to vitamin D deficiency and the development, as well as progression, of HF. In patients with established HF, the presence of diabetes mellitus has been associated with increased hospitalization rates and overall increased mortality [34]. Several mechanisms have been described regarding the role of vitamin D in glucose metabolism. In an animal model of vitamin D deficiency, decreased protein kinase B/Akt phosphorylation and reduced expression of glucose transporter 4 (GLUT4) in cardiomyocytes was shown. This, in turn, resulted in insulin resistance by the myocardium, a significant downregulation of the endogenous antioxidant enzymes SOD2 and catalase, and diminished LVEF [41]. The high prevalence of vitamin D deficiency throughout the 12-month follow-up shown in our cohort of patients with HF may thus worsen or predispose them to alterations in glucose metabolism. Moreover, a significant correlation between 25(OH)D levels < 43 ng/mL and impaired fasting glucose (OR = 3.40) and low insulin sensitivity (defined as HOMA index for insulin sensitivity < 90%) (OR = 2.65) has been reported in healthy adults with a median age of 39.4 years old [42]. Specifically, in patients with HF, vitamin D deficiency was shown to be correlated with increased rates of diabetes mellitus when compared with a non-vitamin D deficient group (31.4% vs. 22.8%) [43]. Despite these observations, our results show no correlation between 25(OH)D concentrations and fasting glucose level, insulin concentration, or HOMA index. These findings may be attributed to the low LVEF (below 40%) of all our patients, or to the medications they were taking.

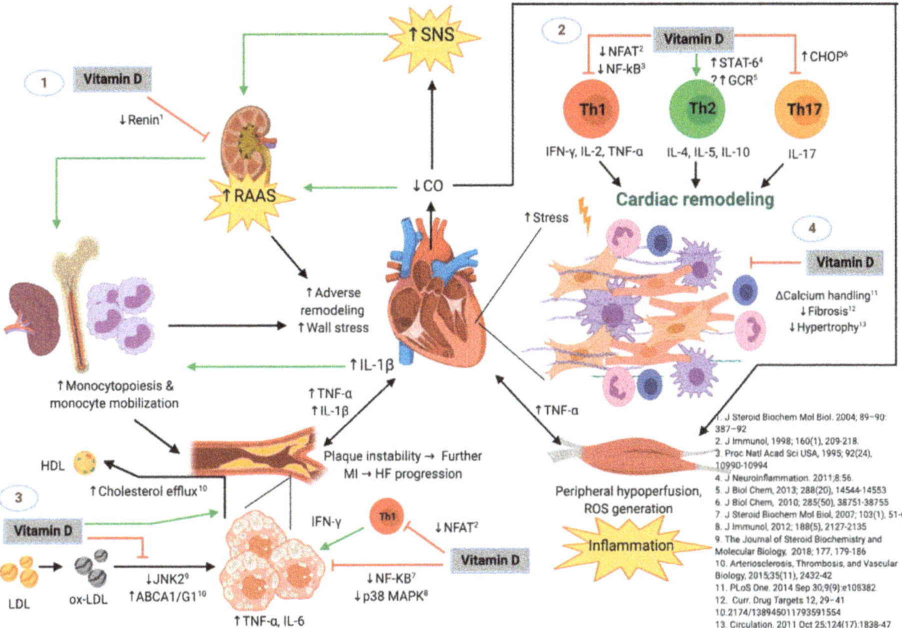

Figure 6. The Possible Immunomodulatory Role of Vitamin D in the Context of Heart Failure (HF) Progression. Green and black lines with arrowheads represent a direct stimulation and a positive correlation, respectively. Red lines ending in perpendicular bars indicate inhibition. (↑) represent an increase in expression, concentration, or activity of the parameter or mechanism. (↓) indicate a decrease in expression, concentration, or activity of the parameter of mechanism. HF is initiated and maintained by chronic maladaptive activation of neurohumoral networks that result from decreased cardiac output (e.g., the sympathetic nervous system [SNS] and the renin-angiotensin-aldosterone system [RAAS]), innate and adaptive immune mechanisms triggered by myocardial damage and stress, and systemic inflammation that results from hypoperfusion of peripheral organs [20,21]. Vitamin D may play a protective role in the progression of HF by interfering with several deleterious pathways [25], which include but are not limited to: (1) inhibiting the release of renin [44], which leads to attenuated RAAS-mediated adverse cardiac remodeling and release of monocytes from bone marrow and lymphoid tissue (an event that is also favored by IL-1β and perhaps other inflammatory cytokines from damaged myocardium) [21]. Monocytes are known to infiltrate damaged myocardium [45] and atheromatous plaques [35], where they exacerbate the local inflammatory response and promote further adverse ventricular remodeling as well as growth and instability of atherosclerotic plaques, respectively. Hence, vitamin D may indirectly attenuate monocyte/macrophage-mediated myocardial damage. (2) Vitamin D may also favor an anti-inflammatory T-helper lymphocyte phenotype by decreasing Th1 [46–48] and Th17 [9,23] cytokine production and favoring the Th2 phenotype [49], which would theoretically promote adequate remodeling and decreased fibrosis, and reduce cardiomyocyte hypertrophy and dysfunction. (3) Vitamin D could also slow the progression of atherosclerotic plaques by inhibiting the

proinflammatory activation of foam cells (macrophages that have phagocytosed ox-LDL particles) through diminished ox-LDL capture [38], increased cholesterol efflux [37], and decreased activation of transcription factors involved in the expression of proinflammatory cytokines such as TNF-α and IL-6 [22,50]. Like the mechanism described above, vitamin D may also act on T-lymphocytes present in atherosclerotic plaques, decreasing IFN-γ production and hence proinflammatory polarization of macrophages/foam cells. This is of relevance, as growth of atherosclerotic plaques results in plaque instability and further myocardial infarction, with consequent exacerbation of the pathogenic loop that characterizes HF. (4) Vitamin D has also been demonstrated to exert direct effects on the cardiomyocyte and interstitium, including modulation of intracellular calcium handling [26], expression of procollagen-1 [27], and hypertrophic calcineurin/NFAT signaling [28]. Abbreviations: HF = heart failure; MI = myocardial infarction; ABCA1/G1 = ATP-binding cassette transporter types A1 and G1; CHOP = C/EBP homologous protein; CO = cardiac output; GCR = glucocorticoid receptor; IFN = interferon; IL = interleukin; JNK2 = c-Jun N-terminal kinase 2; LDL = low-density lipoprotein; HDL = high-density lipoprotein; NFAT = nuclear factor of activated T-cells; NF-κB = nuclear factor kappa-B; ox-LDL = oxidized low-density lipoprotein; p38 MAPK = p38 mitogen-activated protein kinases; RAAS = renin-angiotensin-aldosterone system; SNS = sympathetic nervous system; STAT6 = signal transducer and activator of transcription 6; Th = T-helper; TNF = tumor necrosis factor-alpha. Original image created with BioRender® (BioRender, Toronto, ON, Canada; website URL: https://biorender.com/; accessed on 9 October 2019).

Contrary to what was expected, no correlation was seen between serum 25(OH)D concentrations and LVEF in our cohort. This could be explained by the fact that all of our patients presented abnormally low values of LVEF. Therefore, we could not analyze the whole spectrum of the relationship between vitamin D levels and LVEF.

3.3. *Inflammation Pathways Associating Vitamin D Deficiency with HFrEF*

Figure 6 shows the possible immunomodulatory role of Vitamin D regarding inflammation in the context of HF progression. HFrEF has long been recognized as a systemic proinflammatory state, with pathways involving activation of both innate and adaptive immunity mechanisms [19,20]. Most likely, a crosstalk between the heart and the peripheral organs in which HF promotes inflammation and vice versa might be present [21]. Hemodynamic stress and volume overload have been described to increment wall stress and cell mechanical damage, which stimulates cardiomyocyte release of proinflammatory cytokines, such as TNF-α, IL-1, MCP-1, and IL-6. On the other hand, peripheral organs respond to these inflammatory signals inducing more deleterious effects, such as: skeletal muscle inflammation (which may itself be secondary to chronic vasoconstriction and hypoperfusion due to decreased cardiac output), adipose tissue inflammation, increased atherogenic progression in the endothelium, augmented monocyte production by the bone marrow, and increased bacterial translocation from the gut [21]. This chronic proinflammatory cycle has been shown to contribute to the deterioration of ventricular function by inducing myocardial contractile dysfunction, hypertrophy, apoptosis, and fibrosis, with subsequent progression of HF [51]. Although systemic inflammation is a central pathogenic hallmark of HF [20], it is currently unclear whether vitamin D deficiency plays a specific role in the pathogenesis of HF through the attributed immunomodulatory functions [52]. However, it has been shown that vitamin D deficiency is indeed associated with HFrEF and increased markers of ongoing inflammation [17,53].

In the context of inflammation, the relationship between vitamin D and the cytokine milieu has been shown to represent a core aspect that leads to a proinflammatory state through several mechanisms. Our results in this cohort of patients with HF show serum 25(OH)D levels to be inversely correlated with the proinflammatory cytokines IL-1β, TNF-α, IL-6, and IL-8, although no correlation with the anti-inflammatory cytokine IL-10 was noted. Vitamin D has been found to inhibit the secretion of proinflammatory cytokines TNF-α and IL-6 by monocytes and macrophages through inhibiting p38 MAPK via induction of MKP1, which results in dephosphorylation of p38 [22]. The MAPK pathway is stimulated by diverse stressors to induce expression of proinflammatory cytokines, not only in immune cells but also in the myocardium itself [54]. Since a high prevalence of vitamin D deficiency was found in our study population, induction of this metabolic pathway is highly probable, which might aggravate the inflammatory status that these patients with HFrEF are already in.

Other mechanisms by means of which vitamin D has been associated with inflammation are related to the vitamin D-activated vitamin D receptor/retinoid X receptor (VDR/RXR) heterodimer. VDR/RXR has been reported to interact with transcription factors such as NF-κB, NFAT, and the glucocorticoid receptor (GCR), all of which participate in the transcription of genes involved in inflammatory processes [52]. NF-κB is involved in the transcription of many proinflammatory cytokines, including TNF-α, IL-1β, and IL-6, among others [55], which are direct mediators of cardiac dysfunction [56]. NF-κB expression and activation has been shown to be downregulated by vitamin D [48,50], which may function as a protective mechanism that would not be present in the setting of vitamin D deficiency. In its own way, NFAT upregulates the expression of IL-2, a crucial cytokine for T-cell replication, activation, and induction of Th1-mediated inflammatory responses [47], which appear to be relevant in HF, as evidenced by the higher number of circulating Th1 cells seen in this condition [57]. NFAT signaling has been shown to be inhibited by vitamin D, preventing this transcription factor from binding to its response elements [46], which further indicates another immunological pathway by which vitamin D deficiency may be deleterious in HF. On the other hand, GCR upregulates the expression of anti-inflammatory cytokines, while downregulating proinflammatory ones [58]. Vitamin D has been found to enhance these GCR-mediated anti-inflammatory activities [59], which may represent an anti-inflammatory mechanism that would be absent in circumstances of vitamin D deficiency. Considering the lines of evidence of the mechanisms and pathways provided, low 25(OH)D levels are expected to be accompanied by increased circulating proinflammatory cytokines, as observed in our cohort, which may possibly contribute to clinical deterioration in these patients.

Regarding one of the main pathways of adaptive immunity, our results show serum 25(OH)D concentrations to be negatively correlated with IL-17A as well. Interestingly, vitamin D has also been shown to inhibit the secretion of the proinflammatory cytokine IL-17 by Th17 cells through a post-transcriptional mechanism [23]. Others have shown that vitamin D suppresses autoimmunity, not only by decreasing the production of IL-17, but also by inhibiting the ability of naïve CD4+ T-cells to commit to the Th17 lineage [9]. The anti-Th17 effects of vitamin D could be of pathophysiological relevance in HFrEF, as IL-17 appears to have a role in adverse cardiac remodeling and myocardial fibrosis [60]. Vitamin D seems to induce a shift from a Th1 towards a Th2 phenotype, as well as to suppress Th17 responses [52]. Hence, it is reasonable to assume that vitamin D deficiency exacerbates the deleterious generalized proinflammatory state in our patients with HFrEF. In opposition to a previous study in our group with healthy adults [6], no significant seasonal variation of any cytokine was observed, which is consistent with the permanent state of inflammation that occurs in the context of HF [19,20].

3.4. Cluster Analysis

In order to characterize the cytokine profile and the 25(OH)D levels in our patients, we performed a cluster analysis, identifying three clusters. Patients from cluster 1 presented the highest circulating levels of 25(OH)D and the highest dietary intake of vitamin D. They showed the lowest significant levels of the proinflammatory cytokines TNF-α and IL-12p70, and the highest levels of the anti-inflammatory cytokine IL-10, compared with patients from the other two clusters, further reflecting the protective role that vitamin D plays in the cytokine milieu. These patients also showed the lowest concentration of total cholesterol, triglycerides, and Apo-B, representing the possible defensive role of vitamin D in atherosclerosis. Patients from cluster 2 had relatively intermediate levels of vitamin D concentration and vitamin D intake, along with intermediate levels of TNF-α, IL-17A, and IL-10 compared with patients from clusters 1 and 3. Finally, and in opposition, patients from cluster 3 presented the lowest serum 25(OH)D levels. These patients showed the highest concentrations of TNF-α, IL-8, and IL-17A, (significant for TNF-α), and the lowest significant levels of IL-10. They also had the highest levels of total cholesterol, LDL cholesterol, Apo-B, and triglycerides, indicating the proinflammatory and atherosclerotic mechanistic roles of vitamin D deficiency described previously. The role of dietary intake as the main lifestyle factor associated with vitamin D deficiency in this setting of patients may also be considered.

3.5. Correlation of Vitamin D Deficiency with Anthropometric and Lifestyle Parameters in Patients with HFrEF

The relationship between vitamin D and anthropometric parameters in the setting of HF has not been thoroughly explored. In the general population, there seems to be a consensus of a negative correlation between serum 25(OH)D concentrations and BMI, WC, and BF% [33,61–64], mainly attributed to sequestration of vitamin D by adipocytes. However, controversy has been found in patients with HF, as no correlation has been reported in this context between serum 25(OH)D levels and body weight, BMI, mid-arm muscle circumference [65], arm lean mass, leg lean mass, and grip strength (rather related to muscle mass and strength) [66]. Similarly, we found no correlation between 25(OH)D levels and BMI, BF%, and fat mass in kg. On the other hand, in a study with 14 HF patients vs. 14 controls, an inverse relationship was found between 25(OH)D levels and BMI [67]. Our results could be explained by the fact that most of our patients belonged to the overweight BMI and BF% category, and WC was within normal limits. Our sample of patients, like most with HF, are rather not obese because of a higher metabolic waste status due to the disease, and hence do not fully represent each of the BMI or BF% categories.

Likewise, lifestyle factors related to vitamin D status in patients with HF have been understudied. In a questionnaire-based study, patients with HF were less exposed to summer outdoor activity at least once a year compared with controls, but no difference was found in fish consumption (the only food source investigated) or vitamin D supplementation. However, serum 25(OH)D levels were not measured [67]. In the previously mentioned cross-sectional case-control study, that included 14 patients with HF vs. 14 controls, no significant differences in 25(OH)D concentration was reported. HF patients were shown to present less weekly sun exposure, but no differences in daily dietary vitamin D intake between the groups was found [67]. In addition, using data from a large population study, no differences were seen among various amounts of vitamin D intake and the incidence (new cases) of HF, but measurement of 25(OH)D levels was not performed either [68]. Our results during the 12-month follow up showed a significant positive correlation between dietary intake of vitamin D-rich foods and serum 25(OH)D concentrations. Regarding sun exposure, when the total population was considered, no correlation was found between serum 25(OH)D levels and sun exposure. Nevertheless, in a sub-analysis in which patients were separated into two groups, those with any amount of sun exposure versus those with no exposure at all, the prevalence of vitamin D sufficiency was significantly greater in patients exposed to the sun compared with those that had no exposure. Our group previously studied the seasonal variation of serum 25(OH)D concentration in a cohort of 23 healthy subjects with a follow-up of 12 months, in which seasonal variation was observed [6]. However, no seasonal variation of 25(OH)D levels was found in our present study, suggesting that sun exposure may not play a dominant role in vitamin D status (as opposed to healthy controls), but rather vitamin D intake seems to play a greater role. Consideration has to be given to our patients' skin phototype. Skin pigmentation represents a significant independent risk factor for vitamin D deficiency because UVB is needed in the initial conversion of 7-dehydrocholesterol into cholecalciferol, which is less available for this reaction in darker skin phototypes, as melanin acts as a light-absorbent protein. Thus, our cohort might be prone to vitamin D deficiency, given that the majority of the subjects belonged to phototypes IV and V.

The recommended daily allowance (RDA) of vitamin D for adults 51–70 years old is 600 IU/day, while that of adults >70 years old is 800 IU/day [69]. Therefore, none of the patients in our cohort met the nutritional requirements for vitamin D intake, as all of them had an average intake throughout the year of less than 400 IU/day. Although the time of sun exposure required for vitamin D synthesis varies considerably depending on the living latitude and the season of the year, it has been estimated that 5 to 30 min of sun exposure between 10 a.m. and 3 p.m. at least two times per week may be a reasonable amount [70]. Most of the patients in our cohort (82.3% during summer, autumn, and spring, and 52.9% during winter) obtained at least this level of sun exposure while living at an adequate latitude. Therefore, vitamin D ingestion might be a lifestyle factor as important or even more significant than sun exposure in determining 25(OH)D levels in the setting of our cohort of patients with HF.

The study has some limitations. The sample size is both small and from a Hispanic ethnicity; thus, our results might not be extrapolated to other populations. Hence, results should be cautiously interpreted. Even though data regarding vitamin D-rich foods intake and sun exposure was obtained through a personal interview, the information given by the patients might still be subjective. However, our study has several strengths. This is a 12-month follow up to approach seasonal variation of vitamin D and cytokines. In-vitro and animal model studies concerning the molecular mechanisms were identified to describe and apply these mechanisms to explain our results, given that studies in humans hardly explain them in such detail. We also emphasize these molecular mechanisms in an original figure presented in this article, which could offer insight for further research.

4. Materials and Methods

4.1. Study Population

We conducted a longitudinal study in a sample of HFrEF patients (49 to 82 years old) following standard treatment. Patients (n = 17) were followed during 12 months. Every three months, corresponding to each of the four seasons of the year, every patient was personally interviewed and evaluated for lifestyle, anthropometric, and laboratory parameters. All of the patients live in northeastern Mexico (city of Monterrey, latitude 25°40′0′′N) and are of Hispanic origin. The Fitzpatrick classification of phototypes, the standardized clinical method to evaluate a person's level of melanin pigmentation, was used to determine the level of skin pigmentation of the patients. Phototype I individuals have fair skin, which burns and peels easily with sun exposure, but does not tan; they usually have light eye and hair color. Phototype VI individuals are dark-skinned; when exposed to the sun, they always tan and never burn [71]. The inclusion criteria were a proved diagnosis of HFrEF (LVEF $< 40\%$) and a New York Heart Association (NYHA) functional class of III or IV [29]. The exclusion criteria were chronic kidney disease, liver disease, LVEF $\geq 40\%$, chronic use of corticosteroids, ingestion of vitamin D supplements, and being institutionalized. Written informed consent was explained, accepted, and signed by all of the patients. The study was approved on May 14, 2013, by the Ethics and Research Commissions of the School of Medicine of Tecnologico de Monterrey and the Santos y de la Garza Evia Foundation, as well as by the Mexican Secretariat of Health with the project identification code "ESVDIC".

4.2. Vitamin D Intake

Every 3 months, each of the patients were interviewed about regular ingestion of vitamin D-rich foods. Standardized portions of these foods were shown to the patients in order to quantify in a more accurate way the amount of vitamin D ingested per source. The patients were asked about consumption of milk (124 IU/portion), yogurt (120 IU/portion), cheese (20 IU/portion), egg (20 IU/portion), fish (150 IU/portion), and cereal (50 IU/portion). Estimated vitamin D content per food source and portions were determined according to the USDA National Nutrient Database for Standard Reference Release [72].

4.3. Sun Exposure

During each visit, patients were asked for the total amount of time being exposed to the sun (minutes per day [mpd] and days per week [dpw]). Activities involving and not involving physical exercise were both taken into account. Skin characteristics according to the Fitzpatrick classification of phototypes were registered [71].

4.4. Anthropometric Parameters

For each of the seasons, the following anthropometric parameters were measured: height (m), weight (kg), BMI (kg/m^2), WC (cm), BF% (%), and fat mass (kg) according to standardized protocols [73]. The body fat percentage was measured by bioimpedance using a TANITA's BF-350 (Tanita Corporation of America, Inc., Arlington Heights, IL, USA).

4.5. Vitamin D, Biochemical Parameters, and Cytokines

The patients had a blood sample withdrawn by venipuncture for each of the seasons. Calcium concentration was obtained by spectrophotometry with the Calcium 3L79-21 kit (06753/R05) (Abbott Laboratories Diagnostics Division, Chicago, IL, USA), while serum 25(OH)D and PTH levels were measured by chemiluminescence with the Vit D25OH 5P02-25 kit (G5-9160/R01) and the iPTH 8K25-25 kit (G6-5257/-R05), respectively, on the Architect iSystem (Abbott Laboratories Diagnostics Division, Chicago, IL, USA). Serum and plasma were also obtained from these samples, centrifuged, and frozen at −80 °C afterward. Vitamin D status was determined according to the Endocrine Clinical Society Guidelines: sufficiency as serum 25(OH)D ≥ 30 ng/mL, insufficiency as serum 25(OH)D ≥ 20 and < 30 ng/mL, and deficiency as serum 25(OH)D < 20 ng/mL [30]. High-sensitivity C-reactive protein levels were measured by quantitative immunoturbidimetric assay with a CRP Vario 6K26-30 and 6K26-41 kits (Abbott Laboratories Diagnostic Division, Chicago, IL, USA). Fasting glucose levels were determined by the hexokinase (HK)/glucose-6-phosphate dehydrogenase (G-6-PD) method with the Glucose 3L82 reagent (DENKA SEIKEN Co. Ltd., Tokyo, Japan). Insulin concentrations were measured by chemiluminescence, using the ARCHITECT Insulin Reagent 8K41-27 kit (Abbot Laboratories Diagnostic Division, IL, USA). Total cholesterol was determined with the Cholesterol 7D62-21 reagent kit (307166/R04) (Abbot Laboratories Diagnostic Division, Chicago, IL, USA) on the Architect cSystem. HDL cholesterol was measured by the accelerator selective detergent method with the Ultra HDL 3K33-21 (307177-R04) assay (Abbot Laboratories Diagnostic Division, Chicago, IL, USA) on the Architect cSystem. Triglycerides were measured via the glycerol-phosphate-oxidase reaction with the Triglyceride 7D74-21 kit (307170/R03) (Abbot Laboratories Diagnostic Division, Chicago, IL, USA) on the Architect cSystem. LDL cholesterol was calculated using the total cholesterol, HDL cholesterol, and triglyceride determinations previously mentioned [74]. Apo-A and Apo-B were obtained via immunoturbidimetric assays with the Apolipoprotein A1 9D92-21 kit (306769/R03) and the Apolipoprotein B 9D93-21 kit (306770/R03), respectively, using the Architect cSystem (Abbot Laboratories Diagnostic Division, Chicago, IL, USA). BNP was measured by chemiluminescence with the BNP 8K28-28 kit (616-010-R01) (Abbott Laboratories Diagnostics Division, Chicago, IL, USA) on the Architect iSystem. Serum was used to obtain a cytokine profile using the Legendplex Human Inflammation Panel through a multianalyte flow cytometry assay (BioLegend, San Diego, CA, USA). This panel allows determination of 13 cytokines: IL-1β, IFN-α, IFN-γ, TNF-α, MCP-1, IL-6, IL-8, IL-10, IL-12p70, IL-17A, IL-18, IL-23, and IL-33. As recommended by Biolegend's instructions, each experiment was performed in triplicate. The flow cytometer used was FACS-Canto II (Becton Dickinson, Franklin Lakes, NJ, USA). Using the standard curves and the Legendplex Data Analysis software version 7.1 provided by BioLegend (BioLegend, San Diego, CA, USA), the analyte concentration and the serum concentrations were calculated for each of the cytokines.

4.6. Subgrouping Analysis by Cluster

In order to better identify which variables may represent a given profile regarding vitamin D deficiency in patients with HFrEF, a cluster analysis by partitioning around medoids (PAM) was performed. A total of 4 observations per patient were included in this analysis (corresponding to each of the seasons). The 25(OH)D levels, inflammatory cytokines, lifestyle factors, biochemical, and anthropometric parameters were included. Differences among clusters were assessed by ANOVA with Tukey's post-hoc test.

4.7. Statistical Analysis

Principal component analysis (PCA) was performed to explore continuous demographic, anthropometric, and lifestyle variables. Seasonal variation was tested with repeated-measures ANOVA. Categorical variables were evaluated with Fisher's exact test. Univariate association with 25(OH)D concentration was calculated with Spearman's rank correlation test. A conditional inference tree was constructed for vitamin D status prediction according to parameters with statistically significant association in univariate analysis. The R platform (R package version 2.0.7-1; The R Foundation, Vienna, Austria), along with the cluster [75] and party [76] packages, were used.

5. Conclusions

Several lines of evidence establish the mechanistic and metabolic roles of vitamin D deficiency contributing to the inflammatory, immunomodulatory, and atherosclerotic status of patients with HF. Our results in this 12-month follow-up of patients with HFrEF are consistent with previous observations of a high prevalence of vitamin D deficiency in these type of patients. We found no seasonal variation of 25(OH)D levels, nor of the proinflammatory cytokine profile. Serum 25(OH)D concentrations correlated with dietary intake of vitamin D-rich foods. When analyzing our total population, no correlation was seen between 25(OH)D levels and sun exposure; nonetheless, when subgrouping the patients into positive (any amount of sun exposure) and negative (no sun exposure), there was a higher prevalence of vitamin D sufficiency in those belonging to the positive group. Considering the inflammatory state observed in the setting of HF, 25(OH)D levels were inversely correlated with the proinflammatory cytokines IL-1β, TNF-α, IL-6, IL-8, IL-17A, IL-18, and IL-33. Total cholesterol, LDL cholesterol, and triglycerides were also negatively correlated with 25(OH)D concentrations. Cluster analysis to better identify different patient profiles related to vitamin D status revealed that patients with the highest levels of 25(OH)D showed the highest dietary intake of vitamin D-rich foods, along with the lowest concentration of TNF-α, IL-17A, total cholesterol, triglycerides, and Apo-B. Overall, these results point towards the lifestyle factors and lipid parameters related to adequate levels of 25(OH)D. In addition, the protective anti-inflammatory role that vitamin D has in the setting of HFrEF was shown in this group of patients. Some of the pathways underlying these mechanisms have been described mostly in in-vitro and animal models: inhibition of renin biosynthesis; less detrimental ventricular remodeling by inhibition of procollagen-1 expression; increased myocardial contractility; and downregulation of JNK2, calcineurin/NFAT, and NF-κB cellular pathways. Induction of cholesterol efflux from foam cells; increased cholesterol uptake by hepatocytes; adequate regulation of GLUT4 expression in cardiomyocytes; and inhibition of the secretion of proinflammatory cytokines by downregulation of the MAPK pathway, have been attributed to vitamin D. Future studies are necessary in patients with HF to establish the molecular mechanisms involved in the pathogenesis of this disease.

Author Contributions: Conceptualization, L.E.-M. and G.G.-R.; methodology, L.E.-M., G.G.-R., C.E.; software, J.R.V.-C.; validation, L.E.-M., G.G.-R., C.E., A.M.G.-G., J.R.V.-C., E.C.C., A.S.H.-D., D.N.R.-V.; formal analysis, J.R.V.-C.; investigation, A.M.G.-G., E.C.C., A.S.H.-D., D.N.R.-V.; resources, L.E.-M., G.G.-R., C.E., A.M.G.-G.; data curation, A.S.H.-D., D.N.R.-V., A.M.G.-G., E.C.C.; writing—original draft preparation, D.N.R.-V., A.S.H.-D., A.M.G.-G.; writing—review and editing, L.E.-M., D.N.R.-V., A.S.H.-D., A.M.G.-G., J.R.V.-C.; visualization, L.E.-M., D.N.R.-V., A.S.H.-D.; supervision, L.E.-M., G.G.-R., E.C.C.; project administration, L.E.-M., G.G.-R.; funding acquisition, L.E.-M., G.G.-R.

Funding: This research was funded by the Center for Research in Obesity and Clinical Nutrition, Tecnologico de Monterrey (L.E.-M.), by the XIGNUX foundation and by CONACYT grant number 256577, a1-s-48883 (G.G.-R.).

Acknowledgments: The authors would like to acknowledge the invaluable fieldwork of Mariana Peschard-Franco [1], Gustavo Gutierrez-DelBosque [1], Gerardo Mendoza-Lara [1], Bianca Nieblas [1], and Sofia Tenorio-Martínez [1]. [1] Tecnologico de Monterrey, Center for Research in Clinical Nutrition, Escuela de Medicina, 64710 Monterrey, N.L., Mexico.

Conflicts of Interest: The authors declare no conflict of interest.

Abbreviations

HF	Heart failure
LDL	Low density lipoprotein
Apo	Apolipoprotein
IL	Interleukin
TNF	Tumor necrosis factor
BMI	Body mass index
UVB	Ultraviolet B-type
25(OH)D	25-hydroxyvitamin D
1,25(OH)2D3	1,25-dihydroxyvitamin D (calcitriol)
PTH	Parathyroid hormone
FGF-23	Fibroblast growth factor 23
WC	Waist circumference
IU	International units
LVEF	Left ventricular ejection fraction
HFrEF	Heart failure with reduced ejection fraction
HFpEF	Heart failure with preserved ejection fraction
HFmrEF	Heart failure with mid-range ejection fraction
RAAS	Renin angiotensin aldosterone system
PK	Protein kinase
NF-κB	Nuclear-factor kappa B
MCP-1	Monocyte chemoattractant protein-1
MAPK	Mitogen-activated protein kinase
MMP	Matrix metalloproteinase
NYHA	New York Heart Association
USDA	United States Department of Agriculture
IFN	Interferon
PCA	Principal component analysis
HSD	Honest significance difference
PMA	Partitioning around medoids
HDL	High density lipoprotein
HOMA	Homeostatic model assessment
BNP	Brain natriuretic peptide
ANOVA	Analysis of variance
cAMP	Cyclic adenosine monophosphate
NFAT	Nuclear factor of activated T-cells
oxLDL	Oxidized low-density lipoprotein
APC	Antigen-presenting cells
JNK2	c-Jun N-terminal protein kinase 2
GLUT4	Glucose transporter 4
VDR/RXR	Vitamin D receptor/Retinoid X receptor
GCR	Glucocorticoid receptor

Appendix A

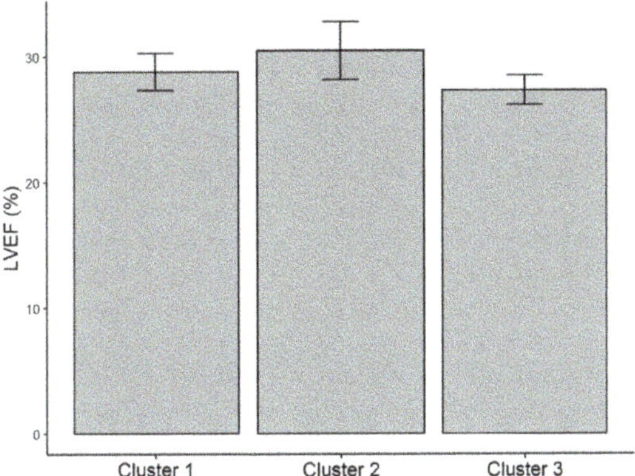

Figure A1. LVEF (%) values for each of the clusters. Mean and standard error (vertical bars) are shown. There was no significant difference of this parameter among clusters. LVEF = left ventricular ejection fraction.

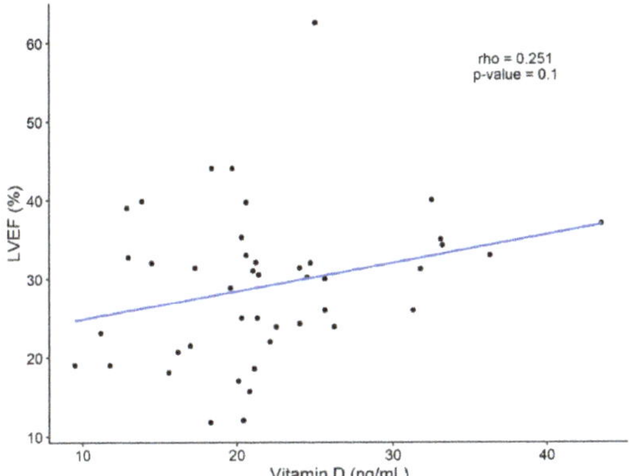

Figure A2. Linear correlation between LVEF (%) and 25(OH)D concentration. A weak positive correlation was found (R = 0.251), but no statistical significance was seen. LVEF = left ventricular ejection fraction; 25(OH)D = 25-hydroxivitamin D.

References

1. Cashman, K.D. Vitamin D Deficiency: Defining, Prevalence, Causes, and Strategies of Addressing. *Calcif. Tissue Int.* **2019**, 1–16. [CrossRef] [PubMed]
2. Jeon, S.M.; Shin, E.A. Exploring vitamin D metabolism and function in cancer. *Exp. Mol. Med.* **2018**, *50*, 20. [CrossRef] [PubMed]
3. Veldurthy, V.; Wei, R.; Campbell, M.; Lupicki, K.; Dhawan, P.; Christakos, S. 25-Hydroxyvitamin D_3 24-Hydroxylase: A Key Regulator of $1,25(OH)_2D_3$ Catabolism and Calcium Homeostasis. *Vitam. Horm.* **2016**, *100*, 137–150. [CrossRef] [PubMed]
4. Dhawan, P.; Peng, X.; Sutton, A.L.; MacDonald, P.N.; Croniger, C.M.; Trautwein, C.; Centrella, M.; McCarthy, T.L.; Christakos, S. Functional cooperation between CCAAT/enhancer-binding proteins and the vitamin D receptor in regulation of 25-hydroxyvitamin D3 24-hydroxylase. *Mol. Cell. Biol.* **2005**, *25*, 472–487. [CrossRef] [PubMed]
5. Holick, M.F. Vitamin D deficiency. *N. Engl. J. Med.* **2007**, *357*, 266–281. [CrossRef]

6. Elizondo-Montemayor, L.; Castillo, E.C.; Rodríguez-López, C.; Villarreal-Calderón, J.R.; Gómez-Carmona, M.; Tenorio-Martínez, S.; Nieblas, B.; García-Rivas, G. Seasonal Variation in Vitamin D in Association with Age, Inflammatory Cytokines, Anthropometric Parameters, and Lifestyle Factors in Older Adults. *Mediat. Inflamm.* **2017**, *2017*, 5719461. [CrossRef]
7. Herrick, K.A.; Storandt, R.J.; Afful, J.; Pfeiffer, C.M.; Schleicher, R.L.; Gahche, J.J.; Potischman, N. Vitamin D status in the United States, 2011–2014. *Am. J. Clin. Nutr.* 2019. [CrossRef]
8. Peterson, C.A.; Heffernan, M.E. Serum tumor necrosis factor-alpha concentrations are negatively correlated with serum 25(OH)D concentrations in healthy women. *J. Inflamm. (Lond.)* **2008**, *5*, 10. [CrossRef]
9. Tang, J.; Zhou, R.; Luger, D.; Zhu, W.; Silver, P.B.; Grajewski, R.S.; Su, S.B.; Chan, C.C.; Adorini, L.; Caspi, R.R. Calcitriol suppresses antiretinal autoimmunity through inhibitory effects on the Th17 effector response. *J. Immunol.* **2009**, *182*, 4624–4632. [CrossRef]
10. Rai, V.; Agrawal, D.K. Role of Vitamin D in Cardiovascular Diseases. *Endocrinol. Metab. Clin. N. Am.* **2017**, *46*, 1039–1059. [CrossRef]
11. Benjamin, E.J.; Muntner, P.; Alonso, A.; Bittencourt, M.S.; Callaway, C.W.; Carson, A.P.; Chamberlain, A.M.; Chang, A.R.; Cheng, S.; Das, S.R.; et al. Heart Disease and Stroke Statistics-2019 Update: A Report From the American Heart Association. *Circulation* **2019**, *139*, e56–e528. [CrossRef] [PubMed]
12. Savarese, G.; Lund, L.H. Global Public Health Burden of Heart Failure. *Card. Fail. Rev.* **2017**, *3*, 7–11. [CrossRef]
13. Ciapponi, A.; Alcaraz, A.; Calderón, M.; Matta, M.G.; Chaparro, M.; Soto, N.; Bardach, A. Burden of Heart Failure in Latin America: A Systematic Review and Meta-analysis. *Rev. Esp. Cardiol. (Engl. Ed.)* **2016**, *69*, 1051–1060. [CrossRef] [PubMed]
14. Ponikowski, P.; Voors, A.A.; Anker, S.D.; Bueno, H.; Cleland, J.G.; Coats, A.J.; Falk, V.; González-Juanatey, J.R.; Harjola, V.P.; Jankowska, E.A.; et al. 2016 ESC Guidelines for the diagnosis and treatment of acute and chronic heart failure: The Task Force for the diagnosis and treatment of acute and chronic heart failure of the European Society of Cardiology (ESC). Developed with the special contribution of the Heart Failure Association (HFA) of the ESC. *Eur. J. Heart Fail.* **2016**, *18*, 891–975. [CrossRef] [PubMed]
15. Ziaeian, B.; Fonarow, G.C. Epidemiology and aetiology of heart failure. *Nat. Rev. Cardiol.* **2016**, *13*, 368–378. [CrossRef] [PubMed]
16. Saponaro, F.; Saba, A.; Frascarelli, S.; Frontera, C.; Clerico, A.; Scalese, M.; Sessa, M.R.; Cetani, F.; Borsari, S.; Pardi, E.; et al. Vitamin D measurement and effect on outcome in a cohort of patients with heart failure. *Endocr. Connect.* **2018**, *7*, 957–964. [CrossRef] [PubMed]
17. D'Amore, C.; Marsico, F.; Parente, A.; Paolillo, S.; De Martino, F.; Gargiulo, P.; Ferrazzano, F.; De Roberto, A.M.; La Mura, L.; Marciano, C.; et al. Vitamin D deficiency and clinical outcome in patients with chronic heart failure: A review. *Nutr. Metab. Cardiovasc. Dis.* **2017**, *27*, 837–849. [CrossRef]
18. Shirazi, L.F.; Bissett, J.; Romeo, F.; Mehta, J.L. Role of Inflammation in Heart Failure. *Curr. Atheroscler. Rep.* **2017**, *19*, 27. [CrossRef]
19. Mann, D.L. Innate immunity and the failing heart: The cytokine hypothesis revisited. *Circ. Res.* **2015**, *116*, 1254–1268. [CrossRef]
20. Sánchez-Trujillo, L.; Vázquez-Garza, E.; Castillo, E.C.; García-Rivas, G.; Torre-Amione, G. Role of Adaptive Immunity in the Development and Progression of Heart Failure: New Evidence. *Arch. Med. Res.* **2017**, *48*, 1–11. [CrossRef]
21. Van Linthout, S.; Tschöpe, C. Inflammation—Cause or Consequence of Heart Failure or Both? *Curr. Heart Fail. Rep.* **2017**, *14*, 251–265. [CrossRef] [PubMed]
22. Zhang, Y.; Leung, D.Y.; Richers, B.N.; Liu, Y.; Remigio, L.K.; Riches, D.W.; Goleva, E. Vitamin D inhibits monocyte/macrophage proinflammatory cytokine production by targeting MAPK phosphatase-1. *J. Immunol.* **2012**, *188*, 2127–2135. [CrossRef] [PubMed]
23. Chang, S.H.; Chung, Y.; Dong, C. Vitamin D suppresses Th17 cytokine production by inducing C/EBP homologous protein (CHOP) expression. *J. Biol. Chem.* **2010**, *285*, 38751–38755. [CrossRef] [PubMed]
24. Riek, A.E.; Oh, J.; Bernal-Mizrachi, C. 1,25(OH)2 vitamin D suppresses macrophage migration and reverses atherogenic cholesterol metabolism in type 2 diabetic patients. *J. Steroid Biochem. Mol. Biol.* **2013**, *136*, 309–312. [CrossRef]
25. Wu, M.; Xu, K.; Wu, Y.; Lin, L. Role of Vitamin D in Patients with Heart Failure with Reduced Ejection Fraction. *Am. J. Cardiovasc. Drugs.* **2019**, 1–12. [CrossRef]

26. Choudhury, S.; Bae, S.; Ke, Q.; Lee, J.Y.; Singh, S.S.; St-Arnaud, R.; Monte, F.D.; Kang, P.M. Abnormal calcium handling and exaggerated cardiac dysfunction in mice with defective vitamin d signaling. *PLoS ONE* **2014**, *9*, e108382. [CrossRef]
27. Meems, L.M.; van der Harst, P.; van Gilst, W.H.; de Boer, R.A. Vitamin D biology in heart failure: Molecular mechanisms and systematic review. *Curr. Drug Targets* **2011**, *12*, 29–41. [CrossRef]
28. Chen, S.; Law, C.S.; Grigsby, C.L.; Olsen, K.; Hong, T.T.; Zhang, Y.; Yeghiazarians, Y.; Gardner, D.G. Cardiomyocyte-specific deletion of the vitamin D receptor gene results in cardiac hypertrophy. *Circulation* **2011**, *124*, 1838–1847. [CrossRef]
29. The Joint Commission. *New York Heart Association (NYHA) Classification*; Measures, J.C.N.Q., Ed.; The Joint Commission: Washington, DC, USA, 2016.
30. Holick, M.F.; Binkley, N.C.; Bischoff-Ferrari, H.A.; Gordon, C.M.; Hanley, D.A.; Heaney, R.P.; Murad, M.H.; Weaver, C.M.; Society, E. Evaluation, treatment, and prevention of vitamin D deficiency: An Endocrine Society clinical practice guideline. *J. Clin. Endocrinol. Metab.* **2011**, *96*, 1911–1930. [CrossRef]
31. Anderson, J.L.; May, H.T.; Horne, B.D.; Bair, T.L.; Hall, N.L.; Carlquist, J.F.; Lappé, D.L.; Muhlestein, J.B.; Group, I.H.C.I.S. Relation of vitamin D deficiency to cardiovascular risk factors, disease status, and incident events in a general healthcare population. *Am. J. Cardiol.* **2010**, *106*, 963–968. [CrossRef]
32. Belen, E.; Sungur, A.; Sungur, M.A. Vitamin D levels predict hospitalization and mortality in patients with heart failure. *Scand. Cardiovasc. J.* **2016**, *50*, 17–22. [CrossRef] [PubMed]
33. Cartier, J.L.; Kukreja, S.C.; Barengolts, E. Lower Serum 25-Hydroxyvitamin D Is Associated with Obesity but Not Common Chronic Conditions: An Observational Study of African American and Caucasian Male Veterans. *Endocr. Pract.* **2017**, *23*, 271–278. [CrossRef] [PubMed]
34. Bozkurt, B.; Aguilar, D.; Deswal, A.; Dunbar, S.B.; Francis, G.S.; Horwich, T.; Jessup, M.; Kosiborod, M.; Pritchett, A.M.; Ramasubbu, K.; et al. Contributory Risk and Management of Comorbidities of Hypertension, Obesity, Diabetes Mellitus, Hyperlipidemia, and Metabolic Syndrome in Chronic Heart Failure: A Scientific Statement From the American Heart Association. *Circulation* **2016**, *134*, e535–e578. [CrossRef] [PubMed]
35. Libby, P.; Buring, J.E.; Badimon, L.; Hansson, G.K.; Deanfield, J.; Bittencourt, M.S.; Tokgözoğlu, L.; Lewis, E.F. Atherosclerosis. *Nat. Rev. Dis. Primers.* **2019**, *5*, 56. [CrossRef]
36. Gimbrone, M.A.; García-Cardeña, G. Endothelial Cell Dysfunction and the Pathobiology of Atherosclerosis. *Circ. Res.* **2016**, *118*, 620–636. [CrossRef]
37. Yin, K.; You, Y.; Swier, V.; Tang, L.; Radwan, M.M.; Pandya, A.N.; Agrawal, D.K. Vitamin D Protects Against Atherosclerosis via Regulation of Cholesterol Efflux and Macrophage Polarization in Hypercholesterolemic Swine. *Arterioscler. Thromb. Vasc. Biol.* **2015**, *35*, 2432–2442. [CrossRef]
38. Oh, J.; Riek, A.E.; Zhang, R.M.; Williams, S.A.S.; Darwech, I.; Bernal-Mizrachi, C. Deletion of JNK2 prevents vitamin-D-deficiency-induced hypertension and atherosclerosis in mice. *J. Steroid Biochem. Mol. Biol.* **2018**, *177*, 179–186. [CrossRef]
39. Faridi, K.F.; Zhao, D.; Martin, S.S.; Lupton, J.R.; Jones, S.R.; Guallar, E.; Ballantyne, C.M.; Lutsey, P.L.; Michos, E.D. Serum vitamin D and change in lipid levels over 5 y: The Atherosclerosis Risk in Communities study. *Nutrition* **2017**, *38*, 85–93. [CrossRef]
40. Forrest, K.Y.; Stuhldreher, W.L. Prevalence and correlates of vitamin D deficiency in US adults. *Nutr. Res.* **2011**, *31*, 48–54. [CrossRef]
41. Nizami, H.L.; Katare, P.; Prabhakar, P.; Kumar, Y.; Arava, S.K.; Chakraborty, P.; Maulik, S.K.; Banerjee, S.K. Vitamin D Deficiency in Rats Causes Cardiac Dysfunction by Inducing Myocardial Insulin Resistance. *Mol. Nutr. Food Res.* **2019**, *63*, e1900109. [CrossRef]
42. Mayer, O.; Seidlerová, J.; Černá, V.; Kučerová, A.; Karnosová, P.; Hronová, M.; Wohlfahrt, P.; Fuchsová, R.; Filipovský, J.; Cífková, R.; et al. Serum Vitamin D Status, Vitamin D Receptor Polymorphism, and Glucose Homeostasis in Healthy Subjects. *Horm. Metab. Res.* **2018**, *50*, 56–64. [CrossRef] [PubMed]
43. Cubbon, R.M.; Lowry, J.E.; Drozd, M.; Hall, M.; Gierula, J.; Paton, M.F.; Byrom, R.; Kearney, L.C.; Barth, J.H.; Kearney, M.T.; et al. Vitamin D deficiency is an independent predictor of mortality in patients with chronic heart failure. *Eur. J. Nutr.* **2019**, *58*, 2535–2543. [CrossRef] [PubMed]
44. Li, Y.C.; Qiao, G.; Uskokovic, M.; Xiang, W.; Zheng, W.; Kong, J. Vitamin D: A negative endocrine regulator of the renin-angiotensin system and blood pressure. *J. Steroid Biochem. Mol. Biol.* **2004**, *89*, 387–392. [CrossRef] [PubMed]

45. Epelman, S.; Liu, P.P.; Mann, D.L. Role of innate and adaptive immune mechanisms in cardiac injury and repair. *Nat. Rev. Immunol.* **2015**, *15*, 117–129. [CrossRef] [PubMed]
46. Takeuchi, A.; Reddy, G.S.; Kobayashi, T.; Okano, T.; Park, J.; Sharma, S. Nuclear factor of activated T cells (NFAT) as a molecular target for 1alpha,25-dihydroxyvitamin D3-mediated effects. *J. Immunol.* **1998**, *160*, 209–218.
47. Hermann-Kleiter, N.; Baier, G. NFAT pulls the strings during CD4± T helper cell effector functions. *Blood* **2010**, *115*, 2989–2997. [CrossRef]
48. Yu, X.P.; Bellido, T.; Manolagas, S.C. Down-regulation of NF-kappa B protein levels in activated human lymphocytes by 1,25-dihydroxyvitamin D3. *Proc. Natl. Acad. Sci. USA* **1995**, *92*, 10990–10994. [CrossRef]
49. Sloka, S.; Silva, C.; Wang, J.; Yong, V.W. Predominance of Th2 polarization by vitamin D through a STAT6-dependent mechanism. *J. Neuroinflamm.* **2011**, *8*, 56. [CrossRef]
50. Stio, M.; Martinesi, M.; Bruni, S.; Treves, C.; Mathieu, C.; Verstuyf, A.; d'Albasio, G.; Bagnoli, S.; Bonanomi, A.G. The Vitamin D analogue TX 527 blocks NF-kappaB activation in peripheral blood mononuclear cells of patients with Crohn's disease. *J. Steroid Biochem. Mol. Biol.* **2007**, *103*, 51–60. [CrossRef]
51. Zhang, Y.; Bauersachs, J.; Langer, H.F. Immune mechanisms in heart failure. *Eur. J. Heart Fail.* **2017**, *19*, 1379–1389. [CrossRef]
52. Wöbke, T.K.; Sorg, B.L.; Steinhilber, D. Vitamin D in inflammatory diseases. *Front. Physiol.* **2014**, *5*, 244. [CrossRef] [PubMed]
53. Liu, L.C.; Voors, A.A.; van Veldhuisen, D.J.; van der Veer, E.; Belonje, A.M.; Szymanski, M.K.; Sillјé, H.H.; van Gilst, W.H.; Jaarsma, T.; de Boer, R.A. Vitamin D status and outcomes in heart failure patients. *Eur. J. Heart Fail.* **2011**, *13*, 619–625. [CrossRef] [PubMed]
54. Li, M.; Georgakopoulos, D.; Lu, G.; Hester, L.; Kass, D.A.; Hasday, J.; Wang, Y. p38 MAP kinase mediates inflammatory cytokine induction in cardiomyocytes and extracellular matrix remodeling in heart. *Circulation* **2005**, *111*, 2494–2502. [CrossRef] [PubMed]
55. Liu, T.; Zhang, L.; Joo, D.; Sun, S.C. NF-κB signaling in inflammation. *Signal Transduct. Target Ther.* **2017**, *2*. [CrossRef] [PubMed]
56. Gullestad, L.; Ueland, T.; Vinge, L.E.; Finsen, A.; Yndestad, A.; Aukrust, P. Inflammatory cytokines in heart failure: Mediators and markers. *Cardiology* **2012**, *122*, 23–35. [CrossRef] [PubMed]
57. Fukunaga, T.; Soejima, H.; Irie, A.; Sugamura, K.; Oe, Y.; Tanaka, T.; Kojima, S.; Sakamoto, T.; Yoshimura, M.; Nishimura, Y.; et al. Expression of interferon-gamma and interleukin-4 production in CD4± T cells in patients with chronic heart failure. *Heart Vessels* **2007**, *22*, 178–183. [CrossRef]
58. Baschant, U.; Tuckermann, J. The role of the glucocorticoid receptor in inflammation and immunity. *J. Steroid Biochem. Mol. Biol.* **2010**, *120*, 69–75. [CrossRef]
59. Zhang, Y.; Leung, D.Y.; Goleva, E. Vitamin D enhances glucocorticoid action in human monocytes: Involvement of granulocyte-macrophage colony-stimulating factor and mediator complex subunit 14. *J. Biol. Chem.* **2013**, *288*, 14544–14553. [CrossRef]
60. Feng, W.; Li, W.; Liu, W.; Wang, F.; Li, Y.; Yan, W. IL-17 induces myocardial fibrosis and enhances RANKL/OPG and MMP/TIMP signaling in isoproterenol-induced heart failure. *Exp. Mol. Pathol.* **2009**, *87*, 212–218. [CrossRef]
61. Cheng, S.; Massaro, J.M.; Fox, C.S.; Larson, M.G.; Keyes, M.J.; McCabe, E.L.; Robins, S.J.; O'Donnell, C.J.; Hoffmann, U.; Jacques, P.F.; et al. Adiposity, cardiometabolic risk, and vitamin D status: The Framingham Heart Study. *Diabetes* **2010**, *59*, 242–248. [CrossRef]
62. Pantovic, A.; Zec, M.; Zekovic, M.; Obrenovic, R.; Stankovic, S.; Glibetic, M. Vitamin D Is Inversely Related to Obesity: Cross-Sectional Study in a Small Cohort of Serbian Adults. *J. Am. Coll. Nutr.* **2019**, *38*, 405–414. [CrossRef] [PubMed]
63. Gu, J.K.; Charles, L.E.; Millen, A.E.; Violanti, J.M.; Ma, C.C.; Jenkins, E.; Andrew, M.E. Associations between adiposity measures and 25-hydroxyvitamin D among police officers. *Am. J. Hum Biol.* **2019**, *31*, e23274. [CrossRef] [PubMed]
64. Hannemann, A.; Thuesen, B.H.; Friedrich, N.; Völzke, H.; Steveling, A.; Ittermann, T.; Hegenscheid, K.; Nauck, M.; Linneberg, A.; Wallaschofski, H. Adiposity measures and vitamin D concentrations in Northeast Germany and Denmark. *Nutr. Metab. (Lond.)* **2015**, *12*, 24. [CrossRef] [PubMed]
65. Price, R.J.; Witham, M.D.; McMurdo, M.E. Defining the nutritional status and dietary intake of older heart failure patients. *Eur. J. Cardiovasc. Nurs.* **2007**, *6*, 178–183. [CrossRef] [PubMed]

66. Loncar, G.; Bozic, B.; von Haehling, S.; Düngen, H.D.; Prodanovic, N.; Lainscak, M.; Arandjelovic, A.; Dimkovic, S.; Radojicic, Z.; Popovic, V. Association of adiponectin with peripheral muscle status in elderly patients with heart failure. *Eur. J. Intern. Med.* **2013**, *24*, 818–823. [CrossRef]
67. DiCarlo, C.; Schmotzer, B.; Vest, M.; Boxer, R. Body mass index and 25 hydroxyvitamin D status in patients with and without heart failure. *Congest. Heart Fail.* **2012**, *18*, 133–137. [CrossRef]
68. Robbins, J.; Petrone, A.B.; Gaziano, J.M.; Djoussé, L. Dietary vitamin D and risk of heart failure in the Physicians' Health Study. *Clin. Nutr.* **2016**, *35*, 650–653. [CrossRef]
69. National Academy of Sciences. *Dietary Reference Intakes for Adequacy: Calcium and Vitamin D*; Institute of Medicine of the National Academies: Washington, DC, USA, 2011. Available online: https://www.ncbi.nlm.nih.gov/books/NBK56070/pdf/Bookshelf_NBK56070.pdf (accessed on 12 September 2019).
70. National Institutes of Health; ODS. *Vitamin D: Fact Sheet for Health Professionals*; Office of Dietary Supplements, National Institutes of Health: Bethesda, MD, USA, 2019. Available online: https://ods.od.nih.gov/factsheets/VitaminD-HealthProfessional/ (accessed on 12 September 2019).
71. U.S. Department of Health & Human Services. *USDA National Nutrient Database for Standard Reference Release 28*; National Institutes of Health, Office of Dietary Supplements: Bethesda, MD, USA, 2015. Available online: https://ods.od.nih.gov/pubs/usdandb/VitaminD-Content.pdf (accessed on 13 September 2019).
72. Fitzpatrick, T.B. The validity and practicality of sun-reactive skin types I through VI. *Arch. Dermatol.* **1988**, *124*, 869–871. [CrossRef]
73. Centers for Disease Control and Prevention. *National Health and Nutrition Examination Survey (NHANES): Anthropometry Procedures Manual*; National Center for Health Statistics: Hyattsville, MD, USA, 2017. Available online: https://wwwn.cdc.gov/nchs/data/nhanes/2017-2018/manuals/2017_Anthropometry_Procedures_Manual.pdf (accessed on 12 September 2019).
74. Krishnaveni, P.; Gowda, V.M.N. Assessing the Validity of Friedewald's Formula and Anandraja's Formula For Serum LDL-Cholesterol Calculation. *J. Clin. Diagn. Res.* **2015**, *9*, BC01–BC04. [CrossRef]
75. Maechler, M.; Rousseeuw, P.; Struyf, A.; Hubert, M.; Hornik, K. *Cluster: Cluster Analysis Basics and Extensions*, R package version 2.0.7-1; The R Foundation: Vienna, Austria, 2018.
76. Hothorn, T.; Hornik, K.; Zeileis, A. Unbiased Recursive Partitioning: A Conditional Inference Framework. *J. Comput. Graph. Stat.* **2006**, *15*, 651–674. [CrossRef]

International Journal of
Molecular Sciences

Article

Vitamin D Attenuates Loss of Endothelial Biomarker Expression in Cardio-Endothelial Cells

Chi-Cheng Lai [1,2,3,†], Wang-Chuan Juang [4,5,†], Gwo-Ching Sun [6,7], Yu-Kai Tseng [8,9], Rong-Chang Jhong [10], Ching-Jiunn Tseng [10,11], Tzyy-Yue Wong [10,12,*] and Pei-Wen Cheng [10,11,*]

1 Cardiovascular Center, Kaohsiung Veterans General Hospital, Kaohsiung 81362, Taiwan; llccheng@gmail.com
2 Department of Cardiology, Kaohsiung Municipal United Hospital, Kaohsiung 80457, Taiwan
3 Department of Pharmacy and Graduate Institute of Pharmaceutical Technology, Tajen University, Pingtung 90741, Taiwan
4 Department of Emergency, Kaohsiung Veterans General Hospital, Kaohsiung 81362, Taiwan; wcchuang@vghks.gov.tw
5 Department of Business Management, National Sun Yat-Sen University, Kaohsiung 80424, Taiwan
6 Department of Anesthesiology, Kaohsiung Medical University Hospital, Kaohsiung 80756, Taiwan; gcsun39@yahoo.com.tw
7 Faculty of Medicine, College of Medicine, Kaohsiung Medical University, Kaohsiung 80708, Taiwan
8 Department of Medical Research, Chang Bing Show Chwan Health Memorial Hospital, Changhua 50544, Taiwan; iamtsengyukai@gmail.com
9 Department of Orthopedics, Show Chwan Memorial Hospital, Changhua 50008, Taiwan
10 Department of Medical Education and Research, Kaohsiung Veterans General Hospital, Kaohsiung 81362, Taiwan; jrchang@vghks.gov.tw (R.-C.J.); cjtseng@vghks.gov.tw (C.-J.T.)
11 Department of Biomedical Science, National Sun Yat-Sen University, Kaohsiung 80424, Taiwan
12 International Center for Wound Repair and Regeneration, National Cheng Kung University, Tainan 70101, Taiwan
* Correspondence: wongtzyyyue@gmail.com (T.-Y.W.); pwcheng@vghks.gov.tw (P.-W.C.); Tel.: +886-04-725-6166 (T.-Y.W.); +886-7-3422121 (ext. 1593) (P.-W.C.)
† These authors contributed equally to this work.

Received: 3 February 2020; Accepted: 20 March 2020; Published: 22 March 2020

Abstract: Vitamin D is associated with cardiovascular health through activating the vitamin D receptor that targets genes related to cardiovascular disease (CVD). The human cardiac microvascular endothelial cells (HCMECs) were used to develop mechanically and TGF-β1-induced fibrosis models, and the rat was used as the isoproterenol (ISO)-induced fibrosis model. The rats were injected with ISO for the first five days, followed by vitamin D injection for the consecutive three weeks before being sacrificed on the fourth week. Results showed that mechanical stretching reduced endothelial cell marker CD31 and VE-cadherin protein expressions, as well as increased α-smooth muscle actin (α-SMA) and fibronectin (FN). The transforming growth factor-β1 (TGF-β1) reduced CD31, and increased α-SMA and FN protein expression levels. Vitamin D presence led to higher protein expression of CD31, and lower protein expressions of α-SMA and FN compared to the control in the TGF-β1-induced fibrosis model. Additionally, protein expression of VE-cadherin was increased and fibroblast-specific protein-1 (FSP1) was decreased after vitamin D treatment in the ISO-induced fibrosis rat. In conclusion, vitamin D slightly inhibited fibrosis development in cell and animal models. Based on this study, the beneficial effect of vitamin D may be insignificant; however, further investigation of vitamin D's effect in the long-term is required in the future.

Keywords: cardiovascular disease; vitamin D; stretching; TGF-β1; fibrosis

1. Introduction

Acute cardiomyopathy patients are commonly found to have low vitamin D [1], which has been associated with increased mortality in critically ill patients [2]. Two key enzymes, 25-hydroxylase in the liver and 1α-hydroxylase in the kidney, are essential to the activation of vitamin D. The administration of active vitamin D has been reported to confer cardiovascular protection and increased survival in clinical [3] and experimental researches [4,5]. Paricalcitol (25-dihydroxyvitamin D2), a vitamin D receptor (VDR) activator, is reported to improve cardiovascular health or survival in patients with advanced kidney disease who tend to have low vitamin D [6,7]. Deficiency in 25-hydroxyvitamin D ($1{,}25(OH)_2D$, or vitamin D) has been associated with cardiovascular disease (CVD) risk factors, such as age, high blood pressure, obesity and diabetes [8]. The association of vitamin D with CVD risk factors have been attributed to hypertension, congestive heart failure, myocardial infarction and stroke [9]. However, the mechanism for CVD involving vitamin D is not clear.

The risk factors for CVD vary differently, therefore vitamin D's effect is observed from different aspects. For example, atherosclerosis accumulates plaque of fats in the blood vessels, blocking the blood flow. This blockade hinders normal blood flow, increasing blood pressure as the blood pushes through the thickening walls, leading to hypertension development as the blockade continues. As atherosclerosis progresses, the effect of pressure built on the thickened walls of the blood vessels leads to extracellular matrix (ECM) secretion, such as fibronectin, for remodeling [10,11]. ECM accumulation gradually leads to fibrosis in the vascular cells. Vitamin D is involved in the renin–angiotensin II system, which regulates blood pressure through vasoconstriction when blood pressure is increased [12]. Vitamin D deficiency is also linked to fibrosis [13–16] and immune T-cell activation [17,18]. In addition, vitamin D prevents liver fibrosis through inhibition of fibrosis markers TGF-β1, collagen I and α-SMA [19]. Therefore, our previous studies demonstrated that paricalcitol ameliorated isoproterenol (ISO)-induced cardiac dysfunction and fibrosis via regulation of endothelial-to-mesenchymal cell transition (EndoMT) [20]. Vitamin D plays a role in modulating fibrosis; however, the role of vitamin D in fibrosis of cardiovascular cells is not clear.

Vitamin D binds to its receptor with high affinity, the vitamin D receptor (VDR), which belongs to the nuclear receptor family [21] and is expressed in cardiovascular cells, including vascular smooth muscle cells, endothelial cells, cardiomyocytes and immune cells [22]. The VDR is activated by VDR activators, such as calcitriol, paricalcitol and alfacalcidol, to prevent cardiac fibrosis through inhibition of left ventricular hypertrophy [23]. In this study, we proposed that vitamin D is responsible for suppressing fibrosis in cardiovascular cells.

2. Results

2.1. *Effect of Vitamin D on the Mechanically Induced Fibrosis Model*

Increased blood pressure modulates vascular cell remodeling, which consequently affects cellular function and responses. To correctly determine the impact of mechanical stress on human cardiac microvascular endothelial cells (HCMECs), the HCMECs were mechanically stretched at 15% elongation with a 1 Hz frequency for 6 h. The HCMECs orientated and aligned perpendicularly to the direction of the uniaxial force under the given mechanical strain (Figure 1A,B). The immunofluorescence assay showed that 6h stretching increased α-SMA and α-tubulin (Figure 1C–F). Mechanical stretching significantly decreased CD31 and increased α-SMA protein levels (Figure 1G–J). Vitamin D was administered to the dynamic culture to study its effect on fibrogenesis. Mechanical stimulation reduced endothelial marker VE-Cad, and vitamin D did not restore its protein expression (Figure 2A,B). However, the release of TGF-β1 was decreased in the presence of vitamin D compared to the control (Figure 2C). In addition, vitamin D slightly attenuated the decrease of VE-cad compared to the control; the increase of fibronectin (FN) in the presence of vitamin D might be due to an onset of remodeling as pressure is built up by mechanical stretching (Figure 2D). Furthermore, the vitamin D receptor VD3R was increased in vitamin D presence; the Akt pathway that is associated with VD3R activation was not affected when vitamin D was present (Figure 2E).

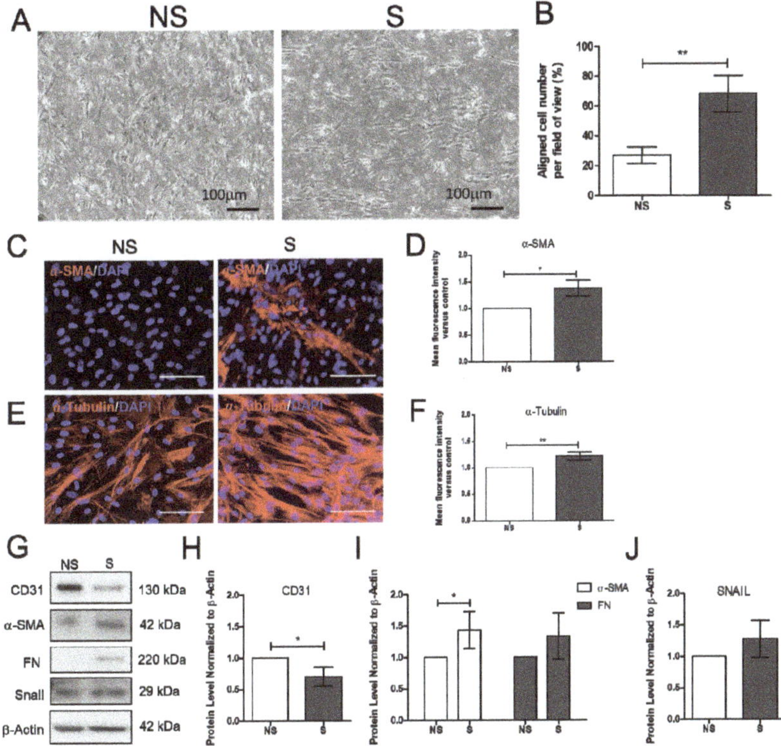

Figure 1. Mechanical stretching reduced CD31, but increased α-smooth muscle actin and fibronectin. (**A**,**B**) Cells orientated perpendicular to the direction of uniaxial stretching after 6 h of stretching. (**C**–**F**) Expression of α-smooth muscle actin (α-SMA), and α-tubulin analyzed by immunofluorescence assay. Scale bar = 100 μm. (**G**–**J**) Expression of endothelial cell marker CD31, fibrosis markers α-SMA and FN and the endothelial-to-mesenchymal transition marker Snail. Data are represented as at least three independent experiments. ** $p < 0.01$; * $p < 0.05$.

2.2. *Effect of Vitamin D on the TGF-β1-Induced Fibrosis Model*

To further determine the effect of vitamin D on fibrosis, the TGF-β1-induced fibrosis model was established. The HCMECs were induced with 5 ng/mL TGF-β1 for 24 h. Results showed that CD31 was significantly decreased (Figure 3A,B), but Snail significantly increased at 5 ng/mL TGF-β1, 24 h TGF-β1 treatment (Figure 3C,D). The HCMECs were further induced with 5 ng/mL TGF-β1 for 2 and 5 days because the expression of the fibrosis marker was not significantly increased when compared to the control in the 24 h TGF-β1 treatment. α-SMA and FN were significantly increased in a time-dependent manner at 5 ng/mL after 5 days of TGF-β1 treatment (Figure 3A,B). In addition, Snail was significantly increased at 5 ng/mL TGF-β1 on Day 5 (Figure 3C,D). The immunofluorescence assay showed that vitamin D slightly increased CD31 expression compared to the TGF-β1-treated group (Figure 4A,B). In the presence of vitamin D, the CD 31 decrease was not attenuated, FN and α-SMA were not suppressed (Figure 4C,D). Furthermore, the vitamin D receptors VD3R, VEGFR and TGFβR1, receptors which have been reported to interact with vitamin D, were increased in the presence of TGF-β1 and vitamin D (Figure 4E,F).

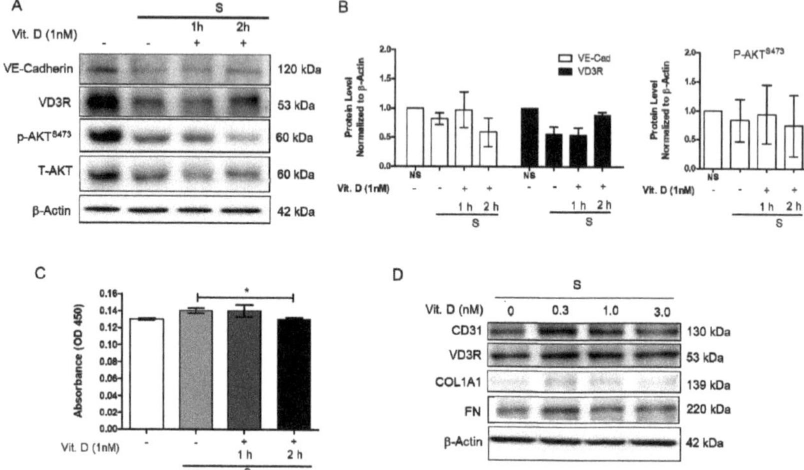

Figure 2. Vitamin D attenuated decrease of endothelial cell marker vascular E-cadherin protein expression. (**A**,**B**) Protein expression of endothelial marker VE-cadherin, VD3R, p-Akt^{S473} and T-Akt. (**C**) The release of TGF-β1 measured by ELISA in presence of vitamin D at 1 nM. (**D**) Protein expression of CD31, vitamin D3 receptor VD3R and fibrosis marker COL1A1, as well as fibronectin FN when 1nM vitamin D was added during the stretching. Data are represented as at least three independent experiments. * $p < 0.05$.

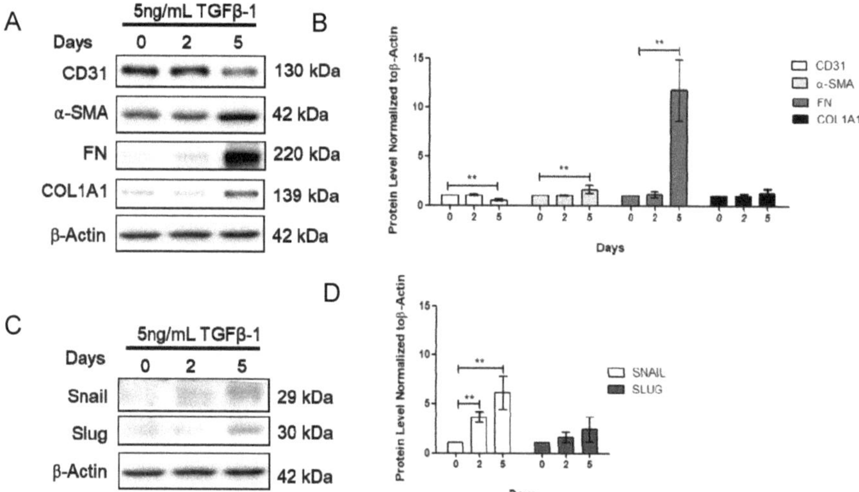

Figure 3. Transforming growth factor β1 reduced CD31 but increases α-smooth muscle actin and fibronectin. (**A**,**B**) Expression of CD31, α-SMA, FN and COL1A1 after 0, 2, and 5 days of TGF-β1 treatment. (**C**,**D**) Expression of Snail and Slug after 0, 2, and 5 days of TGF-β1 treatment. Data are represented as at least three independent experiments. ** $p < 0.01$; * $p < 0.05$.

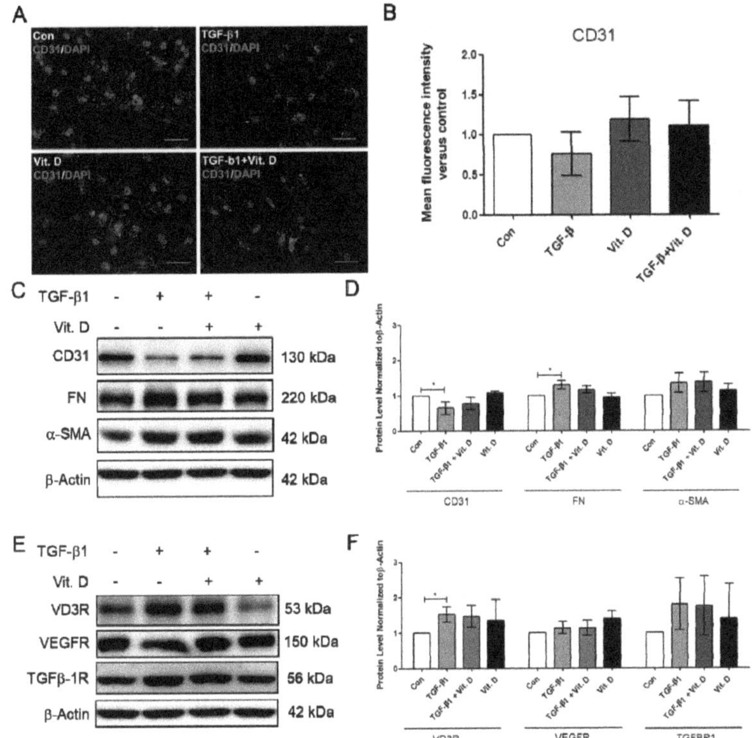

Figure 4. Vitamin D slightly attenuated decrease of endothelial cell marker CD31. (**A**,**B**) Immunofluorescence assay for CD31 in the TGF-β1-induced fibrosis. Scale bar = 100 μm. (**C**,**D**) Protein expression of CD31, FN, and α-SMA in presence of vitamin D at 1nM. (**E**,**F**) Protein expression of VD3R, growth factor receptors VEGFR and TGF-β1R. Data are represented as at least three independent experiments. * $p < 0.05$.

2.3. Vitamin D Slightly Attenuated Fibrosis Biomarker in an ISO-Induced Fibrosis Model

Further investigation of vitamin D's effect on CVD led to the analysis of myocardial heart tissue. Results showed that the myocardial infarction heart expressed the fibrosis marker FSP1, which was reduced when vitamin D was administered (Figure 5A,B). The effects of vitamin D are summarized in Table 1, thereby confirming the fundamental effect of vitamin D on fibrogenesis in cardiovascular cells.

Table 1. Summary of vitamin D's effect in the mechanically, chemically and isoproterenol-induced cardiovascular fibrosis models.

Approaches	**+S**	**+ TGFβ-1**	**+ISO**
Dynamics	Cyclic stretching	Static	Rat, in vivo
Duration for induction	6 Hours	5 Days	4 Weeks
Fibrosis markers	FN, α-SMA	FN, α-SMA	FSP-1
Vitamin D effect	VE-Cad decrease attenuated	CD31 decrease attenuated	FSP-1 decreased

+S: Induction of fibrosis markers by mechanical stimulation, stretching. +TGF-β1: Induction of fibrosis markers by adding TGF-β1. +ISO: Induction of fibrosis markers by isoproterenol injection into rat, a myocardial infarction model.

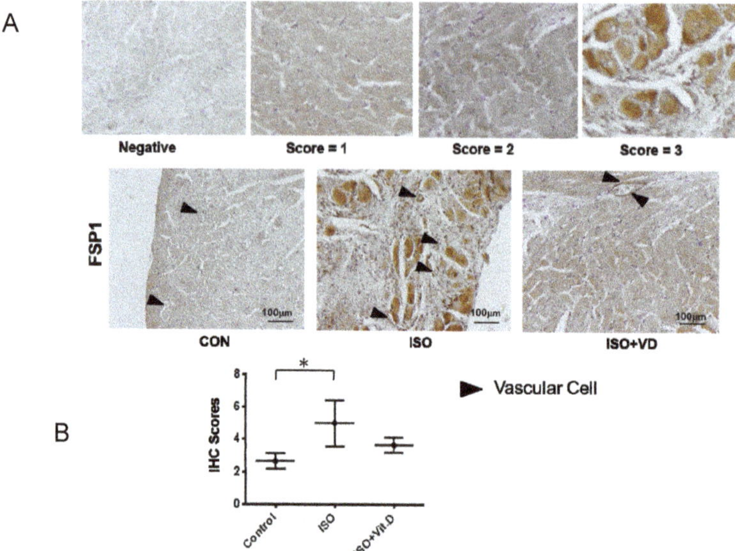

Figure 5. Vitamin D attenuated fibronectin specific protein-1 protein expression in fibrosis animal model. (**A**,**B**) Immunohistological analysis for fibronectin-specific protein-1 (FSP1) in the heart section of ISO-induced rats. For scoring the FSP1 expression, the score was divided into 0, 1, 2 and 3 from negative to positive expressions. The black arrow showed the blood vessels. Scale bar = 100 μm. Data are represented as at least three independent experiments. * $p < 0.05$.

3. Discussions

Fibrosis is a state when tissue become fibrous as myofibroblasts accumulate excess ECM components, such as fibronectin, α-smooth muscle actin (α-SMA) and collagen I. Mechanistically, fibroblasts secrete the transforming growth factor-β1 (TGF-β1) to promote myofibroblast differentiation [24,25]. Apart from activation of fibroblast and epithelial-to-mesenchymal transition (EMT), endothelial-to-mesenchymal transition (EndoMT) is also a source of myofibroblasts [26]. The secreted TGF-β1 remains inactive in the ECM until being activated by mechanical cues in the ECM, such as the contraction force of a myofibroblast [27–29]. The mechanical cues can alter ECM arrangement [30], cell adhesion [31], morphology, intra-cellular cytoskeletal organization [32], gene expression, and so forth. The accumulating ECM components pose a strain on the cardiac blood vessels that lead to pressure overload. Moreover, the contraction and relaxation of the heart exert pressure on endothelial cells at the luminal surface, and a pressure overload can lead to cardiac fibrosis [33].

The role of vitamin D on fibrosis is observed to be beneficial; however, statistical data showed no significance. The reason might be that vitamin D affects the CVD-related symptoms, in particular, hypertension, which is discussed in this study. Based on our observations, the endothelial cell markers CD31 and VE-Cad were slightly increased when vitamin D is present (Figures 1, 3 and 5). Consistent with the data on fibrosis markers was that vitamin D slightly suppressed the increase of α-SMA and FN in mechanically and TGF-β-induced models, as well as FSP1 in a rat model (Figures 1, 3 and 5).

Recently, vascular cells were shown to require the mechanically activated ion channel to maintain endothelial cell function and vascular architecture in hypertension-associated vascular remodeling [34,35]. A breakthrough finding in 2007 showed that 27–33% of total cardiac fibroblasts are of endothelial origin in the heart failure murine model [36]. Evidence has shown that EndMT contributes to cardiac fibrosis in streptozotocin-induced diabetes mice with cells expressing the α-SMA and FSP1 [37]. As observed from our data, the fibrosis markers expressed in HCMEC were α-SMA and FN, and FSP1 was expressed in the in vivo model (Figures 1G, 3E and 5D). The effect of vitamin D on the fibrosis marker increase may be resulted from the progressive state of the fibrosis. Overall, vitamin D is observed to suppress fibrosis. The accumulation of ECM during hypertension activates the remodeling process in vascular smooth muscle cells to allow blood vessel to withstand the increased blood pressure. At the moment of blood-vessel wall remodeling, the effect of vitamin D on the cardiovascular cells is not yet clarified. In our data, the vitamin D receptor VD3R was increased in presence of vitamin D (Figure 2C,D and Figure 4E,F). In addition, the VD3R increase was not significant, indicating that the vitamin D receptor was not activated.

The limitation of this study is that the signaling pathway of vitamin D has not been examined due to the insignificance of vitamin D's effect on cardiovascular cell fibrosis. However, the effect of vitamin D on our health remains an interesting topic, and further investigating how vitamin D is involved in cardiac fibrogenesis or before the onset of fibrogenesis would help us understand the mechanism of vitamin D on CVD.

4. Materials and Methods

4.1. Cell Culture

The cells HCMECs were purchased from ScienCell Research Laboratories, Inc. (Catalogue #6000, Carlsbad, CA, USA). The HCMECs were maintained in endothelial cell medium (Catalogue #1001; ScienCell Research Laboratories, Inc., Carlsbad, CA, USA), supplemented with 10% fetal bovine serum (Catalogue #0025; ScienCell Research Laboratories, Inc., Carlsbad, CA, USA), 100 IU/mL penicillin and 100 µg/mL streptomycin. The HCMECs were incubated at 37 °C, with 5% CO_2, passaged every four days, and passages from 1 to 7 were used for this study.

4.2. Chemicals

Recombinant human TGF-β1 (GFH39; CellGS) was prepared in double-distilled water at stock concentration of 0.1 mg/mL. Paricalcitol or vitamin D (Zemplar) was purchased from the pharmacy counter as a 5 µg/mL stock solution containing propylene glycol (30% *v*/*v*) and ethanol (20% *v*/*v*), and was diluted in a medium to a final concentration of 1 nM.

4.3. In Vitro Stretching Device

The stretching device (ATMS Boxer) was provided by the manufacturer TAIHOYA (Kaohsiung, Taiwan) for this study. The cells were cultured on polydimethylsiloxane (PDMS) membrane coated with collagen I for one day before being subjected to stretching. The PDMS with the attached cells were mounted to the stretch device using clamps that were fixed on both ends of the PDMS membrane. The cells were stretched to 15% elongation, with a frequency of 1 Hz for 6 h.

4.4. Immunofluorescence Microscopy

Cells were fixed with 4% paraformaldehyde for 20 min, washed with PBS, and then incubated with Triton X-100 for 10 min (0.5% *v*:*v*). Blocking was done by incubating with 5% (*w*:*v*) bovine serum albumin (BSA) for 30 min, washed with PBS, and then incubated with primary antibody overnight at 4 °C. Anti-α-SMA (Abcam, ab5694, 1:100, Cambridge, UK), anti-CD31 (Santa Cruz, SC1506, 1:100, Dallas, TX, USA) and vE-cadherin (Arigo-ARG11036, 1:100, Hsinchu, Taiwan) primary antibodies were used. After binding to the primary antibodies, cells were further incubated for 1h with Alexa Fluor-conjugated anti-rabbit and anti-mouse (Jackson Immunoresearch, 1:500, West Grove, PA, USA). Finally, cells were stained and mounted using the Prolong®Diamond Antifade Mounting Medium containing DAPI (Life Technology, Carlsbad, CA, USA). Cells were analyzed microscopically with an Olympus DP71 device by magnification of 100 times and 200 times, or the Re-scan confocal microscope (Confocal.nl, Amsterdam, The Netherlands) at 600 times magnification.

4.5. Western Blot

To collect cell lysates, cells were lysed in lysis buffer, and protein concentration was quantified by the Bio-Rad assay kit (Bio-Rad Laboratories, Hercules, CA, USA). The protein was separated by sodium dodecyl sulfate-polyacrylamide gel electrophoresis. After gel electrophoresis, separated proteins were transferred to the PVDF membrane and immunoblotted with primary antibodies: Collagen I GeneTex, GTX26308, CD31 Santa Cruz, SC1506, fibronectin Abcam, ab2413, Snail Cellsignal, #3879, Slug Abcam, ab27568, vE-cadherin Arigo-ARG11036, VDR3 Cellsignal, #12550s, TGF-β1R Abcam, ab31013, VEGFR Abcam, ab32152, p-Akt^{S473} Cellsignal, #3879, Akt Cellsignal, #9276s and β-actin. The signals were visualized by enhanced chemiluminescence using ECL (ThermoFisher Scientific, Waltham, MA, USA).

4.6. Immunohistochemistry Assay

The paraffin-embedded sections of heart were cut at 5 μm, followed by dewaxing, and then quenching in 3% hydrogen peroxide methanol solution. Antigen retrieval in citric buffer (10 mM, pH 6.0) was blocked with 5% goat serum. The sections were further incubated in a solution of primary antibody FSP-1 (1:100; Arigo-ARG55205, Hsinchu, Taiwan) for 4 °C overnight. The next day, the sections were incubated in biotinylated secondary antibody (1:200; Vector Laboratories, Burlingame, CA, USA) for 1 h. For visualization, the tissue was incubated in DAB (Vector Laboratories, Burlingame, CA, USA) solution, counterstained with hematoxylin.

4.7. ELISA

The secreted TGF-β1 was measured by the ELISA kit from Cloud-Clone Corporation (Katy, TX, USA) according to the manufacturer's instruction. Briefly, cell culture supernatant was collected from cells subjected to stretching for 6 h. The readings were measured by a microplate reader at 450 nm.

4.8. Animal

To analyze the effect of vitamin D on fibrosis, a rat model was established by isoproterenol (ISO) induction. The animal preparation is approved by the Kaohsiung Veterans General Hospital institutional review board, approved serial number was VGHKS103-029, from January 2015 until December 2015, following standard guidelines for animal care, and is based on a previous method [20]. In brief, 8-week-old male Wistar–Kyoto (WKY) rats, which weighed between 250 and 350 g, were obtained from the National Science Council Animal Facility (Taipei, Taiwan), housed in the animal center at Kaohsiung Veterans General Hospital, and were provided normal rat chow (Purina, St. Louis, MO, USA) and tap water ad libitum. The control rats were given normal saline with 0.1% ascorbic acid. The ISO rats were given 2 mg/kg per day once every day for 5 days, and ISO with 200 ng vitamin D or paricalcitol thrice in a week for 3 weeks. The control, ISO and ISO + vitamin D rats were monitored for tail systolic blood pressure and echocardiography. The rats were sacrificed after 4 weeks; their hearts were extracted and sectioned for immunohistochemistry analysis.

4.9. Statistical Analysis

All measurements were produced at least three times under independent conditions. The results are shown as the mean ± standard error of the mean (SEM). A two-sided Student's t test was used to compare the mean values obtained from two independent conditions: * $p < 0.05$ indicates a significant result; ** $p < 0.01$ indicates a very significant result; and *** $p < 0.05$ indicates a highly significant result.

5. Conclusions

In conclusion, vitamin D suppressed the decrease of CD31 protein expression in the TGF-β-induced fibrosis models, and slightly suppressed the decrease of VE-Cad protein expression in the mechanically and ISO-induced fibrosis models. Our data showed that vitamin D is beneficial for suppressing fibrogenesis. Therefore, vitamin D can be a long-term supplement for individuals having high CVD risk.

Author Contributions: Conceptualization of this study was contributed by G.-C.S., Y.-K.T., W.-C.J. and C.-C.L.; data organization, writing and editing were performed by T.-Y.W.; data collection was performed by R.-C.J. and T.-Y.W.; supervision and interpretation were contributed by C.-J.T. and P.-W.C. All authors have read and agreed to the published version of the manuscript.

Funding: This work was supported by funding from the National Science Council (MOST108-2320-B-075B-002, MOST107-2320-B-075B-002-MY2) and Kaohsiung Veterans General Hospital (VGHKS108-195, VGHKS109-158).

Conflicts of Interest: The authors of this study declared no conflicts of interest.

References

1. Ng, L.L.; Sandhu, J.K.; Squire, I.B.; Davies, J.E.; Jones, D.J. Vitamin D and prognosis in acute myocardial infarction. *Int. J. Cardiol.* **2013**, *168*, 2341–2346. [CrossRef]
2. Quraishi, S.A.; Bittner, E.A.; Blum, L.; McCarthy, C.M.; Bhan, I.; Camargo, C.A., Jr. Prospective study of vitamin D status at initiation of care in critically ill surgical patients and risk of 90-day mortality. *Crit. Care. Med.* **2014**, *42*, 1365–1371. [CrossRef] [PubMed]
3. Tamez, H.; Zoccali, C.; Packham, D.; Wenger, J.; Bhan, I.; Appelbaum, E.; Pritchett, Y.; Chang, Y.; Agarwal, R.; Wanner, C.; et al. Vitamin D reduces left atrial volume in patients with left ventricular hypertrophy and chronic kidney disease. *Am. Heart J.* **2012**, *164*, 902–909. [CrossRef] [PubMed]
4. Meems, L.M.; Cannon, M.V.; Mahmud, H.; Voors, A.A.; van Gilst, W.H.; Sillje, H.H.; Ruifrok, W.P.; de Boer, R.A. The vitamin D receptor activator paricalcitol prevents fibrosis and diastolic dysfunction in a murine model of pressure overload. *J. Steroid. Biochem. Mol. Biol.* **2012**, *132*, 282–289. [CrossRef] [PubMed]
5. Bae, S.; Yalamarti, B.; Ke, Q.; Choudhury, S.; Yu, H.; Karumanchi, S.A.; Kroeger, P.; Thadhani, R.; Kang, P.M. Preventing progression of cardiac hypertrophy and development of heart failure by paricalcitol therapy in rats. *Cardiovasc. Res.* **2011**, *91*, 632–639. [CrossRef] [PubMed]
6. Teng, M.; Wolf, M.; Lowrie, E.; Ofsthun, N.; Lazarus, J.M.; Thadhani, R. Survival of patients undergoing hemodialysis with paricalcitol or calcitriol therapy. *N. Engl. J. Med.* **2003**, *349*, 446–456. [CrossRef] [PubMed]
7. Thadhani, R.; Appelbaum, E.; Chang, Y.; Pritchett, Y.; Bhan, I.; Agarwal, R.; Zoccali, C.; Wanner, C.; Lloyd-Jones, D.; Cannata, J.; et al. Vitamin D receptor activation and left ventricular hypertrophy in advanced kidney disease. *Am. J. Nephrol.* **2011**, *33*, 139–149. [CrossRef] [PubMed]
8. Gouni-Berthold, I.; Krone, W.; Berthold, H.K. Vitamin D and cardiovascular disease. *Current. Vascular. Pharmacol.* **2009**, *7*, 414–422. [CrossRef]
9. Danik, J.S.; Manson, J.E. Vitamin d and cardiovascular disease. *Curr. Treat. Options Cardiovasc. Med.* **2012**, *14*, 414–424. [CrossRef]
10. Harvey, A.; Montezano, A.C.; Lopes, R.A.; Rios, F.; Touyz, R.M. Vascular Fibrosis in Aging and Hypertension: Molecular Mechanisms and Clinical Implications. *Can. J. Cardiol.* **2016**, *32*, 659–668. [CrossRef]
11. Lan, T.H.; Huang, X.Q.; Tan, H.M. Vascular fibrosis in atherosclerosis. *Cardiovasc. Pathol. Off. J. Soc. Cardiovasc. Pathol.* **2013**, *22*, 401–407. [CrossRef] [PubMed]
12. Selvin, E.; Najjar, S.S.; Cornish, T.C.; Halushka, M.K. A comprehensive histopathological evaluation of vascular medial fibrosis: Insights into the pathophysiology of arterial stiffening. *Atherosclerosis* **2010**, *208*, 69–74. [CrossRef] [PubMed]
13. Shi, Y.; Liu, T.; Yao, L.; Xing, Y.; Zhao, X.; Fu, J.; Xue, X. Chronic vitamin D deficiency induces lung fibrosis through activation of the renin-angiotensin system. *Sci. Rep.* **2017**, *7*, 3312. [CrossRef] [PubMed]
14. Repo, J.M.; Rantala, I.S.; Honkanen, T.T.; Mustonen, J.T.; Koobi, P.; Tahvanainen, A.M.; Niemela, O.J.; Tikkanen, I.; Rysa, J.M.; Ruskoaho, H.J.; et al. Paricalcitol aggravates perivascular fibrosis in rats with renal insufficiency and low calcitriol. *Kidney Int.* **2007**, *72*, 977–984. [CrossRef] [PubMed]
15. Tan, X.; Li, Y.; Liu, Y. Paricalcitol attenuates renal interstitial fibrosis in obstructive nephropathy. *J. Am. Soc. Nephrol. JASN* **2006**, *17*, 3382–3393. [CrossRef]
16. Park, J.W.; Bae, E.H.; Kim, I.J.; Ma, S.K.; Choi, C.; Lee, J.; Kim, S.W. Renoprotective effects of paricalcitol on gentamicin-induced kidney injury in rats. *Am. J. Physiol. -Ren. Physiol.* **2009**, *298*, 301–313. [CrossRef]
17. Pincikova, T.; Paquin-Proulx, D.; Sandberg, J.K.; Flodström-Tullberg, M.; Hjelte, L. Vitamin D treatment modulates immune activation in cystic fibrosis. *Clin. Exp. Immunol.* **2017**, *189*, 359–371. [CrossRef]

18. González-Mateo, G.T.; Fernández-Míllara, V.; Bellón, T.; Liappas, G.; Ruiz-Ortega, M.; López-Cabrera, M.; Selgas, R.; Aroeira, L.S. Paricalcitol Reduces Peritoneal Fibrosis in Mice through the Activation of Regulatory T Cells and Reduction in IL-17 Production. *PLoS ONE* **2014**, *9*, e108477. [CrossRef]
19. Abramovitch, S.; Sharvit, E.; Weisman, Y.; Bentov, A.; Brazowski, E.; Cohen, G.; Volovelsky, O.; Reif, S. Vitamin D inhibits development of liver fibrosis in an animal model but cannot ameliorate established cirrhosis. *Am. J. Physiol. -Gastrointest. Liver Physiol.* **2015**, *308*, 112–120. [CrossRef]
20. Lai, C.C.; Liu, C.P.; Cheng, P.W.; Lu, P.J.; Hsiao, M.; Lu, W.H.; Sun, G.C.; Liou, J.C.; Tseng, C.J. Paricalcitol Attenuates Cardiac Fibrosis and Expression of Endothelial Cell Transition Markers in Isoproterenol-Induced Cardiomyopathic Rats. *Crit. Care Med.* **2016**, *44*, 866–874. [CrossRef]
21. Norman, P.E.; Powell, J.T. Vitamin D and Cardiovascular Disease. *Circ. Res.* **2014**, *114*, 379. [CrossRef] [PubMed]
22. Ni, W.; Watts, S.W.; Ng, M.; Chen, S.C.; Glenn, D.J.; Gardner, D.G. Elimination of vitamin D receptor in vascular endothelial cells alters vascular function. *Hypertension* **2014**, *64*, 1290–1298. [CrossRef] [PubMed]
23. Panizo, S.; Barrio-Vazquez, S.; Naves-Diaz, M.; Carrillo-Lopez, N.; Rodriguez, I.; Fernandez-Vazquez, A.; Valdivielso, J.M.; Thadhani, R.; Cannata-Andia, J.B. Vitamin D receptor activation, left ventricular hypertrophy and myocardial fibrosis. *Nephrol. Dial. Transplant.: Off. Publ. Eur. Dial. Transpl. Assoc. - Eur. Ren. Assoc.* **2013**, *28*, 2735–2744. [CrossRef] [PubMed]
24. Meredith, A.; Boroomand, S.; Carthy, J.; Luo, Z.; McManus, B. 1,25 Dihydroxyvitamin D3 Inhibits TGFbeta1-Mediated Primary Human Cardiac Myofibroblast Activation. *PLoS ONE* **2015**, *10*, e0128655. [CrossRef]
25. Kasabova, M.; Joulin-Giet, A.; Lecaille, F.; Gilmore, B.F.; Marchand-Adam, S.; Saidi, A.; Lalmanach, G. Regulation of TGF-beta1-driven differentiation of human lung fibroblasts: Emerging roles of cathepsin B and cystatin C. *J. Biol. Chem.* **2014**, *289*, 16239–16251. [CrossRef]
26. Piera-Velazquez, S.; Li, Z.; Jimenez, S.A. Role of Endothelial-Mesenchymal Transition (EndoMT) in the Pathogenesis of Fibrotic Disorders. *Am. J. Pathol.* **2011**, *179*, 1074–1080. [CrossRef]
27. Hinz, B. The extracellular matrix and transforming growth factor-β1: Tale of a strained relationship. *Matrix Biol.* **2015**, *47*, 54–65. [CrossRef]
28. Wipff, P.J.; Rifkin, D.B.; Meister, J.J.; Hinz, B. Myofibroblast contraction activates latent TGF-beta1 from the extracellular matrix. *J. Cell Biol.* **2007**, *179*, 1311–1323. [CrossRef]
29. Zhou, Y.; Hagood, J.S.; Lu, B.; Merryman, W.D.; Murphy-Ullrich, J.E. Thy-1-integrin alphav beta5 interactions inhibit lung fibroblast contraction-induced latent transforming growth factor-beta1 activation and myofibroblast differentiation. *J. Biol. Chem.* **2010**, *285*, 22382–22393. [CrossRef]
30. Tse, J.M.; Cheng, G.; Tyrrell, J.A.; Wilcox-Adelman, S.A.; Boucher, Y.; Jain, R.K.; Munn, L.L. Mechanical compression drives cancer cells toward invasive phenotype. *Proc. Natl. Acad. Sci. USA* **2012**, *109*, 911–916. [CrossRef]
31. Bershadsky, A.D.; Balaban, N.Q.; Geiger, B. Adhesion-dependent cell mechanosensitivity. *Ann. Rev. Cell Dev. Biol.* **2003**, *19*, 677–695. [CrossRef] [PubMed]
32. Ingber, D.E. Tensegrity-based mechanosensing from macro to micro. *Prog. Biophys. Mol. Biol.* **2008**, *97*, 163–179. [CrossRef] [PubMed]
33. Chen, W.-Y.; Hong, J.; Gannon, J.; Kakkar, R.; Lee, R.T. Myocardial pressure overload induces systemic inflammation through endothelial cell IL-33. *Proc. Natl. Acad. Sci. USA* **2015**, *112*, 7249. [CrossRef] [PubMed]
34. Retailleau, K.; Duprat, F.; Arhatte, M.; Ranade, S.S.; Peyronnet, R.; Martins, J.R.; Jodar, M.; Moro, C.; Offermanns, S.; Feng, Y.; et al. Piezo1 in Smooth Muscle Cells Is Involved in Hypertension-Dependent Arterial Remodeling. *Cell Rep.* **2015**, *13*, 1161–1171. [CrossRef] [PubMed]
35. Ranade, S.S.; Qiu, Z.; Woo, S.-H.; Hur, S.S.; Murthy, S.E.; Cahalan, S.M.; Xu, J.; Mathur, J.; Bandell, M.; Coste, B.; et al. Piezo1, a mechanically activated ion channel, is required for vascular development in mice. *Proc. Natl. Acad. Sci. USA* **2014**, *111*, 10347–10352. [CrossRef]

36. Zeisberg, E.M.; Tarnavski, O.; Zeisberg, M.; Dorfman, A.L.; McMullen, J.R.; Gustafsson, E.; Chandraker, A.; Yuan, X.; Pu, W.T.; Roberts, A.B.; et al. Endothelial-to-mesenchymal transition contributes to cardiac fibrosis. *Nature Med.* **2007**, *13*, 952–961. [CrossRef]
37. Widyantoro, B.; Emoto, N.; Nakayama, K.; Anggrahini, D.W.; Adiarto, S.; Iwasa, N.; Yagi, K.; Miyagawa, K.; Rikitake, Y.; Suzuki, T.; et al. Endothelial cell-derived endothelin-1 promotes cardiac fibrosis in diabetic hearts through stimulation of endothelial-to-mesenchymal transition. *Circulation* **2010**, *121*, 2407–2418. [CrossRef]

Review

Role of Magnesium Deficiency in Promoting Atherosclerosis, Endothelial Dysfunction, and Arterial Stiffening as Risk Factors for Hypertension

Krasimir Kostov [1,*] and Lyudmila Halacheva [2]

1 Department of Pathophysiology, Medical University-Pleven, 1 Kliment Ohridski Str., 5800 Pleven, Bulgaria
2 Department of Physiology, Medical University-Pleven, 1 Kliment Ohridski Str., 5800 Pleven, Bulgaria; l_halacheva@abv.bg
* Correspondence: dr.krasi_kostov@abv.bg; Tel.: +359-889-257-459

Received: 18 May 2018; Accepted: 8 June 2018; Published: 11 June 2018

Abstract: Arterial hypertension is a disease with a complex pathogenesis. Despite considerable knowledge about this socially significant disease, the role of magnesium deficiency (MgD) as a risk factor is not fully understood. Magnesium is a natural calcium antagonist. It potentiates the production of local vasodilator mediators (prostacyclin and nitric oxide) and alters vascular responses to a variety of vasoactive substances (endothelin-1, angiotensin II, and catecholamines). MgD stimulates the production of aldosterone and potentiates vascular inflammatory response, while expression/activity of various antioxidant enzymes (glutathione peroxidase, superoxide dismutase, and catalase) and the levels of important antioxidants (vitamin C, vitamin E, and selenium) are decreased. Magnesium balances the effects of catecholamines in acute and chronic stress. MgD may be associated with the development of insulin resistance, hyperglycemia, and changes in lipid metabolism, which enhance atherosclerotic changes and arterial stiffness. Magnesium regulates collagen and elastin turnover in the vascular wall and matrix metalloproteinase activity. Magnesium helps to protect the elastic fibers from calcium deposition and maintains the elasticity of the vessels. Considering the numerous positive effects on a number of mechanisms related to arterial hypertension, consuming a healthy diet that provides the recommended amount of magnesium can be an appropriate strategy for helping control blood pressure.

Keywords: magnesium deficiency; arterial hypertension; vascular tone; arterial stiffness; vascular remodeling; insulin resistance; magnesium supplementation; dietary magnesium intake

---

1. Introduction

Magnesium (Mg^{2+}) is the fourth most common mineral in the human body after calcium (Ca^{2+}), potassium (K^+), and sodium (Na^+), and should be continuously replenished by food and water intake [1]. Mg^{2+} is the second richest intracellular cation after K^+ and is a cofactor in more than 325 enzyme systems in cells [2]. Mg^{2+} is abundant in all green leafy vegetables, cereal, nuts, and legumes. Chocolate products, fruits, meat, and fish contain moderate amounts of Mg^{2+}, and dairy products are low in Mg^{2+}. Drinking water can be an important source of Mg^{2+} when it contains up to 30 mg/L of Mg^{2+} [3]. Chronic inadequate intake of Mg^{2+} over a long period of time can manifest as latent magnesium deficiency (MgD) [1]. Chronic MgD is associated with an increased risk of multiple preclinical and clinical manifestations including hypertension (HTN), atherosclerosis, cardiac arrhythmias, stroke, changes in lipid metabolism, insulin resistance, metabolic syndrome (MetS), type 2 diabetes (T2D), osteoporosis, depression, and other neuropsychiatric disorders [4] (Table 1). The assessment of Mg^{2+} status in the body is difficult because most is found in the cells or in the bones. Only 1% of the total Mg^{2+} in the body is present in extracellular fluids, and only 0.3% is found in the serum. The normal reference range for Mg^{2+} in the serum is 0.76–1.15 mmol/L [4]. In order to maintain these levels, the Recommended Dietary Allowance (RDA) for Mg^{2+} is 420 mg/day for men and 320 mg/day for women [5]. The hypomagnesaemia is defined as a condition where the serum concentration of Mg^{2+} in the body is <0.75 mmol/L [4]. When the Mg^{2+} intake is poor, the kidney can compensate by increasing reabsorption to >99% of the filtered amount, mainly in Henle's loop and further in the distal tubules [1]. Signs and symptoms of hypomagnesemia usually occur when serum Mg^{2+} is decreased below 0.5 mmol/L. Many patients with hypomagnesemia remain asymptomatic [6,7].

Table 1. Negative effects of MgD on the body and organs.

General: anxiety, agitation, irritability, headache, loss of appetite, nausea.
Musculature: muscle spasm and tetany.
CNS/Nerves: nervousness, migraine, depression, poor memory, low stress tolerance, paraesthesia, tremor, and seizures.
Cardiovascular system: HTN, risk of arrhythmias, coronary spasm, atherosclerosis, endothelial dysfunction, low-grade vascular inflammation, arterial stiffness, vascular ECM remodeling, arterial calcification, vascular aging, increased platelet aggregation, potentiates Ca^{2+}-mediated vasoconstriction, potentiates the vasoconstrictor effects of ATII, ET-1, NA, Adr, and TxA_2.
Electrolytes: sodium retention, hypokalemia, and hypocalcemia.
Metabolism: dyslipoproteinemia, insulin resistance, pancreatic β-cell dysfunction, decreased glucose tolerance, increased risk of MetS and T2D, disorders of vitamin D metabolism, resistance to PTH, and osteoporosis.
Pregnancy: pregnancy complications (e.g., eclampsia).
Gastrointestinal tract: constipation.

Over the last eight decades, nutritional and serum Mg^{2+} levels have received more attention and have been the subject of comprehensive cardiovascular health studies. MgD is considered an important risk factor for various types of cardiovascular diseases (CVD) [8] (Table 1). Dietary studies in Europe and the United States of America (USA) reveal that Mg^{2+} intake is lower than recommended [4]. The often recommended daily intake for adults is 320–400 mg/day (or 6 mg/kg/bodyweight for both sexes) [1]. Epidemiological studies in Europe and North America show that people who consume Western-style diets have low Mg^{2+} content, <30–50% of the RDA for Mg^{2+}. It is assumed that the Mg^{2+} intake in the USA has decreased over the past 100 years from about 500 mg/day to 175–225 mg/day. This is probably the result of the increasing use of fertilizers and processed foods [4]. Refining or processing of food may deplete Mg^{2+} content by nearly 85%. Especially boiling of Mg^{2+}-rich foods can result in a significant loss of Mg^{2+} [3]. Furthermore, the incidence rate of MgD can vary considerably in different regions due to the large differences of Mg^{2+} content in drinking water, which can provide up to 30% of daily needs [1].

Therefore, it seems reasonable to assume that MgD is mainly related to the low intake of Mg^{2+} in food and drinking water, including the use of purified salt to cook, which may lead to a negative balance over time [1]. The processing and cooking of food may therefore explain the apparently high prevalence of low Mg^{2+} intake in many populations [3]. In the USA, the prevalence of insufficient Mg^{2+} intake in adults is about 64% among men and 67% among women. Among people over 71 years of age, the figure increases to 81% for men and 82% for women respectively [8]. In the USA the RDA, set at 320 and 420 mg/day for females and males respectively, is higher than the Reference Nutrient Intake of the United Kingdom (UK). Reported data for Mg^{2+} intake in UK compared to the USA RDA are significantly lower than recommended for all population groups. The UK general population's mean intake is 270 mg and 221.4 mg for males and females respectively, representing only 64% and 69% of the US RDA [9]. Dietary intake of Mg^{2+} has also been shown to be insufficient in middle aged French adults, as 77% of women and 72% of men consume less than the recommended levels [10]. Furthermore, MgD may be potentiated by malabsorption or medication intake, such as diuretics (loop and thiazide), proton pump inhibitors, some antibiotics, and chemotherapeutic agents. With age, the absorption of Mg^{2+} from the intestine is reduced and its loss from the body in the urine increases in both sexes [1].

A number of literature reviews and editorials have focused on the importance of Mg^{2+} for the occurrence of CVD. These examinations have shown that the prevalence of cardiovascular disease events caused by an inadequate intake of Mg^{2+} and low serum concentrations of Mg^{2+}, have been underestimated and cardiovascular health may be associated with the intake of Mg^{2+}. A significant association was found between dietary Mg^{2+} intake and total cardiovascular disease risk. The greatest risk reduction was observed when dietary Mg^{2+} intake increased from 150 mg/day to 400 mg/day. Higher serum Mg^{2+} concentrations with 0.1 mmol/L were associated with a 9% lower risk for total cardiovascular events [8].

A significant part of the experimental studies link MgD and CVD, such as HTN, and atherosclerosis [4]. Given the increasing incidence of HTN, the identification of effective and safe preventative measures that offer even modest lowering of blood pressure (BP), could have a significant public health impact. In this regard, Mg^{2+} may have beneficial health effects for the primary prevention of HTN [11]. HTN is the largest contributor to the global burden of cardiovascular disease. The World Health Organization (WHO) estimates that the number of adults with high BP will increase from 1 billion to 1.5 billion worldwide by 2020 [12]. HTN is a major risk factor for heart disease and stroke [4]. According to WHO, 62% of all strokes and 49% of coronary heart disease events are attributable to high BP. Furthermore, experimental and epidemiological studies considered that HTN may serve as an effect modifier of the Mg^{2+} and CVD association [8].

2. Mechanisms Connecting MgD and Arterial Hypertension

Mg^{2+} is involved in BP regulation. Every modification of the endogenous Mg^{2+} status leads to changes in vascular tonus, and as a consequence, changes in arterial BP [4]. Mg^{2+} plays an important role in BP regulation by modulating vascular tone and reactivity [13]. Small changes in both extracellular and intracellular Mg^{2+} concentrations have significant effects on vascular tone, contractility, reactivity, and growth [14,15]. MgD can increase BP by affecting multiple molecular and cellular mechanisms [16] (Figure 1).

2.1. Disturbances of Mg^{2+} Transport and Its Endocrine Control

Mg^{2+} transport occurs through two main pathways—transcellular and paracellular [5]. Transcellular transport includes influx and efflux transport systems. Mg^{2+} influx is controlled by a number of transporters such as mitochondrial RNA splicing 2 protein (Mrs2p), human solute carrier family 41, members 1 and 2 (SLC41A1 and SLC41A2), ancient conserved domain protein 2 (ACDP2), and Mg^{2+} transporter 1 (MagT1), as well as specialized cationic channels—transient receptor potential melastatin-6 and -7 (TRPM6 and TRPM7) cation channels [17]. TRPM6 channels are predominantly expressed in the kidneys and caecum, where they regulate Mg^{2+} reabsorption. TRPM7 channels are ubiquitously expressed and their absence is lethal [18]. Mg^{2+} efflux is accomplished by Na^+-dependent and Na^+-independent pathways [17]. Na^+-dependent transport involves Na^+/Mg^{2+} pump. The following participate in Na^+-independent transport. The Mg^{2+}/Ca^{2+} pump and Mn^{2+}/Mg^{2+} antiporter Cl^-/Mg^{2+} co-transporter. Paracellular Mg^{2+} transport is a passive process and takes place through dense intercellular contacts of the epithelial cells in the intestinal tract and the kidneys and depends on the specific structural proteins—claudins. Intestinal Mg^{2+} absorption is connected with the relatively low expression of 'tightening' claudins 1, 3, 4, 5, and 8. In the kidney paracellular Mg^{2+} transport depends mainly of claudin 16 (paracellin-1) and claudin 19 [5]. Disturbances of Mg^{2+} transport may predispose to the development of HTN and subsequent CVD [17].

The data reported up to now indicates the potential regulatory role of cation TRPM7 channels in maintenance of vascular integrity [19]. In vascular cells Mg^{2+} influx is mainly determined by the TRPM7 channels and disturbances in vascular smooth muscle cells (VSMCs) function in HTN can be partially linked to defective TRPM7 expression or activity [18]. TRPM7, implicated as a signaling kinase, is involved in a number of processes affecting VSMCs—growth, apoptosis, adhesion, cohesion, contraction, cytoskeletal organization, and migration. TRPM7 channels expressed in vasculature are regulated by vasoactive agents such as bradykinin, aldosterone (ALDO), endothelin-1 (ET-1), and angiotensin II (ATII) and different effects on the vascular wall such as pressure, tension, and osmotic changes. Thus these channels alter intracellular Mg^{2+} concentration by influencing the influx and efflux and may be associated with the onset and maintenance of HTN [17,18,20]. Experimental studies with Spontaneously Hypertensive (SHR) and Wistar-Kyoto (WKY) rats showed that TRPM6 and TRPM7 are differently regulated in their VSMCs. Inhibition of TRPM7, but not of TRPM6, may play a role in the altered Mg^{2+} homeostasis in VSMCs of SHR rats [21]. TRPM6 and TRPM7 are expressed in the endothelium, where they play an important role in intracellular Mg^{2+} homeostasis [22]. Because these channels have different effects on cell endothelial function it is suggested that they have potential importance, especially TRPM7, in the regulation of angiogenesis and vascular remodeling [23]. Mechanisms regulating vascular Mg^{2+} in health and disease remain unclear but TRPM7 could be important in HTN [20]. TRPM7 channels modulate endothelial behavior and any condition that leads to their increased expression, e.g., MgD or oxidative stress can damage endothelial function [19]. TRPM7 plays an important role in modulating VSMCs Mg^{2+} homeostasis, a major determinant of VSMCs function and vascular tone. Antunes et al. showed that in heterozygous TRPM7 kinase-deficient mice, ATII induces an exaggerated hypertensive response. These observations indicate that TRPM7 kinase differentially regulates vascular function in HTN. TRPM7 kinase is a modulator of BP regulation, which, when downregulated, promotes severe HTN and worsening of cardiovascular function. Moreover, ATII is a negative regulator of TRPM7. The study defines a novel TRPM7 kinase-sensitive mechanism involved in ATII-induced HTN. Based on these results, it can be concluded that aberrant TRPM7 expression/activity may contribute to impaired intracellular-free Mg^{2+} concentration and VSMCs contraction, proliferation, inflammation, and fibrosis, important determinants of vascular dysfunction and remodeling in HTN [20].

It has been proven that estrogens and epidermal growth factor (EGF) are magnesiotropic hormones [24–27]. Estrogens stimulate TRPM6 activity in short-term treatment and have long-term regulatory effects on TRPM6 transcription. The stimulation of a specific EGF receptor (EGFR) promotes TRPM6 trafficking to the plasma membrane. Long-term treatment with EGF regulates TRPM6 expression [28]. Considering the role of TRPM6 channels for changes of intracellular Mg^{2+} concentrations could be assumed that they are related to the modifications of the electrical and mechanical activity of cardiac myocytes or VSMCs and hence to CVD such as cardiac arrhythmias or HTN [29]. Thus, estrogen deficiency and the lack of their regulatory effects on Mg^{2+} exchange may be directly linked to increased loss of Mg^{2+} in postmenopausal women, which can lead to the development of HTN, cardiac arrhythmias, increased neuro-muscular excitability, and osteoporosis.

It is known that 1,25-dihydroxyvitamin D [1,25$(OH)_2$D] can stimulate intestinal Mg^{2+} absorption [30]. In addition, conversion of vitamin D by hepatic 25-hydroxylation and renal 1α-hydroxylation into the active form and the vitamin D-binding protein are Mg^{2+}-dependent. MgD leads to reduced 1,25$(OH)_2$D and impaired parathyroid hormone (PTH) response. Absorption of Mg^{2+} and Ca^{2+} is inter-related. MgD impairs hypocalcemic-induced PTH release, which can be corrected after infusion of Mg^{2+}. Mg^{2+} is also an important factor for the sensitivity of the target tissues to PTH. On the other hand, PTH has significant effects on Mg^{2+} homeostasis. For example, PTH release enhances Mg^{2+} reabsorption in the kidneys and absorption in the gut. PTH influences Mg^{2+} absorption, however, hypercalcemia antagonizes this effect [4]. The findings indicate a possible link between the vasculature and mineral metabolism. Epidemiological studies have shown that both PTH and Ca^{2+} are associated with high BP [31], which may be due to Mg^{2+} homeostasis disturbances.

2.2. Mg^{2+} as a Regulator of Vascular Tone and Reactivity

Mg^{2+} is one of the important physiological regulators of blood vessel tone. It improves vascular relaxation responses and attenuates agonist-induced vasoconstriction, thus mitigating the increased peripheral vascular resistance. The effects of Mg^{2+} as a regulator of vascular tone are most often the result of a competitive relationship with Ca^{2+}. MgD increases the contractile response to various agonists and changes vascular responses to a number of vasoactive substances (Figure 1).

2.2.1. Mg^{2+} as a Natural Calcium Antagonist

In the dissolved state, Mg^{2+} binds hydration water tighter than Ca^{2+}. Thus, the hydrated Mg^{2+} is more difficult to dehydrate. The radius of hydrated Mg^{2+} is ~400 times larger than its dehydrated radius. This difference explains a lot of the biological properties of Mg^{2+}, including its often antagonistic behavior to Ca^{2+}, despite similar chemical reactivity and charge. For example, it is almost impossible for Mg^{2+} to pass through narrow channels in biological membranes, compared to Ca^{2+}, because it cannot easily lose its hydration shell [32]. Because of its unique chemical properties, Mg^{2+} is linked to the modulation of intracellular Ca^{2+} homeostasis and decreased extracellular or intracellular Mg^{2+} can be combined with an increase in Ca^{2+} levels [16]. Ca^{2+} concentration in the cytosol is one of the principal factors determining the contractile properties of the VSMCs. Mg^{2+} counteracts Ca^{2+} and functions as physiological Ca^{2+} blocker, like synthetic Ca^{2+} antagonists [33–35]. Both extracellular and intracellular free Mg^{2+} can modulate VSMCs tone by voltage-dependent L-type Ca^{2+} channels. Extracellular Mg^{2+} inhibits Ca^{2+} current in VSMCs by two main mechanisms. First, extracellular Mg^{2+} effectively neutralizes the fixed negative charges on the external surface of the cell membrane. This stabilizes the excitable membranes and raises the excitation threshold which diminishes current via the voltage-gated Ca^{2+} channels. Second, some evidence suggests that extracellular Mg^{2+} can decrease Ca^{2+} current by directly binding to the Ca^{2+} channels. Binding of Mg^{2+} may either mechanically block the channel pore or may cause an allosteric modulation of the channel gating, resulting in its closure [36]. Elevated levels of extracellular Mg^{2+} inhibit Ca^{2+} influx, while decreased extracellular Mg^{2+} activates Ca^{2+} influx through Ca^{2+} channels [37]. Intracellular Mg^{2+} concentrations modulate VSMCs tone via its effects on ion channels and signal transduction pathways, especially those involving Ca^{2+}. Changes in intracellular Mg^{2+} are known to influence these channels by affecting its amplitude, its activation/inactivation kinetics, and its modulation by factors such as phosphorylation, ultimately leading to decreased Ca^{2+} entry. Intracellular Mg^{2+} activates sarcoplasmic/endoplasmic reticular Ca^{2+} ATPase pump that sequesters intracellular Ca^{2+} into the sarcoplasmic reticulum. Elevated intracellular Mg^{2+} stimulates inositol-1,4,5-trisphosphate (IP3) breakdown, inhibits IP3-induced Ca^{2+} release from the sarcoplasmic reticulum, and competes with intracellular Ca^{2+} for cytoplasmic and reticular binding sites [36]. Low intracellular Mg^{2+} stimulates IP3 mediated mobilization of Ca^{2+} from the sarcoplasmic reticulum and reduces Ca^{2+} ATPase activity, decreasing Ca^{2+} efflux and reuptake by the sarcoplasmic reticulum. This leads to accumulation of cytosolic Ca^{2+}, increased intracellular Ca^{2+} concentration, which is a crucial factor for vasoconstriction [37]. Mg^{2+} can also block Ca^{2+} release from the sarcoplasmic reticulum via the ryanodine receptor [17]. An important property of Mg^{2+} is to compete with Ca^{2+} for binding sites on regulatory molecule troponin C. Thus, Mg^{2+} regulates the activity of contractile proteins and their dynamics [16]. Lastly, intracellular Mg^{2+} regulates activity of the G-protein-coupled receptors as AT1 (ATII), ETA (ET-1), V1a (vasopressin), and α_1-adrenoceptors (norepinephrine and epinephrine) on VSMCs and intracellular Ca^{2+} signal transduction pathways as translocation of phospholipase C (PLC) and protein kinase C (PKC) activation [36].

Besides the direct effects of Mg^{2+} on VSMCs, Mg^{2+} also modulates endothelial function, which in turn contributes to its vasodilatory actions. Normal endothelium plays a fundamental role in regulating vasomotor tone by synthesizing vasodilatory prostacyclin (PGI_2) and nitric oxide (NO). Mg^{2+} has been shown to increase endothelial release of PGI_2 in cultured human endothelial cells and in healthy human volunteers [17,36].

2.2.2. MgD and Vascular Reactivity

Other Mg^{2+} effects could be due to altered binding of agonists to their specific cell membrane receptors and/or the production of vasoactive peptides, such as ATII and ET-1, which are powerful vasoconstrictors [38]. ATII, ET-1, vasopressin, and epinephrine/norepinephrine exert their vasoconstrictor effect via stimulation of AT1, ETA, V1a, and α_1 receptors, respectively, on VSMCs. Activation of these Gq–protein-coupled receptors initiates the PLC, IP3, diacylglycerol, Ca^{2+}, and PKC signal transduction pathway. Evidence suggests that following receptor–ligand interaction intracellular Mg^{2+} is also altered and that it too functions as a second messenger to modulate signal transduction [36]. For example, elevated levels of Mg^{2+} decrease ET-1 induced contraction, whereas decreased Mg^{2+} levels increase it [38]. Kharitonova et al. have explored the role of various Mg^{2+} compounds for the development of systemic inflammation and endothelial dysfunction in rats fed on a low Mg^{2+} diet for 74 days. Low Mg^{2+} diet reduces Mg^{2+} concentration in the plasma and red blood cells (RBCs), which is accompanied by a decreased concentration of endothelial NO synthase (eNOS), elevated levels of ET-1 in serum and impaired endothelial-dependent vasodilation. ET-1 produces multiple effects in the blood vessels: causes significant vasoconstriction, has proinflammatory effects, possesses mitogenic and proliferative properties, stimulates the formation of free radicals, and activates platelets. The analysis of the activity of the inorganic Mg^{2+} salts showed that supplementation with Mg-sulfate did not significantly reduce pathologically elevated levels of ET-1, but Mg-chloride completely restored the concentration of ET-1 to a normal value. Tested Mg-organic compounds Mg-oxybutyrate and Mg-*L*-aspartate reduce the concentration of ET-1 to a normal level, whereas treatment with Mg-*N*-acetyltaurate leads only to partial reduction of ET-1 in MgD groups [39]. Mg^{2+} is an essential cofactor of the enzyme δ-6-desaturase, which is the rate determining conversion of linoleic acid to gamma-linolenic, which in turn is converted to dihomo-γ-linoleic acid. The latter is a precursor of prostaglandin E_1 (PGE_1), which is both a vasodilator and inhibitor of platelet aggregation. MgD disrupts production of PGE_1, which leads to vasoconstriction and increase in BP [17].

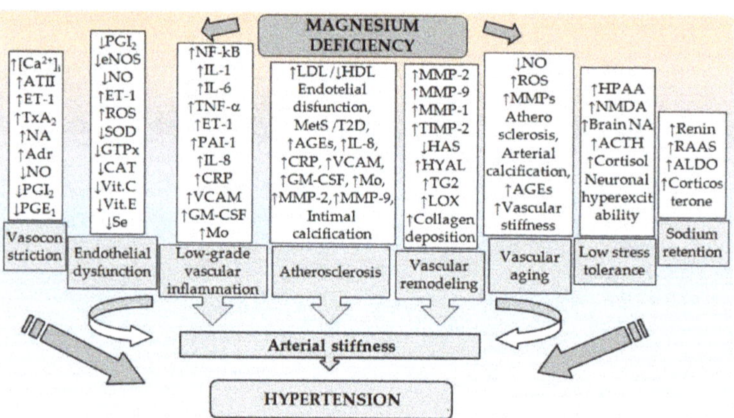

Figure 1. Pathogenetic relationships between MgD and arterial hypertension. (Abbreviations: ↑: increased; ↓: decreased; $[Ca^{2+}]_i$: intracellular calcium in VSMCs; ATII: angiotensin II; ET-1: endothelin-1; TxA_2: thromboxane A_2; NO: nitric oxide; PGE_1: prostaglandin E_1; PGI_2: prostaglandin I_2; ROS: reactive oxygen species; NF-kB: nuclear factor kappa B; TNF-α: tumor necrosis factor-α; IL-1: interleukin-1; IL-6: interleukin-1; IL-8: interleukin-8; PAI-1: plasminogen activator inhibitor-1; VCAM: vascular cell adhesion molecule-1; GM-CSF: granulocyte-macrophage colony-stimulating factor; Mo: monocytes; CRP: C-reactive protein; SOD: superoxide dismutase; GTPx: glutathione peroxidase; CAT: catalase; Vit.C: vitamin C; Vit.E: vitamin E; Se: selenium; LDL: low-density lipoproteins; HDL: high-density lipoproteins; MetS: metabolic syndrome; T2D: type 2 diabetes; AGEs: advanced glycation end products; MMPs: matrix metalloproteinases; MMP-1: matrix metalloproteinase-1; MMP-2: matrix metalloproteinase-2; MMP-9: matrix metalloproteinase-9; TIMP-2: tissue inhibitor of metalloprotease-2; HAS: hyaluronan synthases; HYAL: hyaluronidase; TG2: transglutaminase; LOX: lysyl oxidase; HPAA: hypothalamic pituitary adrenal axis; NMDA: N-methyl-D-aspartate receptor; NA: noradrenaline; Adr: adrenaline; ACTH: adrenocorticotropic hormone; RAAS: renin-angiotensin-aldosterone system; ALDO: aldosterone).

2.2.3. MgD and the Renin-Angiotensin-Aldosterone System (RAAS)

Few studies have investigated the effects of MgD on the hormonal systems which control BP. The RAAS plays an essential role in humoral and hemodynamic regulation [40]. It was established, that in hypertensive patients with high plasma renin activity, serum Mg^{2+} was much lower than in normotensive persons [16]. MgD increases the ATII-induced plasma concentration of ALDO, as well as the production of thromboxane A_2 and vasoconstrictor prostaglandins [3]. Mg^{2+} has some direct effects on ALDO synthesis rather than indirect effects via the RAAS. ALDO secretion from the zona glomerulosa of the adrenal gland is Ca^{2+} dependent process. Mg^{2+} infusion in humans decreases the production of ALDO, by inhibiting the cellular Ca^{2+} influx. MgD facilitates cellular Ca^{2+} influx, which may stimulate the production and release of ALDO [40]. It was proved that MgD rats have increased plasma renin activity and circulating levels of ALDO and corticosterone. Additionally, other authors indicate that Mg^{2+} supplementation decreases ATII stimulated production and release of ALDO from the adrenal cortex of normotensive subjects [16]. Rats fed with an MgD diet showed a continuous increment of the juxtaglomerular granulation index and width of the zona glomerulosa of the adrenal cortex, whereas the thickness of the inner zones diminished slightly. In Mg^{2+}-recovering rats juxtaglomerular granulation index the width of the zona glomerulosa returned to normal [41]. Thus, one could suggest that MgD, by facilitating cellular Ca^{2+} entry, may promote ALDO production and release. Experimental studies have reported that MgD-induced HTN in rats is associated with increased vascular total Ca^{2+} content, and increased vasoconstrictor activity to endogenous agonists such as ATII and noradrenaline (NA) [40]. Mg^{2+} supplementation can reduce the pressor effect of ATII and stimulate the production of the vasodilator PGI_2 [3].

2.2.4. MgD and Catecholamines (CA)

Ca^{2+} exerts a major role in CA release from the adrenal gland and adrenergic nerve terminals in response to sympathetic stimulation. Because Mg^{2+} competes with Ca^{2+} for membrane channels, it can modify these types of Ca^{2+}-mediated responses. The ability of Mg^{2+} to inhibit the release of CA from both the adrenal gland and peripheral adrenergic nerve terminals is well established in laboratory experiments. On the basis of these effects, Mg^{2+} can be used in patients where an excess of CA is prevalent, such as in phaeochromocytoma [42]. Acetylcholine (ACh) evoked CA release from adrenal glands. Ca^{2+} stimulates the secretion of ACh and there exists a direct relationship between Ca^{2+} concentration and the rate of CA release. Mg^{2+} antagonizes the stimulant effects of ACh on adrenal chromaffin cells; this effect can be overcome by the addition of Ca^{2+} [43]. Mg^{2+} is required for the catalytic action of adenylate cyclase (ADCY). For example, the decreased activity of the ADCY9 in the absence of Mg^{2+} results in increased secretion of ACh from preganglionic nerves, which in turn stimulates further release of CA from the adrenal glands [44]. On the other hand, Mg^{2+} together with ATP can greatly stimulate the release of CA from adrenal medullary granules. Neither ATP nor Mg^{2+} alone may have any effect on the release of CA. The effects of Ca^{2+}, which cause the release of CA from the granules, are inhibited in the presence of ATP and Mg^{2+}. The uptake of ^{14}C-adrenaline (Adr) into the granules can also be stimulated by ATP and Mg^{2+} [45].

There are evidences that high concentrations of Mg^{2+} prevent the release of NA in some arteries by blocking *N*-type Ca^{2+} channels of nerve endings, which counteracts the rise in BP. Also rats fed with a MgD diet showed an increase in CA excretion [16]. The effects of Ca^{2+} and Mg^{2+} concentrations on responses to periarterial nervous sympathetic stimulation, NA, and tyramine have been investigated on the isolated rabbit central ear artery. An increase in Ca^{2+} concentrations potentiates responses to sympathetic stimulation until complete removal of Ca^{2+} inhibits these responses. The addition of the Mg^{2+} solution greatly reduced the responses to sympathetic stimulation, NA, and tyramine. These actions of Mg^{2+} on sympathetic transmission are important in determining the responsiveness of arterial smooth muscle to direct and indirectly acting sympathomimetic amines [46]. The effects of Adr infusion, sufficient to achieve its pathophysiological levels, and of therapeutic intravenous infusion of salbutamol, a β_2-agonist, on plasma Mg^{2+} levels, were studied in a placebo-controlled design in healthy subjects. Plasma Mg^{2+} levels fell significantly during the Adr infusion and also during the salbutamol infusion, though more slowly. It is propose that intracellular shifts of Mg^{2+} occur as a result of β-adrenergic stimulation [47]. In another study, infusion of Adr in man significantly reduced the plasma Mg^{2+} levels in healthy males. This effect was abolished by simultaneous infusion of propranolol. NA had no such effect. These results suggest that the β-adrenergic system affect Mg^{2+} homeostasis [48]. On the other hand, dose-dependent increase in circulating Mg^{2+} was observed in rats infused with isoproterenol (ISO). Pretreatment with butoxamine or propranolol has prevented the ISO-induced increase in serum Mg^{2+} levels, whereas administration of atenolol has minimal effects. This evidence demonstrates the existence of a pool of Mg^{2+} that is mobilized into the circulation in response to selective β_2-adrenergic stimulation [49].

2.3. *MgD and Arterial Stiffness*

Normally conduit arteries adapt pressure and blood flow during cardiac systole to facilitate perfusion to tissues during diastole. This is determined in large part by the elasticity, distensibility, and compliance of the arterial system. Loss of elasticity and increased stiffness demand greater force to accommodate blood flow, leading to increased systolic BP and consequent increased cardiac work load. Multiple interacting factors at the systemic (BP, hemodynamics), vascular (vascular contraction/dilatation, extracellular matrix remodeling), cellular (cytoskeletal organization and inflammatory responses), and molecular (oxidative stress, intracellular signaling, and mechanotransduction) levels contribute to arterial stiffness in HTN. Dysregulation of endothelial cells, VSMCs, and adaptive immune responses are also implicated in HTN [12]. Changes in Mg^{2+} concentrations play an important role in many of these processes.

2.3.1. MgD, Low-Grade Inflammation, and Oxidative Stress at the Vascular Wall

A number of immunopathological mechanisms may be at the basis of HTN. There are evidence in animal models and humans that link HTN with changes in both humoral and cellular immunity, and in particular with the key role of low-grade vascular inflammation [50]. One study showed that the total intracellular Mg^{2+} is considerably lower in lymphocytes of the hypertensive patients, compared with healthy subjects, whereas serum Mg^{2+}, erythrocyte Mg^{2+} and ionized platelet Mg^{2+} were not significantly different [51]. MgD leads to inflammation and increased production of free radicals [52] in the vascular wall, and they in turn contribute to the development of endothelial dysfunction and vascular remodeling. Low Mg^{2+} intake is associated with a higher probability of increased serum C-reactive protein (CRP) levels in children [53]. There is also an association between the dietary intake of Mg^{2+} and elevated CRP levels in the adult population. Insufficient dietary intake of Mg^{2+} may be associated with an increased inflammatory response resulting in more frequent occurrence of cardiovascular accidents [54]. Intake of Mg^{2+} is also inversely related to the level of hs-CRP, interleukin 6 (IL-6), and fibrinogen [55]. MgD significantly increases production of various proinflammatory molecules such as, interleukin 1 (IL-1), IL-6, tumor necrosis factor α (TNF-α), vascular cell adhesion molecule-1 (VCAM), plasminogen activator inhibitor-1 (PAI-1), and decreases expression and activity of the antioxidant enzymes such as glutathione peroxidase, superoxide dismutase, and catalase. Cellular and tissue levels of important antioxidants such as glutathione, vitamin C, vitamin E and selenium are also reduced [16]. This shows that MgD can increase cytotoxicity of the free radicals to endothelial cells [56]. For example Mg^{2+} deficit, in rats leads to an increase in the inflammatory mediators, such as histamine, IL-1, IL-6, TNF-α, and ET-1, which is associated with leukocytosis and generation of free radicals [34] (Figure 1).

2.3.2. MgD, Vascular Structure and Remodeling

Vascular remodeling is a permanent process of structural changes in the vessel wall in response to a number of hemodynamic stimulus [57]. In HTN, resistance arteries undergo an inward remodeling, while larger arteries show outward hypertrophic remodeling [58–60]. The vascular extracellular matrix (ECM) comprises multiple structural proteins, including collagens, elastin, fibronectin, and proteoglycans.

The absolute and relative quantities of collagen and elastin determine biomechanical properties of vessels. Excessive ECM protein deposition, in particular collagen and fibronectin, contributes to increased intima–media thickening, vascular fibrosis, and stiffening leading to the development of HTN [12]. Mg^{2+} regulates collagen and elastin turnover and the structure of the ECM. MgD leads to a delay in the synthesis of all structural molecules (collagen, elastin, proteoglycans and glycosaminoglycans) (Figure 1). Hyaluronan synthases (HAS)—HAS1, HAS2 and HAS3, contain Mg^{2+} ion in their active centers. On the other hand it is known that inhibitors of the hyaluronidase (HYAL) depend on the concentration of Mg^{2+} ions. Thus, low levels of Mg^{2+} could lead to decreased activity of HAS, and at the same time to an increased activity of HYAL. Tissue transglutaminase (TG2) is an enzyme of the transglutaminase superfamily that is ubiquitously expressed in the vasculature [61]. TG2 is associated with a wide range of CVD and processes, including the development of HTN, and the progression of atherosclerosis, regulating vascular permeability, and angiogenesis. TG2 activity is associated with arterial stiffening in humans and rats. TG2 forms glutamine-lysine cross-links between variety of extracellular proteins, including collagen and elastin [62]. TG2 is secreted through a Golgi-independent mechanism to the ECM, where it can be activated to a Ca^{2+}-bound open conformation to catalyze the formation of isopeptide bonds [61]. Thus, Ca^{2+} may be limiting TG2 activity in the ECM [62]. TG2 is activated by Ca^{2+}, and inhibited by Mg^{2+} [63]. A disturbance in lysyl oxidase (LOX) expression has also been reported in CVD, and an increase in vascular LOX activity has been described in experimental models of HTN. Mg^{2+} can inhibit LOX, which is also associated with crosslinking of chains of elastin and/or collagen [63,64]. Additionally, MgD could lead to the production of defective collagen, elastin, and fibronectin by fibroblasts [65].

Fundamental to many of the processes underlying ECM reorganization and fibrosis in HTN is activation of matrix metalloproteinases (MMPs) and tissue inhibitors of metalloproteinases (TIMPs). ECM MMPs and TIMPs may contribute to the profibrotic phenotype in HTN. Activated MMPs degrade collagen, elastin, and other ECM proteins, resulting in a modified ECM, often associated with a proinflammatory microenvironment that triggers a shift of endothelial and VSMCs to a more secretory, migratory, and proliferative phenotype, which contributes to fibrosis, calcification, endothelial dysfunction, and increased intima–media thickness, further impacting on vascular remodeling and arterial stiffness [12]. In endothelial cells cultured in MgD medium, a significant increase in expression and activity of MMP-2 and MMP-9 has been reported. Also, MMP-2 and MMP-9 have been associated with alterations of the vascular wall in Mg^{2+}-deficient rats [66]. In addition, there is evidence that the addition of Mg-sulfate effectively attenuated MMP-9 activity in a human umbilical cord vein endothelial cell line [67]. These data are confirmed by K. Kostov et al. who find that in patients with essential HTN there was a moderate negative correlation of serum Mg^{2+} with MMP-2 ($r = -0.318$, $p = 0.013$). There was a similar correlation of Mg^{2+} with MMP-9 in patients with HTN and T2D ($r = -0.376$, $p = 0.003$). The results show that lower and higher serum Mg^{2+} levels correlate inversely with MMP-2 and MMP-9 levels in HTN [68]. It is noteworthy that in Mg^{2+}-deficient endothelial cells, MMP-2 and MMP-9 activity overrides the inhibitory effect of TIMP-2, which probably is induced as an attempt to counterbalance the effects of the proteases [66]. A nuclear factor (NF)-κB-binding site is present in the promoter of the MMP-9 gene. It is therefore possible that low Mg^{2+} availability might directly increase MMP-9 expression via NF-κB [66,69]. In cultured rat VSMCs, Mg^{2+} significantly reduced the production of MMP-2 under basal and platelet-derived growth factor-stimulated conditions in a dose-dependent manner, while neither verapamil nor nifedipine showed any effect under the same conditions. These data suggest that the beneficial effect of Mg^{2+} supplementation on vascular disease processes may be due, at least in part, to the inhibitory effect of Mg^{2+} on the production of MMP-2 in VSMCs [70]. Evidence supporting this data is that in cultured rat cardiac fibroblasts, Mg^{2+} significantly reduced the production of MMP-2 in a dose-dependent manner [71]. MgD may increase the activity of MMPs, including collagenases, which begin to degrade the extracellular vascular matrix and primarily collagen with an increased speed. The degradation of elastin fibers can significantly increase (up to 2–3 times) in the presence of Mg^{2+}. MgD is associated with low elastase activity and an increased number of elastic fibers [63]. Altura et al. describe and other possible mechanisms by which MgD can affect vascular remodeling processes. They present new evidence for effects on platelet-activating factor, proto-oncogenes, and sphingolipids, e.g., ceramide and sphingosine with upstream regulation in both VSMCs and cardiac muscle cells. These findings will be helpful in explaining many of the known cardiovascular manifestations of MgD, especially vascular remodeling seen in atherosclerosis and HTN [72].

2.3.3. MgD, Endothelial Dysfunction and Atherosclerosis

MgD may potentiate the development of endothelial dysfunction via activation of NF-κB, which includes the transcriptional program leading to development of the proinflammatory phenotype [69]. Low extracellular Mg^{2+} slows endothelial cell proliferation, stimulates the adhesion of monocytes, and affect the synthesis of vasoactive molecules, such as NO and PGI_2. Endothelial function is significantly impaired in a model of familial hypomagnesemia in mice. Compared to controls, in the aortas of these animals were found reduced amounts of eNOS and increased expression of proinflammatory molecules, such as VCAM, PAI-1, as well as of the TRPM7 channel [19]. Endothelial dysfunction is an early event in the process of atherogenesis and precedes the angiographic and ultrasound evidences of damage to the arterial wall [66]. The pathogenesis of atherosclerotic changes and disturbances in endothelial function are complex and multifactorial. In this context, Mg^{2+} deficit is too important [73]. This mineral is especially important because of its antiatherosclerotic effects [74]. Endothelial function correlates to the levels of Mg^{2+} and results of Mg^{2+} supplementation have showed significantly improved endothelial function in patients with ischemic heart disease and diabetes. These results in humans have also been observed in different experimental models in which Mg^{2+} deficit affects vascular structure and function. Low levels of extracellular Mg^{2+} favor and increase endothelial permeability. More specifically, MgD enhance the transport of low-density lipoproteins (LDL) through the endothelial layer [66]. Several studies have reported beneficial effects of Mg^{2+} supplementation on plasma LDL levels, as well as on high-density lipoproteins (HDL) levels, which are increased [75]. Another possibility by which Mg^{2+} contributes to the development of atherogenesis is through the effect on triglyceride levels which are increased in MgD. Accumulation of triglyceride-rich lipoproteins is accompanied by decreased concentration of HDL and increased plasma concentration of apolipoprotein B. Since the oxidation of lipoproteins play a key role in the development of atherosclerosis, it could be another mechanism by which Mg^{2+} influences. It is also possible proatherogenic lipoprotein changes found in MgD to be a consequence of the inflammatory response [76]. A central role for Mg^{2+} mediated effects on endothelial cells has IL-1α, which is regulated by NF-κB and may be inducer of the NF-κB. IL-1α increases significantly in the environment of low Mg^{2+} content. IL-1α also induces the production of various chemokines and adhesion molecules in vascular endothelial cells by activation of NF-κB, and thus favors aggregation, adhesion, and diapedesis of monocytes. In particular, low concentrations of Mg^{2+} stimulate the secretion of interleukin 8 (IL-8), and chemokines, which are overexpressed in human atherosclerotic lesions. IL-8 is essential for chemotaxis and adhesion of monocytes to endothelial cells, which is a fundamental event in the initiation of atherogenesis and also stimulates proliferation and migration of VSMCs. By induction of IL-1α, low serum Mg^{2+} may also stimulate overexpression of VCAM-1 on the surface of endothelial cells which assists in the migration of leukocytes. In addition, the secretion of granulocyte-macrophage colony-stimulating factor is significantly higher in endothelial cells with Mg^{2+} deficit [66]. All these date indicate that the MgD is a potential factor for accumulation of monocytes/macrophages in the arterial wall during the early stages of atherosclerosis (Figure 1).

2.3.4. MgD, MetS, and T2D

The presence of MetS is also associated with altered Mg^{2+} metabolism [77]. Usually, the triad consisting of obesity, HTN, and impaired glucose tolerance/insulin resistance is denoted as MetS [78]. Furthermore, HTN is present in a high proportion of patients with T2D [79,80]. A common feature in patients with T2D, HTN, and low levels of HDL is MgD (Table 1).

Currently, there is little data on serum Mg^{2+} levels in people with MetS. The relationship between the intake of Mg^{2+} and MetS was investigated prospectively in 5115 young Americans (aged 18–30 years), initially without MetS and diabetes, which were enrolled in Coronary Artery Risk Development in Young Adults (CARDIA) study from 1985 to 1986. The total number of participants included in the analysis was 4637, and 74% were evaluated at the 15-year period in 2000–2001. During this interval 608 cases of MetS were diagnosed. The findings showed that young people with a high Mg^{2+} intake had a lower risk of developing MetS and that risk was dose-dependent [78]. Guerrero-Romero et al. found a link between Mg^{2+} levels, inflammation, and oxidative stress, as risk factors for the development of MetS. Mg^{2+} intake is inversely proportional to the components of MetS and fasting insulin levels, suggesting that higher Mg^{2+} intake may have a protective role against the risk of developing MetS [77]. The results of several clinical studies have shown that increased synthesis and release of proinflammatory cytokines trigger the process of chronic inflammation, which may be the link between obesity and MetS. On the other hand hypomagnesemia triggers low-grade chronic inflammation and Mg^{2+} deficit may be associated with the development of MetS. These findings support the hypothesis that MgD can play an important role in the pathophysiology of MetS, and the actuation of the inflammatory reaction caused by the shortage of Mg^{2+} could be the link between MgD and MetS [81]. Some studies linked decreased Mg^{2+} levels with chronic inflammatory stress in obese people. Obesity affects over 35% of the adult population of the USA and is a main risk factor for chronic diseases, associated with a lower Mg^{2+} status, such as atherosclerosis and T2D. MgD is often found in people with MetS and T2D, which are connected with higher plasma concentrations of CRP [82].

Mg^{2+} plays a very important role in the development of T2D [83]. Corica et al. have recently shown that patients with T2D having lipid profile with high risk, high BP, and abdominal obesity, have lower levels of Mg^{2+}, compared with patients without metabolic risk factors. Furthermore, plasma triglycerides and waist circumference were independently associated with hypomagnesaemia [77]. T2D is often linked with hypomagnesaemia, as has been reported at an occurrence rate of 13.5–47.7% [78]. The relationship between insulin and Mg^{2+} is bipartite. Insulin regulates Mg^{2+} homeostasis, but on the other hand Mg^{2+} is a major factor determining insulin and glucose metabolism. Extracellular Mg^{2+} acts as Ca^{2+} antagonist and inhibits Ca^{2+} influx, required for insulin secretion. Thus a decreased concentration of extracellular free Mg^{2+} results in an increased Ca^{2+} influx and increased concentration of intracellular free Ca^{2+}. The increased intracellular Ca^{2+} stimulates insulin secretion by beta-cells, as was demonstrated in experiments with an insulinoma cell line [84]. The effect of extracellular Mg^{2+} on insulin secretion was found in healthy human subjects. In subjects with 0.79 mmol/L plasma Mg^{2+}, fasting plasma insulin was 23 μU/mL, while in those with plasma Mg^{2+} 0.87 or 1.00 mmol/L, fasting plasma insulin amounted to 11 μU/mL [85].

There are growing evidences that highlight the clinical significance of altered Mg^{2+} metabolism for the occurrence of peripheral insulin resistance. MgD can lead to disturbances of the tyrosine kinase activity of the insulin receptor (IR), associated with the development of post receptor insulin resistance and reduced cellular glucose utilization, as a lower Mg^{2+} concentration, requires a greater amount of insulin for glucose metabolism [77]. The effects of MgD on glucose-stimulated insulin secretion and insulin action on skeletal muscle were studied in experimental animals. The hypothesis that changes in Mg^{2+} metabolism induce insulin resistance is confirmed by data showing that lower dietary intake of Mg^{2+} is associated with insulin resistance. Rats fed on a low Mg^{2+} diet showed a significant increase of blood glucose and triglycerides. The insulin resistance, observed in the skeletal muscle of rats with MgD is partially associated with a defect of tyrosine kinase activity of the IR [86–88]. Insulin action begins with the binding of insulin to an IR on the cell membrane of the target cells. The IR is a transmembrane glycoprotein with tyrosine kinase activity [89]. Activation of the receptor is an important step in transmembrane signaling for insulin action. The activated kinase promotes autophosphorylation of receptor tyrosine residues. The insulin–receptor complex is internalized and phosphorylates IR substrates 1–6 (IRS 1–6) and other kinases in the insulin signaling cascade [90]. When the intrinsic tyrosine kinase activity of the receptor is triggered by insulin binding, two major signaling pathways have been activated: (1) Ras-mitogen-activated protein kinase (MAPK) pathway, which controls cell growth and differentiation; (2) Phosphoinositide 3-kinase/Akt (PI3K/Akt) pathway. Binding of IRS to the regulatory subunit of phosphoinositide 3-kinase (PI3K) results in activation of PI3K, which phosphorylates membrane phospholipids and phosphatidylinositol 4,5-bisphosphate (PIP2). This complex activates the 3-phosphoinositide-dependent protein kinases (PDK-1 and PDK-2) resulting in activation of Akt/protein kinase B and atypical protein kinase [91,92]. Activated Akt phosphorylates its 160 kDa substrate, which stimulates the translocation of insulin-mediated glucose transporter type 4 (GLUT4) from intracellular vesicles to the plasma membrane [93]. The PI3K/Akt pathway is a key component of the insulin signaling cascade, which is necessary for the metabolic effects of insulin and GLUT4 translocation [94]. Since Mg^{2+} is a necessary cofactor in all ATP transfer reactions, intracellular Mg^{2+} concentration is critical in the phosphorylation of the IR and other kinases [95]. In all these reactions Mg^{2+} operates together with ATP as a kinase substrate. Additionally Mg^{2+} is bound to a regulatory site of the IR tyrosine kinase (IRTK). The apparent affinity of the IRTK for Mg^{2+}-ATP increased as the concentration of free Mg^{2+} increased, and the apparent affinity of the IRTK for free Mg^{2+} increased as the concentration of Mg^{2+}-ATP increased [84]. There are evidences that show a link between decreased Mg^{2+} concentration and reduction of tyrosine-kinase activity at the IR level, which results in the impairment of insulin action and development of insulin resistance [96]. Studies in multiple insulin resistant cell models have demonstrated that an impaired response of the tyrosine kinase to insulin stimulation is one potential mechanism causing insulin-resistant state in T2D [97]. Nadler et al. have reported that insulin sensitivity decreases even in nondiabetic individuals after induction of MgD [98]. Finally, Mg^{2+} can also be a limiting factor in carbohydrate metabolism, since many of the enzymes in this process require Mg^{2+} as a cofactor during reactions that utilize the phosphorus bond [99].

Inadequate dietary intake of Mg^{2+} is an independent risk factor for the development of T2D. Lopez-Ridaura et al., evaluating 37,309 participants free of cardiovascular disease and T2D, found a significant inverse association between Mg^{2+} intake and diabetes risk [100]. Van Dam et al. reported a similar relationship. Their findings indicated that a diet high in Mg^{2+}-rich foods, particularly whole grains, is associated with a substantially lower risk of T2D [101]. Benefits of Mg^{2+} supplementation in diabetic subjects have been found in some clinical studies. Rodriguez-Moran et al. reported that Mg^{2+} supplementation improves insulin sensitivity and secretion as well as metabolic control in patients with T2D [96]. Mooren et al. have shown beneficial effect of oral Mg^{2+} supplementation on insulin sensitivity in overweight, nondiabetic subjects [102].

2.3.5. MgD and Vascular Calcification

Vascular calcification is the extracellular deposition of Ca^{2+} in the arterial wall and is intimately linked with the HTN. On the other hand, HTN was considered a risk factor for atherosclerosis and associated calcification. Two types of extracellular vascular calcification are recognized, intimal and medial. Intimal calcification is exclusively associated with atherosclerosis. Medial calcification may contribute to increasing BP by decreasing the elasticity of the media. Decreased elasticity results in arterial stiffening which accelerates pulse wave velocity and widening the pulse pressure, leading ultimately to HTN. Intimal and, especially, medial vascular calcification are associated with arterial stiffening, the major cause of isolated systolic HTN in the elderly [103]. The first in vitro evidence in human aortic VSMCs for a protective role of Mg^{2+} on vascular calcification was based on the observation that living cells are necessary for Mg^{2+} ions to exert its protective effect. These studies suggested a potentially active intracellular role for Mg^{2+} ions in attenuating the vascular calcification process [33,35,104]. In confirmation of this, Hruby et al. reported on favorable associations between dietary and supplemental Mg^{2+} intake and lower calcification of the coronary arteries [105]. Furthermore, it was found that higher Mg^{2+} levels prevented calcification of bovine VSMCs, and further progression of the already established calcification. Inhibition of the Wnt/β -catenin signaling pathway was identified as one of the possible intracellular mechanisms by which Mg^{2+} achieved its anti-calcifying effect [104].

2.3.6. MgD and Vascular Aging

Aging represents a major risk factor for MgD. The total body Mg^{2+} content and intracellular Mg^{2+} tend to decrease with age. Aging is often associated with a Mg^{2+} deficiency due to reduced intake and/or absorption, increased renal wasting and/or reduced tubular reabsorption, as well as age-related illnesses and their treatment with certain drugs [106]. The aging process is associated with alterations in the properties of all the elements of the vascular wall, including endothelium, VSMCs, and ECM. This increases vascular stiffness and leads to the development of isolated systolic HTN. "Aging"-associated arterial changes and those associated with HTN (and early atherosclerosis and diabetes) are fundamentally intertwined at the cellular and molecular levels [107]. At the molecular and cellular levels, arterial aging and HTN-associated vascular changes are characterized by reduced NO production, increased generation of reactive oxygen species (oxidative stress), activation of transcription factors, induction of "aging" genes, stimulation of proinflammatory and profibrotic signaling pathways, reduced collagen turnover, calcification, VSMCs proliferation, and ECM remodeling. These processes contribute to increased fibrosis, which is further promoted by prohypertensive vasoactive agents, such as ATII, ET-1, and ALDO [12]. The cellular and molecular proinflammatory mechanisms that underlie arterial aging are novel putative candidates to be targeted by interventions aimed at attenuating arterial aging, and thus possibly attenuating the major risk factors for HTN and atherosclerosis [107]. Targeted interventions aimed at correcting MgD and maintaining an optimal Mg^{2+} balance may prove to be an appropriate strategy against arterial aging due to its positive effects on low-grade inflammation and oxidative stress associated with aging process (Figure 1).

2.4. MgD and Stress Response

Stress is among the potential psychological risk factors for HTN. Acute stressful events have no consistent association with the HTN. Chronic stress on the other hand, particularly the non-adaptive response to stress, may be a more likely cause of sustained elevation of BP. The mechanisms underlying the association between psychosocial stress and HTN can be divided into behavioral, psychological and pathophysiological. The latter involves neuro-endocrine activation mediated by the hypothalamic pituitary adrenal axis (HPAA) [108]. Mg^{2+} plays a key role in the activity of psychoneuroendocrine systems. For example, all elements of the limbic-HPAA are sensitive to the action of Mg^{2+}. Mg^{2+} modulates activity of the HPAA which is a central substrate of the stress response system. Activation of the HPAA instigates adaptive autonomic, neuroendocrine, and behavioral responses to cope with the demands of the stressor [10]. MgD induced an increase in the transcription of the corticotropin releasing hormone in the paraventricular hypothalamic nucleus, and elevated adrenocorticotropic hormone (ACTH) plasma levels, indicating an enhanced set-point of the HPAA [109]. MgD results in a stressor effect and increases susceptibility to the physiological damage produced by stress (Figure 1). Mg^{2+} supply has been shown to attenuate the development of adverse stress reactions. Stress activates the HPAA and the sympathetic nervous system. The innervation of the kidney may result in the overproduction of renin, which in turn activates the production of ATII, a powerful vasoconstrictor that elevates the BP [110]. Additionally, Mg^{2+} deficient mice are more sensitive to anxiety-provoking situations. Dysregulation of the HPAA evoked by MgD is normalizes by chronic desipramine or diazepam treatment. These data indicate that dysregulation in the HPAA may contribute to hyper-emotionality in response to dietary induced hypomagnesaemia [109].

Mg^{2+} ions also have a key role in the modulation of neurotransmission. Numerous studies have confirmed that the function of the native *N*-methyl-D-aspartate (NMDA) receptor is the result of equilibrium between extracellular and intracellular concentration of Mg^{2+}. The blockade of the ion channel of the NMDA receptor is the most well-known and established way in which Mg^{2+} affects the functioning of the central nervous system (CNS) [109]. Mg^{2+} reduces neuronal hyperexcitability by inhibiting NMDA receptor activity and also is essential for the activity of metabotropic glutamate receptors (mGluRs) in the brain. The mGluRs play a key modulatory role in glutamatergic activity, secretion, and presynaptic release of glutamate, activity of the gamma-aminobutyric acid (GABA)-ergic system, and regulation of the neuroendocrine system. Mg^{2+} may additionally modulate anxiety via increasing GABA availability by decreasing presynaptic glutamate release. GABA is a primary inhibitory transmitter in the CNS that counterbalances the excitatory action of glutamate [111]. The state of acute and chronic stress leads to depletion of intracellular Mg^{2+} and its loss in the urine, because in stressful situations secreted elevated amounts of Adr and NA help to remove Mg^{2+} from the cells. Intracellular RBCs Mg^{2+} depletion is found in patients with HTN. MgD affects the balance of monoamines, such as CA and serotonin in the brain. CA released into the blood is rapidly inactivated by the enzyme catechol-*O*-methyl transferase (COMT). The latter is activated by Mg^{2+} and is inhibited by Ca^{2+}. MgD leads to decreased activity of COMT, which in turn increases the concentration of circulating CA [44]. Brain NA was determined in adult male mice with genetically low (MGL) or high (MGH) blood Mg^{2+} levels. NA levels were significantly higher in MGL than in MGH mice. These data together with the higher urinary NA excretion observed in the MGL line might account for the higher sensitivity and/or reactivity of MGL animals to stress [112]. Moreover, in stressful conditions, MGL mice displayed a more aggressive behavior than the control MGH strain. Altogether, MGL mice showed a more restless behavior, and much higher brain and urine NA levels than the MGH animals [113]. An analysis of the literature suggests the possible role of MgD in the susceptibility to CVD, observed among subjects displaying a type A behavior pattern. Type A subjects are more sensitive to stress and produce more CA than type B subjects. This, in turn, seems to induce an intracellular Mg^{2+} loss. In the long run, type A individuals would develop a state of MgD, which may promote a greater sensitivity to stress, and ultimately to the development of CVD [114], including HTN. Hypomagnesemia usually involves cellular Mg^{2+} depletion, but acute stress that increase serum concentrations of CA may lower serum Mg^{2+} concentration, which does not always imply depleted tissue Mg^{2+} stores [115].

The potential effect of Mg^{2+} in attenuating psychological response to stress merits further investigation since stress is a ubiquitous feature of modern life. The modulation of HPAA by Mg^{2+}, which has been shown to reduce central (ACTH), peripheral (cortisol) endocrine responses [111], and reduces neuronal hyperexcitability by NMDA, mGluRs and GABA-effects suggests that behavioral responses to stress exposure may be attenuated by Mg^{2+} supplementation in patients with HTN.

3. MgD, Groups at Risk, Replacement Therapy and Prevention

3.1. Mg^{2+} Supplements in Hypertensive Subjects

HTN is a complex, heterogeneous disorder whose etiology, pathogenesis, and treatment still raises some unresolved questions. Maintenance of optimal Mg^{2+} status in the human body may help prevent or treat HTN. Although most epidemiological and experimental studies support a role of MgD in the pathophysiology of HTN, data from clinical studies have been less convincing [38]. In some studies the inverse association between Mg^{2+} and BP remained inconclusive, but not in others [116]. Ultimately, the view is that MgD in patients with HTN is linking with significant adverse effects on BP. In the Atherosclerosis Risk in Communities (ARIC) study, serum Mg^{2+} level in hypertensive patients was inversely proportional to the systolic BP. The study examined a cohort of 15248 participants aged 45–64 years [117]. In another meta-analysis of 34 randomized, double-blind, placebo-controlled trials involving a total of 2028 subjects, it was found that oral administration of Mg^{2+} resulted in a significant reduction in both systolic and diastolic BP (2.00 mmHg and 1.78 mmHg respectively) [118]. An analytical review of 44 studies in humans have shown that low doses of Mg^{2+} supplementation, for example 243 mg/day can significantly lower BP in patients with uncomplicated HTN, treated six months or longer with antihypertensive drugs [119]. Moreover, the researchers reported that patients with MgD require higher doses of antihypertensive drugs compared to those with normal Mg^{2+} concentration [118]. The evidence supporting the cause–consequence antihypertensive effect of Mg^{2+} in adults suggest that oral Mg^{2+} supplements may be recommended for the prevention of arterial HTN or as adjuvant antihypertensive therapy [11]. It should be noted that MgD is not found in all patients with HTN. On the other hand, not all people with hypomagnesemia have high BP. These differences are probably due to the fact that patients with high BP do not constitute a homogenous group [78]. This may be one of the possible causes for the discrepancy between epidemiological and clinical data. Despite these discrepancies concerning Mg^{2+} status and high BP, some hypertensive patients constantly demonstrate hypomagnesemia. Among them are patients with obesity, insulin resistance, hypertriglyceridemia, severe forms of HTN, hyperaldosteronism (volume-dependent HTN), pregnancy induced HTN, and patients of African-American origin [120]. In view of the still ill-defined role of Mg^{2+} in clinical HTN, Mg^{2+} supplementation is advised in those hypertensive patients who are receiving diuretics, who have resistant or secondary HTN or who have frank MgD [121]. In the USA, the Estimated Average Requirement (EAR) and RDA of Mg^{2+} for adult women are set at 255–265 mg and 310–320 mg/day, respectively. The EAR and RDA of Mg^{2+} for adult men are set at 330–350 mg and 410–420 mg/day, respectively. These dietary reference intakes are based on data from 16 men and 18 women who have consumed self-selected diets and have had a decreased Mg^{2+} intake during the balance periods, which could have affected balance values [122,123]. The current RDA for Mg^{2+} ranges from 80 mg/day for children 1–3 years of age to 130 mg/day for children 4–8 years of age. For older males, the RDA for Mg^{2+} ranges from as low as 240 mg/day (range, 9–13 years of age) and increases to 420 mg/day for males 31–70 years of age and older. For females, the RDA ranges from 240 mg/day (9–13 years of age) to 360 mg/day for females 14–18 years of age. The RDA for females 31–70 years of age and older is 320 mg/day. Many nutritional experts feel the ideal intake for Mg^{2+} should be based on the body weight (e.g., 4–6 mg per kg/day) [4]. Intravenous Mg^{2+} supplementation may be more effective, but this treatment has the disadvantage that it requires regular hospital visits. The treatment regimen of intravenous Mg^{2+} supplementation normally consists of 8–12 g of Mg-sulfate in the first 24 h followed by 4–6 g/day for 3 or 4 days. When serum Mg^{2+} levels are extremely low or are accompanied by hypokalemia, Mg^{2+} supplementation may not be sufficient to restore normal Mg^{2+} levels. In that case, patients are often cosupplemented with K^+ or receive amiloride to prevent K^+ secretion [124]. Recent reports indicate that individuals with serum Mg^{2+} concentrations >0.75 mmol/L, or high as 0.85 mmol/L, could be Mg^{2+}-deficient. Thus, to assess Mg^{2+} status of an individual with a serum Mg^{2+} concentration between 0.75 and 0.85 mmol/L is requires additional measures of status. A urinary excretion of <80 mg (3.29 mmol)/day and/or dietary intake history showing a Mg^{2+} intake of <250 mg/day would support the presence of MgD [122]. In the treatment of MgD are recommended organic bound Mg^{2+} salts, such as Mg^{2+} citrate, gluconate, orotate, or aspartate, because of their high bioavailability [4,125].

Hypermagnesemia is a rare condition and is seen most often in patients with renal impairment who take medicines containing Mg^{2+}. Excessive intake of supplemental Mg^{2+} can result in adverse effects, especially in impaired renal function. Serum concentrations >8 mmol/L cause drowsiness, vasodilation, slowing of atrioventricular conduction, and hypotension [126–128].

3.2. *Food and Water Sources of Mg^{2+}*

3.2.1. Mg^{2+} Intake from Food

In the Western World, dietary intake of Mg^{2+} is subnormal, with shortfalls of between 65 and 225 mg of Mg^{2+}/day, depending upon geographic region. Several epidemiologic studies in North America and Europe have shown that children and adults, some which are pregnant women, consuming Western-type diets are low in Mg^{2+} content (i.e., 30–50% of the RDA for these populations) [129]. Epidemiological observations suggest a negative correlation between dietary Mg^{2+} intake and BP [11]. Overall, the current evidence supports the importance of adequate dietary Mg^{2+} intake for the reduction of BP and total CVD risk. These findings support the importance of increasing the consumption of Mg^{2+}-rich foods, including fruits, vegetables, nuts, and whole grains in the treatment and prevention of high BP [125].

A Mg^{2+}-rich diet should be encouraged in hypertensive subjects as well as in predisposed communities because of the advantages of such a diet in the prevention of HTN [121]. The Dietary Approaches to Stop HTN (DASH) diet (originally termed the "combination diet"4) contains larger amounts of Mg^{2+}, K^+, Ca^{2+}, dietary fiber, and protein and smaller amounts of total and saturated fat and cholesterol than the typical diet [130]. The Mediterranean diet is also rich in Mg^{2+}, dietary fiber, antioxidant capacity, and polyphenolic compounds [131]. In trials of vegetarian diets, replacing animal products with vegetable products reduced BP in normotensive and hypertensive people. Aspects of vegetarian diets believed to reduce BP include their high levels of fiber and minerals (such as Mg^{2+} and K^+) and their reduced fat content. In observational studies, significant inverse associations of BP with intake of Mg^{2+}, K^+, Ca^{2+}, and fiber have also been reported [132].

3.2.2. Mg^{2+} Intake from Water

Water is a variable source of Mg^{2+} intake. Typically, water with increased "hardness" has a higher concentration of Mg^{2+} salts. Since this varies depending on the area from which water comes, Mg^{2+} intake from water is usually not estimated to a sufficient extent. This omission may lead to impaired assessment and underestimation of total intake of Mg^{2+} in certain regions [123]. The modern processed food diet, which is low in Mg^{2+} and is spreading globally, makes this well-researched potential of drinking-water Mg^{2+} worth serious consideration, especially in areas where insufficient dietary intake of Mg^{2+} is prevalent. It would be wise and forward-thinking for public health officials to consider how high-Mg^{2+} drinking water might be made available to communities, i.e., water with Mg^{2+} levels of at least 10 mg/L and ideally 25–100 mg/L [133].

4. Conclusions

The enhancing effect of MgD on BP should be considered in the context of total intake and loss of Mg^{2+} in each individual patient with HTN. Special attention should be given to the risk groups in which serum Mg^{2+} levels should be monitored periodically. Considering the numerous positive effects of Mg^{2+} on a number of mechanisms related to HTN, consuming a healthy diet that provides the recommended amount of Mg^{2+} can be an appropriate strategy for helping control BP.

Author Contributions: K.K. designed the review, interpreted the data, and wrote the manuscript; K.K. and L.H. performed the literature research. The authors approved the final version of the manuscript.

Acknowledgments: This review was accomplished with the financial support from the Medical University-Pleven, Bulgaria.

Conflicts of Interest: The authors declare no conflict of interest.

Abbreviations

MgD	Magnesium deficiency
Mg^{2+}	Magnesium
Ca^{2+}	Calcium
K^+	Potassium
Na^+	Sodium
HTN	Hypertension
MetS	Metabolic syndrome
T2D	Type 2 diabetes
RDA	Recommended Dietary Allowance
EAR	Estimated Average Requirement
CVD	Cardiovascular diseases
BP	Blood pressure
WHO	World Health Organization
TRPM6	Transient receptor potential melastatin-6 channel
TRPM7	Transient receptor potential melastatin-7 channel

VSMCs	Vascular smooth muscle cells
ET-1	Endothelin-1
ATII	Angiotensin II
ALDO	Aldosterone
SHR	Spontaneously hypertensive rats
WKY	Wistar-Kyoto rats
EGF	Epidermal growth factor
PTH	Parathyroid hormone
IP3	Inositol-1,4,5-trisphosphate
PIP2	Phosphatidylinositol 4,5-bisphosphate
AT1	Angiotensin II receptor type 1
ETA	Endothelin A receptor
V1a	Vasopressin receptor 1a
PLC	Phospholipase C
PKC	Protein kinase C
PGI_2	Prostacyclin
NO	Nitric oxide
RBCs	Red blood cells
eNOS	Endothelial nitric oxide synthase
PGE_1	Prostaglandin E_1
RAAS	Renin-Angiotensin-Aldosterone System
NA	Noradrenaline
Adr	Adrenaline
CA	Catecholamines
ACh	Acetylcholine
ADCY	Adenylate cyclase
ISO	Isoproterenol
CRP	C-reactive protein
IL	Interleukin
TNF-α	Tumor necrosis factor-α
VCAM	Vascular cell adhesion molecule-1
PAI-1	Plasminogen activator inhibitor-1
ECM	Extracellular matrix
HAS	Hyaluronan synthases
HYAL	Hyaluronidase
TG2	Transglutaminase
LOX	Lysyl oxidase
MMPs	Matrix metalloproteinases
TIMPs	Tissue inhibitors of metalloproteinases
NF-κB	Nuclear factor kappa B
LDL	Low-density lipoproteins
HDL	High-density lipoproteins
IR	Insulin receptor
MAPK	Mitogen-activated protein kinase
PI3K	Phosphoinositide 3-kinase
PIP2	Phosphatidylinositol 4,5-bisphosphate
GLUT4	Glucose transporter type 4
IRTK	Insulin receptor tyrosine kinase
CNS	Central nervous system
HPAA	Hypothalamic Pituitary Adrenal Axis
ACTH	Adrenocorticotropic hormone
NMDA	*N*-methyl-D-aspartate receptor
mGluRs	Metabotropic glutamate receptors
GABA	Gamma-aminobutyric acid
COMT	Catechol-*O*-methyl transferase
MGL	Magnesium low blood levels
MGH	Magnesium high blood levels

References

1. Ismail, A.A.; Ismail, N.A. Magnesium: A mineral essential for health yet generally underestimated or even ignored. *J. Nutr. Food Sci.* **2016**, *6*, 2. [CrossRef]
2. Choi, M.K.; Bae, Y.J. Association of magnesium intake with high blood pressure in Korean adults: Korea national health and nutrition examination survey 2007–2009. *PLoS ONE* **2015**, *10*, e0130405. [CrossRef] [PubMed]
3. Swaminathan, R. Magnesium metabolism and its disorders. *Clin. Biochem. Rev.* **2003**, *24*, 47. [PubMed]
4. Gröber, U.; Schmidt, J.; Kisters, K. Magnesium in prevention and therapy. *Nutrients* **2015**, *7*, 8199–8226. [CrossRef] [PubMed]
5. De Baaij, J.H.; Hoenderop, J.G.; Bindels, R.J. Regulation of magnesium balance: Lessons learned from human genetic disease. *Clin. Kidney J.* **2012**, *5*, i15–i24. [CrossRef] [PubMed]
6. Seo, J.W.; Park, T.J. Magnesium metabolism. *Electrolyte Blood Press.* **2008**, *6*, 86–95. [CrossRef] [PubMed]
7. Anand, G.; Konrad, M.; Heuss, L.T.; Schorn, R. A Case of Severe Hypomagnesaemia. *Open J. Nephrol.* **2013**, *3*, 66. [CrossRef]
8. Qu, X.; Jin, F.; Hao, Y.; Li, H.; Tang, T.; Wang, H.; Yan, W.; Dai, K. Magnesium and the risk of cardiovascular events: A meta-analysis of prospective cohort studies. *PLoS ONE* **2013**, *8*, e57720. [CrossRef] [PubMed]
9. Kass, L.; Sullivan, K.R. Low Dietary Magnesium Intake and Hypertension. *World J. Cardiovasc. Dis.* **2016**, *6*, 447. [CrossRef]
10. Galan, P.; Preziosi, P.; Durlach, V.; Valeix, P.; Ribas, L.; Bouzid, D.; Favier, A.; Hercberg, S. Dietary magnesium intake in a French adult population. *Magnes. Res.* **1997**, *10*, 321–328. [PubMed]
11. Zhang, X.; Li, Y.; Del Gobbo, L.C.; Rosanoff, A.; Wang, J.; Zhang, W.; Song, Y. Effects of magnesium supplementation on blood pressure: A meta-analysis of randomized double-blind placebo-controlled trials. *Hypertension* **2016**, *68*, 324–333. [CrossRef] [PubMed]
12. Harvey, A.; Montezano, A.C.; Lopes, R.A.; Rios, F.; Touyz, R.M. Vascular fibrosis in aging and hypertension: Molecular mechanisms and clinical implications. *Can. J. Cardiol.* **2016**, *32*, 659–668. [CrossRef] [PubMed]
13. Landau, R.; Scott, J.A.; Smiley, R.M. Magnesium-induced vasodilation in the dorsal hand vein. *BJOG* **2004**, *111*, 446–451. [CrossRef] [PubMed]
14. Cunha, A.R.; Medeiros, F.; Umbelino, B.; Oigman, W.; Touyz, R.M.; Neves, M.F. Altered vascular structure and wave reflection in hypertensive women with low magnesium levels. *J. Am. Soc. Hypertens.* **2013**, *7*, 344–352. [CrossRef] [PubMed]
15. Touyz, R.M.; Yao, G. Modulation of vascular smooth muscle cell growth by magnesium-role of mitogen-activated protein kinases. *J. Cell. Physiol.* **2003**, *197*, 326–335. [CrossRef] [PubMed]
16. Belin, R.J.; He, K. Magnesium physiology and pathogenic mechanisms that contribute to the development of the metabolic syndrome. *Magnes. Res.* **2007**, *20*, 107–129. [PubMed]
17. Houston, M. The role of magnesium in hypertension and cardiovascular disease. *J. Clin. Hypertens. (Greenwich)* **2011**, *13*, 843–847. [CrossRef] [PubMed]
18. Paravicini, T.M.; Chubanov, V.; Gudermann, T. TRPM7: A unique channel involved in magnesium homeostasis. *Int. J. Biochem. Cell Biol.* **2012**, *44*, 1381–1384. [CrossRef] [PubMed]
19. Baldoli, E.; Castiglioni, S.; Maier, J.A. Regulation and function of TRPM7 in human endothelial cells: TRPM7 as a potential novel regulator of endothelial function. *PLoS ONE* **2013**, *8*, e59891. [CrossRef] [PubMed]
20. Antunes, T.T.; Callera, G.E.; He, Y.; Yogi, A.; Ryazanov, A.G.; Ryazanova, L.V.; Zhai, A.; Stewart, D.J.; Shrier, A.; Touyz, R.M. Transient receptor potential melastatin 7 cation channel kinase: New player in angiotensin II-induced hypertension. *Hypertension* **2016**, *67*, 763–773. [CrossRef] [PubMed]
21. Touyz, R.M.; He, Y.; Montezano, A.C.; Yao, G.; Chubanov, V.; Gudermann, T.; Callera, G.E. Differential regulation of transient receptor potential melastatin 6 and 7 cation channels by ANG II in vascular smooth muscle cells from spontaneously hypertensive rats. *Am. J. Physiol. Regul. Integr. Comp. Physiol.* **2006**, *290*, R73–R78. [CrossRef] [PubMed]
22. Baldoli, E.; Maier, J.A. Silencing TRPM7 mimics the effects of magnesium deficiency in human microvascular endothelial cells. *Angiogenesis* **2012**, *15*, 47–57. [CrossRef] [PubMed]
23. Di, A.; Malik, A.B. TRP channels and the control of vascular function. *Curr. Opin. Pharmacol.* **2010**, *10*, 127–132. [CrossRef] [PubMed]

24. Groenestege, W.M.T.; Thébault, S.; van der Wijst, J.; van den Berg, D.; Janssen, R.; Tejpar, S.; van den Heuvel, L.P.; van Cutsem, E.; Hoenderop, J.G.; Knoers, N.V.; et al. Impaired basolateral sorting of pro-EGF causes isolated recessive renal hypomagnesemia. *J. Clin. Investig.* **2007**, *117*, 2260–2267. [CrossRef] [PubMed]
25. Thebault, S.; Alexander, R.T.; Groenestege, W.M.T.; Hoenderop, J.G.; Bindels, R.J. EGF increases TRPM6 activity and surface expression. *J. Am. Soc. Nephrol.* **2009**, *20*, 78–85. [CrossRef] [PubMed]
26. Groenestege, W.M.T.; Hoenderop, J.G.; van den Heuvel, L.; Knoers, N.; Bindels, R.J. The epithelial Mg^{2+} channel transient receptor potential melastatin 6 is regulated by dietary Mg^{2+} content and estrogens. *J. Am. Soc. Nephrol.* **2006**, *17*, 1035–1043. [CrossRef] [PubMed]
27. Cao, G.; van der Wijst, J.; van der Kemp, A.; van Zeeland, F.; Bindels, R.J.; Hoenderop, J.G. Regulation of the epithelial Mg^{2+} channel TRPM6 by estrogen and the associated repressor protein of estrogen receptor activity (REA). *J. Biol. Chem.* **2009**, *284*, 14788–14795. [CrossRef] [PubMed]
28. Van Der Wijst, J.; Hoenderop, J.G.; Bindels, R.J. Epithelial Mg^{2+} channel TRPM6: Insight into the molecular regulation. *Magnes. Res.* **2009**, *22*, 127–132. [PubMed]
29. Mubagwa, K.; Gwanyanya, A.; Zakharov, S.; Macianskiene, R. Regulation of cation channels in cardiac and smooth muscle cells by intracellular magnesium. *Arch. Biochem. Biophys.* **2007**, *458*, 73–89. [CrossRef] [PubMed]
30. Ritchie, G.; Kerstan, D.; Dai, L.J.; Kang, H.S.; Canaff, L.; Hendy, G.N.; Quamme, G.A. 1,25$(OH)_2D3$ stimulates Mg^{2+} uptake into MDCT cells: Modulation by extracellular Ca^{2+} and Mg^{2+}. *Am. J. Physiol. Ren. Physiol.* **2001**, *280*, F868–F878. [CrossRef] [PubMed]
31. Hagström, E.; Ahlström, T.; Ärnlöv, J.; Larsson, A.; Melhus, H.; Hellman, P.; Lind, L. Parathyroid hormone and calcium are independently associated with subclinical vascular disease in a community-based cohort. *Atherosclerosis* **2015**, *238*, 420–426. [CrossRef] [PubMed]
32. Jahnen-Dechent, W.; Ketteler, M. Magnesium basics. *Clin. Kidney J.* **2012**, *5*, i3–i14. [CrossRef] [PubMed]
33. Louvet, L.; Bazin, D.; Büchel, J.; Steppan, S.; Passlick-Deetjen, J.; Massy, Z.A. Characterisation of calcium phosphate crystals on calcified human aortic vascular smooth muscle cells and potential role of magnesium. *PLoS ONE* **2015**, *10*, e0115342. [CrossRef] [PubMed]
34. Chakraborti, S.; Chakraborti, T.; Mandal, M.; Mandal, A.; Das, S.; Ghosh, S. Protective role of magnesium in cardiovascular diseases: A review. *Mol. Cell. Biochem.* **2002**, *238*, 163–179. [CrossRef] [PubMed]
35. Villa-Bellosta, R. Impact of magnesium: Calcium ratio on calcification of the aortic wall. *PLoS ONE* **2017**, *12*, e0178872. [CrossRef] [PubMed]
36. Kolte, D.; Vijayaraghavan, K.; Khera, S.; Sica, D.A.; Frishman, W.H. Role of magnesium in cardiovascular diseases. *Cardiol. Rev.* **2014**, *22*, 182–192. [CrossRef] [PubMed]
37. Cunha, A.R.; Umbelino, B.; Correia, M.L.; Neves, M.F. Magnesium and vascular changes in hypertension. *Int. J. Hypertens.* **2012**, *2012*, 754250. [CrossRef] [PubMed]
38. Sontia, B.; Touyz, R.M. Role of magnesium in hypertension. *Arch. Biochem. Biophys.* **2007**, *458*, 33–39. [CrossRef] [PubMed]
39. Kharitonova, M.; Iezhitsa, I.; Zheltova, A.; Ozerov, A.; Spasov, A.; Skalny, A. Comparative angioprotective effects of magnesium compounds. *J. Trace Elem. Med. Biol.* **2015**, *29*, 227–234. [CrossRef] [PubMed]
40. Laurant, P.; Dalle, M.; Berthelot, A.; Rayssiguier, Y. Time-course of the change in blood pressure level in magnesium-deficient Wistar rats. *Br. J. Nutr.* **1999**, *82*, 243–251. [PubMed]
41. Cantin, M. Relationship of juxtaglomerular apparatus and adrenal cortex to biochemical and extracellular fluid volume changes in magnesium deficiency. *Lab. Investig.* **1970**, *22*, 558–568. [PubMed]
42. James, M.F.; Beer, R.E.; Esser, J.D. Intravenous magnesium sulfate inhibits catecholamine release associated with tracheal intubation. *Anesth. Analg.* **1989**, *68*, 772–776. [CrossRef] [PubMed]
43. Douglas, W.W.; Rubin, R.P. The mechanism of catecholamine release from the adrenal medulla and the role of calcium in stimulus-secretion coupling. *J. Physiol.* **1963**, *167*, 288–310. [CrossRef] [PubMed]
44. Torshin, I.Y.; Gromova, O.A.; Gusev, E.I. Mechanisms of antistress and antidepressant action of magnesium and pyridoxine. *Korsakov's J. Neurol. Psychiatry* **2009**, *109*, 107–111.
45. Oka, M.; Ohuchi, T.; Yoshida, H.; Imaizumi, R. Stimulatory effect of adenosine triphosphate and magnesium on the release of catecholamines from adrenal medullary granules. *Jpn. J. Pharmacol.* **1967**, *17*, 199–207. [CrossRef] [PubMed]
46. Farmer, J.B.; Campbell, I.K. Calcium and magnesium ions: Influence on the response of an isolated artery to sympathetic nerve stimulation, noradrenaline and tyramine. *Br. J. Pharmacol.* **1967**, *29*, 319–328. [CrossRef]

47. Whyte, K.F.; Addis, G.J.; Whitesmith, R.; Reid, J.L. Adrenergic control of plasma magnesium in man. *Clin. Sci.* **1987**, *72*, 135–138. [CrossRef] [PubMed]
48. Joborn, H.; Akerström, G.; Ljunghall, S. Effects of exogenous catecholamines and exercise on plasma magnesium concentrations. *Clin. Endocrinol.* **1985**, *23*, 219–226. [CrossRef]
49. Keenan, D.; Romani, A.; Scarpa, A. Differential regulation of circulating Mg^{2+} in the rat by β_1-and β_2-adrenergic receptor stimulation. *Circ. Res.* **1995**, *77*, 973–983. [CrossRef] [PubMed]
50. Boos, C.J.; Lip, G.Y. Is hypertension an inflammatory process? *Curr. Pharm. Des.* **2006**, *12*, 1623–1635. [CrossRef] [PubMed]
51. Arnaud, M.J. Update on the assessment of magnesium status. *Br. J. Nutr.* **2008**, *99*, S24–S36. [CrossRef] [PubMed]
52. Barbagallo, M.; Belvedere, M.; Dominguez, L.J. Magnesium homeostasis and aging. *Magnes. Res.* **2009**, *22*, 235–246. [PubMed]
53. King, D.E.; Mainous, A.G., III; Geesey, M.E.; Ellis, T. Magnesium intake and serum C-reactive protein levels in children. *Magnes. Res.* **2007**, *20*, 32–36. [PubMed]
54. King, D.E.; Mainous, A.G., III; Geesey, M.E.; Woolson, R.F. Dietary magnesium and C-reactive protein levels. *J. Am. Coll. Nutr.* **2005**, *24*, 166–171. [CrossRef] [PubMed]
55. Kim, D.J.; Xun, P.; Liu, K.; Loria, C.; Yokota, K.; Jacobs, D.R.; He, K. Magnesium intake in relation to systemic inflammation, insulin resistance, and the incidence of diabetes. *Diabetes Care* **2010**, *33*, 2604–2610. [CrossRef] [PubMed]
56. Maier, J.A.; Malpuech-Brugère, C.; Zimowska, W.; Rayssiguier, Y.; Mazur, A. Low magnesium promotes endothelial cell dysfunction: Implications for atherosclerosis, inflammation and thrombosis. *Biochim. Biophys. Acta Mol. Basis Dis.* **2004**, *1689*, 13–21. [CrossRef] [PubMed]
57. Friese, R.S.; Rao, F.; Khandrika, S.; Thomas, B.; Ziegler, M.G.; Schmid-Schönbein, G.W.; O'Connor, D.T. Matrix metalloproteinases: Discrete elevations in essential hypertension and hypertensive end-stage renal disease. *Clin. Exp. Hypertens.* **2009**, *31*, 521–533. [CrossRef] [PubMed]
58. Heagerty, A.M.; Aalkjaer, C.; Bund, S.J.; Korsgaard, N.; Mulvany, M.J. Small artery structure in hypertension. Dual processes of remodeling and growth. *Hypertension* **1993**, *21*, 391–397. [CrossRef] [PubMed]
59. Mulvany, M.J.; Baumbach, G.L.; Aalkjaer, C.; Heagerty, A.M.; Korsgaard, N.; Schiffrin, E.L.; Heistad, D.D. Vascular remodeling. *Hypertension* **1996**, *28*, 505–506. [PubMed]
60. Mulvany, M.J. Small artery remodeling and significance in the development of hypertension. *Physiology* **2002**, *17*, 105–109. [CrossRef]
61. Steppan, J.; Bergman, Y.; Viegas, K.; Armstrong, D.; Tan, S.; Wang, H.; Melucci, S.; Hori, D.; Park, S.Y.; Barreto, S.F.; et al. Tissue transglutaminase modulates vascular stiffness and function through crosslinking-dependent and crosslinking-independent functions. *J. Am. Heart. Assoc.* **2017**, *6*, e004161. [CrossRef] [PubMed]
62. Bains, W. Transglutaminse 2 and EGGL, the protein cross-link formed by transglutaminse 2, as therapeutic targets for disabilities of old age. *Rejuvenation Res.* **2013**, *16*, 495–517. [CrossRef] [PubMed]
63. Torshin, I.Y.; Gromova, O.A. Connective tissue dysplasia, cell biology and molecular mechanisms of magnesium exposure. *RMZh* **2008**, *16*, 230–238.
64. Martínez-Revelles, S.; García-Redondo, A.B.; Avendaño, M.S.; Varona, S.; Palao, T.; Orriols, M.; Roque, F.R.; Fortuño, A.; Touyz, R.M.; Martínez-González, J.; et al. Lysyl oxidase induces vascular oxidative stress and contributes to arterial stiffness and abnormal elastin structure in hypertension: Role of p38MAPK. *Antioxid. Redox Signal.* **2017**, *27*, 379–397. [CrossRef] [PubMed]
65. Avtandilov, A.G.; Pukhaeva, A.A.; Manizer, E.D. Magnesium and mitral valve prolapse. Efficiency and application points. *Ration. Pharmacother. Card.* **2010**, *6*, 677–684. [CrossRef]
66. Maier, J.A. Endothelial cells and magnesium: Implications in atherosclerosis. *Clin. Sci.* **2012**, *122*, 397–407. [CrossRef] [PubMed]
67. Dolinsky, B.M.; Ippolito, D.L.; Tinnemore, D.; Stallings, J.D.; Zelig, C.M.; Napolitano, P.G. The effect of magnesium sulfate on the activity of matrix metalloproteinase-9 in fetal cord plasma and human umbilical vein endothelial cells. *Am. J. Obstet. Gynecol.* **2010**, *203*, 371e1–371e5. [CrossRef] [PubMed]
68. Kostov, K. Pathophysiological Role of Endothelin-1, Matrix Metalloproteinases-2, -9 and Magnesium in the Development and Dynamics of Vascular Changes at Various Degrees of Arterial Hypertension. Doctoral Disertation, Medical University of Pleven, Pleven, Bulgaria, 2 December 2016. Available online: http://www.mu-pleven.bg/procedures/59/Avtoreferat.pdf (accessed on 1 December 2016).

69. Ferrè, S.; Baldoli, E.; Leidi, M.; Maier, J.A. Magnesium deficiency promotes a pro-atherogenic phenotype in cultured human endothelial cells via activation of NFkB. *Biochim. Biophys. Acta Mol. Basis Dis.* **2010**, *1802*, 952–958. [CrossRef] [PubMed]
70. Yue, H.; Lee, J.D.; Shimizu, H.; Uzui, H.; Mitsuke, Y.; Ueda, T. Effects of magnesium on the production of extracellular matrix metalloproteinases in cultured rat vascular smooth muscle cells. *Atherosclerosis* **2003**, *166*, 271–277. [CrossRef]
71. Yue, H.; Uzui, H.; Lee, J.D.; Shimizu, H.; Ueda, T. Effects of magnesium on matrix metalloproteinase-2 production in cultured rat cardiac fibroblasts. *Basic Res. Cardiol.* **2004**, *99*, 257–263. [CrossRef] [PubMed]
72. Altura, B.M. Insights into the possible mechanisms by which platelet-activating factor and PAF-receptors function in vascular smooth muscle in magnesium deficiency and vascular remodeling: Possible links to atherogenesis, hypertension and cardiac failure. *Int. J. Cardiol. Res.* **2016**, *3*, 1–3.
73. Kisters, K.; Gröber, U. Magnesium in health and disease. *Plant Soil* **2013**, *368*, 155–165. [CrossRef]
74. Sakaguchi, Y.; Fujii, N.; Shoji, T.; Hayashi, T.; Rakugi, H.; Isaka, Y. Hypomagnesemia is a significant predictor of cardiovascular and non-cardiovascular mortality in patients undergoing hemodialysis. *Kidney Int.* **2014**, *85*, 174–181. [CrossRef] [PubMed]
75. Mishra, S.; Padmanaban, P.; Deepti, G.N.; Sarkar, G.; Sumathi, S.; Toora, B.D. Serum magnesium and dyslipidemia in type 2 diabetes mellitus. *Biomed. Res.* **2012**, *23*, 295–300.
76. King, D.E. Inflammation and elevation of C-reactive protein: Does magnesium play a key role? *Magnes. Res.* **2009**, *22*, 57–59. [PubMed]
77. Barbagallo, M.; Dominguez, L.J. Magnesium metabolism in type 2 diabetes mellitus, metabolic syndrome and insulin resistance. *Arch. Biochem. Biophys.* **2007**, *458*, 40–47. [CrossRef] [PubMed]
78. Geiger, H.; Wanner, C. Magnesium in disease. *Clin. Kidney J.* **2012**, *5*, i25–i38. [CrossRef] [PubMed]
79. Kostov, K.; Blazhev, A.; Atanasova, M.; Dimitrova, A. Serum concentrations of endothelin-1 and matrix metalloproteinases-2,-9 in pre-hypertensive and hypertensive patients with type 2 diabetes. *Int. J. Mol. Sci.* **2016**, *17*, 1182. [CrossRef] [PubMed]
80. De Boer, I.H.; Bangalore, S.; Benetos, A.; Davis, A.M.; Michos, E.D.; Muntner, P.; Rossing, P.; Zoungas, S.; Bakris, G. Diabetes and hypertension: A position statement by the American Diabetes Association. *Diabetes Care* **2017**, *40*, 1273–1284. [CrossRef] [PubMed]
81. Guerrero-Romero, F.; Bermudez-Peña, C.; Rodríguez-Morán, M. Severe hypomagnesemia and low-grade inflammation in metabolic syndrome. *Magnes. Res.* **2011**, *24*, 45–53. [PubMed]
82. Nielsen, F.H. Magnesium, inflammation, and obesity in chronic disease. *Nutr. Rev.* **2010**, *68*, 333–340. [CrossRef] [PubMed]
83. Chaudhary, D.P.; Sharma, R.; Bansal, D.D. Implications of magnesium deficiency in type 2 diabetes: A review. *Biol. Trace Elem. Res.* **2010**, *134*, 119–129. [CrossRef] [PubMed]
84. Günther, T. The biochemical function of Mg^{2+} in insulin secretion, insulin signal transduction and insulin resistance. *Magnes. Res.* **2010**, *23*, 5–18. [PubMed]
85. Rosolova, H.; Mayer, O., Jr.; Reaven, G.M. Insulin-mediated glucose disposal is decreased in normal subjects with relatively low plasma magnesium concentrations. *Metabolism* **2000**, *49*, 418–420. [CrossRef]
86. Venu, L.; Padmavathi, I.J.; Kishore, Y.D.; Bhanu, N.V.; Rao, K.R.; Sainath, P.B.; Ganeshan, M.; Raghunath, M. Long-term effects of maternal magnesium restriction on adiposity and insulin resistance in rat pups. *Obesity* **2008**, *16*, 1270–1276. [CrossRef] [PubMed]
87. Suarez, A.; Pulido, N.; Casla, A.; Casanova, B.; Arrieta, F.J.; Rovira, A. Impaired tyrosine-kinase activity of muscle insulin receptors from hypomagnesaemic rats. *Diabetologia* **1995**, *38*, 1262–1270. [CrossRef] [PubMed]
88. Barbagallo, M.; Dominguez, L.J. Magnesium and the cardiometabolic syndrome. *Curr. Nutr. Rep.* **2012**, *1*, 100–108. [CrossRef]
89. Ward, C.W.; Lawrence, M.C. Ligand-induced activation of the insulin receptor: A multi-step process involving structural changes in both the ligand and the receptor. *Bioessays* **2009**, *31*, 422–434. [CrossRef] [PubMed]
90. Jensen, M.; Hansen, B.; De Meyts, P.; Schäffer, L.; Ursø, B. Activation of the insulin receptor by insulin and a synthetic peptide leads to divergent metabolic and mitogenic signaling and responses. *J. Biol. Chem.* **2007**, *282*, 35179–35186. [CrossRef] [PubMed]
91. Farese, R.V.; Sajan, M.P.; Standaert, M.L. Insulin-sensitive protein kinases (atypical protein kinase C and protein kinase B/Akt): Actions and defects in obesity and type II diabetes. *Exp. Biol. Med.* **2005**, *230*, 593–605. [CrossRef]

92. Sale, E.M.; Sale, G.J. Protein kinase B: Signalling roles and therapeutic targeting. *Cell. Mol. Life Sci.* **2008**, *65*, 113–127. [CrossRef] [PubMed]
93. Sano, H.; Kane, S.; Sano, E.; Mîinea, C.P.; Asara, J.M.; Lane, W.S.; Garner, C.W.; Lienhard, G.E. Insulin-stimulated phosphorylation of a Rab GTPase-activating protein regulates GLUT4 translocation. *J. Biol. Chem.* **2003**, *278*, 14599–14602. [CrossRef] [PubMed]
94. Siddle, K. Signalling by insulin and IGF receptors: Supporting acts and new players. *J. Mol. Endocrinol.* **2011**, *47*, R1–R10. [CrossRef] [PubMed]
95. Barbagallo, M.; Gupta, R.K.; Bardicef, O.; Bardicef, M.; Resnick, L.M. Altered ionic effects of insulin in hypertension: Role of basal ion levels in determining cellular responsiveness. *J. Clin. Endocrinol. Metab.* **1997**, *82*, 1761–1765. [CrossRef] [PubMed]
96. Rodríguez-Morán, M.; Guerrero-Romero, F. Oral magnesium supplementation improves insulin sensitivity and metabolic control in type 2 diabetic subjects: A randomized double-blind controlled trial. *Diabetes Care* **2003**, *26*, 1147–1152. [CrossRef] [PubMed]
97. Häring, H.U. The insulin receptor: Signalling mechanism and contribution to the pathogenesis of insulin resistance. *Diabetologia* **1991**, *34*, 848–861. [CrossRef] [PubMed]
98. Nadler, J.L.; Rude, R.K. Disorders of magnesium metabolism. *Endocrinol. Metab. Clin. N. Am.* **1995**, *24*, 623–641.
99. Chaudhary, D.P.; Boparai, R.K.; Sharma, R.; Bansal, D.D. Studies on the development of an insulin resistant rat model by chronic feeding of low magnesium high sucrose diet. *Magnes. Res.* **2004**, *17*, 293–300. [PubMed]
100. Lopez-Ridaura, R.; Willett, W.C.; Rimm, E.B.; Liu, S.; Stampfer, M.J.; Manson, J.E.; Hu, F.B. Magnesium intake and risk of type 2 diabetes in men and women. *Diabetes Care* **2004**, *27*, 134–140. [CrossRef] [PubMed]
101. Van Dam, R.M.; Hu, F.B.; Rosenberg, L.; Krishnan, S.; Palmer, J.R. Dietary calcium and magnesium, major food sources, and risk of type 2 diabetes in US black women. *Diabetes Care* **2006**, *29*, 2238–2243. [CrossRef] [PubMed]
102. Mooren, F.C.; Krüger, K.; Völker, K.; Golf, S.W.; Wadepuhl, M.; Kraus, A. Oral magnesium supplementation reduces insulin resistance in non-diabetic subjects—A double-blind, placebo-controlled, randomized trial. *Diabetes Obes. Metab.* **2011**, *13*, 281–284. [CrossRef] [PubMed]
103. Kalra, S.S.; Shanahan, C.M. Vascular calcification and hypertension: Cause and effect. *Ann. Med.* **2012**, *44*, S85–S92. [CrossRef] [PubMed]
104. Floege, J. Magnesium in CKD: More than a calcification inhibitor? *J. Neurol.* **2015**, *28*, 269–277. [CrossRef] [PubMed]
105. Hruby, A.; O'Donnell, C.J.; Jacques, P.F.; Meigs, J.B.; Hoffmann, U.; McKeown, N.M. Magnesium intake is inversely associated with coronary artery calcification: The Framingham Heart Study. *JACC Cardiovasc. Imaging* **2014**, *7*, 59–69. [CrossRef] [PubMed]
106. Barbagallo, M.; Dominguez, L.J. Magnesium, oxidative stress, and aging muscle. *Aging* **2014**, 157–166. [CrossRef]
107. Lakatta, E.G. The reality of aging viewed from the arterial wall. *Artery Res.* **2013**, *7*, 73–80. [CrossRef] [PubMed]
108. Agyei, B.; Nicolaou, M.; Boateng, L.; Dijkshoorn, H.; Agyemang, C. Relationship between psychosocial stress and hypertension among Ghanaians in Amsterdam, the Netherlands–the GHAIA study. *BMC Public Health* **2014**, *14*, 692. [CrossRef] [PubMed]
109. Pochwat, B.; Szewczyk, B.; Sowa-Kucma, M.; Siwek, A.; Doboszewska, U.; Piekoszewski, W.; Gruca, P.; Papp, M.; Nowak, G. Antidepressant-like activity of magnesium in the chronic mild stress model in rats: Alterations in the NMDA receptor subunits. *Int. J. Neuropsychopharmacol.* **2014**, *17*, 393–405. [CrossRef] [PubMed]
110. Rayssiguier, Y.; Libako, P.; Nowacki, W.; Rock, E. Magnesium deficiency and metabolic syndrome: Stress and inflammation may reflect calcium activation. *Magnes. Res.* **2010**, *23*, 73–80. [PubMed]
111. Boyle, N.B.; Lawton, C.; Dye, L. The effects of magnesium supplementation on subjective anxiety and stress–a systematic review. *Nutrients* **2017**, *9*, 429. [CrossRef] [PubMed]
112. Amyard, N.; Leyris, A.; Monier, C.; Frances, H.; Boulu, R.G.; Henrotte, J.G. Brain catecholamines, serotonin and their metabolites in mice selected for low (MGL) and high (MGH) blood magnesium levels. *Magnes. Res.* **1995**, *8*, 5–9. [PubMed]
113. Henrotte, J.G.; Franck, G.; Santarromana, M.; Francès, H.; Mouton, D.; Motta, R. Mice selected for low and high blood magnesium levels: A new model for stress studies. *Physiol. Behav.* **1997**, *61*, 653–658. [CrossRef]
114. Henrotte, J.G. Type A behavior and magnesium metabolism. *Magnesium* **1986**, *5*, 201–210. [PubMed]
115. Ryzen, E.; Servis, K.L.; Rude, R.K. Effect of intravenous epinephrine on serum magnesium and free intracellular red blood cell magnesium concentrations measured by nuclear magnetic resonance. *J. Am. Coll. Nutr.* **1990**, *9*, 114–119. [CrossRef] [PubMed]

116. Han, H.; Fang, X.; Wei, X.; Liu, Y.; Jin, Z.; Chen, Q.; Fan, Z.; Aaseth, J.; Hiyoshi, A.; He, J.; et al. Dose-response relationship between dietary magnesium intake, serum magnesium concentration and risk of hypertension: A systematic review and meta-analysis of prospective cohort studies. *Nutr. J.* **2017**, *16*, 26. [CrossRef] [PubMed]
117. Ma, J.; Folsom, A.R.; Melnick, S.L.; Eckfeldt, J.H.; Sharrett, A.R.; Nabulsi, A.A.; Hutchinson, R.G.; Metcalf, P.A. Associations of serum and dietary magnesium with cardiovascular disease, hypertension, diabetes, insulin, and carotid arterial wall thickness: The ARIC study. *J. Clin. Epidemiol.* **1995**, *48*, 927–940. [CrossRef]
118. Whang, R.; Chrysant, S.; Dillard, B.; Smith, W.; Fryer, A. Hypomagnesemia and hypokalemia in 1,000 treated ambulatory hypertensive patients. *J. Am. Coll. Nutr.* **1982**, *1*, 317–322. [CrossRef] [PubMed]
119. Rosanoff, A. Magnesium supplements may enhance the effect of antihypertensive medications in stage 1 hypertensive subjects. *Magnes. Res.* **2010**, *23*, 27–40. [PubMed]
120. Touyz, R.M. Magnesium in clinical medicine. *Front. Biosci.* **2004**, *9*, 1278–1293. [CrossRef] [PubMed]
121. Touyz, R.M. Role of magnesium in the pathogenesis of hypertension. *Mol. Aspects Med.* **2003**, *24*, 107–136. [CrossRef]
122. Nielsen, F.H. Guidance for the determination of status indicators and dietary requirements for magnesium. *Magnes. Res.* **2016**, *29*, 154–160. [PubMed]
123. Standing Committee on the Scientific Evaluation of Dietary Reference Intakes; Food and Nutrition Board; Institute of Medicine (IOM). *Dietary Reference Intakes for Calcium, Phosphorus, Magnesium, Vitamin D, and Fluoride*; National Academies Press: Washington, DC, USA, 1997.
124. De Baaij, J.H.; Hoenderop, J.G.; Bindels, R.J. Magnesium in man: Implications for health and disease. *Physiol. Rev.* **2015**, *95*, 1–46. [CrossRef] [PubMed]
125. Rosique-Esteban, N.; Guasch-Ferré, M.; Hernández-Alonso, P.; Salas-Salvadó, J. Dietary Magnesium and Cardiovascular Disease: A Review with Emphasis in Epidemiological Studies. *Nutrients* **2018**, *10*, 168. [CrossRef] [PubMed]
126. Crosby, V.; Elin, R.J.; Twycross, R.; Mihalyo, M.; Wilcock, A. Magnesium. *J. Pain Symptom Manag.* **2013**, *45*, 137–144. [CrossRef] [PubMed]
127. Nishikawa, M.; Shimada, N.; Kanzaki, M.; Ikegami, T.; Fukuoka, T.; Fukushima, M.; Asano, K. The characteristics of patients with hypermagnesemia who underwent emergency hemodialysis. *Acute Med. Surg.* **2018**, *5*, 1–2. [CrossRef]
128. Celi, L.A.; Scott, D.J.; Lee, J.; Nelson, R.; Alper, S.L.; Mukamal, K.J.; Mark, R.G.; Danziger, J. Association of hypermagnesemia and blood pressure in the critically ill. *J. Hypertens.* **2013**, *31*, 2136. [CrossRef] [PubMed]
129. Altura, B.M.; Li, W.; Zhang, A.; Zheng, T.; Shah, N.C. Sudden cardiac death in infants, children and young adults: Possible roles of dietary magnesium intake and generation of platelet-activating factor in coronary arteries. *J. Heart Health* **2016**, *2*, 1–5.
130. Sacks, F.M.; Svetkey, L.P.; Vollmer, W.M.; Appel, L.J.; Bray, G.A.; Harsha, D.; Obarzanek, E.; Conlin, P.R.; Miller, E.R., 3rd; Simons-Morton, D.G.; et al. Effects on blood pressure of reduced dietary sodium and the Dietary Approaches to Stop Hypertension (DASH) diet. *N. Engl. J. Med.* **2001**, *344*, 3–10. [CrossRef] [PubMed]
131. Park, Y.M.; Steck, S.E.; Fung, T.T.; Zhang, J.; Hazlett, L.J.; Han, K.; Lee, S.H.; Kwon, H.S.; Merchant, A.T. Mediterranean diet, Dietary Approaches to Stop Hypertension (DASH) style diet, and metabolic health in US adults. *Clin. Nutr.* **2017**, *36*, 1301–1309. [CrossRef] [PubMed]
132. Appel, L.J.; Moore, T.J.; Obarzanek, E.; Vollmer, W.M.; Svetkey, L.P.; Sacks, F.M.; Bray, G.A.; Vogt, T.M.; Cutler, J.A.; Windhauser, M.M.; et al. A clinical trial of the effects of dietary patterns on blood pressure. *N. Engl. J. Med.* **1997**, *336*, 1117–1124. [CrossRef] [PubMed]
133. Rosanoff, A. The high heart health value of drinking-water magnesium. *Med. Hypotheses* **2013**, *81*, 1063–1065. [CrossRef] [PubMed]

International Journal of
Molecular Sciences

Article

Standardized *Aronia melanocarpa* Extract as Novel Supplement against Metabolic Syndrome: A Rat Model

Vladimir Jakovljevic [1,2,*], Petar Milic [3], Jovana Bradic [4], Jovana Jeremic [4], Vladimir Zivkovic [1], Ivan Srejovic [1], Tamara Nikolic Turnic [4], Isidora Milosavljevic [4], Nevena Jeremic [4], Sergey Bolevich [2], Milica Labudovic Borovic [5], Miroslav Mitrovic [6] and Vesna Vucic [7]

1 Department of Physiology, Faculty of Medical Sciences, University of Kragujevac, Svetozara Markovica 69, 34000 Kragujevac, Serbia; vladimirziv@gmail.com (V.Z.); ivan_srejovic@hotmail.com (I.S.)
2 Department of Human Pathology, 1st Moscow State Medical, University IM Sechenov, Trubetskaya street 8, Moscow 119991, Russia; bolevich2011@yandex.ru
3 Department of Pharmacy, High Medical School of Professional Studies in Cuprija, Lole Ribara 1/2, 35000 Cuprija, Serbia; milic.milic@gmail.com
4 Department of Pharmacy, Faculty of Medical Sciences, University of Kragujevac, Svetozara Markovica 69, 34000 Kragujevac, Serbia; jovanabradickg@gmail.com (J.B.); jovana.jeremic@medf.kg.ac.rs (J.J.); tamara.nikolic@medf.kg.ac.rs (T.N.T.); isidora.stojic@medf.kg.ac.rs (I.M.); nbarudzic@hotmail.com (N.J.)
5 Institute of Histology and Embryology "Aleksandar Dj. Kostic", Faculty of Medicine, University of Belgrade, Dr Subotic 8, 11000 Belgrade, Serbia; milica.borovic@supremalab.rs
6 Pharmanova, Generala Anrija 6, 11010 Belgrade, Serbia; miroslav.mitrovic@pharmanova.com
7 Institute for Medical Research, Centre of Research Excellence in Nutrition and Metabolism, University of Belgrade, Tadeusa Koscuska 1, 11129 Belgrade, Serbia; vesna.vucic.imr@gmail.com
* Correspondence: drvladakgbg@yahoo.com; Tel.: + 381-34-306-800

Received: 30 November 2018; Accepted: 12 December 2018; Published: 20 December 2018

Abstract: The aim of our study was to examine the effects of different dietary strategies, high-fat (HFd) or standard diet (Sd) alone or in combination with standardized oral supplementation (0.45 mL/kg/day) of Aronia melanocarpa extract (SAE) in rats with metabolic syndrome (MetS). SAE is an official product of pharmaceutical company Pharmanova (Belgrade, Serbia); however, the procedure for extraction was done by EU-Chem company (Belgrade, Serbia). Rats were divided randomly into six groups: control with Sd, control with Sd and SAE, MetS with HFd, MetS with HFd and SAE, MetS with Sd and MetS with Sd and SAE during 4 weeks. At the end of the 4-week protocol, cardiac function and liver morphology were assessed, while in the blood samples glucose, insulin, iron levels and systemic redox state were determined. Our results demonstrated that SAE had the ability to lower blood pressure and exert benefits on in vivo and ex vivo heart function. Moreover, SAE improved glucose tolerance, attenuated pathological liver alterations and oxidative stress present in MetS. Obtained beneficial effects of SAE were more prominent in combination with changing dietary habits. Promising potential of SAE supplementation alone or in combination with different dietary protocols in triggering cardioprotection should be further examined in future.

Keywords: metabolic syndrome; *Aronia melanocarpa*; standardized extract; dietary strategies; supplementation

1. Introduction

Metabolic syndrome (MetS) represents one of the metabolic disorders characterized by abdominal obesity, dyslipidemia, hypertension, insulin resistance and diabetes mellitus (DM) type 2 [1,2]. There is a great concern since MetS directly promotes the development of cardiovascular disorders, possibly because it results in increased oxidative stress and low-grade inflammation [3]. Therapeutic approaches involve dietary restriction or a combination of synthetic antidiabetic and hypolipidemic drugs [4]. However, increasing incidence of MetS associated with the undesirable side—effects and high cost of available drugs indicates the need to discover new, less harmful herbal medicines efficient in controlling both blood glucose and lipids [4,5]. Therefore, a lot of plant extracts as well as plant-derived biomoleculs such as polyphenols, have been under research for the prevention and therapy of MetS [1]. It has been well documented that polyphenols, especially anthocyanins and quercetin, exert the potential to enhance the glucose uptake by muscle and adipocyte cells, thus exerting antidiabetic effect [6].

Aronia melanocarpa (*A. melanocarpa*) or black chokeberry is a fruit/plant which belongs to the *Rosaceae* family and is native to North America [7]. However, it has been commonly used in Europe as ingredient for juices, wine, jams, teas and cordial liqueurs [8,9]. *A. melanocarpa* represents one of the richest sources of polyphenols among fruits, with anthocyanins and flavonoids identified as major components responsible for its therapeutic potential [10,11]. Recent researches have focused attention on *A. melanocarpa* due to its numerous health benefits in a broad range of pathological conditions [12]. It has been reported that fruit and extracts of *A. melanocarpa* exert gastroprotective, hepatoprotective, antiinflammatory and antiproliferative activity [12,13]. Furthermore, the health-promoting effects of extracts of this plant involve antiatherosclerotic, antiplatelet and hypoglycemic properties [7,14]. Moreover, it was previously confirmed that *A. melanocarpa* extract may reduce systolic and diastolic pressure and be useful in the management of DM [5,13]. However, to our best knowledge, the effectiveness of *A. melanocarpa* extract in combination with different diet regimens in the treatment of MetS has been not investigated so far.

Therefore, the aim of our study was to examine different dietary strategies, high-fat (HFd) or standard diet (Sd) alone or in combination with standardized *A. melanocarpa* extract (SAE) supplementation, and their potential benefits in the prevention and treatment of various complications in rats with MetS.

2. Results

2.1. Body Mass Index (BMI) of Healthy and Rats with MetS after 4-Weeks of Dietary Changes

BMI was statistically higher in groups with MetS (MetS + HFd, MetS + HFd + SAE, MetS + Sd, MetS + Sd + SAE) than in healthy rats (CTRL, SAE). However, 4 weeks after dietary regime, MetS + Sd and MetS + Sd + SAE groups had significantly lower BMI levels than MetS + HFd and MetS + HFd + SAE, while MetS + Sd + SAE had significantly lower BMI level than MetS + Sd (Figure 1).

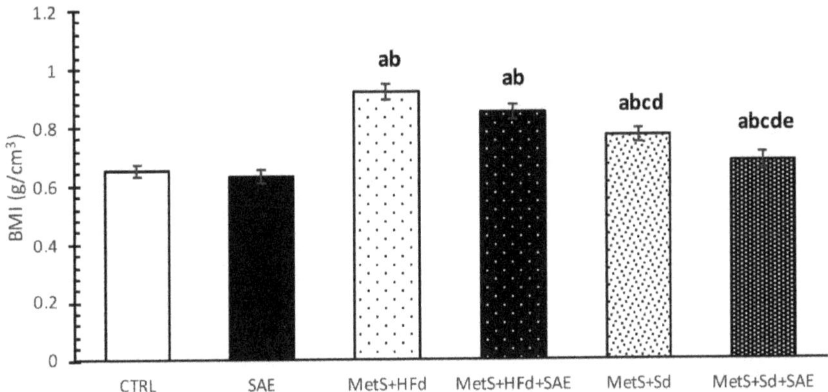

Figure 1. BMI in examined groups 4 weeks after dietary changes. Values are expressed as mean ± standard deviation for 10 animals, for each group. For statistical significance were considered values $p < 0.05$. [a] Statistical significance in relation to control (CTRL) group; [b] Statistical significance in relation to standardized *A. melanocarpa* extract (SAE) group; [c] Statistical significance in relation to MetS + HFd group; [d] Statistical significance in relation to MetS + HFd + SAE group; [e] Statistical significance in relation to MetS + Sd group.

2.2. Changes in Blood Pressure and Heart Rate in Healthy and Rats with MetS on the Different Diet Regime

Systolic and diastolic blood pressure (SBP and DBP), as well as heart rate (HR) were increased in rats with MetS compared to healthy rats, as excepted (Figure 2A–C). More interesting was that with the addition of SAE and Sd in rats with MetS, SBP was statistically decreased compared to MetS + Sd and MetS + HFd + SAE (Figure 2A). DBP was significantly increased in MetS + HFd in comparison with MetS + HFd + SAE, MetS + Sd and MetS + Sd + SAE, while in MetS + Sd + SAE group DBP was significantly lower than in other MetS groups (Figure 2B). SAE treatment in MetS + HFd + SAE group induced significant increase of HR compared to MetS + HFd and MetS + Sd groups (Figure 2C).

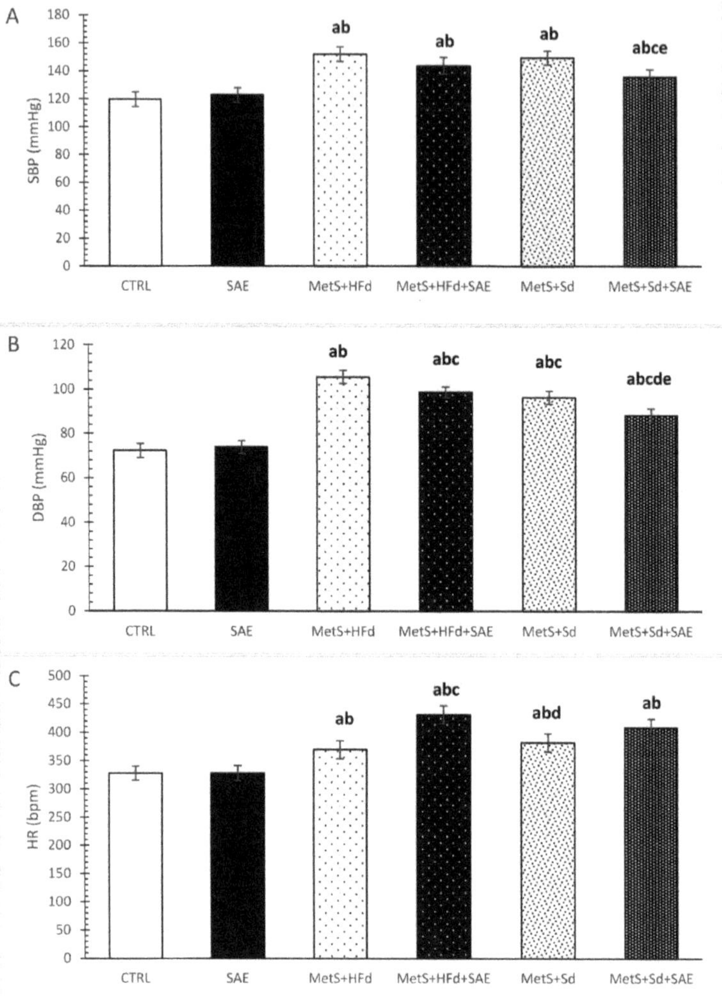

Figure 2. Changes in blood pressure and heart rate in healthy and rats with MetS on the different diet regime: (**A**) systolic blood pressure (SBP, mmHg); (**B**) diastolic blood pressure (DBP, mmHg); (**C**) heart rate (HR, bpm). Values are expressed as mean ± standard deviation for 10 animals, for each group. For statistical significance were considered values $p < 0.05$. [a] Statistical significance in relation to CTRL group; [b] Statistical significance in relation to SAE group; [c] Statistical significance in relation to MetS + HFd group; [d] Statistical significance in relation to MetS + HFd + SAE group; [e] Statistical significance in relation to MetS + Sd group.

2.3. Effect of Dietary Changes in Healthy and Rats with MetS in In Vivo Cardiac Function

SAE supplementation in healthy rats significantly increased interventricular septal wall thickness at end diastole (IVSd), left ventricle posterior wall thickness at end diastole (LVPWd), interventricular septal wall thickness at end systole (IVSs), left ventricle posterior wall thickness at end systole (LVPWs), fractional shortening (FS) and reduced left ventricle internal dimension at end systole (LVIDs) and left ventricle internal dimension at end diastole (LVIDd) compared to control. On the other hand, IVSd and IVSs were significantly decreased while LVIDd was significantly increased in MetS + HFd compared to SAE group. More importantly, FS was statistically decreased in MetS + HFd compared to CTRL and SAE, as well as in MetS + HFd + SAE and MetS + Sd compared to SAE. SAE supplementation in rats with MetS fed with Sd significantly increased LVIDd, IVSs and LVIDs compared to MetS + Sd group (Table 1).

Table 1. Effect of dietary changes in healthy and rats with MetS on in vivo cardiac function.

Milimeters (mm)	CTRL	SAE	MetS + HFd	MetS + HFd + SAE	MetS + Sd	MetS + Sd + SAE
IVSd	1.22 ± 0.2	1.89 ± 0.2 [a]	1.28 ± 0.3 [b]	1.34 ± 0.2	1.58 ± 0.2	1.59 ± 0.2
LVIDd	6.80 ± 0.4	5.73 ± 0.2 [a]	6.23 ± 0.4	6.70 ± 0.4	5.53 ± 0.3	7.93 ± 0.4 [e]
LVPWd	1.94 ± 0.4	3.45 ± 0.4 [a]	2.22 ± 0.1	1.72 ± 0.1 [b]	1.88 ± 0.2 [b]	1.93 ± 0.2 [b]
IVSs	2.48 ± 0.3	3.64 ± 0.3 [a]	2.33 ± 0.1 [b]	2.52 ± 0.2	2.45 ± 0.2 [b]	3.24 ± 0.1 [ace]
LVIDs	3.20 ± 0.5	2.03 ± 0.2 [a]	3.24 ± 0.2 [b]	3.16 ± 0.1	2.48 ± 0.2 [acd]	3.45 ± 0.2 [be]
LVPWs	2.98 ± 0.2	4.97 ± 0.1 [a]	3.16 ± 0.2	3.26 ± 0.2	2.76 ± 0.2 [b]	3.45 ± 0.3
FS (%)	53.2 ± 4.18	65.7 ± 5.01 [a]	49.0 ± 3.99 [ab]	52.2 ± 4.32 [b]	51.4 ± 4.19 [b]	56.3 ± 4.21

Values are expressed as mean ± standard deviation for 10 animals, for each group. For statistical significance were considered values $p < 0.05$. [a] Statistical significance in relation to CTRL group; [b] Statistical significance in relation to SAE group; [c] Statistical significance in relation to MetS + HFd group; [d] Statistical significance in relation to MetS + HFd + SAE group; [e] Statistical significance in relation to MetS + Sd group.

2.4. *Effect of Dietary Changes in Healthy and Rats with MetS on Ex Vivo Cardiac Function*

Figure 3 shows the values of ex vivo measured cardiac function parameters and coronary flow, during pressure changing protocols (PCPs) on the Langendorff apparatus. To examine the potential difference due to various dietary habits, we compared the percentage of decrease (−) or increase (+) between PCP 1 and PCP 2 in the group (Table 2).

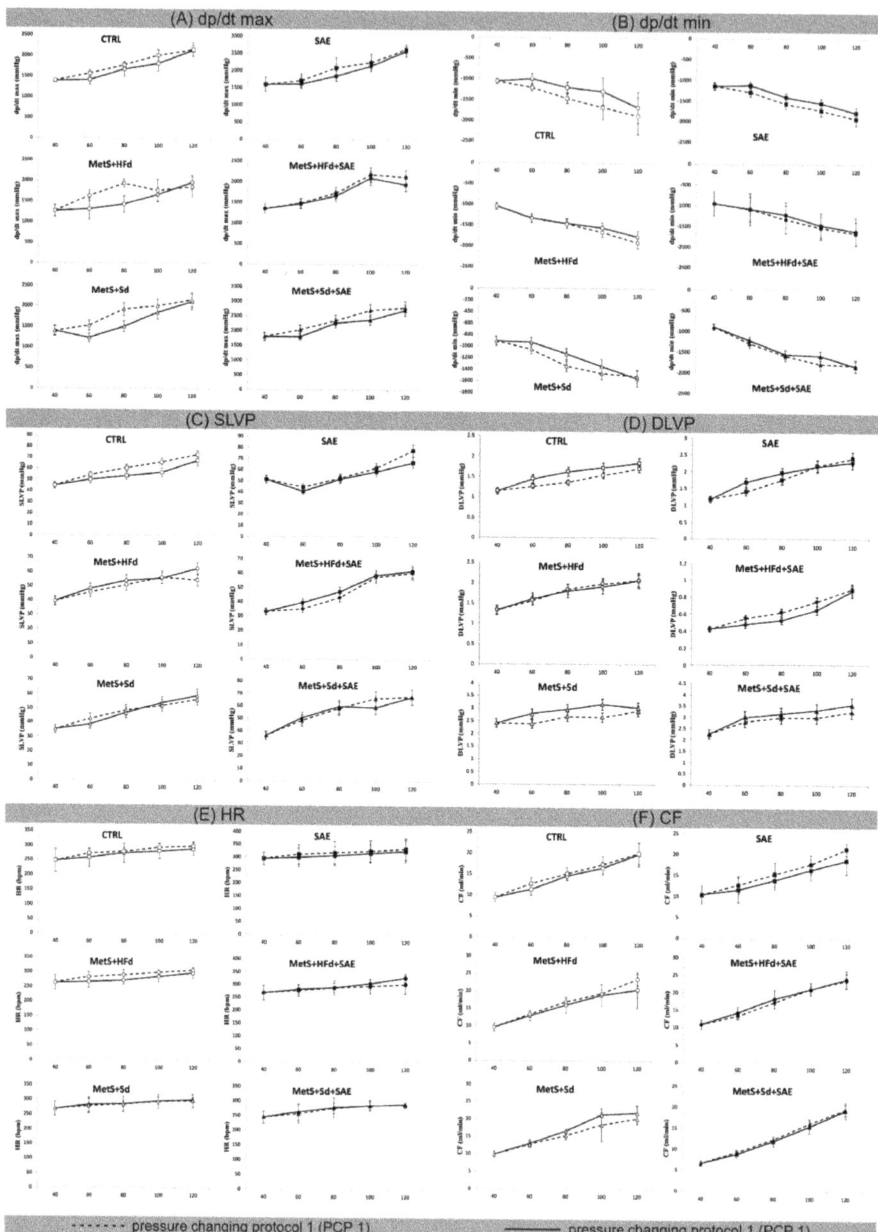

Figure 3. Effect of dietary changes in healthy and rats with MetS on cardiac function parameters, measured ex vivo: (**A**) values of dp/dt max within each of 6 groups during pressure changing protocol 1 (PCP 1) and pressure changing protocol 2 (PCP 2); (**B**) values of dp/dt min within each of 6 groups during PCP 1 and PCP 2; (**C**) values of SLVP within each of 6 groups during PCP 1 and PCP 2; (**D**) values of diastolic left ventricular pressure (DLVP) within each of 6 groups during PCP 1 and PCP 2; (**E**) values of HR within each of 6 groups during PCP 1 and PCP 2; (**F**) values of CF within each of 6 groups during PCP 1 and PCP 2. All values are expressed as mean ± standard deviation for each group.

Table 2. Percentage differences between PCP 1 and PCP 2 during ex vivo perfusion on Langendorff aparatus.

CPP	CTRL	SAE	MetS + HFd	MetS + HFd + SAE	MetS + Sd	MetS + Sd + SAE
				dp/dt max		
60	−11.05	−5.99	−24.77	−1.07	−24.23	−13.13
80	−6.11	−12.90	−35.30	−4.41	−28.11	−3.77
100	−10.88	−4.79	−6.22	−4.08	−8.80	−14.02
120	−0.63	−1.91	5.62	−9.39	−2.06	−3.24
				dp/dt min		
60	−21.25	−14.81	−0.46	−0.62	−13.64	−6.36
80	−22.03	−11.53	−0.80	−9.03	−18.90	−2.32
100	−29.70	−11.43	−7.04	−3.81	−9.14	−12.42
120	−12.04	−8.14	−8.08	−3.15	2.63	1.16
				SLVP		
60	−8.35	−9.49	5.37	10.80	−10.58	4.06
80	−13.11	−1.66	5.70	8.55	−4.18	2.44
100	−16.81	−5.81	−1.22	2.15	3.72	−12.46
120	−7.79	−16.29	12.60	1.99	4.12	−0.07
				DLVP		
60	12.50	16.67	2.50	−15.79	14.41	6.67
80	16.05	10.20	−2.22	−19.05	9.32	4.76
100	10.47	−0.93	−4.21	−15.38	15.87	9.09
120	7.61	−5.26	0.00	−2.86	4.13	8.45
				HR		
60	−5.29	−4.01	−6.26	1.59	1.39	2.51
80	−1.96	−3.85	−6.82	0.23	0.41	1.10
100	−4.58	−2.77	−5.04	3.66	0.31	−0.37
120	−3.41	−2.63	−3.21	7.85	1.02	0.61
				CF		
60	−12.37	−10.07	−3.11	6.74	1.93	−4.55
80	−3.57	−10.40	−5.78	5.48	7.88	−3.36
100	−5.10	−8.33	−2.55	−0.57	13.44	−4.55
120	−1.21	−14.66	−14.96	1.51	7.59	−1.04

Major changes in maximum rate of pressure development in the left ventricle (dp/dt max) in PCP 1 versus PCP 2 were observed at coronary perfusion pressure (CPP) = 60 cm H_2O and 80 cm H_2O in MetS + HFd (−24.77; −35.3) and MetS + Sd (−24.23; −28.11) groups. In MetS groups that were fed with combination of mentioned diets and SAE, this parameter was not changed during pressure changing protocols. On the other hand, the most significant differences in minimum rate of pressure development in the left ventricle (dp/dt min) were observed in CTRL group at CPP = 60 cm H_2O, 80 cm H_2O and 100 cm H_2O (−21.25; −22.03; −29.7), while during the PCPs there were no significant changes in systolic left ventricular pressure (SLVP), HR, and coronary flow (CF) in any of the examined groups (Figure 3, Table 2).

2.5. Effect of Dietary Changes in Healthy and Rats with MetS on Glucose and Insulin Levels during Oral Glucose Tolerance Test (OGTT)

2.5.1. Effects on Glucose Levels during OGTT

The average blood glucose values during an OGTT were present in Figure 4. Fasting blood glucose concentrations were significantly increased in all MetS groups compared to healthy groups, except in MetS + HFd + SAE where glucose level was the lowest among the MetS groups. In addition, glucose level was lower in MetS + Sd + SAE than in MetS + Sd. The similar trend was maintained during 30, 60 and 120 min, while in 180′ the highest level was in MetS + HFd.

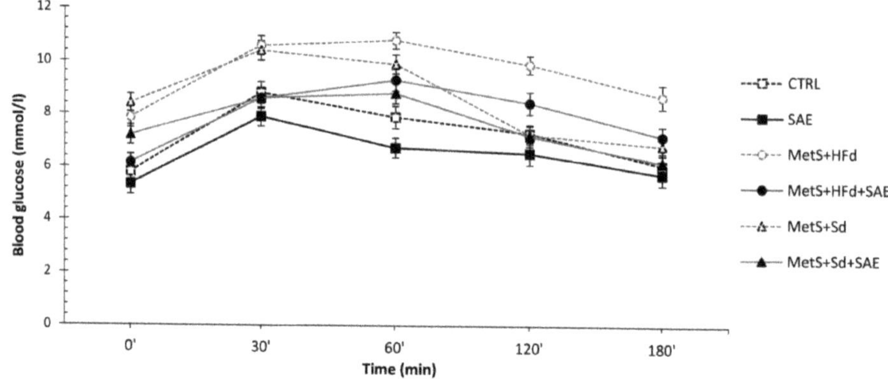

	0′	30′	60′	120′	180′
CTRL	5.81±0.34	8.82±0.41	7.93±0.42	7.29±0.34	6.08±0.39
SAE	5.35±0.40	7.91±0.36	6.75±0.37^{a}	6.55±0.44	5.75±0.41
MetS+HFd	7.85±0.38^{ab}	10.61±0.36^{ab}	10.77±0.32^{ab}	9.91±0.34^{ab}	8.71±0.45^{ab}
MetS+HFd+SAE	6.15±0.32^{c}	8.59±0.40^{c}	9.32±0.41^{abc}	8.45±0.42^{abc}	7.23±0.38^{c}
MetS+Sd	8.42±0.33^{abd}	10.43±0.38^{abd}	9.90±0.36^{ab}	7.25±0.35^{cd}	6.85±0.37^{c}
MetS+Sd+SAE	7.19±0.37^{abdc}	8.60±0.39^{ce}	8.78±0.38^{bce}	7.18±0.39^{cd}	6.22±0.36^{c}

Figure 4. Effect of dietary changes in healthy and rats with MetS on glucose levels during OGTT. Values are expressed as mean ± standard deviation for 10 animals, for each group. For statistical significance were considered values $p < 0.05$. ^{a} Statistical significance in relation to CTRL group; ^{b} Statistical significance in relation to SAE group; ^{c} Statistical significance in relation to MetS + HFd group; ^{d} Statistical significance in relation to MetS + HFd + SAE group; ^{e} Statistical significance in relation to MetS + Sd group.

2.5.2. Effects on Insulin Levels during OGTT

Table 3 shows the insulin concentration measured during the OGTT. The SAE group had the lowest, while MetS + Sd group had the highest insulin concentration measured fasting (0′), as well as 3 h after glucose administration (180′). In MetS groups fasting insulin concentration was significantly higher than in CTRL group. Moreover, insulin concentration was significantly lower in MetS + Sd + SAE than in MetS + Sd group, in both measured moments of interest.

Table 3. Effect of dietary changes in healthy and rats with MetS on insulin levels during OGTT.

Groups	0′	180′
CTRL	122.9 ± 6.76	123.7 ± 6.61
SAE	106.9 ± 6.04 ^{a}	113.8 ± 6.51
MetS + HFd	185.1 ± 7.78 ^{ab}	129.3 ± 6.38
MetS + HFd + SAE	180.3 ± 8.02 ^{ab}	131.1 ± 7.13 ^{b}
MetS + Sd	205.8 ± 9.87 ^{abcd}	145.2 ± 7.65 ^{b}
MetS + Sd + SAE	182.1 ± 9.32 ^{abcde}	127.3 ± 6.72 ^{e}

Values are expressed as mean ± standard deviation for 10 animals, for each group. For statistical significance were considered values $p < 0.05$. ^{a} Statistical significance in relation to CTRL group; ^{b} Statistical significance in relation to SAE group; ^{c} Statistical significance in relation to MetS + HFd group; ^{d} Statistical significance in relation to MetS + HFd + SAE group; ^{e} Statistical significance in relation to MetS + Sd group.

2.6. Effect of Dietary Changes in Healthy and Rats with MetS on Serum Iron Levels

SAE supplementation significantly increased iron levels in serum of healthy and rats with MetS fed with HFd or Sd than in non-treated groups. CTRL group had the lowest values of iron in relation to all other examined groups (Figure 5).

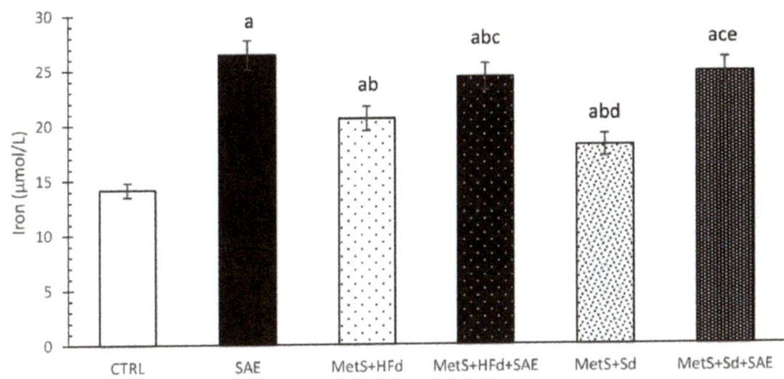

Figure 5. Effect of dietary changes in healthy and rats with MetS on serum iron levels. Values are expressed as mean ± standard deviation for 10 animals, for each group. For statistical significance were considered values $p < 0.05$. [a] Statistical significance in relation to CTRL group; [b] Statistical significance in relation to SAE group; [c] Statistical significance in relation to MetS + HFd group; [d] Statistical significance in relation to MetS + HFd + SAE group; [e] Statistical significance in relation to MetS + Sd group.

2.7. Evaluation of Systemic Redox State

Level of nitrites (NO_2^-) was significantly decreased in SAE group compared to CTRL group, and significantly increased in MetS groups compare to CTRL and SAE groups. The highest values of NO_2^- were observed in MetS + HFd group, while with the addition of SAE in HFd or Sd these values drop dramatically. Interestingly, MetS + Sd had significantly higher NO_2^- levels compare to MetS + HFd + SAE group (Figure 6A).

The highest level of superoxide anion radical (O_2^-) was noticed in MetS groups untreated with SAE extract. Moreover, SAE supplementation in healthy and rats with MetS significantly reduced O_2^- levels (Figure 6B).

Hydrogen peroxide (H_2O_2) levels were significantly increased in all examined group (except in MetS + HFd + SAE) compared to CTRL (Figure 6C).

Superoxide dismutase (SOD) activity was significantly reduced in rats with MetS, compare to healthy rats. The transition to a standard food with or without SAE supplementation led to a significant increase of SOD activity (Figure 6D).

Catalase (CAT) activity was significantly higher in SAE group compared to other examined groups, except MetS + Sd + SAE. On the other hand, in CTRL group this parameter was significantly increased compared to MetS + HFd and MetS + HFd + SAE and significantly decreased compared to MetS + Sd + SAE. With the addition of SAE in diet of rats with MetS, we observed significant increasement of CAT activity in comparison to MetS rats untreated with SAE (Figure 6E).

Reduced glutathione (GSH) levels were significantly increased in SAE and MetS + HFd + SAE compared to other observed groups (Figure 6F).

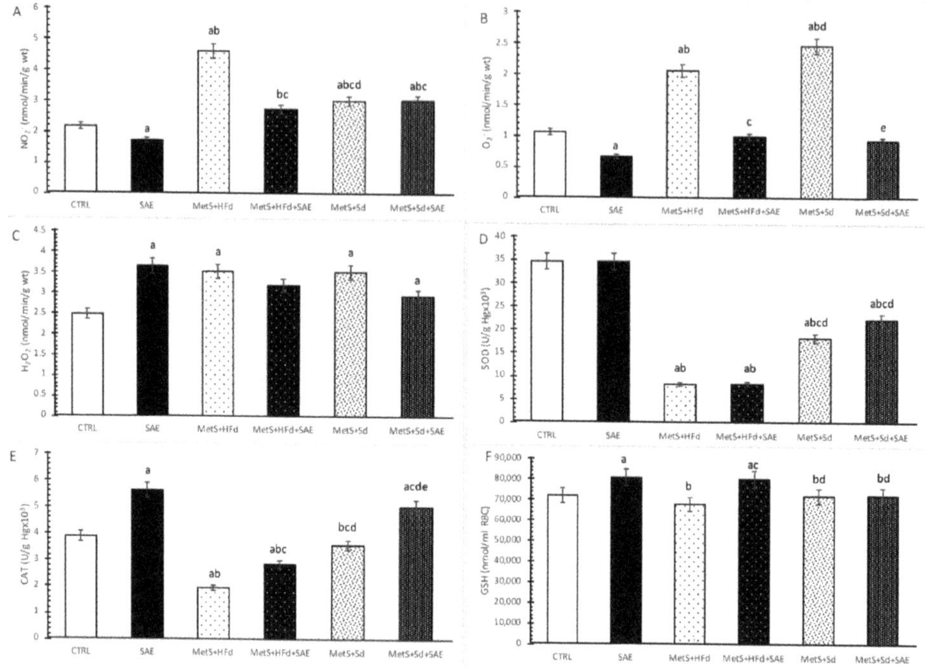

Figure 6. Effect of dietary changes in healthy and rats with MetS on systemic oxidative stress parameters: (**A**) NO_2^-; (**B**) O_2^-; (**C**) H_2O_2; (**D**) SOD; (**E**) CAT; (**F**) GSH. Values are expressed as mean ± standard deviation for 10 animals, for each group. For statistical significance were considered values $p < 0.05$. [a] Statistical significance in relation to CTRL group; [b] Statistical significance in relation to SAE group; [c] Statistical significance in relation to MetS + HFd group; [d] Statistical significance in relation to MetS + HFd + SAE group; [e] Statistical significance in relation to MetS + Sd group.

2.8. Histological Analysis of Liver Tissue

As shown in Figure 7, in CTRL, SAE, MetS + HFd + SAE, MetS + Sd and MetS + Sd + SAE liver tissue is a common feature. Liver lobulus is fully preserved. Hepatocytes are correctly arranged in the liver plates, without change. Sinusoidal capillaries are common characteristics, also without change. Fibrosis as well as inflammation have not been detected. On the other hand, in MetS + HFd there was a microvesicular steatosis (fatty change).

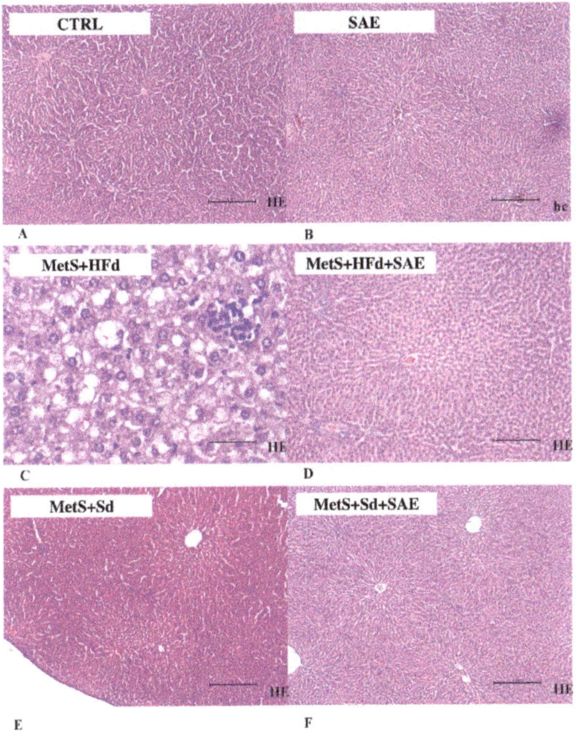

Figure 7. Representative hematoxylin and eosin (H&E) staining photos of liver tissue (for sub-figures A, B, D, E and F original magnification is 100× and scale bar is 2 mm, while in sub-figure C magnification is 400× and scale bar is 2 mm) in: (**A**) CTRL, (**B**) SAE, (**C**) MetS + HFd, (**D**) MetS + HFd + SAE, (**E**) MetS + Sd and (**F**) MetS + Sd + SAE groups.

3. Discussion

Several epidemiologic studies have implicated visceral fat as a major risk factor for insulin resistance, type 2 diabetes mellitus, cardiovascular disease, stroke, metabolic syndrome and death [15]. Taking into consideration increasing incidence of MetS and its related complications we wanted to estimate the effectiveness of HFd cessation and introduction of polyphenol-rich extract (SAE) on weight gain, cardiac function, glucose tolerance, serum insulin and iron levels, as well as systemic redox state and morphological characteristic of the liver.

Our results clearly show that the highest increase in BMI was observed in MetS + HFd, which was expected as many data suggest weight gain in rats exposed to HFd for different periods [16–18]. Introduction of Aronia extract in MetS + HFd group suppressed the body weight gain and decreased BMI; however changing dietary habits from high-fat to standard food had better anti-obesity effect when compared to MetS + HFd group. Moreover, the most prominent reduction in body weight and BMI was achieved by standard diet associated with consumption of SAE extract.

Taking into account, that cardiovascular complications in MetS are very common, we wanted to examine cardiovascular effects such as their ability to affect blood pressure as well as in vivo and ex vivo cardiac function, after dietary changes. Transition from a high caloric to normal fat diet-induced a decline in diastolic blood pressure. Nevertheless, the highest hypotensive effect in rats with MetS, evidenced with a drop in both systolic and diastolic pressure was reached when this regimen was combined with Aronia extract. Extract of Aronia was able to induce a drop in diastolic pressure even in rats who were fed with high-fat food continuously. On the contrary, in healthy rats SAE did not affect blood pressure. This is in line with the data that both Aronia berries and Aronia polyphenol extracts reduce quite effectively both SBP and DBP in spontaneously hypertensive rats [19]. The proposed mechanism might be through inhibition of the kidney renin-angiotensin system [20]. Others also found that *A. melanocarpa* extract decreases blood pressure in experimental model of hypertension [21,22]. Growing evidence suggests that the flavonoid-rich foods intake is related with decline in SBP and DBP, so we assume that the blood pressure-lowering effect of SAE is attributed to polyphenols [22,23]. Hypotensive effect was confirmed in patients with metabolic syndrome as well [24].

During blood pressure measuring, HR was also registered. Increase in HR in groups with MetS compared to healthy rats may be explained by an increase in sympathetic nervous system activity induced by HFd [25]. In that sense obtained decrease in HR after the transition to standard diet in MetS + Sd group appears to be logical. However, there was no change in HR when SAE was added to a dietary regimen in rats with MetS who were on a Sd, while we noticed an increase in rats on HFd. Moreover, Aronia extract did not alter heart rate in healthy rats, reflecting preserved myocardial function and contractility. Similar results were found in previous investigations regarding the effects of polyphenol-containing extracts on HR [26].

An echocardiographic examination illustrated that the highest impact of Aronia extract on in vivo myocardial functions was found in healthy rats, where we observed significant increase in IVSd, LVPWd, IVSs, LVPWs, FS and decrease in LVIDs compared to CTRL. On the other hand, addition of Aronia extract in rats with MetS during both HFd and Sd did not significantly affect cardiac function compared to MetS + HFd group. The greatest benefit of SAE involves improvement in systolic function, manifested as a significant increase in fractional shortening (FS) in healthy rats relative to almost all other groups. Similar values in healthy and MetS + Sd + SAE and MetS + HFd + SAE group suggest that Aronia extract was capable of improving fractional shortening during both high-calorie and standard dietary conditions in the presence of metabolic syndrome. In accordance with our findings, it has been previously reported that polyphenolics and plants rich in polyphenolics had effect in lessening the pathological alterations in FS promoted by MetS [27–29]. A decrease in systemic blood pressure after Aronia extract treatment may increase fractional shortenings, resulting in increased myocardial contractility [30].

Similar results were obtained during ex vivo, retrograde perfusion on Langendorff. Cardiac contractility, estimated by maximum and minimum rate of left ventricle pressure development, (dp/dt mSAE and dp/dt min), was preserved in MetS groups treated with SAE in combination with HFd or Sd, especially in normoxic conditions (CPP = 60 and 80 cm H_2O). Furthermore, addition of Aronia extract in healthy rats significantly improved heart relaxation (for the CPP values from 60–100 cm H_2O) compared to CTRL group. These results confirm the assumption that Aronia extract triggers cardioprotection, most probably because of its antioxidant, antiinflammatory, vasorelaxant and antithrombotic effects [31].

To estimate if SAE extract might improve glucose tolerance, which is strongly related to insulin resistance and insulin secretion, we performed oral glucose tolerance (OGT) test. Our results highlighted that SAE extract did not affect fasting glucose level in healthy rats, while it exerted hypoglycemic effect in animals with MetS on both HFd and Sd. The similar trend was noticed during 30, 60, 120 and 180 min, except the fact that in 120 and 180 min there was no difference in glucose level between MetS + Sd and MetS + Sd + SAE. Regarding the concentration of insulin, it was the lower in group of healthy rats receiving SAE group compared to healthy untreated group in 0′. Moreover, higher insulin concentration in MetS + HFd group wasn't diminished after adding SAE extract. On the other hand, this extract potentiates effect of standard food on lowering insulin, as evidenced by a decrease in insulin concentration in MetS + Sd + SAE compared to MetS + Sd group in 0′ and 180′. In line with our observation, better glucose tolerance achieved by treatment with SAE extract was confirmed by several papers. Other authors showed the beneficial effects of *A. melanocarpa* extract on attenuating insulin resistance and improving insulin sensitivity in HFd-induced obese mice [32]. Furthermore, glucose lowering potential was confirmed in patients with DM as well [33]. Proposed mechanism though which Aronia exerts hypoglicemic effect involve inhibition of dipeptidyl peptidase IV and α-glucosidase activities [34].

Since iron is an essential trace element that has been involved in maintenance of regular homeostasis, understanding the influence of SAE alone or in combination with different dietary regiments its serum levels would be of a great importance [35]. We found the highest level of iron in group of healthy animals receiving SAE, while increased level of iron was found in all groups with MetS compared to healthy rats. Previous study indicates that increased iron stores have been associated with MetS [36]. Changing dietary habits from high-fat to standard food did not result in change of iron, while addition of SAE in this group induced significant rise when compared to MetS group. This is in agreement with other research which showed that supplementation with Aronia juice increased serum level of iron [34]. In fact, certain flavonoids have potential to chelate iron and decrease iron absorption through mechanism independent of the hepcidin, a hormone included in iron homeostasis [35,37]. On the other hand, other flavonoids may decrease the activity of hepcidin resulting in increased iron uptake and serum iron levels, which may be an explanation for the SAE extract-induced increase in our research [35,38].

Increased oxidative stress has been linked with pathogenesis of MetS, thus indicating the need for consuming natural antioxidants from food sources in treatment of MetS-related diseases [39]. In that sense, in order to test if 4-week SAE supplementation alters systemic redox homeostasis we determined levels of pro-oxidants, as well as capacity of antioxidant defense system in blood samples. Our results demonstrated that SAE consumption led to a drop in NO_2^- and O_2^- and rise in CAT and GSH in healthy rats. Increase in CAT activity which catalyzes the decomposition of hydrogen peroxide to water and oxygen support unchanged values of H_2O_2. As it was expected increased generation of pro-oxidants and decreased activity of antioxidant enzymes SOD and CAT were noticed in MetS + HFd group compared to control. Introduction of three different dietary strategies such as consumption of SAE or transition to standard diet or its combination induced decline only in NO_2^- compared to MetS + HFd. Regarding the antioxidant status, the highest impact of increase in activities of antioxidant enzymes was noticed when standard diet was combined with SAE treatment. Striking evidence indicate that polyphenols might increase antioxidant capacity via rise in activities of SOD, CAT and GSH-peroxidase and act as direct free radical scavengers as well [40,41]. Chelation of iron ions which catalyze several free radical-generating reactions is one of the mechanisms underlying antioxidant effects of polyphenols. Nevertheless, rise in iron level induced by SAE in our research lead us to a hypothesis that enhanced activity of antioxidant enzymes and direct scavenging rather than iron chelation were responsible for antioxidant potential of applied extract. However, poor bioavailability of polyphenols through food intake suggests necessity for polyphenol-enriched foods or supplements treatment such as our extract [42].

MetS in combination with high-fat altered structure of liver tissue manifested as microvesicular hepatic steatosis. Nevertheless, transition from high-fat to standard food and combined approach which involved SAE extract consumption associated with both dietary protocols significantly normalized liver changes in MetS groups. Obtained positive effects in those groups are evidenced by the absence of fibrosis and inflammation. Ability of anthocyanins in Aronia to diminish liver steatosis induced by MetS was documented before, so we may hypothesize that these bioactive compounds are responsible for the beneficial effects in our study [43,44]. Some results show beneficial effects of *A. melanocarpa* against hepatic lipid accumulation through the inhibition of peroxisome proliferator-activated receptor $\gamma 2$ (PPARγ2) expression along with improvements in body weight, liver functions, lipid profiles and antioxidant capacity suggesting the potential therapeutic efficacy of *A. melanocarpa* on nonalcoholic fatty liver disease [45]. Recently, it was showed clearly an increase in acetylcholinesterase (AChE) and butyryl cholinesterase activity and disruption of lipid metabolism in patients with MetS. After supplementation of MetS patients with *A. melanocarpa* extract, a decrease in AChE activity and oxidative stress was noted [46].

4. Material and Methods

4.1. Ethical Approval

This research was carried out in the laboratory for cardiovascular physiology of the Faculty of Medical Sciences, University of Kragujevac, Serbia. The study protocol was approved (number: 119-01-5/14/2017-09, date: 30 June 2017) by the Ethical Committee for the welfare of experimental animals of the Faculty of Medical Sciences, University of Kragujevac, Serbia. All experiments were performed according to EU Directive for welfare of laboratory animals (86/609/EEC) and principles of Good Laboratory Practice.

4.2. Animals

Sixty *Wistar albino* rats (males, six weeks old, body-weight 200 ± 30 g, on beginning of experiments) were included in the study. They were housed at temperature of 22 ± 2 °C, with 12 h of automatic illumination daily. The rats were randomly divided into two groups: healthy animals (n = 20), fed with Sd which contains 9% fat, 20% protein, 53% starch, 5% fiber and animals with MetS (n = 40), fed with HFd which contains 25% fat, 15% protein, 51% starch and 5% fiber during 4 weeks. After one month on their respective diets, rats from MetS group after 6–8 h of starvation received one dose of streptozotocin intraperitoneally. Streptozotocin was prepared *ex tempore* by dissolving in citrate buffer and, depending on the body weight, it was administered in a dose of 25 mg/kg body weight [47]. Three days (72 h) after streptozotocin treatment e and 12 h after starvation fasting glucose and insulin level as well as blood pressure, were measured. Animals with systolic blood pressure greater than 140 mmHg, diastolic blood pressure more than 85 mmHg, fasting glucose level above 7.0 mmol/L and fasting insulin level over 150 µLU/mL were included in the study and were used in the study as rats with MetS.

Healthy and rats with MetS were divided into 6 groups as follows: CTRL—healthy rats, fed with a Sd for 4 weeks; SAE—healthy rats, fed with a Sd and treated with highly concentrated Aronia extract standardized with polyphenol content—SAE in the dose 0.45 mL/kg/day per os for 4 weeks; MetS + HFd—rats with MetS, fed with HFd for 4 weeks; MetS + HFd + SAE—rats with MetS, fed with HFd and treated with SAE (0.45 mL/kg/day, per os) for 4 weeks; MetS + Sd—rats with MetS, fed with a Sd for 4 weeks; MetS + Sd + SAE—rats with MetS, fed with a Sd for 4 weeks and treated with SAE (0.45 mL/kg/day, per os) for 4 weeks.

Standardized Aronia extract (SAE) is official product of pharmaceutical company Pharmanova (Belgrade, Serbia); however procedure of extraction was done by EU-Chem company (Belgrade, Serbia).

4.3. Measurement of BMI

At the end of the study protocol, body weight and body length were measured. Body length represents nose-anus length. Those parameters were used to calculate the BMI of the rats as follows:

$$\text{Body mass index (BMI)} = \text{body weight (g)}/\text{length}^2\ (\text{cm}^2) \quad (1)$$

4.4. Evaluation of Blood Pressure and Heart Rate

A day before sacrificing animals, the blood pressure and heart rate were measured by a tail-cuff noninvasive method BP system (Rat Tail Cuff Method Blood Pressure Systems (MRBP-R), IITC Life Science Inc., Los Angeles, CA, USA) [48].

4.5. Evaluation of in vivo Cardiac Function

After accomplishing 4-week treatment, transthoracic echocardiograms were performed. Mixture of ketamine—50 mg/kg and xylazine—10 mg/kg intraperitoneally was used as anesthesia. Echocardiograms were performed using a Hewlett-Packard Sonos 5500 (Andover, MA, USA) sector scanner equipped with a 15.0-MHz phased-array transducer as previously described [49]. From the parasternal long-axis view in 2-dimensional mode, and M-mode cursor was positioned perpendicularly to the interventricular septum and posterior wall of the left ventricle (LV) at the level of the papillary muscles and M-mode images were obtained. Interventricular septal wall thickness at end diastole (IVSd), LV internal dimension at end diastole (LVIDd), LV posterior wall thickness at end diastole (LVPWd), interventricular septal wall thickness at end systole (IVSs), LV internal diameter at end systole (LVIDs) and LV posterior wall thickness at end systole (LVPWs) were recorded with M-mode. Fractional shortening percentage (FS%) was calculated from the M-mode LV diameters using the equation: $[(LVEDd-LVESd)/LVEDd] \times 100\%$. Where LVEDd is left ventricular end diastolic diameter and LVESd is left ventricular end systolic diameter.

4.6. Evaluation of ex vivo Cardiac Function

Following 4-week protocol, after short-term narcosis induced by intraperitoneal application of ketamine (10 mg/kg) and xylazine (5 mg/kg) and premedication with heparin as an anticoagulant, animals were sacrificed by decapitation. Then the chest was opened via midline thoracotomy, hearts were immediately removed and immersed in cold saline and aortas were cannulated and retrogradely perfused according to *Langendorff* technique, under gradually increasing coronary perfusion pressure (CPP) from 40 to 120 cm H_2O [50]. The composition of Krebs-Henseleit buffer used for retrograde perfusion was as follows (mmol/L): NaCl 118 mmol/L, KCl 4.7 mmol/L, $MgSO_4 \times 7H_2O$ 1.7 mmol/L, $NaHCO_3$ 25 mmol/L, KH_2PO_4 1.2 mmol/L, $CaCl_2 \times 2H_2O$ 2.5 mmol/L, glucose 11 mmol/L, pyruvate 2 mmol/L, equilibrated with 95% O_2 plus 5% CO_2 and warmed to 37 °C (pH 7.4).

After placing the sensor (transducer BS473-0184, Experimetria Ltd., Budapest, Hungary) in the left ventricle, the following parameters of myocardial function have been measured: maximum rate of pressure development in the left ventricle (dp/dt max), minimum rate of pressure development in the left ventricle (dp/dt min), systolic left ventricular pressure (SLVP), diastolic left ventricular pressure (DLVP), heart rate (HR). Coronary flow (CF) was measured flowmetrically. Following the establishment of heart perfusion, the hearts were stabilized within 30 min with a basal coronary perfusion pressure of 70 cm H_2O. To examine the heart function, after stabilization period, the perfusion pressure was gradually decreased to 60, and then increased to 80, 100 and 120 cm H_2O and reduced to 40 cm H_2O (pressure changing protocol 1, PCP 1) and again gradually increased from 40 to 120 cm H_2O (pressure changing protocol 1, PCP 1).

4.7. Oral Glucose Tolerance Test

Oral glucose tolerance test (OGTT) was performed at the end of 4-week protocol and a day before sacrificing animals. After overnight (12–14h) fasted animals, the blood sample was taken by tail bleeding to determine the fasting blood glucose and insulin level (0 min) and then glucose was administered orally in a dose of 2 g/kg body weight and blood samples were taken at 30, 60, 120 and 180 min after glucose loading. Glucose levels were determined in 0, 30, 60, 120 and 180 min, using glucometer (Accu-Chek, Roche Diagnostics, Indianapolis, IN, USA) with its corresponding strips. At 0 and 180 min, insulin levels were assessed in plasma samples by the enzyme-linked immunosorbent assay (ELISA) method as previously described [51].

4.8. Evaluation of Serum Iron Levels and Systemic Redox State

In the moment of sacrificing animals blood samples were collected from jugular vein in order to estimate serum iron levels and systemic oxidative stress response. The levels of serum iron (SI) was determined on a biochemical analyzer (Dimension, Dade Behring, Milton Keynes, UK, USA) and the results were expressed in μg/L.

In plasma the following pro-oxidants were determined: the levels of nitrites (NO_2^-), superoxide anion radical (O_2^-) and hydrogen peroxide (H_2O_2). Parameters of antioxidative defence system, such as activities of superoxide dismutase (SOD) and catalase (CAT) and level of reduced glutathione (GSH) were determined in erythrocytes samples.

NO_2^- was determined as an index of NO production with Griess reagent. The method for detection nitrate in plasma is based on the Green and coworkers proposal and measured spectrophotometrically at a wavelength of 543 nm [52]. The concentration of O_2^- in plasma was measured at 530 nm, after the reaction of nitro blue tetrazolium in TRIS buffer [53]. The measurement of H_2O_2 is based on the oxidation of phenol red by hydrogen peroxide, in a reaction catalyzed by horseradish peroxidase (HRPO) as previously described by Pick and colleagues. The level of H_2O_2 was measured at 610 nm [54].

For determination of antioxidant parameters, isolated erytrocytes were prepared according to McCord and Fridovich [54]. CAT activity were determined at 360 nm toward to Beutler [55]. Lysates were diluted with distilled water (1:7 v/v) and treated with chloroform-ethanol (0.6:1 v/v) to remove hemoglobin [51]. SOD activity was determined by the epinephrine method of Misra and Fridovich. Detection was performed at 470 nm [56]. Level of GSH is based on GSH oxidation via 5,5-dithiobis-6,2-nitrobenzoic acid as previously described by Beutler. Measuring was performed at 420 nm [57].

4.9. Histological Analysis of Liver Tissue

Liver tissue samples were fixed in 4% buffered paraformaldehyde solution and immersed in paraffin. Afterwards 4-micrometre-thick sections were stained with hematoxylin and eosin (H&E) [43].

4.10. Statistical Analysis

IBM SPSS Statistics 20.0 Desktop for Windows was used for statistical analysis. Distribution of data was checked by Shapiro–Wilk test. Where distribution between groups was normal, statistical comparisons were performed using the one-way analysis of variance (ANOVA) tests with a Tukey's post hoc test for multiple comparisons. Kruskal–Wallis was used for comparison between groups where the distribution of data was different than normal. Values of $p < 0.05$ were considered to be statistically significant.

5. Conclusions

Our results highlighted cardioprotective potential of SAE in treatment of MetS, involving lowering blood pressure and favorable effects on heart function. Furthermore, SAE effectively suppressed the body weight gain, improved glucose tolerance and attenuated liver steatosis and oxidative stress present in MetS, thus indicating its promising role in management of MetS-related diseases. Moreover, increase in iron concentration indicates its health benefits in iron deficiency. Obtained beneficial effects would be more prominent in c combination with changing dietary habits. This research may be a starting point for further experimental and clinical investigations which would fully evaluate the effects of SAE alone or in combination with different dietary protocols in various models of chronic diseases.

Author Contributions: Conceptualization, V.J. and M.M.; methodology, I.M.; validation, N.J. and T.N.T.; formal analysis, J.J. and J.B.; investigation, I.S., M.L.B. and J.J.; writing—original draft preparation, J.B. and J.J.; writing—review and editing, V.Z.; visualization, P.M. and S.B.; supervision, V.J. and V.V.; project administration, V.Z. and I.S.

Funding: This research received no external funding.

Conflicts of Interest: The authors declare no conflict of interest.

References

1. Tabatabaei-Malazy, O.; Larijani, B.; Abdollahi, M. Targeting metabolic disorders by natural products. *J. Diabetes Metab. Disord.* **2015**, 14, 57. [CrossRef] [PubMed]
2. Moore, J.X.; Chaudhary, N.; Akinyemiju, T. Metabolic Syndrome Prevalence by Race/Ethnicity and Sex in the United States, National Health and Nutrition Examination Survey, 1988–2012. *Prev. Chronic Dis.* **2017**, 14, E24. [CrossRef] [PubMed]

3. Bernabé, J.; Mulero, J.; Marhuenda, J.; Cerdá, B.; Avilés, F.; Abellán, J.; Zafrilla, P. Changes in Antioxidant Enzymes in Metabolic Syndrome Patients after Consumption a Citrus-Based Juice Enriched with Aronia Melanocarpa. *J. Nutr. Disorders Ther.* **2015**, *5*, 4.
4. Nammi, S.; Sreemantula, S.; Roufogalis, B.D. Protective effects of ethanolic extract of Zingiber officinale rhizome on the development of metabolic syndrome in high-fat diet-fed rats. *Basic Clin. Pharmacol. Toxicol.* **2009**, *104*, 366–373. [CrossRef] [PubMed]
5. Ranasinghe, P.; Mathangasinghe, Y.; Jayawardena, R.; Hills, A.P.; Misra, A. Prevalence and trends of metabolic syndrome among adults in the asia-pacific region: A systematic review. *BMC Public Health* **2017**, *17*, 101. [CrossRef]
6. De la Iglesia, R.; Loria-Kohen, V.; Zulet, M.A.; Martinez, J.A.; Reglero, G.; Ramirez de Molina, A. Dietary Strategies Implicated in the Prevention and Treatment of Metabolic Syndrome. *Int. J. Mol. Sci.* **2016**, *17*, 1877. [CrossRef] [PubMed]
7. Daskalova, E.; Delchev, S.; Peeva, Y.; Vladimirova-Kitova, L.; Kratchanova, M.; Kratchanov, C.; Denev, P. Antiatherogenic and Cardioprotective Effects of Black Chokeberry (*Aronia melanocarpa*) Juice in Aging Rats. *Evid. Based Complement Alternat. Med.* **2015**, *2015*. [CrossRef]
8. Kokotkiewicz, A.; Jaremicz, Z.; Luczkiewicz, M. Aronia plants: A review of traditional use, biological activities, and perspectives for modern medicine. *J. Med. Food* **2010**, *13*, 255–269. [CrossRef]
9. Ochmian, I.; Grajkowski, J.; Smolik, M. Comparison of some morphological features, quality and chemical content of four cultivars of chokeberry fruits (*Aronia melanocarpa*). *Not. Bot. Horti. Agrobot. Cluj Napoca* **2012**, *40*, 253–260. [CrossRef]
10. Tolić, M.T.; Landeka Jurčević, I.; Panjkota Krbavčić, I.; Marković, K.; Vahčić, N. Phenolic Content, Antioxidant Capacity and Quality of Chokeberry (*Aronia melanocarpa*) Products. *Food Technol. Biotechnol.* **2015**, *53*, 171–179. [CrossRef]
11. Cebova, M.; Klimentova, J.; Janega, P.; Pechanova, O. Effect of Bioactive Compound of *Aronia melanocarpa* on Cardiovascular System in Experimental Hypertension. *Oxid. Med. Cell. Longev.* **2017**, *2017*. [CrossRef] [PubMed]
12. Jurikova, T.; Mlcek, J.; Skrovankova, S.; Sumczynski, D.; Sochor, J.; Hlavacova, I.; Snopek, L.; Orsavova, J. Fruits of Black Chokeberry *Aronia melanocarpa* in the Prevention of Chronic Diseases. *Molecules* **2017**, *22*, 944. [CrossRef] [PubMed]
13. Lupascu, N.; Cadar, E.; Cherim, M.; Erimia, C.L.; Sîrbu, R. Research Concerning the Efficiency of Aronia Melanocarpa for Pharmaceutical Purpose. *Eur. J. Interdiscip. Stud.* **2016**, *4*, 115–121. [CrossRef]
14. Borowska, S.; Brzóska, M.M. Chokeberries (*Aronia melanocarpa*) and Their Products as a Possible Means for the Prevention and Treatment of Noncommunicable Diseases and Unfavorable Health Effects Due to Exposure to Xenobiotics. *Compr. Rev. Food Sci. Food Saf.* **2016**, *15*, 982–1017. [CrossRef]
15. Finelli, C.; Sommella, L.; Gioia, S.; La Sala, N.; Tarantino, G. Should visceral fat be reduced to increase longevity? *Ageing Res. Rev.* **2013**, *12*, 996–1004. [CrossRef] [PubMed]
16. Yin, Y.; Yu, Z.; Xia, M.; Luo, X.; Lu, X.; Ling, W. Vitamin D attenuates high fat diet-induced hepatic steatosis in rats by modulating lipid metabolism. *Eur. J. Clin. Investig.* **2012**, *42*, 1189–1196. [CrossRef] [PubMed]
17. Wang, Y.; Wang, P.Y.; Qin, L.Q.; Davaasambuu, G.; Kaneko, T.; Xu, J.; Murata, S.; Katoh, R.; Sato, A. The development of diabetes mellitus in Wistar rats kept on a high-fat/low-carbohydrate diet for long periods. *Endocrine* **2003**, *22*, 85–92. [CrossRef]
18. Auberval, N.; Dal, S.; Bietiger, W.; Pinget, M.; Jeandidier, N.; Maillard-Pedracini, E.; Schini-Kerth, V.; Sigrist, S. Metabolic and oxidative stress markers in Wistar rats after 2 months on a high-fat diet. *Diabetol. Metab. Syndr.* **2014**, *6*, 130. [CrossRef] [PubMed]
19. Hellström, J.K.; Shikov, A.N.; Makarova, M.N.; Pihlanto, A.M.; Pozharitskaya, O.N.; Ryhänen, E.L.; Kivijärvi, P.; Makarov, V.G.; Mattila, P.H. Blood pressure-lowering properties of chokeberry (*Aronia mitchurinii* var. Viking). *J. Funct. Foods* **2010**, *2*, 163–169. [CrossRef]
20. Yamane, T.; Kozuka, M.; Imai, M.; Yamamoto, Y.; Ohkubo, I.; Sakamoto, T.; Nakagaki, T.; Nakano, Y. Reduction of blood pressure by aronia berries through inhibition of angiotensin-converting enzyme activity in the spontaneously hypertensive rat kidney. *Funct. Food Health Dis.* **2017**, *7*, 280–290.
21. Ciocoiu, M.; Badescu, M.; Badulescu, O.; Tutunaru, D.; Badescu, L. Polyphenolic extract association with renin inhibitors in experimental arterial hypertension. *J. Biomed. Sci. Eng.* **2013**, *6*, 493–497. [CrossRef]

22. Naruszewicz, M.; Laniewska, I.; Millo, B.; Dłuzniewski, M. Combination therapy of statin with flavonoids rich extract from chokeberry fruits enhanced reduction in cardiovascular risk markers in patients after myocardial infraction (MI). *Atherosclerosis* **2007**, *194*, e179–e184. [CrossRef] [PubMed]
23. Tjelle, T.E.; Holtung, L.; Bøhn, S.K.; Aaby, K.; Thoresen, M.; Wiik, S.Å.; Paur, I.; Karlsen, A.S.; Retterstøl, K.; Iversen, P.O.; Blomhoff, R. Polyphenol-rich juices reduce blood pressure measures in a randomised controlled trial in high normal and hypertensive volunteers. *Br. J. Nutr.* **2015**, *114*, 1054–1063. [CrossRef] [PubMed]
24. Broncel, M.; Kozirog, M.; Duchnowicz, P.; Koter-Michalak, M.; Sikora, J.; Chojnowska-Jezierska, J. *Aronia melanocarpa* extract reduces blood pressure, serum endothelin, lipid, and oxidative stress marker levels in patients with metabolic syndrome. *Med. Sci. Monit.* **2010**, *16*, CR28–CR34. [PubMed]
25. Kanthe, P.S.; Patil, B.S.; Bagali, S.C.; Reddy, R.C.; Aithala, M.R.; Das, K.K. Protective effects of Ethanolic Extract of *Emblica officinalis* (amla) on Cardiovascular Pathophysiology of Rats, Fed with High Fat Diet. *J. Clin. Diagn. Res.* **2017**, *11*, CC05–CC09. [CrossRef] [PubMed]
26. El-Sayyad, S.M.; Makboul, M.A.; Ali, R.M.; Farag, S.F. Hypotensive effect of *Ficus sycomorus* L. on the arterial blood pressure of rabbits. *J. Pharmacogn. Phytochem.* **2016**, *4*. Available online: https://www.researchgate.net/publication/309610756_Hypotensive_effect_of_Ficus_sycomorus_L_on_the_arterial_blood_pressure_of_rabbits (accessed on 28 November 2018).
27. Cherniack, E.P. Polyphenols: Planting the seeds of treatment for the metabolic syndrome. *Nutrition* **2011**, *27*, 617–623. [CrossRef]
28. Leung, L.; Martin, J.B.; Lawmaster, T.; Arthur, K.; Broderick, T.L.; Al-Nakkash, L. Sex-Dependent Effects of Dietary Genistein on Echocardiographic Profile and Cardiac GLUT4 Signaling in Mice. *Evid. Based Complement Alternat. Med.* **2016**, *2016*, 1796357. [CrossRef]
29. Razmaraii, N.; Babaei, H.; Mohajjel Nayebi, A.; Assadnassab, G.; Ashrafi-Helan, J.; Azarmi, Y. Cardioprotective Effect of Grape Seed Extract on Chronic Doxorubicin-Induced Cardiac Toxicity in Wistar Rats. *Adv. Pharm. Bull.* **2016**, *6*, 423–433. [CrossRef]
30. Yadegari, M.; Khamesipour, F.; Talebiyan, R.; Katsande, S. Echocardiography findings after intravenous injection of *Achillea millefolium* (Yarrow) extract in the dog. *Malays. Appl. Biol.* **2015**, *44*, 85–91.
31. Parzonko, A.; Naruszewicz, M. Cardioprotective effects of *Aronia melanocarpa* anthocynanins. From laboratory experiments to clinical practice. *Curr. Pharm. Des.* **2016**, *22*, 174–179. [CrossRef]
32. Kim, N.H.; Jegal, J.; Kim, Y.N.; Heo, J.D.; Rho, J.R.; Yang, M.H.; Jeong, E.J. Chokeberry Extract and Its Active Polyphenols Suppress Adipogenesis in 3T3-L1 Adipocytes and Modulates Fat Accumulation and Insulin Resistance in Diet-Induced Obese Mice. *Nutrients* **2018**, *10*, 1734. [CrossRef]
33. Simeonov, S.B.; Botushanov, N.P.; Karahanian, E.B.; Pavlova, M.B.; Husianitis, H.K.; Troev, D.M. Effects of *Aronia melanocarpa* juice as part of the dietary regimen in patients with diabetes mellitus. *Folia Med.* **2002**, *44*, 20–23.
34. Yamane, T.; Kozuka, M.; Konda, D.; Nakano, Y.; Nakagaki, T.; Ohkubo, I.; Ariga, H. Improvement of blood glucose levels and obesity in mice given aronia juice by inhibition of dipeptidyl peptidase IV and α-glucosidase. *J. Nutr. Biochem.* **2016**, *31*, 106–112. [CrossRef] [PubMed]
35. Imam, M.U.; Zhang, S.; Ma, J.; Wang, H.; Wang, F. Antioxidants Mediate Both Iron Homeostasis and Oxidative Stress. *Nutrients* **2017**, *9*, 671. [CrossRef]
36. Lee, H.J.; Jang, H.B.; Park, J.E.; Park, K.H.; Kang, J.H.; Park, S.I.; Song, J. Relationship between Serum Levels of Body Iron Parameters and Insulin Resistance and Metabolic Syndrome in Korean Children. *Osong Public Health Res. Perspect.* **2014**, *5*, 204–210. [CrossRef] [PubMed]
37. Skarpańska-Stejnborn, A.; Basta, P.; Sadowska, J.; Pilaczyńska-Szcześniak, L. Effect of supplementation with chokeberry juice on the inflammatory status and markers of iron metabolism in rowers. *J. Int. Soc. Sports Nutr.* **2014**, *11*, 48. [CrossRef]
38. Mu, M.; An, P.; Wu, Q.; Shen, X.; Shao, D.; Wang, H.; Zhang, Y.; Zhang, S.; Yao, H.; Min, J.; et al. The dietary flavonoid myricetin regulates iron homeostasis by suppressing hepcidin expression. *J. Nutr. Biochem.* **2016**, *30*, 53–61. [CrossRef] [PubMed]
39. Gregório, B.M.; De Souza, D.B.; de Morais Nascimento, F.A.; Pereira, L.M.; Fernandes-Santos, C. The potential role of antioxidants in metabolic syndrome. *Curr. Pharm. Des.* **2016**, *22*, 859–869. [CrossRef]
40. Louis, X.L.; Thandapilly, S.J.; Kalt, W.; Vinqvist-Tymchuk, M.; Aloud, B.M.; Raj, P.; Yu, L.; Le, H.; Netticadan, T. Blueberry polyphenols prevent cardiomyocyte death by preventing calpain activation and oxidative stress. *Food Funct.* **2014**, *5*, 1785–1794. [CrossRef]

41. Mattera, R.; Benvenuto, M.; Giganti, M.G.; Tresoldi, I.; Pluchinotta, F.R.; Bergante, S.; Tettamanti, G.; Masuelli, L.; Manzari, V.; Modesti, A.; et al. Effects of Polyphenols on Oxidative Stress-Mediated Injury in Cardiomyocytes. *Nutrients* **2017**, *9*, 523. [CrossRef] [PubMed]
42. Brglez Mojzer, E.; Knez Hrnčič, M.; Škerget, M.; Knez, Ž.; Bren, U. Polyphenols: Extraction Methods, Antioxidative Action, Bioavailability and Anticarcinogenic Effects. *Molecules* **2016**, *21*, 901. [CrossRef] [PubMed]
43. Kotronen, A.; Westerbacka, J.; Bergholm, R.; Pietiläinen, K.H.; Yki-Järvinen, H. Liver fat in the metabolic syndrome. *J. Clin. Endocrinol. Metab.* **2007**, *92*, 3490–3497. [CrossRef] [PubMed]
44. Bhaswant, M.; Shafie, S.R.; Mathai, M.L.; Mouatt, P.; Brown, L. Anthocyanins in chokeberry and purple maize attenuate diet-induced metabolic syndrome in rats. *Nutrition* **2017**, *41*, 24–31. [CrossRef] [PubMed]
45. Park, C.H.; Kim, J.H.; Lee, E.B.; Hur, W.; Kwon, O.J.; Park, H.J.; Yoon, S.K. *Aronia melanocarpa* extract ameliorates hepatic lipid metabolism through PPARγ2 downregulation. *PLoS ONE* **2017**, *12*, e0169685. [CrossRef] [PubMed]
46. Duchnowicz, P.; Ziobro, A.; Rapacka, E.; Koter-Michalak, M.; Bukowska, B. Changes in Cholinesterase Activity in Blood of Adolescent with Metabolic Syndrome after Supplementation with Extract from *Aronia melanocarpa*. *BioMed Res. Int.* **2018**. [CrossRef]
47. Pari, L.; Chandramohan, R. Modulatory effects of naringin on hepatic key enzymes of carbohydrate metabolism in high-fat diet/low-dose streptozotocin-induced diabetes in rats. *Gen Physiol Biophys* **2017**, *36*, 343–352. [CrossRef]
48. Feng, M.; Whitesall, S.; Zhang, Y.; Beibel, M.; D'Alecy, L.; DiPetrillo, K. Validation of volume-pressure recording tail-cuff blood pressure measurements. *Am. J. Hypertens.* **2008**, *21*, 1288–1291. [CrossRef]
49. Joffe, I.I.; Travers, K.E.; Perreault-Micale, C.L.; Hampton, T.; Katz, S.E.; Morgan, J.P.; Douglas, P.S. Abnormal cardiac function in the streptozotocin-induced non-insulin-dependent diabetic rat: Noninvasive assessment with Doppler echocardiography and contribution of the nitric oxide pathway. *J. Am. Coll. Cardiol.* **1999**, *34*, 2111–2119. [CrossRef]
50. Zivkovic, V.; Jakovljevic, V.; Djordjevic, D.; Vuletic, M.; Barudzic, N.; Djuric, D. The effects of homocysteine-related compounds on cardiac contractility, coronary flow, and oxidative stress markers in isolated rat heart. *Mol. Cell. Biochem.* **2012**, *370*, 59–67. [CrossRef]
51. Green, L.C.; Wagner, D.A.; Glogowski, J.; Skipper, P.L.; Wishnok, J.S.; Tannenbaum, S.R. Analysis of nitrate, nitrite, and [15N]nitrate in biological fluids. *Anal. Biochem.* **1982**, *126*, 131–138. [CrossRef]
52. Auclair, C.; Voisin, E. Nitroblue tetrazolium reduction. In *Handbook of Methods for Oxygen Radical Research*; Greenvvald, R.A., Ed.; CRC Press: Boca Raton, FL, USA, 1985; pp. 123–132.
53. Pick, E.; Keisari, Y. A simple colorimetric method for the measurement of hydrogen peroxide produced by cells in culture. *J. Immunol. Methods.* **1980**, *38*, 161–170. [CrossRef]
54. McCord, J.M.; Fridovich, I. The utility of superoxide dismutase in studying free radical reactions. I. Radicals generated by the interaction of sulfite, dimethyl sulfoxide, and oxygen. *J. Biol. Chem.* **1969**, *244*, 6056–6063. [PubMed]
55. Beutler, E. Catalase. In *Red Cell Metabolism, a Manual of Biochemical Methods*; Beutler, E., Ed.; Grune and Stratton: New York, NY, USA, 1982; pp. 105–106.
56. Misra, H.P.; Fridovich, I. The role of superoxide-anion in the autooxidation of epinephrine and a simple assay for superoxide dismutase. *J. Biol. Chem.* **1972**, *247*, 3170–3175.
57. Beutler, E. Reduced glutathione (GSH). In *Red Cell Metabolism, a Manual of Biochemical Methods*; Beutler, E., Ed.; Grune and Stratton: New York, NY, USA, 1975; pp. 112–114.

International Journal of
Molecular Sciences

MDPI

Review

Acute and Chronic Effects of Cocaine on Cardiovascular Health

Sung Tae Kim [1] and Taehwan Park [2,3,*]

1 Department of Pharmaceutical Engineering, Inje University, Gimhae 50834, Korea; stkim@inje.ac.kr
2 Pharmacy Administration, St. Louis College of Pharmacy, St. Louis, MO 63110, USA
3 Center for Health Outcomes Research and Education, St. Louis College of Pharmacy, St. Louis, MO 63110, USA
* Correspondence: Taehwan.Park@stlcop.edu; Tel.: +1-314-446-8193

Received: 8 January 2019; Accepted: 27 January 2019; Published: 29 January 2019

Abstract: Cardiac complications resulting from cocaine use have been extensively studied because of the complicated pathophysiological mechanisms. This study aims to review the underlying cellular and molecular mechanisms of acute and chronic effects of cocaine on the cardiovascular system with a specific focus on human studies. Studies have consistently reported the acute effects of cocaine on the heart (e.g., electrocardiographic abnormalities, acute hypertension, arrhythmia, and acute myocardial infarction) through multifactorial mechanisms. However, variable results have been reported for the chronic effects of cocaine. Some studies found no association of cocaine use with coronary artery disease (CAD), while others reported its association with subclinical coronary atherosclerosis. These inconsistent findings might be due to the heterogeneity of study subjects with regard to cardiac risk. After cocaine use, populations at high risk for CAD experienced coronary atherosclerosis whereas those at low risk did not experience CAD, suggesting that the chronic effects of cocaine were more likely to be prominent among individuals with higher CAD risk. Studies also suggested that risky behaviors and cardiovascular risks may affect the association between cocaine use and mortality. Our study findings highlight the need for education regarding the deleterious effects of cocaine, and access to interventions for cocaine abusers.

Keywords: cocaine; cardiovascular health; heart disease; acute effects; chronic effects

1. Introduction

Cocaine is a tropane alkaloid compound that can be extracted from the leaves of an Andean shrub, *Erythroxylon coca*, in South America. Cocaine was originally used for local surgeries as an anesthetic agent in the 1880s, but it became a recreational drug in the 1970s. In the 1980s, there was an epidemic of cocaine use, with the number of cocaine users in the US estimated at 5.8 million in 1985 [1]. In 2016, the total number of cocaine users was estimated to be 18.2 million worldwide [2]. Approximately 34% of these cocaine users resided in North America, and 20% resided in Western and Central Europe. In the US, there were 1.5 million cocaine users aged 12 or older, representing 0.6% of the population [3]. Young adults aged 18 to 25 were the most common cocaine users (1.4%).

Cocaine may be administered by smoking, intravenous injection, nasal inhalation, or oral application. Pharmacokinetics vary by route of administration, with time to peak blood concentration ranging from 1–5 min (smoking or intravenous injection) to 60–90 min (oral administration) [4,5]. The duration of pharmacological action ranges between 5–60 min following smoking or intravenous administration, and up to 180 min following oral administration. In addition to nasal mucous membranes, cocaine absorption through other mucous membranes such as intravaginal or intrarectal mucus membranes is also possible [6]. Cocaine administration through mucous membranes results in slower onset of action, later peak concentration, and longer duration of action compared with that of smoking or intravenous administration, but faster onset of action, earlier peak concentration, and shorter duration of action than that of oral administration. Cocaine is converted into two major metabolites by plasma and liver cholinesterases: benzoylecgonine and ecgonine methyl ester. These water-soluble metabolites are excreted in the urine and are detectable in the urine for 24 to 36 h after intake.

Cocaine is categorized as a Schedule II substance under the Controlled Substances Act. Drugs or substances in this schedule have a high potential for abuse, which may lead to severe psychological or physical dependence. Cocaine abuse can result in a range of adverse health outcomes. About 0.9 million U.S. adults had a cocaine use disorder in 2014 [7]. Approximately 40% of all emergency department visits related to drug misuse and abuse were attributed to cocaine [8].

Prior studies have consistently reported the deleterious effects of cocaine use/abuse on the cardiovascular system. Cocaine-related cardiac complications include acute conditions such as arrhythmia and acute myocardial infarction (MI), as well as chronic conditions such as cardiomyopathy and coronary artery disease (CAD). Cocaine-induced cardiotoxicity can result in sudden death. In addition, previous studies have explored the complicated pathophysiological mechanisms of cocaine cardiotoxicity. Herein, we first review the cellular and molecular mechanisms of cocaine in the cardiovascular system to obtain a better understanding of its acute and chronic effects on the heart and blood vessels. Furthermore, we discuss recent evidence from human studies that examined cocaine-associated changes in the cardiovascular system. As such, our review includes recent clinical studies that have been published in the past 10 years (from September 2008 through September 2018) retrieved from the Medline database, and several other important clinical studies published before September 2008.

2. Pathophysiological Mechanisms of Cocaine on Cardiovascular Health

Cocaine stimulates the sympathetic nervous system by inhibiting reuptake of norepinephrine, dopamine, and serotonin by interacting with each transporter, leading to exaggerated, prolonged sympathetic nervous system activity [9,10]. Cocaine also blocks sodium/potassium channels, which induces abnormal, depressed cardiovascular profiles [11]. In particular, concurrent cocaine and alcohol abuse significantly increases cocaine levels in the blood, leading to increased, prolonged cardiovascular risks [12]. Previous studies have reported that use/abuse of cocaine is associated with increased risk of subsequent cardiovascular complications such as hypertension, coronary spasm, arrhythmias, MI, cardiomyopathy, atherosclerosis, and CAD [13], as summarized in Figure 1. In this section, we summarize the acute and chronic pathophysiological mechanisms of cocaine on cardiovascular health.

2.1. Mechanisms of Acute Toxicity

2.1.1. Acute Hypertension and Coronary Spasm

Acute coronary events usually occur within minutes to hours after cocaine administration. Cocaine stimulates the adrenergic system by binding to norepinephrine transporters, resulting in increased norepinephrine effects at postsynaptic receptor sites. Blocking norepinephrine reuptake induces tachycardia and hypertension, which increases myocardial oxygen demand and reduces myocardial oxygen supply by vasoconstriction [11,14,15]. As such, cocaine induces sympathetic effects on the cardiovascular system by enhanced inotropic and chronotropic effects through increased vasoconstriction. In particular, cocaine induces acute hypertension due to increased vasoconstriction induced by increased endothelin-1 [16], impaired acetylcholine-induced vasorelaxation [17], inhibition of nitric oxide synthase [18], impaired intracellular calcium handling [19], and inhibition of sodium/potassium channels [20] as determined by cellular and molecular analytical approaches [11]. In addition, acute vessel damage induces platelet aggregation/blood clots through increased fibrinogen and von Willebrand factor, leading to acute heart damage due to reduced blood flow [21]. Taken together, cocaine induces acute hypertension, coronary spasm, which may lead to subsequent myocardial infarction.

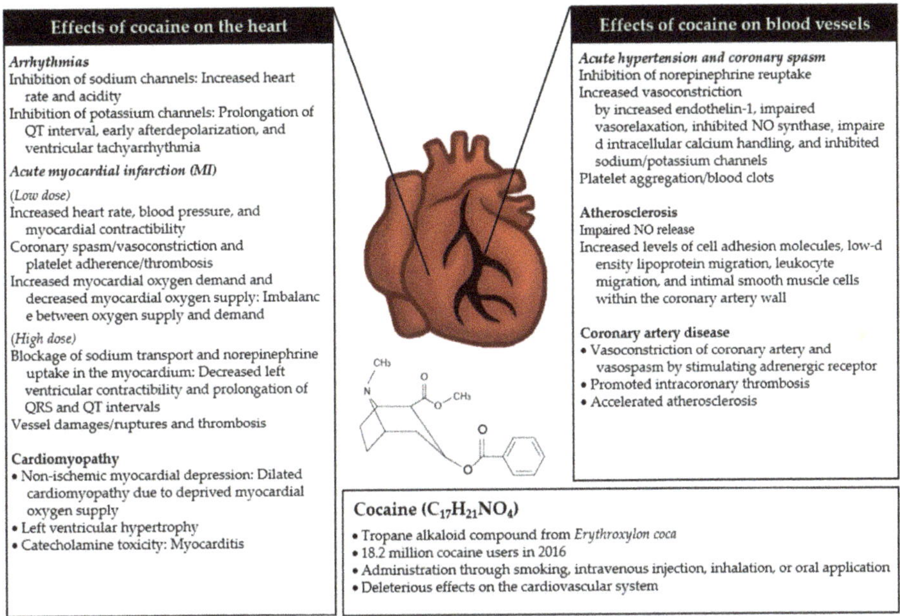

Figure 1. Effects of cocaine on cardiovascular health. Use of cocaine (bottom) results in both acute (italic) and chronic (normal) changes in the heart (left) and blood vessels (right). (Note: Cocaine often induces cardiac condition(s) (e.g., acute myocardial infarction (MI) and coronary artery disease) by affecting the heart and vessels simultaneously).

2.1.2. Arrhythmias

Previous studies have shown that cocaine inhibits cardiac ion channels such as sodium channels and potassium channels [22]. The upstroke of action potential was shown to be delayed in response to sodium channel blockade, which is modulated by heart rate and acidity. Increased heart rate and acidity boost the effect of cocaine on sodium channels [23,24]. Inhibition of sodium channels is intensified when cocaine is abused or when cocaethylene is formed after administration of cocaine with alcohol [25,26]. Cocaine effects on potassium channel blockade result in prolonged QT interval, early afterdepolarization, and ventricular tachyarrhythmia [27,28]. Similar to effects on sodium channels, cocaine abuse or cocaine use with alcohol exacerbates inhibition of potassium channels and QT prolongation [29]. In addition, cocaine administration increases body temperature, resulting in hyperthermia. Cocaine overdose can induce cardiac arrhythmias and result in an impaired electrocardiographic profile, which may be related to the increased prevalence of cocaine-associated mortality in hot weather and in crowded circumstances [30,31]. In addition to these factors, cardiac arrhythmias may be affected by other factors such as catecholamine excess and calcium channel blockade. Acidosis and electrolyte abnormalities can also modulate cardiac arrhythmias. [23]. As such, cocaine-induced cardiac arrhythmias can be generated via many mechanisms in cocaine users.

2.1.3. Acute Myocardial Infarction

Mechanisms of acute MI resulting from cocaine use are multifactorial. Cocaine and its metabolites are sympathomimetic agents [32] and induce local anesthetic effects [11]. At low doses, cocaine-induced sympathetic effects increase heart rate, blood pressure, and myocardial contractibility, leading to increased myocardial oxygen demand [33]. Cocaine also enhances coronary spasm/vasoconstriction and platelet adherence/thrombosis, leading to reduced myocardial oxygen supply [34]. Thus, an imbalance between oxygen supply and demand results in MI [35]. At high doses, cocaine-induced local anesthesia results in decreased left ventricular (LV) contractibility and prolongation of QRS and QT intervals in electrocardiograms by blocking sodium transport and norepinephrine uptake in the myocardium [4]. In vessels, cocaine contributes to MI by increasing endothelin-1 [36] and reducing nitric oxide production in endothelial cells [37]. When vessels are stressed, acute damages/ruptures can occur, which promotes thrombosis by increasing platelet activity/aggregation [38,39] and elevating fibrinogen levels [40] and plasminogen activator inhibitor activity [41,42]. These cellular and molecular cascades result in reduced cardiac blood flow, leading to acute MI and possibly atherosclerosis and coronary thrombosis in the long term [43,44]. As such, cocaine induces acute MI by directly affecting myocardial tissues in the heart and indirectly enhancing thrombosis in vessels.

2.2. Mechanisms of Chronic Toxicity

2.2.1. Cardiomyopathy

Cocaine causes systolic dysfunction or LV failure, which results from reduced ejection fraction and an enlarged left ventricular chamber [45]. Cocaine administration reduces myocardial contractility and ejection fraction [46] and enhances left ventricular end-diastolic pressure and end-systolic volume [47–49]. It may cause non-ischemic myocardial depression, leading to dilated cardiomyopathies such as Takotsubo cardiomyopathy, a type of non-ischemic cardiomyopathy [50]. Previous studies reported that cocaine-induced cardiomyopathy, especially dilated cardiomyopathy [51,52], resulting from deprivation of myocardial oxygen supply despite increased demand for oxygen, leads to reduced coronary blood flow. Dilated cardiomyopathy is the most common consequence of long-term cocaine use and can lead to several complications including heart failure and heart-valve defects [53]. Chronic abuse of cocaine is associated with left ventricular hypertrophy [54]. In addition, catecholamine toxicity from chronic cocaine use was shown to be associated with myocarditis [55], which was related to increased local immune reactions and myocardial necrosis [56].

2.2.2. Atherosclerosis

Coronary atherosclerosis often occurs in young cocaine users [24,57] or cocaine users with other cardiovascular diseases (e.g., MI) [58]. According to previous studies, cocaine impairs nitric oxide release from endothelial cells [59,60]. In addition, cocaine increases levels of cell adhesion molecules (e.g., intracellular adhesion molecule-1 (ICAM-1), cluster of differentiation 54 (CD54), vascular cell adhesion molecule-1 (VCAM-1), endothelial leukocyte adhesion molecule-1 (ELAM-1)), low-density lipoprotein migration, and leukocyte migration in blood vessels [61]. Moreover, intimal smooth muscle cells within the coronary artery wall increase [24,62], presumably leading to progression of atherosclerosis and potential sudden cardiac death [63]. Based on immunological studies, mast cells in plaques may contribute to atherosclerosis, vasospasm, thrombosis, and sudden death [57,59,64]. Briefly, proteolytic substances released from mast cells accelerate atherosclerosis by degrading and facilitating uptake of low-density lipoprotein cholesterol by macrophages [65,66]. Histamine released from mast cells increases endothelial permeability, which leads to leukocyte migration [67]. As such, cocaine has complex effects on endothelial cell dysfunction, facilitates low-density lipoprotein and leukocyte migration, and increases intimal smooth muscle cells, all of which contribute to atherosclerosis in long-term users.

2.2.3. Coronary Artery Diseases

Chronic cocaine use causes repetitive damages to the heart and vessels by interacting with norepinephrine transporters [68]. Alpha-2 adrenergic receptors induce vasoconstriction of coronary arteries through contraction of vascular smooth muscle cells [34], leading to prothrombotic effects caused by increased von Willebrand factor [21]. Cocaine induces vasospasm through stimulation of adrenergic receptors on coronary arteries [69]. Cocaine also promotes intracoronary thrombosis [70,71] through increased von Willebrand factor release, increased levels of endothelial tissue factor, an important factor in pathogenesis of acute coronary syndrome (ACS), decreased levels of tissue factor pathway inhibitor [72], and accelerated atherosclerosis due to endothelial cell dysfunction [60]. In addition, long-term use of cocaine induces endothelial injury, vascular fibrosis [73,74], and subsequent vessel wall weakening [75], resulting in apoptosis of vascular smooth muscle cells and cystic medial necrosis [76,77]. According to previous reports, cocaine sometimes induces coronary and carotid aortic dissections [78–80]. Thus, cocaine causes coronary artery diseases through multifactorial mechanisms including vasoconstriction, intracoronary thrombosis, and accelerated atherosclerosis.

3. Cocaine Cardiotoxicity in Human Studies

Cocaine-induced cardiotoxicity can result in deleterious effects on the heart and vessels through multifactorial pathophysiological mechanisms, as described above. In this section, we focus on recent human studies published in the past 10 years, retrieved from the Medline database. Table 1 presents these studies that examined the association of cocaine use with both acute and chronic cardiovascular diseases and mortality.

3.1. Acute Effects of Cocaine

A number of studies have reported a possible link between cocaine use and acute cardiovascular conditions such as acute hypertension, arrhythmia, coronary artery aneurysms (CAAs), and acute MI. Because the study populations and data sources varied across the studies, the findings of these studies should be interpreted carefully in the context of each individual study.

Kozor et al. [81] in Australia compared blood pressure, aortic stiffness, and LV mass in cocaine users with those in cocaine non-users. The authors recruited 20 regular cocaine users aged 37 ± 7 years (85% male) and 20 control subjects aged 33 ± 7 years (95% male). This study defined regular cocaine use as using cocaine at least monthly during the year prior to when the study was conducted. The study findings showed that cocaine users had higher systolic blood pressure (134 ± 11 vs. 126 ± 11 mm Hg), increased aortic stiffness, and greater LV mass (124 ± 25 vs. 105 ± 16 g) compared with cocaine non-users.

Table 1. Acute and chronic effects of cocaine on the cardiovascular system.

Study (year)	Country	Study Design	Data Source	Study Population (Sample Size)	Male %, Age (mean ± SD)	Outcome(s)	Findings
Acute effects of cocaine							
Kozor et al. (2014) [81]	Australia	Cross-sectional	Study participants	Adults with no coronary disease, no previous MI, no contraindication to CMR imaging, and no cocaine use in the 48 h prior to image acquisition (n = 20 for social cocaine users; n = 20 for cocaine non-users)	85%, 37 ± 7 yrs in the social cocaine users' group; 95%, 33 ± 7 yrs in the cocaine nonusers group	Systolic blood pressure, aortic stiffness, and LV mass	Cocaine use associated with high systolic blood pressure (134 ± 11 vs. 126 ± 11 mmHg), increased aortic stiffness, and greater LV mass (124 ± 25 vs. 105 ± 16 g) compared with no cocaine use
Sharma et al. (2016) [43]	US	Retrospective	ECG recordings in the Atherosclerosis Risk in Communities (ARIC) study from Aug. 2006 to Dec. 2014	Cocaine-dependent subjects (n = 97); non-cocaine-using control subjects (n = 8513)	86%, 50 ± 4 yrs in the cocaine-dependent subjects' group; 46%, 52 ± 5 yrs in the controls group	Resting ECG parameters	Significant effects of cocaine use on early repolarization (OR = 4.92, 95% CI: 2.73–8.87), bradycardia (OR = 3.02, 95% CI: 1.95-4.66), severe bradycardia (OR = 5.11, 95% CI: 2.95-8.84), and heart rate (B weight = −5.84, 95% CI: −7.85 to −3.82)
Kariyanna et al. (2018) [82]	US	Case-report	Patient	A 55-year-old woman presenting with a chest pain after cocaine use (n = 1)	0%, 55 yrs	Second degree Mobitz type II atrioventricular block	Cocaine-induced Mobitz type II second degree atrioventricular block
Satran et al. (2005) [83]	US	Retrospective	Angiographic database at Hennepin County Medical Center in Minnesota	Patients with a history of cocaine use (n = 112); Patients with no history of cocaine use (n = 79)	79%, 44 ± 8 yrs in the cocaine users' group; 61%, 46 ± 5 yrs in the cocaine non-users group	CAA	Significantly higher CAA in cocaine users compared with cocaine nonusers (30.4% vs. 7.6%)
Gupta et al. (2014) [1] [84]	US	Retrospective	Acute Coronary Treatment and Intervention Outcomes Network Registry-Get With The Guidelines (ACTION Registry-GWTG)	Patients admitted within 24 h of acute MI from July 2008 to March 2010 (n = 924 in the cocaine group; n = 102,028 in the non-cocaine group)	80%, 50 (range: 44–56) yrs in the cocaine group; 65%, 64 (range: 54–76) yrs in the non-cocaine group	Acute STEMI, cardiogenic shock, multivessel CAD, and in-hospital mortality	Higher percentages of STEMI (46.3% vs. 39.7%) and cardiogenic shock (13% vs. 4.4%) in the cocaine group, but a lower percentage of multivessel coronary artery disease (53.3% vs. 64.5%). Similar in-hospital mortality between the cocaine group and the non-cocaine group (OR = 1.00, 95% CI: 0.69–1.44)
Salihu et al. (2018) [85]	US	Retrospective	National Inpatient Sample (NIS) from Jan. 2002 to Dec. 2014	Pregnant women aged 13-49 yrs who had pregnancy-related inpatient hospitalizations (n = 153,608 cocaine users; n = 56,882,258 non-drug users)	0%, Age group: 13–24 (21.4%); 25–34 (55.4%); 35–49 (20.5%) in the cocaine users' group; 0%, Age group: 13-24 (34.0%); 25–34 (51.3%); 35–49 (14.7%) in the non-drug users' group	Acute MI or cardiac arrest	Cocaine use associated with acute MI or cardiac arrest (adjusted OR = 1.83, 95% CI: 1.28–2.62)

Table 1. *Cont.*

Study (year)	Country	Study Design	Data Source	Study Population (Sample Size)	Male %, Age (mean ± SD)	Outcome(s)	Findings
Aslibekyan et al. (2008) [86]	US	Retrospective	National Health and Nutrition Examination Survey (NHANES) in 1988–1994 and 2005–2006	Civilian non-institutionalized US adults (a) aged 18-59 (n = 11,993); (b) aged 18-45 (n = 9337)	(a) 46%, 36 yrs (N/R); (b) 39%, 31 yrs (N/R)	Prevalence of MI	(a) No significant association between cocaine use and MI in the 18–59 age group; (b) Significant association between cocaine use of > 10 lifetime instances and MI in the 18–45 age group (aged-adjusted OR = 4.60, 95% CI: 1.12–18.88), but this association was attenuated in the multivariate-adjusted model (OR = 3.84, 95% CI: 0.98–15.07)
Gunja et al. (2018) [2] [87]	US	Retrospective	Veterans Affairs database	Veterans with CAD undergoing cardiac catheterization from Oct. 2007 to Sep. 2014 (n = 3082 in the cocaine group; n = 118,953 in the non-cocaine group)	98.6%, median age: 58 (IQR: 54–62) yrs in the cocaine group; 98.6%, median age: 65 (IQR: 61–72) yrs in the non-cocaine group	MI and 1-year all-cause mortality	With adjustment of basic cardiac risk factors, cocaine use was significantly associated with MI (HR = 1.40, 95% CI: 1.07–1.83) and mortality (HR = 1.23, 95% CI: 1.08–1.39). After adjustment for risky behaviors, cocaine use was associated with mortality (HR = 1.22, 95% CI: 1.04–1.42), but not with MI (HR = 1.17, 95% CI: 0.87–1.56). After adjustment for causal pathway conditions, mortality was no longer significant (HR = 1.15, 95% CI: 0.99–1.33)
Chronic effects of cocaine							
Maceira et al. (2014) [45]	Spain	Prospective	Study participants and a gender and age matched healthy group	Cocaine abusers attending a rehabilitation clinic for the first time (n = 94)	86%, 37 ± 7 yrs	Cocaine cardiotoxicity using a CMR protocol	Increased LV end-systolic volume, LV mass index, and RV end-systolic volume, and decreased LV ejection fraction and RV ejection fraction in cocaine abusers compared with those in the gender and age matched healthy group
Arora et al. (2015) [88]	US	Cross-sectional	Drug treatment center in Florida	Caucasian adults with cocaine use disorder (n = 33)	33%, 37 ± 9 yrs	Presence of subclinical CAD using CIMT	No association between chronic cocaine use and subclinical CAD measured by CIMT
Bamberg et al. (2009) [89]	US	Nested matched cohort	Massachusetts General Hospital	Patients who presented to the emergency department with acute chest pain in May to July, 2005 (n = 44 in the cocaine group; n = 132 in the non-cocaine group)	86%, 46 ± 7 yrs in the cocaine group; 86%, 46 ± 7 yrs in the non-cocaine group	ACS and CAD using coronary CT	Significant association of cocaine use with increased risk of ACS group (OR = 5.79, 95% CI: 1.24–27.02), but no association with coronary stenosis
Chang et al. (2011) [90]	US	Cross-sectional	University of Pennsylvania Hospital	Patients who received coronary CTA for evaluation of CAD in the emergency department from May 2005 to Dec. 2008 (n = 157 in the cocaine group; n = 755 in the non-cocaine group)	58%, 46 ± 6 yrs in the cocaine group; 40%, 48 ± 9 yrs in the non-cocaine group	CAD	No association between recent cocaine use and the presence of coronary lesions ≥ 25% (adjusted RR = 0.92, 95% CI: 0.58–1.45) and coronary lesions ≥ 50% (adjusted RR = 0.96, 95% CI: 0.46–2.01)

Table 1. *Cont.*

Study (year)	Country	Study Design	Data Source	Study Population (Sample Size)	Male %, Age (mean ± SD)	Outcome(s)	Findings
Lai et al. (2016) [91]	US	Cross-sectional	Study participants	African American adults with/without HIV infection in Baltimore (n = 737 in the cocaine group; n = 692 in the non-cocaine group)	60.3%, 45 (IQR: 40–50) yrs in the entire population	Subclinical CAD defined by the presence of CAC detected by noncontrast CT and/or coronary plaque detected by contrast-enhanced coronary CT angiography	Chronic cocaine use associated with high risk for subclinical CAD (propensity score-adjusted prevalence ratio = 1.27, 95% CI: 1.08–1.49), CAC (propensity score-adjusted prevalence ratio=1.26, 95% CI: 1.05–1.52), any coronary stenosis (propensity score-adjusted prevalence ratio = 1.30, 95% CI: 1.08–1.57), and calcified plaques (propensity score-adjusted prevalence ratio = 1.37, 95% CI: 1.10–1.71)
Lucas et al. (2016) [92]	US	Cross-sectional and longitudinal	Study participants	Adults with/without human immunodeficiency virus infection in Baltimore (n = 57 never cocaine users; n = 82 past cocaine users; n = 153 current cocaine users)	67%, 46 (IQR: 41–53) yrs in the never users; 66%, 51 (IQR: 46–54) yrs in the past users; 75%, 49 (IQR:45–52) yrs in the current users	Subclinical CVD: carotid artery plaque	Cocaine use associated with approximately three-fold higher odds of carotid plaques at baseline (OR = 3.3, 95% CI: 1.5–7.3 for past cocaine users vs. cocaine nonusers; OR = 2.7, 95% CI: 1.3–5.5 for the current cocaine users vs. cocaine nonusers)
Cocaine effects on mortality							
DeFilippis et al. (2018) [93]	US	Retrospective cohort	Two academic medical centers (Brigham and Women's Hospital and Massachusetts General Hospital)	Patients presenting with an MI at ≤50 years between 2000 and 2016 (n = 99 in the cocaine group; 1873 in the non-cocaine group)	85%, 44 (range: 40–46) yrs in the cocaine group; 80%, 45 (range: 42–48) yrs in the non-cocaine group	Cardiovascular mortality and all-cause mortality	Significant association of cocaine use with cardiovascular mortality (HR = 2.32, 95% CI: 1.11–4.85) and all-cause mortality (HR = 1.91, 95% CI: 1.11–3.29)
Morentin et al. (2014) [94]	Spain	Case-control retrospective	Forensic autopsy reports in Biscay, Spain	All SCVD in individuals aged 15–49 (n = 311); SnoCVD (n = 126) from Jan. 2003 to Dec. 2009	82%, 41 ± 7 yrs in SCVD; 71%, 39 ± 7 yrs in SnoCVD	Cocaine detected in blood	Cocaine being the risk for SCVD (OR = 4.10; 95% CI: 1.12–15.0)
Qureshi et al. (2014) [95]	US	Retrospective	NHANES in 1988-1994	Civilian non-institutionalized US adults aged 18–45 (n = 7751 cocaine nonusers; n = 730 infrequent cocaine users (1–10 times); n = 354 frequent cocaine users (>10 times); n = 178 regular cocaine users (>100 times))	43%, 31 ± 8 yrs in the cocaine non-users' group; 59%, 31±10 yrs in the infrequent cocaine users group; 65%, 33 ± 9 yrs in the frequent cocaine users group; 70%, 33 ± 7 yrs in the regular cocaine users group	Cardiovascular mortality and all-cause mortality	Regular lifetime cocaine use was associated with high all-cause mortality (RR = 1.9, 95% CI: 1.2–3.0), but not cardiovascular mortality (RR = 0.6, 95% CI: 0.1–4.7) compared with cocaine nonusers
Hser et al. (2012) [96]	US	Prospective cohort	California Treatment Outcome Project (CalTOP) between 2000 and 2002, the National Death Index by 2008, the National Death Register by 2010, and the California Department of Mental Health	Women admitted to 40 drug abuse treatment programs through CalTOP (n = 4,253 for those alive in 2010; n = 194 for those deceased by 2010)	0%, 33 ± 8 yrs for living; 0% 37 ± 7 yrs for the deceased	8 to 10-year mortality	Cocaine was associated with higher mortality relative to methamphetamine (HR = 3.56, 95% CI: 1.95–6.48)

Table 1. *Cont.*

Study (year)	Country	Study Design	Data Source	Study Population (Sample Size)	Male %, Age (mean ± SD)	Outcome(s)	Findings
Atoui et al. (2011) [97]	US	Retrospective chart review	Electronic medical records in Bronx Lebanon Hospital Center	Patients admitted with chest pain to the hospital who had no cardiovascular risk factors from July 2009 to June 2010 (n = 54 in the cocaine group; n = 372 in the non-cocaine group)	59%, 44 ± 10 yrs in the cocaine group; 49%, 43 ± 12 yrs in the non-cocaine group	Length of stay and mortality	No significant difference in length of stay (3.0 vs. 2.4) and in-hospital mortality (0% vs. 1%) between the cocaine group and the non-cocaine group

ACS: Acute coronary syndrome; CAA: Coronary artery aneurysm; CAC: Coronary artery calcium; CAD: Coronary artery disease; CI: Confidence interval; CIMT: Carotid intima media thickness; CMR: Cardiovascular magnetic resonance; CT: Computed tomography; CTA: Computerized tomographic angiography; CVD: Cardiovascular disease; ECG: Electrocardiogram; HR: Hazard ratio; IQR: Interquartile range; LV: Left ventricular; MI: Myocardial infarction; N/R: not reported; OR: Odds ratio; RR: Relative risk; RV: Right ventricular; SCVD: Sudden cardiovascular death; SnoCVD: Sudden death not due to cardiovascular diseases; STEMI: ST elevation myocardial infarction. [1] Including acute effects (i.e., acute STEMI and cardiogenic shock) and chronic effect (i.e., multivessel CAD) of cocaine and mortality as the study outcomes. [2] Including acute effect of cocaine (i.e., MI) and 1-year all-cause mortality as the study outcomes.

In addition, Sharma et al. [43] retrospectively reviewed electrocardiogram (ECG) recordings of cocaine-dependent subjects to examine cardiotoxicity of cocaine use. The ECGs were collected from 97 cocaine-dependent subjects aged 50 $\pm$ 4 years (86% male) in a comprehensive academic health center and 8513 cocaine-non-using subjects aged 52 $\pm$ 5 years (46% male) participating in the Atherosclerosis Risk in Communities (ARIC) study. The authors found significant effects of cocaine use on early repolarization (odds ratio (OR) = 4.92, 95% confidence interval (CI): 2.73–8.87), bradycardia (OR = 3.02, 95% CI: 1.95–4.66), severe bradycardia (OR = 5.11, 95% CI: 2.95–8.84), and heart rate (B weight = −5.84, 95% CI: −7.85 to −3.82). Recently, there was a case report of Mobitz type II atrioventricular (AV) block associated with cocaine use [82]. This case occurred in a 55-year-old female who presented with chest pain after cocaine use.

Satran et al. [83] investigated the prevalence of CAAs among cocaine users undergoing coronary angiography using a database from a medical center in the US. The study population included 112 patients with a history of cocaine use aged 44 $\pm$ 8 years (79% male) and 79 patients with no history of cocaine use aged 46 $\pm$ 5 years (61% male). Based on the finding that cocaine users had a significantly higher CAA compared with cocaine non-users (30.4% vs. 7.6%, respectively), the authors concluded that cocaine users were likely to be at increased risk of acute MI.

Several studies examined the association between cocaine use and MI. Gupta et al. [84] examined the incidence of acute ST elevation myocardial infarction (STEMI), cardiogenic shock, multivessel CAD, and in-hospital mortality in the cocaine group (n = 924) compared with the non-cocaine group (n = 102,028) among patients admitted within 24 h of acute MI. This study used data from the National Cardiovascular Data Registry Acute Coronary Treatment and Intervention Outcomes Network Registry-Get With The Guidelines (ACTION Registry-GWTG) program. Compared with the non-cocaine group, the cocaine group was younger (average age: 50 (44–56) vs. 64 (54–76)), had a higher proportion of men (80% vs. 65%) and African-Americans (45% vs. 9%), and fewer traditional cardiovascular risk factors such as hypertension (65% vs. 71%), dyslipidemia (42% vs. 59%), previous coronary bypass (6% vs. 14%), and previous revascularization (24% vs. 31%). Gupta et al. [84] found higher percentages of STEMI (46% vs. 40%) and cardiogenic shock (13% vs. 4%) in cocaine users although their percentage of multivessel CAD was lower (53% vs. 65%) compared with cocaine non-users. In-hospital mortality was similar between the two groups (OR = 1.00, 95% CI: 0.69–1.44). Another study conducted by Salihu et al. [85] included pregnant women aged 13–49 years to examine the association of cocaine use with incidence of acute MI or cardiac arrest during pregnancy or childbirth. This retrospective study used data from January 2002 through December 2014 from the National Inpatient Sample (NIS), a large public inpatient database in the U.S. The study results showed that cocaine users (n = 153,608) were at higher risk for acute MI or cardiac arrest compared with drug non-users (n = 56,882,258), with adjusted OR of 1.83 (95% CI: 1.28–2.62). Some studies showed that the association between cocaine use and MI was affected by some confounders such as cardiac risk factors and risky behaviors. For example, Aslibekyan et al. [86] conducted a retrospective study examining the prevalence of MI among civilian non-institutionalized US adults. Using data from the National Health and Nutrition Examination Survey (NHANES), their study included two different age groups for their study population: (a) individuals aged 18–59 years (n = 11,993, 46% male) and (b) those aged 18–45 years (n = 9337, 39% male). Although Aslibekyan et al. [86] found no association between cocaine use and MI in the 18–59 age group, cocaine use of >10 lifetime instances was significantly associated with MI in the 18–45 age group after adjusting for age (aged-adjusted OR = 4.60, 95% CI: 1.12–18.88). This association was affected by cardiac risk factors (e.g., smoking status, history of diabetes, hyperlipidemia, and hypertension) in the multivariate-adjusted model (OR = 3.84, 95% CI: 0.98–15.07). Another retrospective study by Gunja et al. [87] examined the association of cocaine use with MI and 1-year all-cause mortality. The study included veterans with CAD who underwent coronary catheterization between October 2007 and September 2014 using the Veterans Affairs database. Compared with the non-cocaine group (n = 118,953), the cocaine group (n = 3082) was younger (median age: 58 vs. 65), more likely to be African-American (59% vs. 11%) and had fewer traditional cardiac risk factors. After adjusting for cardiac risk factors, cocaine use was significantly associated with MI (hazard ratio (HR) = 1.40, 95% CI: 1.07–1.83); however, this association became attenuated after controlling for risky behaviors in the sequential multivariable model (HR = 1.17, 95% CI: 0.87–1.56).

In summary, prior studies have reported that cocaine use was associated with acute cardiovascular conditions such as elevated blood pressure, (severe) bradycardia, CAAs, and acute MI. These findings are consistent with earlier studies documenting cocaine-related MI [59,98]. Of note, the studies in this review suggest that the association between cocaine use and MI might be confounded by cardiac risk factors or risky behaviors. Accordingly, the risk of MI among cocaine users needs to be understood in the context of risk factors and risky behaviors.

3.2. Chronic Effects of Cocaine

Several studies examined whether cocaine use was associated with chronic cardiovascular conditions such as cardiomyopathy (e.g., LV hypertrophy), subclinical atherosclerosis, and CAD. In this section, we present the characteristics of each study along with the study findings. We interpreted the results with consideration of study populations and data sources.

Maceira et al. [45] found that cocaine abusers had increased LV end-systolic volume, LV mass index, and right ventricular (RV) end-systolic volume, with decreased LV ejection fraction and RV ejection fraction. The study participants were 94 cocaine abusers aged 37 $\pm$ 7 years (86% male) attending a rehabilitation clinic for the first time. They were compared with an age- and gender-matched healthy group. As previously mentioned, Kozor et al. [81] also showed greater LV mass among regular cocaine users compared with cocaine nonusers.

Furthermore, several previous studies examined the association between cocaine use and CAD. The effects of cocaine on subclinical CAD were examined using different CAD surrogate markers [84,88–91]. For example, Arora et al. [88] examined the presence of subclinical CAD using carotid intima media thickness (CIMT) as a surrogate marker. This cross-sectional study included 33 Caucasian adults aged 37 $\pm$ 9 years who used cocaine (33% male). Their findings suggested no association between chronic cocaine use and subclinical CAD measured by CIMT. Another study conducted by Bamberg et al. examined the association of cocaine use with CAD and ACS using coronary computed tomography (CT) [89]. The study subjects were patients who presented to the emergency department (ED) with acute chest pain. In this nested matched cohort study, there were 44 patients in the cocaine group aged 46 $\pm$ 7 years (86% male) and 132 patients in the age- and gender-matched non-cocaine group. The authors found no significant association between cocaine use and coronary stenosis, but found a significant association between cocaine use and ACS (OR = 5.79, 95% CI: 1.24–27.02). Chang et al. [90] conducted another cross-sectional study that included patients who received coronary computerized tomographic angiography (CTA) for evaluation of CAD in the ED. The patients were at low- to intermediate-risk for ACS. Of these patients, cocaine users were aged 46$\pm$6 years (n = 157, 58% male) while the non-cocaine group was aged 48 $\pm$ 9 years (n = 755, 40% male). Chang et al. [90] found no association between repetitive cocaine use and coronary calcifications or between recent cocaine use and CAD. As noted previously, Gupta et al. [84] investigated the incidence of multivessel CAD in addition to STEMI and cardiogenic shock between the cocaine group (n = 924) and the non-cocaine group (n = 102,028) among patients admitted within 24 h of acute MI. They found a lower percentage of multivessel CAD among cocaine users than cocaine nonusers (53% vs. 65%), although the percentages of STEMI (46% vs. 40%) and cardiogenic shock (13% vs. 4%) were higher. In contrast to these studies, a study by Lai et al. [91] found a higher risk for subclinical CAD among cocaine users compared with cocaine non-users (propensity score-adjusted prevalence ratio (PR) = 1.27, 95% CI: 1.08–1.49). The subjects in the study by Lai et al. [91] were African Americans aged 45 years (Interquartile range (IQR): 40–50), of whom 60% were males. Approximately 67% of the subjects were HIV-positive. In this cross-sectional study, subclinical CAD was defined by the presence of coronary artery calcium (CAC) detected by non-contrast CT and/or coronary plaque detected by contrast-enhanced CT angiography (CCTA). Chronic cocaine users were at significantly higher risk for the presence of CAC (propensity score-adjusted PR = 1.26, 95% CI: 1.05–1.52), any coronary stenosis (propensity score-adjusted PR = 1.30, 95% CI: 1.08–1.57), and calcified plaques (propensity score-adjusted PR = 1.37, 95% CI: 1.10–1.71), in addition to subclinical CAD. Another study conducted by Lucas et al. [92] showed a significant association between cocaine use and carotid plaque formation. More than 90% of subjects in this study were African Americans. Cocaine non-users were aged 46 years (IQR: 41–53), and 67% were male. Past cocaine users were aged 51 years (IQR: 46–54), and 66% were male. Current cocaine users were aged 49 years (IQR: 45–52), and 75% were male. Of the study subjects, approximately 66% were HIV-positive. Compared with cocaine non-users, both past cocaine users and current users had approximately three-fold higher odds of having carotid plaques at baseline (OR = 3.3, 95% CI: 1.5–7.3 and OR = 2.7, 95% CI: 1.3–5.5, respectively).

In summary, cocaine was reported to be associated with high risk for cardiomyopathy characterized by LV hypertrophy [45,81] and ACS [89]. In particular, one study found an approximately six-fold higher risk for ACS among cocaine users [89]. However, studies have reported inconsistent findings regarding association between cocaine use and subclinical CAD. Some studies found no association of cocaine use with coronary calcifications [88–90]. This result is consistent with findings in the Coronary Artery Risk Development in Young Adults (CARDIA) study which examined the association between cocaine exposure and prevalence of coronary calcification by including over 3000 participants [99]. The CARDIA study reported no relationship between cocaine exposure and coronary calcium after adjusting for age, sex, ethnicity, socioeconomic status, family history, tobacco use, and alcohol use. However, Lai et al. [91] reported that cocaine use was associated with subclinical coronary atherosclerosis. Lai et al. [95–97] also showed this association in their earlier studies. Similarly, Lucas et al. [92] found greater carotid plaque formation at baseline among cocaine users compared with cocaine nonusers. This variability in findings across studies regarding association between cocaine use and subclinical CAD might be explained by different CAD risk factor profiles of the study populations. The studies reporting cocaine-associated plaques included predominantly African American participants, of whom 40% to 100% were HIV-positive [91,92,100–102]. In contrast, all other studies showing no association between cocaine use and coronary calcifications did not include any HIV-positive individuals [84,88–90]. It has been widely known that HIV infection is a risk factor for CAD. Therefore, the study subjects with HIV may have been at higher risk for development of CAD, as was pointed by Arora et al. [88].

3.3. *Effects of Cocaine on Mortality*

Several studies estimated cardiovascular mortality among cocaine users. These studies have shown mixed results with regard to association of cocaine use with cardiovascular mortality. Some studies have reported higher risk for cardiovascular mortality among cocaine users compared with cocaine non-users. For example, DeFilippis et al. [93] retrospectively analyzed records of patients with MI at $\leq$50 years of age between 2000 and 2016 to examine the risk of cocaine use for cardiovascular mortality and all-cause mortality. Patient data were obtained from two large academic medical centers in the US There were 99 individuals in the cocaine-group (mean age: 44 (40–46), 85% male) and 1873 individuals in the non-cocaine group (mean age: 45 (42–48), 80% male). The authors found significant associations of cocaine use with cardiovascular mortality (HR=2.32, 95% CI: 1.11-4.85) and all-cause mortality (HR = 1.91, 95% CI: 1.11–3.29). In Spain, Morentin et al. [94] investigated the prevalence of recent cocaine use in individuals who had died by sudden cardiovascular death (SCVD) between January 2003 and December 2009 (n = 311). The mean age was 41 $\pm$ 7 years, and 82% were male. Individuals who had died by sudden deaths not due to cardiovascular diseases (SnoCVD) served as the control group (n = 126). The average age and percentage of males in the control group were 39 $\pm$ 7 years and 71%. The authors found that recent cocaine use was a significant risk factor for SCVD (OR = 4.10, 95% CI: 1.12–15.0). In contrast, Qureshi et al. [95] found that regular cocaine use was not associated with cardiovascular mortality (relative risk (RR) = 0.6, 95% CI: 0.1–4.7). The study subjects in this retrospective study were civilian non-institutionalized US adults aged 18–45 in the NHANES dataset. The study included 7751 cocaine nonusers (mean age: 31 ± 8 years, 43% males) and 178 regular cocaine users (lifetime cocaine use > 100 times) (mean age: 33 ± 7 years, 70% males). Although the study results showed a significant association between regular cocaine use and all-cause mortality (RR = 1.9, 95% CI: 1.2–3.0), regular cocaine use was not associated with cardiovascular mortality.

Prior studies examining the association of cocaine use with all-cause mortality have also reported inconsistent findings. In some studies, cocaine use was significantly associated with all-cause mortality. As mentioned previously, DeFilippis et al. [93] and Qureshi et al. [95] found an approximately two-fold higher all-cause mortality among cocaine users compared with cocaine nonusers. Similarly, Hser et al. [96] found an elevated mortality risk associated with cocaine use relative to methamphetamine use (HR = 3.56, 95% CI: 1.95–6.48). The subjects in this study were women admitted to drug abuse treatment programs in the US between 2000 and 2002. Contrary to the findings of these studies, some studies have reported no significant association between in-hospital mortality and cocaine use [84,97]. Atoui et al. [97] conducted a retrospective chart review of patients admitted with chest pain to a US-based hospital between July 2009 and June 2010. Of the study population with no risk factors for CAD, 54 were cocaine users (mean age = 44 ± 10 years, 59% males) and 372 were cocaine non-users (mean age = 43 ± 12 years, 49% males). The study results showed no significant differences in length of stay and in-hospital mortality between cocaine users and nonusers. Similarly, in the aforementioned study by Gupta et al. [84] in-hospital mortality was not significantly different between the cocaine group and the non-cocaine group (OR = 1.00, 95% CI: 0.69–1.44). As mentioned previously, in the study by Gunja et al. [87] cocaine use was initially found to be significantly associated with 1-year all-cause mortality after adjusting for cardiac risk factors and risky behaviors among veterans with CAD (HR = 1.22, 95% CI: 1.04–1.42) [87]. However, after controlling for causal pathway conditions, mortality was no longer significantly associated with cocaine use (HR: 1.15, 95% CI: 0.99–1.33).

In summary, some prior studies have reported an association between cocaine use and cardiovascular or all-cause mortality [93,94,96]. However, this association was not observed in other studies [84,97]. Variations in findings across studies may be driven by heterogeneity in patient characteristics (e.g., age), risky behaviors (e.g., smoking, alcohol, or other illicit drug use), and traditional risk factors (e.g., morbidities), all of which are predictors of mortality. Indeed, the study by Gunja et al. [87] showed how the association between cocaine use and mortality was confounded by these factors. In their study, cocaine was initially found to be associated with increased all-cause mortality. However, this association was no longer observed after controlling for causal pathway conditions such as congestive heart failure, cardiogenic shock, dialysis, depression, anxiety, ACS, and clinical status. This finding suggests that the effects of cocaine on mortality are largely dependent on individual clinical risk factors. Notably, mortality risk was not significantly higher among cocaine users if the cocaine users had fewer risk factors compared with cocaine nonusers. For example, subjects in the studies performed by Atoui et al. [97] and Gupta et al. [84] were individuals at low risk for CVD and young adults with few CV risk factors, respectively. Both studies found no association between cocaine use and mortality. In contrast, the subjects in studies reporting an association between cocaine use and mortality were at higher risk. The presence of risk factors is likely to augment the risk of mortality following cocaine use. Furthermore, frequency of cocaine use could be an important factor affecting mortality risk among cocaine users. As observed in the study by Qureshi et al., all-cause mortality was about two times higher among regular cocaine users (lifetime cocaine use > 100 times) compared with cocaine nonusers [95]. However, all cause-mortality of infrequent cocaine users (lifetime cocaine use: 1–10 times) or frequent cocaine users (lifetime cocaine use > 10 times) was not significantly different from that of cocaine nonusers in this study.

4. Cocaine and Nutrition

Cocaine use/abuse often affects food intake behavior and suppresses appetite, which may lead to the disruption of metabolic and neuroendocrine regulation. In addition, cocaine-induced malnutrition may decrease levels of neurotransmitters, and alter amino acid absorption and utilization. As such, chronic exposure to cocaine can result in an increased risk of health conditions such as hypertension, body weight problems, diabetes, and metabolic syndrome.

Cocaine affects appetite and body weight through multifactorial mechanisms. As mentioned previously, cocaine inhibits the reuptake of dopamine by interacting with the dopamine transporter, resulting in increased levels of dopamine in the central nervous system. Subsequently, changes in dopamine levels affect eating behavior and body weight [103–105]. Increased dopaminergic neurotransmission suppresses overall food intake whereas it increases fat-rich food intake [106]. In addition, cocaine blocks the reuptake of serotonin by interacting with the serotonin transporter, inducing leptin-dependent anorexic effect [107,108]. Prior studies demonstrated that cocaine also upregulated neuromodulators such as cocaine- and amphetamine-regulated transcript (CART), which plays an important role in regulating food intake, maintaining body weight, and in endocrine and cardiovascular functions [109,110]. Overexpression of CART has been reported to decrease food intake and change lipid metabolism related to fat storage [111,112].

In accordance with these mechanisms, several pre-clinical studies have shown the effects of cocaine on food consumption and the nutritional status in animals [113–115]. For example, Balopole et al. [113] reported a decrease in food intake after cocaine administration to rats (10, 15, and 25 mg/kg). They found that the cocaine-induced anorexia was transient and dose-dependent. After an hour of anorexic effect, it was shown that animals overconsumed foods. Therefore, total food intake was not significantly different between cocaine- and saline-exposed rats. Another study examined the effects of cocaine on the milk intake and body weight in rats [114]. Findings of this study suggested that cocaine disrupted ingestion primarily by interfering with the appetitive phase of feeding behavior (orientation and approach to food) rather than the consummatory phase (ingestion of food). A study by Church et al. [115] examined the effects of prenatal cocaine exposure on maternal/fetal toxicity in animals. Cocaine treatments in rats (20, 30, 40, and 50 mg/kg) resulted in significant reductions in the maternal weight gain and food consumption in a dose-dependent manner. Undernutrition led to a significant reduction in fetal weight. However, maternal water consumption was significantly increased in the cocaine-exposed animals possibly because of the increased locomotor activity and diuretic effect. Furthermore, cocaine provoked diarrhea in some of animals that received high doses, suggesting that cocaine, as a gastrointestinal irritant, might cause malabsorption and loss of electrolytes and nutrients, which ultimately can lead to malnutrition.

Human studies have also shown cocaine's anorexigenic effects and the resulting weight reduction in cocaine users [116,117]. Low caloric intake, together with abnormal metabolic and gastrointestinal functions, can lead to malnutrition among cocaine users [118]. For example, Escobar et al. [119] found that hemoglobin and hematocrit levels in cocaine users were below normal, indicating protein-energy malnutrition and anemia. As the authors pointed out, anemia in this population might be associated with a diet poor in micronutrients (e.g., iron), inadequate protein consumption, and clinical issues such as decreased intrinsic factor secretion, intestinal perforations, and bacterial or infectious diseases. Indeed, three cases were reported where patients required surgery for their intestinal perforations after cocaine use [120]. Cocaine led to mesenteric vasoconstriction and focal tissue ischemia by blocking the reuptake of norepinephrine, which might lead to intestinal perforations. Cocaine users in the study by Escobar et al. were also found to have altered lipid and glucose profiles, with low levels of high density lipoprotein (HDL) cholesterol and high levels of triglycerides, LDL cholesterol, total cholesterol, and glucose. These findings suggested that cocaine users might be at a high risk for metabolic and cardiovascular problems. Of note, cocaine users did not experience weight gain despite a compensatory increase in fat consumption following the cocaine-induced anorexia [117]. However, the cessation of cocaine use resulted in weight gain [117,121]. In a study by Ersche et al., cocaine users consumed significantly more fatty foods and carbohydrates compared with cocaine nonusers, but there was no concomitant weight increase in the cocaine group [117]. The authors suggested that an imbalance between fat intake and storage could lead to weight gain among cocaine users when they stop using cocaine. This imbalance might result from metabolic alterations from repeated cocaine use. It is well-known that weight gain increases the risk of cardio-metabolic disorders such as diabetes and cardiovascular conditions [122]. Therefore, weight control, as a means to prevent and lessen cardiovascular diseases, has profound implications during cocaine abstinence.

In summary, cocaine use affects eating behavior and suppresses appetite, leading to malnutrition and anorexia through disruption of the metabolic process and neuroendocrine regulation. Also, cocaine uptake in the body can lead to mesenteric vasoconstriction and focal tissue ischemia, and alter lipid as well as glucose profiles, presumably resulting in increased risk for metabolic and cardiovascular problems in cocaine users. Notably, the cessation of cocaine use causes sudden/excess weight gain during the recovery period/process, leading to increased cardiovascular and cardio-metabolic risks. As such, cocaine-induced changes in food intake patterns and the metabolic process can lead to cardiovascular complications during addiction as well as cessation periods.

5. Conclusions

Cocaine use/abuse has been known to make changes in nutrient status and metabolism, which can result in an increased risk of long-term health conditions including eating disorders, metabolic syndrome, and psychological abnormalities. In this review, we focus the deleterious acute and chronic effects of cocaine use particularly on cardiovascular outcomes. We summarized the pathophysiological mechanisms of cocaine on cardiovascular health, which were multifactorial and complex. Compared to chronic effects, acute effects of cocaine have been well-characterized in previous studies. Use of cocaine, a potent cardiovascular stimulant, has been associated with electrocardiographic abnormalities, elevated blood pressure, arrhythmia, and acute MI. The risk of MI among cocaine users was particularly influenced by individuals' cardiac risk factors and risky behaviors. Cocaine use can lead to acute conditions in a multifactorial fashion, for example, by blocking sodium/potassium channels in the heart and enhancing coronary artery spasm/vasoconstriction in vessels. In contrast, chronic effects of cocaine are difficult to determine as evidenced by inconsistent findings across previous studies. Some studies have reported an association of chronic cocaine use with coronary atherosclerosis using coronary calcification as a marker. Conversely, other studies have demonstrated no association between chronic cocaine use and coronary calcification. Of note, the subjects included in studies showing this association were at higher risk for CAD compared with those in the studies that reported no association. Therefore, chronic effects of cocaine may have been more prominent among those with higher CAD risk factor profiles. Contributions of cocaine to chronic conditions were also multifaceted. Long-term exposure to cocaine can exert chronic effects, for example, on the heart through non-ischemic myocardial depression and vessels by inducing endothelial cell injury and intracoronary thrombosis. Furthermore, prior studies suggested that risky behaviors, risk factors for CVD, and frequency of cocaine use may contribute to association between cocaine use and mortality. To evaluate the effects of long-term cocaine use on atherosclerosis and mortality more precisely, large, well-designed longitudinal studies are required with subjects from both low and high-risk populations. β-blocker therapy has often been suggested for cocaine users, in particular, for those with cocaine-associated heart failure. Studies have shown β-blockers lowered blood pressure, improved LV ejection fraction, and reduced the incidence of MI and mortality among cocaine users [123–125]. Understanding the multifactorial pathophysiological mechanisms of cocaine could help clinicians recognize the various symptoms after cocaine use/abuse and improve treatment of patients with either acute or chronic symptoms. The various deleterious CV outcomes resulting from cocaine use highlight the need for education regarding adverse cardiac effects of cocaine use, and access to effective interventions for cocaine abusers. Concurrently, alterations in lifestyle and behaviors (e.g., alcohol abuse or tobacco use) are also important for reducing the harmful adverse cardiac effects that these behavioral factors contribute to among cocaine users.

Author Contributions: Both S.T.K. and T.P. conceptualized and designed the study, extracted and interpreted the data, drafted the manuscript, and approved the final version of the manuscript.

Funding: This research received no external funding.

Conflicts of Interest: The authors declare no conflict of interest.

References

1. Cornish, J.W.; O'Brien, C.P. Crack cocaine abuse: An epidemic with many public health consequences. *Annu. Rev. Public Health* **1996**, *17*, 259–273. [CrossRef] [PubMed]
2. World Drug Report 2018. Available online: https://www.unodc.org/wdr2018/prelaunch/WDR18_Booklet_3_DRUG_MARKETS.pdf (accessed on 29 October 2018).
3. Behavioral Health Trends in the United States: Results from the 2014 National Survey on Drug Use and Health. Available online: https://www.samhsa.gov/data/sites/default/files/NSDUH-FRR1-2014/NSDUH-FRR1-2014.pdf (accessed on 29 October 2018).
4. Egred, M.; Davis, G.K. Cocaine and the heart. *Postgrad. Med. J.* **2005**, *81*, 568–571. [CrossRef]
5. De Giorgi, A.; Fabbian, F.; Pala, M.; Bonetti, F.; Babini, I.; Bagnaresi, I.; Manfredini, F.; Portaluppi, F.; Mikhailidis, D.P.; Manfredini, R. Cocaine and acute vascular diseases. *Curr. Drug Abuse Rev.* **2012**, *5*, 129–134. [CrossRef]
6. Lange, R.A.; Hillis, L.D. Cardiovascular complications of cocaine use. *N. Engl. J. Med.* **2001**, *345*, 351–358. [CrossRef]
7. Trends in Substance Use Disorders among Adults Aged 18 or Older. Available online: https://www.samhsa.gov/data/sites/default/files/report_2790/ShortReport-2790.html (accessed on 11 November 2018).
8. Drug Abuse Warning Network Trends Tables, 2011 Update. Available online: https://www.samhsa.gov/data/sites/default/files/DAWN2k11ED/DAWN2k11ED/rpts/DAWN2k11-Trend-Tables.htm (accessed on 11 November 2018).
9. Vongpatanasin, W.; Mansour, Y.; Chavoshan, B.; Arbique, D.; Victor, R.G. Cocaine stimulates the human cardiovascular system via a central mechanism of action. *Circulation* **1999**, *100*, 497–502. [CrossRef] [PubMed]
10. Howell, L.L.; Carroll, F.I.; Votaw, J.R.; Goodman, M.M.; Kimmel, H.L. Effects of combined dopamine and serotonin transporter inhibitors on cocaine self-administration in rhesus monkeys. *J. Pharmacol. Exp. Ther.* **2007**, *320*, 757–765. [CrossRef] [PubMed]
11. Schwartz, B.G.; Rezkalla, S.; Kloner, R.A. Cardiovascular effects of cocaine. *Circulation* **2010**, *122*, 2558–2569. [CrossRef] [PubMed]
12. Pennings, E.J.; Leccese, A.P.; Wolff, F.A. Effects of concurrent use of alcohol and cocaine. *Addiction* **2002**, *97*, 773–783. [CrossRef]
13. Rezkalla, S.H.; Kloner, R.A. Cocaine-induced acute myocardial infarction. *Clin. Med. Res.* **2007**, *5*, 172–176. [CrossRef]
14. Davies, O.; Ajayeoba, O.; Kurian, D. Coronary artery spasm: An often overlooked diagnosis. *Niger. Med. J.* **2014**, *55*, 356–358. [CrossRef]
15. Talarico, G.P.; Crosta, M.L.; Giannico, M.B.; Summaria, F.; Calo, L.; Patrizi, R. Cocaine and coronary artery diseases: A systematic review of the literature. *J. Cardiovasc. Med.* **2017**, *18*, 291–294. [CrossRef] [PubMed]
16. Wilbert-Lampen, U.; Seliger, C.; Zilker, T.; Arendt, R.M. Cocaine increases the endothelial release of immunoreactive endothelin and its concentrations in human plasma and urine: Reversal by coincubation with sigma-receptor antagonists. *Circulation* **1998**, *98*, 385–390. [CrossRef]
17. Togna, G.I.; Graziani, M.; Russo, P.; Caprino, L. Cocaine toxic effect on endothelium-dependent vasorelaxation: An in vitro study on rabbit aorta. *Toxicol. Lett.* **2001**, *123*, 43–50. [CrossRef]
18. Mo, W.; Singh, A.K.; Arruda, J.A.; Dunea, G. Role of nitric oxide in cocaine-induced acute hypertension. *Am. J. Hypertens.* **1998**, *11*, 708–714. [CrossRef]
19. Perreault, C.L.; Morgan, K.G.; Morgan, J.P. Effects of cocaine on intracellular calcium handling in cardiac and vascular smooth muscle. *NIDA Res. Monogr.* **1991**, *108*, 139–153. [PubMed]
20. Scholz, A. Mechanisms of (local) anaesthetics on voltage-gated sodium and other ion channels. *Br. J. Anaesth.* **2002**, *89*, 52–61. [CrossRef]

21. Siegel, A.J.; Sholar, M.B.; Mendelson, J.H.; Lukas, S.E.; Kaufman, M.J.; Renshaw, P.F.; McDonald, J.C.; Lewandrowski, K.B.; Apple, F.S.; Stec, J.J.; et al. Cocaine-induced erythrocytosis and increase in von Willebrand factor: Evidence for drug-related blood doping and prothrombotic effects. *Arch. Intern. Med.* **1999**, *159*, 1925–1929. [CrossRef] [PubMed]
22. Crumb, W.J., Jr.; Clarkson, C.W. Characterization of cocaine-induced block of cardiac sodium channels. *Biophys. J.* **1990**, *57*, 589–599. [CrossRef]
23. Hoffman, R.S. Treatment of patients with cocaine-induced arrhythmias: Bringing the bench to the bedside. *Br. J. Clin. Pharmacol.* **2010**, *69*, 448–457. [CrossRef]
24. Williams, M.J.; Restieaux, N.J.; Low, C.J. Myocardial infarction in young people with normal coronary arteries. *Heart* **1998**, *79*, 191–194. [CrossRef]
25. Xu, Y.Q.; Crumb, W.J., Jr.; Clarkson, C.W. Cocaethylene, a metabolite of cocaine and ethanol, is a potent blocker of cardiac sodium channels. *J. Pharmacol. Exp. Ther.* **1994**, *271*, 319–325.
26. Goldstein, R.A.; DesLauriers, C.; Burda, A.; Johnson-Arbor, K. Cocaine: History, social implications, and toxicity: A review. *Semin. Diagn. Pathol.* **2009**, *26*, 10–17. [CrossRef]
27. O'Leary, M.E. Inhibition of HERG potassium channels by cocaethylene: A metabolite of cocaine and ethanol. *Cardiovasc. Res.* **2002**, *53*, 59–67. [CrossRef]
28. Ferreira, S.; Crumb, W.J., Jr.; Carlton, C.G.; Clarkson, C.W. Effects of cocaine and its major metabolites on the HERG-encoded potassium channel. *J. Pharmacol. Exp. Ther.* **2001**, *299*, 220–226. [PubMed]
29. O'Leary, M.E.; Hancox, J.C. Role of voltage-gated sodium, potassium and calcium channels in the development of cocaine-associated cardiac arrhythmias. *Br. J. Clin. Pharmacol.* **2010**, *69*, 427–442. [CrossRef] [PubMed]
30. Crandall, C.G.; Vongpatanasin, W.; Victor, R.G. Mechanism of cocaine-induced hyperthermia in humans. *Ann. Intern. Med.* **2002**, *136*, 785–791. [CrossRef]
31. Catravas, J.D.; Waters, I.W. Acute cocaine intoxication in the conscious dog: Studies on the mechanism of lethality. *J. Pharmacol. Exp. Ther.* **1981**, *217*, 350–356. [PubMed]
32. Bachi, K.; Mani, V.; Jeyachandran, D.; Fayad, Z.A.; Goldstein, R.Z.; Alia-Klein, N. Vascular disease in cocaine addiction. *Atherosclerosis* **2017**, *262*, 154–162. [CrossRef]
33. Mouhaffel, A.H.; Madu, E.C.; Satmary, W.A.; Fraker, T.D., Jr. Cardiovascular complications of cocaine. *Chest* **1995**, *107*, 1426–1434. [CrossRef]
34. Lange, R.A.; Cigarroa, R.G.; Yancy, C.W., Jr.; Willard, J.E.; Popma, J.J.; Sills, M.N.; McBride, W.; Kim, A.S.; Hillis, L.D. Cocaine-induced coronary-artery vasoconstriction. *N. Engl. J. Med.* **1989**, *321*, 1557–1562. [CrossRef]
35. Sandoval, Y.; Smith, S.W.; Thordsen, S.E.; Apple, F.S. Supply/demand type 2 myocardial infarction: Should we be paying more attention? *J. Am. Coll. Cardiol.* **2014**, *63*, 2079–2087. [CrossRef]
36. Pradhan, L.; Mondal, D.; Chandra, S.; Ali, M.; Agrawal, K.C. Molecular analysis of cocaine-induced endothelial dysfunction: Role of endothelin-1 and nitric oxide. *Cardiovasc. Toxicol.* **2008**, *8*, 161–171. [CrossRef] [PubMed]
37. Bohm, F.; Pernow, J. The importance of endothelin-1 for vascular dysfunction in cardiovascular disease. *Cardiovasc. Res.* **2007**, *76*, 8–18. [CrossRef] [PubMed]
38. Previtali, E.; Bucciarelli, P.; Passamonti, S.M.; Martinelli, I. Risk factors for venous and arterial thrombosis. *Blood Transfus.* **2011**, *9*, 120–138.
39. Badimon, L.; Padro, T.; Vilahur, G. Atherosclerosis, platelets and thrombosis in acute ischaemic heart disease. *Eur. Heart J. Acute Cardiovasc. Care* **2012**, *1*, 60–74. [CrossRef]
40. Wright, N.M.; Martin, M.; Goff, T.; Morgan, J.; Elworthy, R.; Ghoneim, S. Cocaine and thrombosis: A narrative systematic review of clinical and in-vivo studies. *Subst. Abuse Treat. Prev. Policy* **2007**, *2*, 1–8. [CrossRef] [PubMed]
41. Kalogeris, T.; Baines, C.P.; Krenz, M.; Korthuis, R.J. Cell biology of ischemia/reperfusion injury. *Int. Rev. Cell Mol. Biol.* **2012**, *298*, 229–317.
42. Graziani, M.; Antonilli, L.; Togna, A.R.; Grassi, M.C.; Badiani, A.; Saso, L. Cardiovascular and hepatic toxicity of cocaine: Potential beneficial effects of modulators of oxidative stress. *Oxid. Med. Cell Longev.* **2016**. [CrossRef]
43. Flores, E.D.; Lange, R.A.; Cigarroa, R.G.; Hillis, L.D. Effect of cocaine on coronary artery dimensions in atherosclerotic coronary artery disease: Enhanced vasoconstriction at sites of significant stenoses. *J. Am. Coll. Cardiol.* **1990**, *16*, 74–79. [CrossRef]

44. Howard, R.E.; Hueter, D.C.; Davis, G.J. Acute myocardial infarction following cocaine abuse in a young woman with normal coronary arteries. *JAMA* **1985**, *254*, 95–96. [CrossRef]
45. Maceira, A.M.; Ripoll, C.; Cosin-Sales, J.; Igual, B.; Gavilan, M.; Salazar, J.; Belloch, V.; Pennell, D.J. Long term effects of cocaine on the heart assessed by cardiovascular magnetic resonance at 3T. *J. Cardiovasc. Magn. Reson.* **2014**, *16*, 26. [CrossRef] [PubMed]
46. Hale, S.L.; Alker, K.J.; Rezkalla, S.; Figures, G.; Kloner, R.A. Adverse effects of cocaine on cardiovascular dynamics, myocardial blood flow, and coronary artery diameter in an experimental model. *Am. Heart J.* **1989**, *118*, 927–933. [CrossRef]
47. Pitts, W.R.; Vongpatanasin, W.; Cigarroa, J.E.; Hillis, L.D.; Lange, R.A. Effects of the intracoronary infusion of cocaine on left ventricular systolic and diastolic function in humans. *Circulation* **1998**, *97*, 1270–1273. [CrossRef]
48. Hale, S.L.; Alker, K.J.; Rezkalla, S.H.; Eisenhauer, A.C.; Kloner, R.A. Nifedipine protects the heart from the acute deleterious effects of cocaine if administered before but not after cocaine. *Circulation* **1991**, *83*, 1437–1443. [CrossRef] [PubMed]
49. Gardin, J.M.; Wong, N.; Alker, K.; Hale, S.L.; Paynter, J.; Knoll, M.; Jamison, B.; Patterson, M.; Kloner, R.A. Acute cocaine administration induces ventricular regional wall motion and ultrastructural abnormalities in an anesthetized rabbit model. *Am. Heart J.* **1994**, *128*, 1117–1129. [CrossRef]
50. Restrepo, C.S.; Rojas, C.A.; Martinez, S.; Riascos, R.; Marmol-Velez, A.; Carrillo, J.; Vargas, D. Cardiovascular complications of cocaine: Imaging findings. *Emerg. Radiol.* **2009**, *16*, 11–19. [CrossRef]
51. Cooper, C.J.; Said, S.; Alkhateeb, H.; Rodriguez, E.; Trien, R.; Ajmal, S.; Blandon, P.A.; Hernandez, G.T. Dilated cardiomyopathy secondary to chronic cocaine abuse: A case report. *BMC Res. Notes* **2013**, *6*, 536. [CrossRef]
52. Phillips, K.; Luk, A.; Soor, G.S.; Abraham, J.R.; Leong, S.; Butany, J. Cocaine cardiotoxicity: A review of the pathophysiology, pathology, and treatment options. *Am. J. Cardiovasc. Drugs* **2009**, *9*, 177–196. [CrossRef]
53. Felker, G.M.; Hu, W.; Hare, J.M.; Hruban, R.H.; Baughman, K.L.; Kasper, E.K. The spectrum of dilated cardiomyopathy. The Johns Hopkins experience with 1,278 patients. *Medicine (Baltimore)* **1999**, *78*, 270–283. [CrossRef]
54. Brickner, M.E.; Willard, J.E.; Eichhorn, E.J.; Black, J.; Grayburn, P.A. Left ventricular hypertrophy associated with chronic cocaine abuse. *Circulation* **1991**, *84*, 1130–1135. [CrossRef]
55. Nahas, G.G.; Trouve, R.; Manger, W.M. Cocaine, catecholamines and cardiac toxicity. *Acta. Anaesthesiol. Scand. Suppl.* **1990**, *94*, 77–81. [CrossRef]
56. Kloner, R.A.; Hale, S.; Alker, K.; Rezkalla, S. The effects of acute and chronic cocaine use on the heart. *Circulation* **1992**, *85*, 407–419. [CrossRef] [PubMed]
57. Kolodgie, F.D.; Virmani, R.; Cornhill, J.F.; Herderick, E.E.; Smialek, J. Increase in atherosclerosis and adventitial mast cells in cocaine abusers: An alternative mechanism of cocaine-associated coronary vasospasm and thrombosis. *J. Am. Coll. Cardiol.* **1991**, *17*, 1553–1560. [CrossRef]
58. Patrizi, R.; Pasceri, V.; Sciahbasi, A.; Summaria, F.; Rosano, G.M.; Lioy, E. Evidence of cocaine-related coronary atherosclerosis in young patients with myocardial infarction. *J. Am. Coll. Cardiol.* **2006**, *47*, 2120–2122. [CrossRef] [PubMed]
59. Minor, R.L., Jr.; Scott, B.D.; Brown, D.D.; Winniford, M.D. Cocaine-induced myocardial infarction in patients with normal coronary arteries. *Ann. Intern. Med.* **1991**, *115*, 797–806. [CrossRef] [PubMed]
60. Havranek, E.P.; Nademanee, K.; Grayburn, P.A.; Eichhorn, E.J. Endothelium-dependent vasorelaxation is impaired in cocaine arteriopathy. *J. Am. Coll. Cardiol.* **1996**, *28*, 1168–1174. [CrossRef]
61. Gan, X.; Zhang, L.; Berger, O.; Stins, M.F.; Way, D.; Taub, D.D.; Chang, S.L.; Kim, K.S.; House, S.D.; Weinand, M.; et al. Cocaine enhances brain endothelial adhesion molecules and leukocyte migration. *Clin. Immunol.* **1999**, *91*, 68–76. [CrossRef]
62. Simpson, R.W.; Edwards, W.D. Pathogenesis of cocaine-induced ischemic heart disease. Autopsy findings in a 21-year-old man. *Arch. Pathol. Lab. Med.* **1986**, *110*, 479–484.
63. Shah, P.K. Mechanisms of plaque vulnerability and rupture. *J. Am. Coll. Cardiol.* **2003**, *41* (Suppl. S4), 15S–22S. [CrossRef]
64. Lindstedt, K.A.; Mayranpaa, M.I.; Kovanen, P.T. Mast cells in vulnerable atherosclerotic plaques—A view to a kill. *J. Cell. Mol. Med.* **2007**, *11*, 739–758. [CrossRef]

65. Kokkonen, J.O.; Kovanen, P.T. Stimulation of mast cells leads to cholesterol accumulation in macrophages in vitro by a mast cell granule-mediated uptake of low density lipoprotein. *Proc. Natl. Acad. Sci. USA* **1987**, *84*, 2287–2291. [CrossRef]
66. Kokkonen, J.O.; Kovanen, P.T. Proteolytic enzymes of mast cell granules degrade low density lipoproteins and promote their granule-mediated uptake by macrophages in vitro. *J. Biol. Chem.* **1989**, *264*, 10749–10755.
67. Huang, M.; Pang, X.; Letourneau, R.; Boucher, W.; Theoharides, T.C. Acute stress induces cardiac mast cell activation and histamine release, effects that are increased in Apolipoprotein E knockout mice. *Cardiovasc. Res.* **2002**, *55*, 150–160. [CrossRef]
68. Bohm, M.; La Rosee, K.; Schwinger, R.H.; Erdmann, E. Evidence for reduction of norepinephrine uptake sites in the failing human heart. *J. Am. Coll. Cardiol.* **1995**, *25*, 146–153. [CrossRef]
69. Billman, G.E. Cocaine: A review of its toxic actions on cardiac function. *Crit. Rev. Toxicol.* **1995**, *25*, 113–132. [CrossRef]
70. Hobbs, W.E.; Moore, E.E.; Penkala, R.A.; Bolgiano, D.D.; Lopez, J.A. Cocaine and specific cocaine metabolites induce von Willebrand factor release from endothelial cells in a tissue-specific manner. *Arterioscler. Thromb. Vasc. Biol.* **2013**, *33*, 1230–1237. [CrossRef]
71. Heesch, C.M.; Wilhelm, C.R.; Ristich, J.; Adnane, J.; Bontempo, F.A.; Wagner, W.R. Cocaine activates platelets and increases the formation of circulating platelet containing microaggregates in humans. *Heart* **2000**, *83*, 688–695. [CrossRef]
72. Moons, A.H.; Levi, M.; Peters, R.J. Tissue factor and coronary artery disease. *Cardiovasc. Res.* **2002**, *53*, 313–325. [CrossRef]
73. Gu, X.; Herrera, G.A. Thrombotic microangiopathy in cocaine abuse-associated malignant hypertension: Report of 2 cases with review of the literature. *Arch. Pathol. Lab. Med.* **2007**, *131*, 1817–1820.
74. Fogo, A.; Superdock, K.R.; Atkinson, J.B. Severe arteriosclerosis in the kidney of a cocaine addict. *Am. J. Kidney Dis.* **1992**, *20*, 513–515. [CrossRef]
75. Su, J.; Li, J.; Li, W.; Altura, B.; Altura, B. Cocaine induces apoptosis in primary cultured rat aortic vascular smooth muscle cells: Possible relationship to aortic dissection, atherosclerosis, and hypertension. *Int. J. Toxicol.* **2004**, *23*, 233–237. [CrossRef]
76. Dabbouseh, N.M.; Ardelt, A. Cocaine mediated apoptosis of vascular cells as a mechanism for carotid artery dissection leading to ischemic stroke. *Med. Hypotheses* **2011**, *77*, 201–203. [CrossRef]
77. Ramondo, A.B.; Mistrorigo, F.; Angelini, A. Intimal hyperplasia and cystic medial necrosis as substrate of acute coronary syndrome in a cocaine abuser: An in vivo/ex vivo pathological correlation. *Heart* **2009**, *95*, 82. [CrossRef]
78. Steinhauer, J.R.; Caulfield, J.B. Spontaneous coronary artery dissection associated with cocaine use: A case report and brief review. *Cardiovasc. Pathol.* **2001**, *10*, 141–145. [CrossRef]
79. Kamineni, R.; Sadhu, A.; Alpert, J.S. Spontaneous coronary artery dissection: Report of two cases and a 50-year review of the literature. *Cardiol. Rev.* **2002**, *10*, 279–284. [CrossRef]
80. Meghani, M.; Siddique, M.N.; Bhat, T.; Samarneh, M.; Elsayegh, S. Internal carotid artery redundancy and dissection in a young cocaine abuser. *Vascular* **2013**, *21*, 243–245. [CrossRef]
81. Kozor, R.; Grieve, S.M.; Buchholz, S.; Kaye, S.; Darke, S.; Bhindi, R.; Figtree, G.A. Regular cocaine use is associated with increased systolic blood pressure, aortic stiffness and left ventricular mass in young otherwise healthy individuals. *PLoS ONE* **2014**, *9*, e89710. [CrossRef]
82. Kariyanna, P.T.; Jayarangaiah, A.; Al-Sadawi, M.; Ahmed, R.; Green, J.; Dubson, I.; McFarlane, S.I. A rare case of second degree Mobitz type II AV block associated with cocaine use. *Am. J. Med. Case Rep.* **2018**, *6*, 146–148. [CrossRef] [PubMed]
83. Satran, A.; Bart, B.A.; Henry, C.R.; Murad, M.B.; Talukdar, S.; Satran, D.; Henry, T.D. Increased prevalence of coronary artery aneurysms among cocaine users. *Circulation* **2005**, *111*, 2424–2429. [CrossRef] [PubMed]
84. Gupta, N.; Washam, J.B.; Mountantonakis, S.E.; Li, S.; Roe, M.T.; de Lemos, J.A.; Arora, R. Characteristics, management, and outcomes of cocaine-positive patients with acute coronary syndrome (from the National Cardiovascular Data Registry). *Am. J. Cardiol.* **2014**, *113*, 749–756. [CrossRef]
85. Salihu, H.M.; Salemi, J.L.; Aggarwal, A.; Steele, B.F.; Pepper, R.C.; Mogos, M.F.; Aliyu, M.H. Opioid drug use and acute cardiac events among pregnant women in the United States. *Am. J. Med.* **2018**, *131*, 64–71. [CrossRef]

86. Aslibekyan, S.; Levitan, E.B.; Mittleman, M.A. Prevalent cocaine use and myocardial infarction. *Am. J. Cardiol.* **2008**, *102*, 966–969. [CrossRef]
87. Gunja, A.; Stanislawski, M.A.; Baron, A.E.; Maddox, T.M.; Bradley, S.M.; Vidovich, M.I. The implications of cocaine use and associated behaviors on adverse cardiovascular outcomes among veterans: Insights from the VA Clinical Assessment, Reporting, and Tracking (CART) Program. *Clin. Cardiol.* **2018**, *41*, 809–816. [CrossRef]
88. Arora, S.; Dodani, S.; Kaeley, G.S.; Kraemer, D.F.; Aldridge, P.; Pomm, R. Cocaine use and subclinical coronary artery disease in Caucasians. *J. Clin. Exp. Cardiol.* **2015**, *6*, 1–6.
89. Bamberg, F.; Schlett, C.L.; Truong, Q.A.; Rogers, I.S.; Koenig, W.; Nagurney, J.T.; Seneviratne, S.; Lehman, S.J.; Cury, R.C.; Abbara, S.; et al. Presence and extent of coronary artery disease by cardiac computed tomography and risk for acute coronary syndrome in cocaine users among patients with chest pain. *Am. J. Cardiol.* **2009**, *103*, 620–625. [CrossRef]
90. Chang, A.M.; Walsh, K.M.; Shofer, F.S.; McCusker, C.M.; Litt, H.I.; Hollander, J.E. Relationship between cocaine use and coronary artery disease in patients with symptoms consistent with an acute coronary syndrome. *Acad. Emerg. Med.* **2011**, *18*, 1–9. [CrossRef]
91. Lai, H.; Moore, R.; Celentano, D.D.; Gerstenblith, G.; Treisman, G.; Keruly, J.C.; Kickler, T.; Li, J.; Chen, S.; Lai, S.; et al. HIV infection itself may not be associated with subclinical coronary artery disease among African Americans without cardiovascular symptoms. *J. Am. Heart Assoc.* **2016**, *5*, e002529. [CrossRef]
92. Lucas, G.M.; Atta, M.G.; Fine, D.M.; McFall, A.M.; Estrella, M.M.; Zook, K.; Stein, J.H. HIV, cocaine use, and hepatitis C virus: A triad of nontraditional risk factors for subclinical cardiovascular disease. *Arterioscler. Thromb. Vasc. Biol.* **2016**, *36*, 2100–2107. [CrossRef]
93. DeFilippis, E.M.; Singh, A.; Divakaran, S.; Gupta, A.; Collins, B.L.; Biery, D.; Qamar, A.; Fatima, A.; Ramsis, M.; Pipilas, D.; et al. Cocaine and marijuana use among young adults with myocardial infarction. *J. Am. Coll. Cardiol.* **2018**, *71*, 2540–2551. [CrossRef]
94. Morentin, B.; Ballesteros, J.; Callado, L.F.; Meana, J.J. Recent cocaine use is a significant risk factor for sudden cardiovascular death in 15–49-year-old subjects: A forensic case-control study. *Addiction* **2014**, *109*, 2071–2078. [CrossRef]
95. Qureshi, A.I.; Chaudhry, S.A.; Suri, M.F. Cocaine use and the likelihood of cardiovascular and all-cause mortality: Data from the third national health and nutrition examination survey mortality follow-up study. *J. Vasc. Interv. Neurol.* **2014**, *7*, 76–82. [CrossRef]
96. Hser, Y.I.; Kagihara, J.; Huang, D.; Evans, E.; Messina, N. Mortality among substance-using mothers in California: A 10-year prospective study. *Addiction* **2012**, *107*, 215–222. [CrossRef] [PubMed]
97. Atoui, M.; Fida, N.; Nayudu, S.K.; Glandt, M.; Chilimuri, S. Outcomes of patients with cocaine induced chest pain in an inner city hospital. *Cardiol. Res.* **2011**, *2*, 269–273. [CrossRef]
98. Hollander, J.E.; Hoffman, R.S. Cocaine-induced myocardial infarction: An analysis and review of the literature. *J. Emerg. Med.* **1992**, *10*, 169–177. [CrossRef]
99. Pletcher, M.J.; Kiefe, C.I.; Sidney, S.; Carr, J.J.; Lewis, C.E.; Hulley, S.B. Cocaine and coronary calcification in young adults: The Coronary Artery Risk Development in Young Adults (CARDIA) study. *Am. Heart J.* **2005**, *150*, 921–926. [CrossRef] [PubMed]
100. Lai, S.; Lai, H.; Meng, Q.; Tong, W.; Vlahov, D.; Celentano, D.; Strathdee, S.; Nelson, K.; Fishman, E.K.; Lima, J.A. Effect of cocaine use on coronary calcium among black adults in Baltimore, Maryland. *Am. J. Cardiol.* **2002**, *90*, 326–328. [CrossRef]
101. Lai, S.; Lima, J.A.; Lai, H.; Vlahov, D.; Celentano, D.; Tong, W.; Bartlett, J.G.; Margolick, J.; Fishman, E.K. Human immunodeficiency virus 1 infection, cocaine, and coronary calcification. *Arch. Intern. Med.* **2005**, *165*, 690–695. [CrossRef] [PubMed]
102. Lai, S.; Fishman, E.K.; Lai, H.; Moore, R.; Cofrancesco, J., Jr.; Pannu, H.; Tong, W.; Du, J.; Barlett, J. Long-term cocaine use and antiretroviral therapy are associated with silent coronary artery disease in African Americans with HIV infection who have no cardiovascular symptoms. *Clin. Infect. Dis.* **2008**, *46*, 600–610. [CrossRef] [PubMed]
103. Wang, G.J.; Volkow, N.D.; Logan, J.; Pappas, N.R.; Wong, C.T.; Zhu, W.; Netusil, N.; Fowler, J.S. Brain dopamine and obesity. *Lancet* **2001**, *357*, 354–357. [CrossRef]
104. Volkow, N.D.; Wang, G.J.; Baler, R.D. Reward, dopamine and the control of food intake: Implications for obesity. *Trends Cogn. Sci.* **2011**, *15*, 37–46. [CrossRef] [PubMed]

105. Mahapatra, A. Overeating, obesity, and dopamine receptors. *ACS Chem. Neurosci.* **2010**, *1*, 346–347. [CrossRef]
106. Thanos, P.K.; Cho, J.; Kim, R.; Michaelides, M.; Primeaux, S.; Bray, G.; Wang, G.J.; Volkow, N.D. Bromocriptine increased operant responding for high fat food but decreased chow intake in both obesity-prone and resistant rats. *Behav. Brain Res.* **2011**, *217*, 165–170. [CrossRef] [PubMed]
107. Ramamoorthy, S.; Bauman, A.L.; Moore, K.R.; Han, H.; Yang-Feng, T.; Chang, A.S.; Ganapathy, V.; Blakely, R.D. Antidepressant- and cocaine-sensitive human serotonin transporter: Molecular cloning, expression, and chromosomal localization. *Proc. Natl. Acad. Sci. USA* **1993**, *90*, 2542–2546. [CrossRef]
108. Yadav, V.K.; Oury, F.; Suda, N.; Liu, Z.W.; Gao, X.B.; Confavreux, C.; Klemenhagen, K.C.; Tanaka, K.F.; Gingrich, J.A.; Guo, X.E.; et al. A serotonin-dependent mechanism explains the leptin regulation of bone mass, appetite, and energy expenditure. *Cell* **2009**, *138*, 976–989. [CrossRef]
109. Vicentic, A.; Jones, D.C. The CART (cocaine- and amphetamine-regulated transcript) system in appetite and drug addiction. *J. Pharmacol. Exp. Ther.* **2007**, *320*, 499–506. [CrossRef]
110. Rogge, G.; Jones, D.; Hubert, G.W.; Lin, Y.; Kuha, M.J. CART peptides: Regulators of body weight, reward and other functions. *Nat. Rev. Neurosci.* **2008**, *9*, 747–758. [CrossRef] [PubMed]
111. Wortley, K.E.; Chang, G.Q.; Davydova, Z.; Fried, S.K.; Leibowitz, S.F. Cocaine- and amphetamine-regulated transcript in the arcuate nucleus stimulates lipid metabolism to control body fat accrual on a high-fat diet. *Regul. Pept.* **2004**, *117*, 89–99. [CrossRef] [PubMed]
112. Hunter, R.G.; Philpot, K.; Vicentic, A.; Dominguez, G.; Hubert, G.W.; Kuhar, M.J. CART in feeding and obesity. *Trends Endocrinol. Metab.* **2004**, *15*, 454–459. [CrossRef]
113. Balopole, D.C.; Hansult, C.D.; Dorph, D. Effect of cocaine on food intake in rats. *Psychopharmacology* **1979**, *64*, 121–122. [CrossRef] [PubMed]
114. Wolgin, D.L.; Hertz, J.M. Effects of acute and chronic cocaine on milk intake, body weight, and activity in bottle- and cannula-fed rats. *Behav. Pharmacol.* **1995**, *6*, 746–753. [CrossRef]
115. Church, M.W.; Morbach, C.A.; Subramanian, M.G. Comparative effects of prenatal cocaine, alcohol, and undernutrition on maternal/fetal toxicity and fetal body composition in the Sprague-Dawley rat with observations on strain-dependent differences. *Neurotoxicol. Teratol.* **1995**, *17*, 559–567. [CrossRef]
116. Castro, F.G.; Newcomb, M.D.; Cadish, K. Lifestyle differences between young adult cocaine users and their nonuser peers. *J. Drug Educ.* **1987**, *17*, 89–111. [CrossRef]
117. Ersche, K.D.; Stochl, J.; Woodward, J.M.; Fletcher, P.C. The skinny on cocaine: Insights into eating behavior and body weight in cocaine-dependent men. *Appetite* **2013**, *71*, 75–80. [CrossRef]
118. Quach, L.A.; Wanke, C.A.; Schmid, C.H.; Gorbach, S.L.; Mwamburi, D.M.; Mayer, K.H.; Spiegelman, D.; Tang, A.M. Drug use and other risk factors related to lower body mass index among HIV-infected individuals. *Drug Alcohol Depend.* **2008**, *95*, 30–36. [CrossRef]
119. Escobar, M.; Scherer, J.N.; Soares, C.M.; Guimaraes, L.S.P.; Hagen, M.E.; von Diemen, L.; Pechansky, F. Active Brazilian crack cocaine users: Nutritional, anthropometric, and drug use profiles. *Braz. J. Psychiatry* **2018**, *40*, 354–360. [CrossRef]
120. Muniz, A.E.; Evans, T. Acute gastrointestinal manifestations associated with use of crack. *Am. J. Emerg. Med.* **2001**, *19*, 61–63. [CrossRef]
121. Billing, L.; Ersche, K.D. Cocaine's appetite for fat and the consequences on body weight. *Am. J. Drug Alcohol Abuse* **2015**, *41*, 115–118. [CrossRef]
122. Ludwig, D.S.; Pereira, M.A.; Kroenke, C.H.; Hilner, J.E.; Van Horn, L.; Slattery, M.L.; Jacobs, D.R., Jr. Dietary fiber, weight gain, and cardiovascular disease risk factors in young adults. *JAMA* **1999**, *282*, 1539–1546. [CrossRef]
123. Rangel, C.; Shu, R.G.; Lazar, L.D.; Vittinghoff, E.; Hsue, P.Y.; Marcus, G.M. Beta-blockers for chest pain associated with recent cocaine use. *Arch. Intern. Med.* **2010**, *170*, 874–879. [CrossRef]

124. Lopez, P.D.; Akinlonu, A.; Mene-Afejuku, T.O.; Dumancas, C.; Saeed, M.; Cativo, E.H.; Visco, F.; Mushiyev, S.; Pekler, G. Clinical outcomes of B-blocker therapy in cocaine-associated heart failure. *Int. J. Cardiol.* **2018**. [CrossRef]
125. Dattilo, P.B.; Hailpern, S.M.; Fearon, K.; Sohal, D.; Nordin, C. Beta-blockers are associated with reduced risk of myocardial infarction after cocaine use. *Ann. Emerg. Med.* **2008**, *51*, 117–125. [CrossRef]

International Journal of
Molecular Sciences

Article

The Functional Role of Zinc Finger E Box-Binding Homeobox 2 (Zeb2) in Promoting Cardiac Fibroblast Activation

Fahmida Jahan [1,3], Natalie M. Landry [2,3], Sunil G. Rattan [2,3], Ian M. C. Dixon [2,3] and Jeffrey T. Wigle [1,3,*]

1 Department of Biochemistry and Medical Genetics, University of Manitoba, Winnipeg, MB R3E0J9, Canada; fjaha037@uottawa.ca
2 Department of Physiology and Pathophysiology, University of Manitoba, Winnipeg, MB R3E0J9, Canada; nlandry@sbrc.ca (N.M.L.); srattan@sbrc.ca (S.G.R.); idixon@sbrc.ca (I.M.C.D.)
3 Institute of Cardiovascular Sciences, St. Boniface Hospital Albrechtsen Research Centre, Max Rady College of Medicine, Rady Faculty of Health Sciences, University of Manitoba, Winnipeg, MB R2H2A6, Canada
* Correspondence: jwigle@sbrc.ca; Tel.: +1-204-235-3953

Received: 5 September 2018; Accepted: 5 October 2018; Published: 17 October 2018

Abstract: Following cardiac injury, fibroblasts are activated and are termed as myofibroblasts, and these cells are key players in extracellular matrix (ECM) remodeling and fibrosis, itself a primary contributor to heart failure. Nutraceuticals have been shown to blunt cardiac fibrosis in both in-vitro and in-vivo studies. However, nutraceuticals have had conflicting results in clinical trials, and there are no effective therapies currently available to specifically target cardiac fibrosis. We have previously shown that expression of the zinc finger E box-binding homeobox 2 (Zeb2) transcription factor increases as fibroblasts are activated. We now show that Zeb2 plays a critical role in fibroblast activation. Zeb2 overexpression in primary rat cardiac fibroblasts is associated with significantly increased expression of embryonic smooth muscle myosin heavy chain (SMemb), ED-A fibronectin and α-smooth muscle actin (α-SMA). We found that Zeb2 was highly expressed in activated myofibroblast nuclei but not in the nuclei of inactive fibroblasts. Moreover, ectopic Zeb2 expression in myofibroblasts resulted in a significantly less migratory phenotype with elevated contractility, which are characteristics of mature myofibroblasts. Knockdown of Zeb2 with siRNA in primary myofibroblasts did not alter the expression of myofibroblast markers, which may indicate that Zeb2 is functionally redundant with other profibrotic transcription factors. These findings add to our understanding of the contribution of Zeb2 to the mechanisms controlling cardiac fibroblast activation.

Keywords: Zeb2; cardiac fibroblast; activated myofibroblast; cardiac fibrosis; fibroblast contractility

1. Introduction

Cardiac fibroblasts are essential for normal cardiac development, function and tissue homeostasis [1,2]. In addition, cardiac fibroblasts play a vital role in controlling both the inflammatory response and wound healing following cardiac injury [3]. Cardiac fibroblasts are recruited to the damaged myocardium by cytokines and growth factors that are secreted by circulating inflammatory cells [4,5]. As fibroblasts infiltrate the injured area, they undergo rapid proliferation and phenoconversion into myofibroblasts, a key step in fibrogenesis [6,7]. Myofibroblasts remodel the extracellular matrix (ECM) by the de-novo secretion and organization of matrix proteins such as type I and III fibrillar collagens [7,8] and polymerized fibronectin [9]. They also frequently (but not exclusively) express contractile proteins such as α-smooth muscle actin (α-SMA), filamentous actin (F-actin) in stress fibres, and embryonic smooth muscle myosin heavy chain (SMemb), which facilitate contraction and closure of the wound [2,10]. In most cases of wound healing, the new scar is reduced by means of tissue regeneration and the apoptotic removal of myofibroblasts [11]. However, following myocardial infarction (MI), cardiomyocytes do not proliferate significantly and thereby lost myocytes are not replaced. The presence of continued inflammation causes myofibroblasts to persist for many years in the infarct scar [12], although deactivation of activated myofibroblasts in the heart is known to occur [13]. Persistence of the infarct scar is associated with loss of normal ventricular geometry that leads to eventual cardiac dysfunction including impaired inotropic and lusitropic function [14]. Cardiac fibrosis is a primary contributor to other cardiovascular diseases including hypertension, dilated cardiomyopathy and ischemia. Various factors can lead to conversion of cardiac fibroblast to myofibroblast including mechanical stress, hypoxia, signaling ligands such as transforming growth factor-β (TGF-β), connective tissue growth factor (CCN2/CTGF), platelet-derived growth factor (PDGF), angiotensin II, endothelin I, fibroblast growth factor (FGF) and insulin-like growth factor 1 (IGF1) [2,15–18].

Despite its primary contributions to cardiac dysfunction, there are no effective therapies [2,4]. Common drug classes used to treat cardiovascular disease associated with cardiac fibrosis include statins, angiotensin-converting enzyme inhibition (ACEi) and β-blockade which ameliorate plasma cholesterol levels and hypertension [2,19–21]. Nutraceuticals such as resveratrol and omega-3 fatty acids have been shown to reduce cardiac fibrosis in rodent models of hypertension [22,23]. Similarly, lycopene derived from tomato skins has been shown to reduce the expansion of interstitial fibrosis that occurs in rat models of myocardial infarction [24]. At a cellular level, resveratrol and cyanidin-3-glucoside have been shown to blunt fibroblast activation in vitro [25,26]. However, these approaches have not been effective yet clinically, potentially due to an inability to achieve an effective therapeutic dose in patients [27]. Therefore, it is crucial to better understand the molecular mechanisms leading to chronic fibrosis in order to discover new therapeutic agents that will improve outcomes for patients.

We investigated the role of the zinc finger E box-binding homeobox 2 (Zeb2) transcription factor in modulating fibroblast activation. Zeb2 is a member of the zinc finger transcription factor family that plays a crucial role in embryonic development, particularly during epithelial-to-mesenchymal transition (EMT) [28]. Zeb2 has been previously shown to repress Meox2 expression, a homeodomain transcription factor, by binding to its promoter [29]. We have previously reported that Ski (a Smad repressor) can contribute to myofibroblast deactivation by downregulating Zeb2, which in turn causes an upregulation of the transcription factor Meox2 [15]. The potential functional role of Zeb2 in fibroblast activation has not yet been addressed in the literature. Herein we provide new evidence that loss of Zeb2 did not alter the expression of myofibroblast-specific markers. However, ectopic expression of Zeb2 promoted fibroblast activation as demonstrated by the increased expression of myofibroblast markers, increased contractility and decreased migratory ability.

2. Results

2.1. Subcellular Distribution of Zeb2 during Fibroblast Activation

To investigate the role of Zeb2 in regulating fibroblast activation, we first compared the subcellular localization of the Zeb2 protein in P0 rat cardiac fibroblasts and P1 myofibroblasts. By Western blot analysis, we determined that the Zeb2 protein was highly expressed in the nuclei of P1 myofibroblasts versus its expression in the nuclei of P0 fibroblasts ($n = 4$, $p \leq 0.05$) (Figure 1). Zeb2 protein was not detected in the cytoplasmic fractions from either P0 or P1 cells.

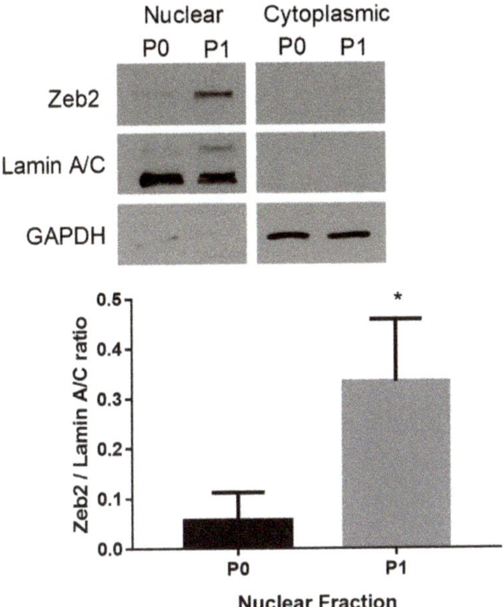

Figure 1. Zeb2 is localized to the nuclei of primary rat cardiac myofibroblasts. Zeb2 protein expression was enriched in the nuclear fraction from P1 rat cardiac myofibroblasts as compared to P0. Lamin and GAPDH were used as nuclear and cytoplasmic loading controls, respectively. The data shown are from $n = 4$ independent experiments, * $p \leq 0.05$ vs. P0. Error bars represent SEM. Data were analyzed by performing Student's *t*-test.

2.2. The Effect of Zeb2 on the Expression of Myofibroblast Markers

To determine the effects of Zeb2 gain of function on the myofibroblast phenotype, we generated an adenoviral vector encoding Zeb2 (Ad-HA-Zeb2). We were able to achieve a five-fold increase in Zeb2 protein levels by infecting myofibroblasts at a MOI of 200 (Figure 2A). We first examined the effect of ectopic Zeb2 expression on the protein levels of three key markers of the myofibroblast phenotype: α-SMA, SMemb and ED-A fibronectin [28]. Fibroblasts were activated by plating on stiff plastic substrates and transduced with either Ad-EGFP (200 MOI) or Ad-HA-Zeb2 (200 MOI) and incubated for 96 h [29]. Our data shows that Zeb2 overexpression significantly increased the levels of expression of α-SMA ($n = 4$, * $p \leq 0.05$), SMemb ($n = 4$, * $p \leq 0.05$) and ED-A fibronectin ($n = 3$, * $p \leq 0.05$) as compared to the Ad-EGFP infected viral control cells. Expression levels were normalized using α-tubulin as a loading control (Figure 2).

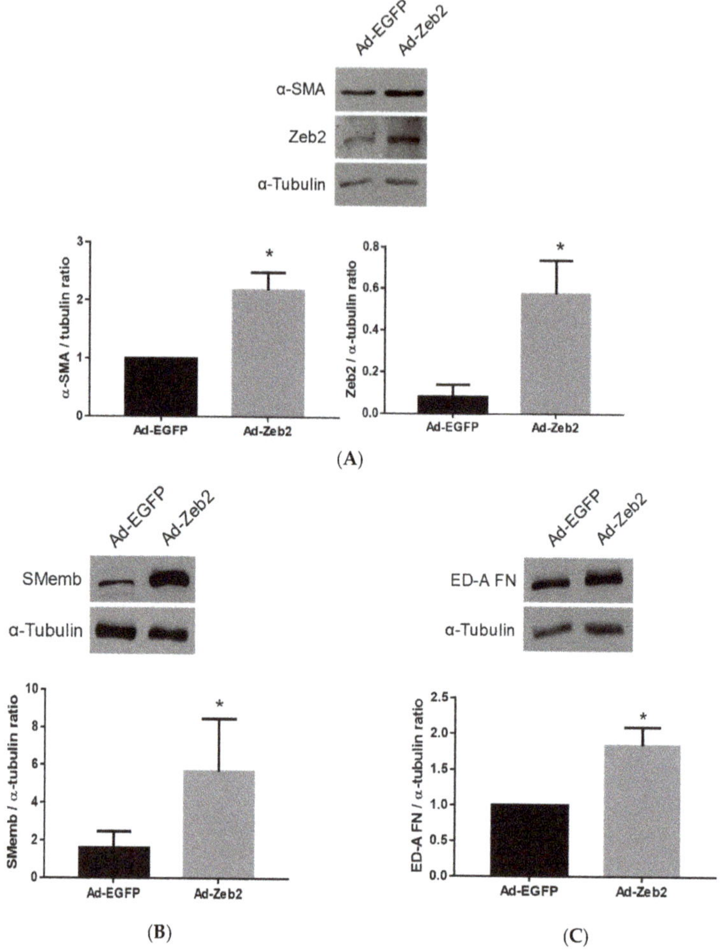

Figure 2. Zeb2 overexpression induces expression of proteins characteristic of the myofibroblast phenotype in primary rat cardiac fibroblasts. P1 cells were transduced with Ad-EGFP or Ad-HA-Zeb2 and incubated for 96 h prior to Western blot analysis. α-Tubulin was used as a loading control. Rabbit polyclonal anti-Zeb2 antibody was used to determine Zeb2 expression levels (Panel A). Ectopic expression of Zeb2 increased the expression of (**A**) α-SMA, (**B**) SMEmb and (**C**) ED-A fibronectin. Right panels are histographic representations of respective expression of markers relative to α-tubulin. The data shown are from $n = 4$ independent experiments for α-SMA and SMemb, and $n = 3$ for ED-A fibronectin. * $p < 0.05$ vs. Ad-EGFP. Error bars represent SEM. Data were analyzed by performing Student's t-tests.

2.3. *Zeb2 Overexpression Inhibits the Migration and Contractility of P1 Myofibroblasts*

Two major functional properties of cardiac myofibroblasts are that (a) they are less motile than fibroblasts, which facilitates their deposition of matrix components within the context of wound healing [28], and (b) they are more contractile than fibroblasts, which allows for the contraction of healing tissue, inherent in wound healing processes in various tissues [4]. Thus, we investigated the effect of Zeb2 on both the migration and contractility of myofibroblasts.

To determine the effect of Zeb2 on cardiac myofibroblast migration, P1 myofibroblasts were transduced with Ad-EGFP or Ad-HA-Zeb2 and incubated for 96 h. After 96 h, culture inserts were removed, medium was changed to 1% FBS-containing medium and images were taken. We found that at 18 h, significantly fewer cells had migrated into the wound area in the Ad-HA-Zeb2 infected plates compared to the Ad-EGFP infected controls (Figure 3).

To determine the effect of Zeb2 on myofibroblast contractility, P1 myofibroblasts were plated on collagen gels and transduced with either Ad-LacZ or Ad-HA-Zeb2. After 72 h, gels were detached from the periphery of the plate and the entire gel allowed to "float" for 12 h. Our results show that Zeb2 overexpression increased P1 myofibroblast contractility, as reflected by the increased relative gel contraction versus control gels seeded with Ad-LacZ infected cells, which is indicated by reduced gel size (Figure 4).

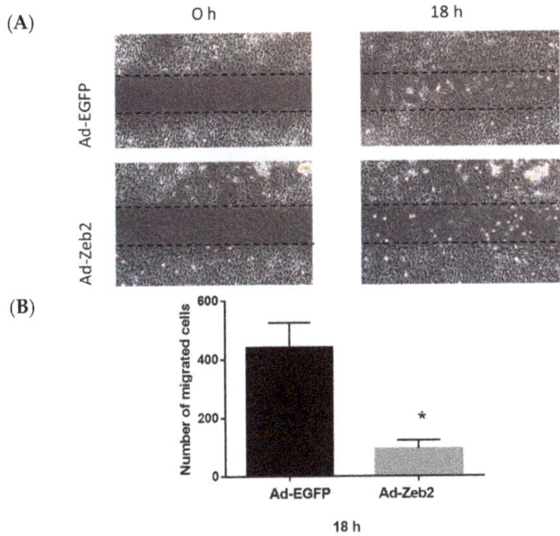

Figure 3. Zeb2 overexpression inhibits the migration of P1 myofibroblasts. (**A**) P1 myofibroblasts were transduced with Ad-EGFP or Ad-HA-Zeb2 and incubated for 96 h. After 96 h, culture inserts were removed and images were taken at the indicated time points (4× objective). (**B**) The number of cells in the wounded area was quantified using ImageJ software. Histographic representation shows the number of migrated cells in Ad-EGFP (200 MOI) and Ad-Zeb2 (200 MOI) infected plates. $n = 3$, * $p \leq 0.05$ vs Ad-EGFP at 18 h. Error bars represent SEM. Data were analyzed by performing a Student's t-test.

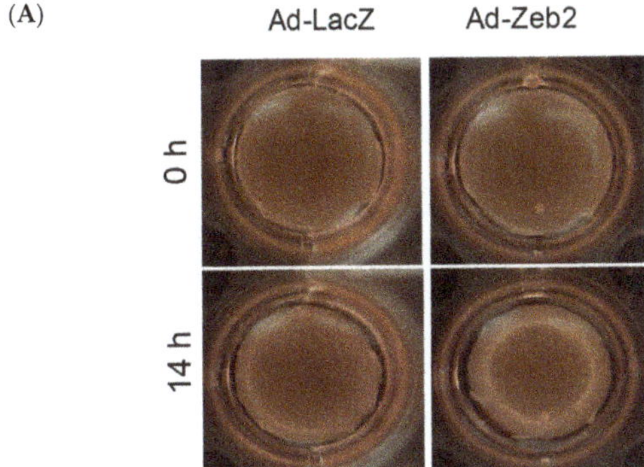

Figure 4. *Cont.*

(B)

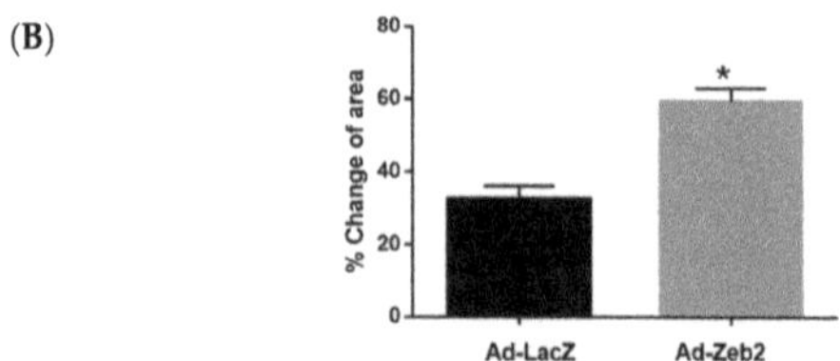

Figure 4. Zeb2 overexpression increases contraction of P1 myofibroblasts—a characteristic of mature myofibroblasts. (**A**) P1 cells were plated on collagen gels and transduced with either Ad-LacZ or Ad-HA-Zeb2 and incubated for 72 h. After 72 h, collagen gels were cut and allowed to contract for 14 h. Images were taken at the 0 h and 14 h timepoints (4× objective). (**B**) The gel size was quantified using Measuregel software. Histographic representation shows the percentage of change of area in case of Zeb2 infected gels compared to Ad-LacZ control. $n = 3$, * $p \leq 0.05$ vs Ad-LacZ. Error bars represent SEM. Data were analyzed by performing a Student's t-test.

2.4. Effect of siRNA-Mediated Zeb2 Knockdown on the Expression of Myofibroblast Markers

To determine the effect of loss of Zeb2 function on the expression of myofibroblast markers, we knocked down Zeb2 in P1 myofibroblasts and then measured the protein expression of α-SMA and SMemb 24 h later. Transfection with siRNA specific to Zeb2 significantly reduced Zeb2 expression as compared to either control or scramble siRNA controls by 88.1% and 82.3%, respectively (Figure 5). There were no differences detected in the levels of either α-SMA or SMemb between the different groups by Western blot analysis (Figure 5).

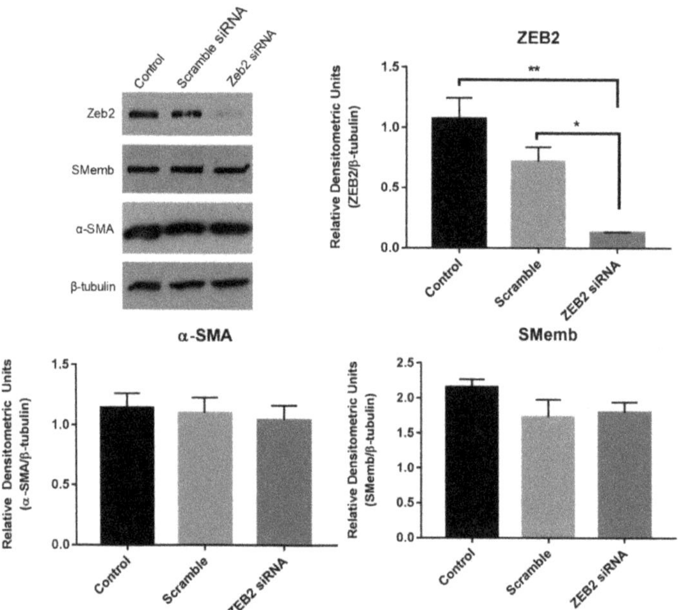

Figure 5. Zeb2 knockdown in primary cardiac myofibroblasts. First-passage (P1) rat cardiac myofibroblasts were subjected to 24 h of serum deprivation prior to treatment with 100 nM of either scramble or FITC-tagged Zeb2-targeted siRNA in serum-free, antibiotic-free DMEM. Untreated cells served as negative controls; transfection efficiency was verified by fluorescent microscopy. Data shown is representative of $n = 3$ biological replicates. ** $p < 0.01$, * $p < 0.05$.

3. Discussion

In this study, we have demonstrated that Zeb2 regulates cardiac fibroblast activation. Zeb2 overexpression was associated with significant increases in the expression levels of three key myofibroblast markers: α-SMA, SMemb and ED-A fibronectin. Ectopic Zeb2 expression was also associated with reduced myofibroblast migration and a markedly contractile phenotype, which is characteristic of mature myofibroblasts [3,28]. Our study indicates that Zeb2 has profibrotic effects in cardiac fibroblasts, but that its function is not required for maintaining the myofibroblast phenotype.

Previously, we have shown that Ski, a negative regulator of TGF-β signaling, deactivates myofibroblasts by downregulating Zeb2 expression, which in turn is associated with upregulation of its target, the Meox2 homeobox transcription factor. We found that ectopic Meox2 expression led to decreased α-SMA and ED-A FN expression levels [15]. Zeb2 has been previously shown to regulate the TGF-β-mediated EMT process [30]. Moreover, we have observed that Zeb2 expression increases in a post-MI rat model [15]. All of these findings point to a possible role of Zeb2 in regulation of cardiac fibroblast phenotype, and a putative molecular regulatory point for controlling cardiac fibroblast function in chronic remodeling of cardiac matrix deposition by activated fibroblasts.

3.1. Zeb2 Regulates Cardiac Myofibroblast Phenotype

Cardiac myofibroblasts characteristically express elevated levels of α-SMA, SMemb and ED-A fibronectin [31–33]. Among these markers, SMemb and α-SMA are the two major contractile proteins expressed by myofibroblasts [31,33]. ED-A fibronectin is considered to be one of the major drivers of myofibroblast phenoconversion that plays a role in the induction of α-SMA expression, collagen deposition and cell contractility [32]. We found that infection with adenovirus encoding Zeb2 induces increased expression of these myofibroblast markers (Figure 2). Thus, it is evident that Zeb2 is sufficient to increase the expression of the major myofibroblast markers, and plays a crucial role in driving the phenoconversion of cardiac fibroblasts to myofibroblasts. The increased expression of the myofibroblast markers may result from either increased expression via a Zeb2/SP1-dependent pathway or via Zeb2-mediated repression of an inhibitor of the myofibroblast phenotype either in a SMAD-dependent or -independent manner [30,34–36]. Although we have used a relatively high MOI, we have achieved a five-fold induction of Zeb2 expression, which is physiologically relevant (Figure 2A). We have previously shown that Ski overexpression leads to Zeb2 downregulation and that increased Meox2 expression leads to downregulation of α-SMA and ED-A fibronectin protein levels [15]. This finding suggests that an increase in Zeb2 protein levels in cardiac myofibroblasts may lead to downregulation of Meox2 expression, which in turn can increase α-SMA and ED-A fibronectin protein levels. However, in Meox2 overexpression studies, we did not observe a decrease in SMemb expression levels, which suggests that Zeb2 has the potential to regulate myofibroblast phenoconversion either directly or via another pathway [15]. Zeb2 is also known as a Smad-interacting protein, which may reflect Zeb2's role in fibroblast activation via binding to the Smad transcriptional complex.

3.2. Zeb2 Regulates Cardiac Myofibroblast Migration

As myofibroblasts are hypersynthetic, their reduced motility results in increased deposition of new matrix at the precise point of where it may be needed in the course of normal wound healing [2]. Our current results using the wound healing assay show that Zeb2 regulates the migratory properties of cardiac myofibroblasts. Zeb2 overexpression is associated with reduced migration of myofibroblasts, as indicated by the decreased number of cells migrating into the cell-free gap region in the wound healing assay (Figure 4). This result is consistent with the findings of our myofibroblast phenotype marker analysis, where Zeb2 overexpression was shown to induce a hypersynthetic mature myofibroblast phenotype.

3.3. Zeb2 Regulates Cardiac Myofibroblast Contraction

Myofibroblasts have a characteristic contractile property that helps in the process of wound closure. Myofibroblasts maintain the matrix in the wound area under tension, which helps reduce wound size and ensures rapid healing [37]. Myofibroblasts synthesize large amounts of contractile proteins such as α-SMA and SMemb, which generate tension by actively contracting to generate force [33,38]. There are also other pathways that can contribute in inducing contractile property of myofibroblasts, for example, Ca^{2+} signaling and Rho/ROCK signaling [39]. We have demonstrated that Zeb2 overexpression in myofibroblasts leads to increased contractility as compared to the Ad-LacZ control (Figure 5). Thus, the finding of increasing myofibroblast marker expression (α-SMA and SMemb) reflects the induction of a more contractile mature cardiac myofibroblast phenotype.

3.4. Zeb2 Is Not Required to Maintain the Myofibroblast Phenotype

Fibroblast activation is controlled by a balance between opposing transcription factor signaling pathways. For example, transcription factors such as Scleraxis promote fibroblast activation, whereas transcription factors such as Ski and Meox2 inhibit fibroblast activation [15,40]. We showed that expression of Zeb2 was sufficient to promote fibroblast activation, as shown by expression of markers and functional changes in migration and contractility. However, siRNA-mediated knockdown of Zeb2 had no effect on the expression of two proteins characteristic of the myofibroblast phenotype. This finding may indicate that Zeb2 is not essential for maintenance of the myofibroblast phenotype. Potentially, Zeb1 may compensate for the loss of Zeb2 in maintaining the myofibroblast phenotype [41,42]. Zeb1 and Zeb2 were shown to be targets of miR200 and miR205. Loss of these miRNAs resulted in the upregulation of both Zeb1 and Zeb2 and the subsequent activation of epithelial-to-mesenchymal transition (EMT). The double knockdown of Zeb1 and Zeb2, but not the individual knockdowns, was able to block this activation of EMT [42]. There may exist a similar degree of functional redundancy for these transcription factors in cardiac fibroblasts. Additionally, the knockdown of Zeb2 was effective but there was still 11.9% of Zeb2 protein left that may be sufficient to maintain expression of these markers. Perhaps there is a threshold that needs to be reached before a decrease in Zeb2 is correlated with decreased expression of myofibroblast markers. Finally, the knockdown of Zeb2 was acute (24 h); perhaps an extended knockdown period may result in a phenotypic change. Future studies will be carried out to investigate the importance and existence of these different mechanisms.

Identification of novel regulators of fibrosis for developing selective antifibrotic strategies is currently moving to the mainstream of cardiovascular research. Nutraceutical-based approaches to reduce cardiac fibrosis have shown promise in preclinical models, but more understanding of the underlying mechanisms will improve their clinical effectiveness. Thus, understanding factors that are involved in cardiac fibroblast activation will enable therapeutic targeting to prevent persistent fibroblast activation and progressive fibrosis in chronic stages of heart disease. Overall, the current data supports the hypothesis that Zeb2 promotes cardiac fibroblast activation. We suggest that Zeb2 promotes fibroblast activation, which is indicated by increased expression of myofibroblast markers—the myofibroblastic phenotype is associated with a less motile and more contractile phenotype. Findings from this study directly contribute to our understanding of the biological role of Zeb2 in modulating cardiac fibroblast phenotype, and underscore its putative role in mediating cardiac matrix remodeling.

4. Materials and Methods

4.1. Cell Isolation and Culture

Approval for experimental protocols for the animal studies was received from the Animal Care Committee of the University of Manitoba, Canada, and the protocols conform to the guidelines established by the Canadian Institutes of Health Research and the Canadian Council on Animal Care (Protocol: 14-049, approved 18 November 2014).

Primary cardiac fibroblasts were isolated from the hearts of adult male Sprague-Dawley rats (150–200 g) [43]. The retrograde Langendorff perfusion method was performed with Dulbucco's Modified Eagle's Medium (DMEM)/F12 (Gibco, Thermo Fisher Scientific, Burlington, ON, Canada) followed by Spinner Minimum Essential Medium (SMEM) (Gibco) [43]. After 10 min of perfusion, hearts were digested with 0.1% *w/v* collagenase type 2 (Worthington Biochemical Corporation, Lakewood, NJ, USA) in SMEM for 20 min at room temperature. Hearts were then transferred to a 10 cm^2 plate and the myocardium teased apart in 10 mL of diluted collagenase solution (0.05% *w/v*) for 15 min. Then, growth medium (DMEM-F12 supplemented with 10% fetal bovine serum (FBS), 100 U/mL penicillin (Gibco), 100 µg/mL streptomycin (Gibco) and 1 µM ascorbic acid (Sigma-Aldrich, Oakville, ON, Canada)) was added. To remove any large tissue pieces, the crude cell suspension was gently passed through a 40 µm sterile cell strainer (Thermo Fisher) and collected in a 50 mL conical tube which was then subjected to centrifugation at 200× *g* for 7 min. Cell pellets were resuspended in growth medium and plated onto 10 cm^2 plates. Cells were allowed to adhere for 3 h at 37 °C in a 5% CO_2 incubator, then washed 2–3 times with 1× phosphate-buffered saline (PBS) prior to replacing the growth medium. The following day, cells were washed with PBS twice and fresh medium was added and cells were allowed to grow for 2–3 days before passaging.

4.2. Nuclear/Cytoplasmic Fractionation

P0 cardiac fibroblasts were allowed to grow for 72 h to achieve 70% confluency and were either harvested or passaged to P1 myofibroblasts, which were allowed to grow for 48 h before harvesting. Fractionation was then carried out using the NE-PER Nuclear and Cytoplasmic Extraction Reagents (Pierce Biotechnology, Waltham, MA, USA) as previously described by us [15]. Protein assays were performed using the DC protein assay [44].

4.3. Total Cell Lysate Preparation

Following incubation, cells were washed twice with PBS, and RIPA lysis buffer (50 mM, 150 mM NaCl, 1 mM EDTA, 1 mM EGTA, 1% Triton X-100, 1% sodium deoxycholate, 1% SDS, pH 7.4), containing complete™ protease inhibitor cocktail (Roche Life Sciences, Laval, QC, Canada), was added to lyse cells. Cells were then mechanically scraped and the lysates were vortexed once, incubated on ice for 45 min, revortexed and centrifuged in a tabletop centrifuge at 14,000 rpm at 4 °C for 10 min. Supernatants were transferred to new tubes and protein assays were performed using the DC protein assay [44].

4.4. Western Blot Analysis

SDS-PAGE of 10–25 µg of protein was performed on either 8% or 10% reducing polyacrylamide gels. Pre-Stained Standard (Bio-Rad Laboratories, Mississauga, ON, Canada) molecular mass markers were used as a standard. Proteins were transferred to a 0.45 µM nitrocellulose membrane (Bio-Rad). Membranes were blocked in 1× TBS containing 5% (*w/v*) skim milk powder for 1 h at room temperature with constant shaking. The following primary antibodies were diluted in 1× TBS with 5% skim milk:rabbit polyclonal anti-Zeb2 (1:1000; Sigma), mouse monoclonal anti-α-tubulin (1:5000; Abcam Inc, Toronto, ON, Canada), rabbit polyclonal anti-β-tubulin (1:5000; Abcam), mouse monoclonal anti-α-smooth muscle actin (1:5000; Sigma), mouse monoclonal anti-ED-A fibronectin (1:1000; Millipore, Bedford, MA, USA), mouse monoclonal anti-SMemb (1:1000; Abcam), mouse monoclonal anti-Lamin A+C (Millipore) and mouse monoclonal anti-glyceraldehyde 3-phosphate dehydrogenase (GAPDH; 1:2000; Abcam). Zeb2, Lamin A+C, GAPDH and SMemb antibodies were incubated overnight at 4 °C while α-SMA, ED-A fibronectin and tubulin antibodies were incubated for 1 h at room temperature. Secondary antibodies were horseradish peroxidase (HRP)-conjugated goat anti-rabbit or goat anti-mouse antibodies (Jackson ImmunoResearch Laboratories Inc., West Grove, PA, USA), which were diluted at 1:5000 in 1× TBS containing 5% skim milk and incubated for 1 h at room temperature with constant shaking. Equal protein loading was confirmed using α- and β-tubulin,

GAPDH or Lamin A+C. Protein bands were detected using Western Blotting Luminol Reagent (Santa Cruz Biotechnology, Santa Cruz, CA, USA) and images were developed on CL-Xposure blue X-ray films using Flour S Max Multi Imager (Bio-rad, Hercules, CA, USA).

4.5. Adenoviral Constructs

The eGFP-expressing control vector (Ad-EGFP) was a gift from Grant Pierce (University of Manitoba) and the LacZ vector (Ad-LacZ) was a gift from Michael Czubryt (University of Manitoba). The HA-tagged human Zeb2 (Ad-HA-Zeb2) virus was constructed using the pAdEasy™ Adenoviral Vector System protocol (Agilent Technologies, Palo Alto, CA, USA). Briefly, adenovirus encoding N-terminal HA-tagged Zeb2 (mouse) was created by excising the Zeb2 cDNA from pcDNA 3.1 (A gift from Anders Lund, University of Copenhagen) and the Zeb2 cDNA was cloned into the pShuttle-CMV vector (Agilent Technologies, Palo Alto, CA, USA). Linearized pShuttle-Zeb2 plasmid DNA and pAdEasy vector were cotransformed into BJ5183-competent *E. coli* cells. Recombined plasmids were amplified in DH5α cells and transfected into HEK293 cells to prepare primary viral stock. Primary stock was then amplified using HEK293 cells and viruses were purified. Finally, Ad-HA-Zeb2 virus titration was performed in HEK293 cells using the Adeno-X™ Rapid Titer Kit (Clontech Laboratories, Mountain View, CA, USA).

4.6. Analysis of Myofibroblast Marker Expression Following Adenoviral Infection

For Western blot analysis, 1.4×10^5 P1 rat cardiac myofibroblasts were plated onto 6 cm^2 plates in 2 mL of 10% FBS-containing DMEM/F-12 medium and infected with either Ad-EGFP or Ad-HA-Zeb2 at a multiplicity of infection (MOI) of 200 and incubated in a 5% CO_2 incubator at 37 °C. The following day, feeding medium was replaced with 1% FBS containing DMEM F-12 medium and incubated for another 72 h with 5% CO_2 and subsequently harvested for protein analysis.

4.7. Wound Healing Migration Assay

P1 cardiac myofibroblasts (2.5×10^4 cells in 70 μL of 10% FBS-containing DMEM-F12 medium) were plated inside culture inserts (ibidi USA Inc., Madison, WI, USA) and grown overnight in a 5% CO_2 incubator at 37 °C. The following day, cells were transduced with either Ad-EGFP (200 MOI) or Ad-HA-Zeb2 (200 MOI) and incubated for 96 h. After 96 h, the culture inserts were carefully removed and cells were washed with 1× PBS and the medium was replaced with 1% FBS-containing DMEM-F12 medium. Images were taken at 0 h and 18 h using a 4× objective. The number of cells in the wounded area was quantified using ImageJ software (version 1.45; National Institutes of Health, Bethesda, MD, USA) [45].

4.8. Collagen Gel Contraction Assay

To initiate this assay, 7 mL of cold collagen I solution (Worthington) was mixed with 2 mL of 5× culture medium (DMEM F-12 without serum and antibiotics) in a 50 mL centrifuge tube and the pH was kept between 7 and 7.5. The volume was then adjusted to 10 mL with double-distilled water (ddH_2O). After that, 500 μL of the mixture was added per well of a 24-well plate. Gels were allowed to solidify by incubating at 37 °C in a 5% CO_2 incubator for minimum 3 h or overnight. P1 myofibroblasts (5×10^4 per well) were then plated onto wells and transduced with either Ad-LacZ (200 MOI) or Ad-HA-Zeb2 (200 MOI) in 10% FBS-containing DMEM-F12 medium. The following day, medium was replaced with 1% FBS-containing DMEM F-12 medium and incubated for another 48 h. After 48 h of incubation, the medium was replaced with serum-free DMEM F-12 medium. Gels were cut around the edges using pipette tips. Images were taken at 0 h and 14 h, and gel surface contraction was measured using IDL Measure Gel software (University of Calgary, Calgary, AB, Canada).

4.9. siRNA-Mediated Gene Knockdown

P1 cardiac myofibroblasts were seeded at 8.0×10^4 cells in each well of a 6-well dish. Cells were left to adhere overnight in DMEM:F-12 (1:1) supplemented with 10% FBS and 100 units/mL of penicillin–streptomycin. The cells were then gently washed twice with 1× PBS, then starved overnight in serum-free, antibiotic-free DMEM. The following day, the myofibroblasts were transfected for 24 h with 100 nM of either scrambled Zeb2 siRNA or Zeb2-targeting FITC-tagged siRNA (Table 1) (Sigma-Aldrich, St. Louis, MO, USA) using Lipofectamine RNAiMax (ThermoFisher, Waltham, MA, USA) as per the manufacturer's protocol; negative control wells were left in DMEM alone. Whole cell lysates were collected and analysed by Western blot.

Table 1. siRNA sequences.

Oligo ID	siRNA Target	Sequences	Modification
rZeb2 sense	Targets Rat Zeb2 mRNA	[Flc]GCAAGAAAUGUAUUGGUUU[dT][dT]	5′FITC
rZeb2 antisense	Targets Rat Zeb2 mRNA	AAACCAAUACAUUUCUUGC[dT][dT]	None
rZeb2 scramble	Scrambled sense rZEB2 oligo	GUACGUUAAGGUUAGAUAU[dT][dT]	None
rZeb2scramble_as	Scrambled Zeb2 antisense oligo	AUAUCUAACCUUAACGUAC[dT][dT]	None

Flc: Fluorescein label; dT: Deoxythymidine.

4.10. Statistical Analysis

All experiments were repeated with 3 biological replicates ($n = 3$). For primary cells, different rat hearts were used as a source of primary fibroblasts and each heart equates to $n = 1$ [46]. Student's t-test was used to compare means between two samples (control and experimental groups) and one-way ANOVA followed by a Tukey's multiple comparison test was used to compare between 3 samples and results, wherein a difference of $p \leq 0.05$ was considered as statistically significant.

Author Contributions: The authors were responsible for the following aspects of this paper: Conceptualization, F.J., N.M.L., I.M.C.D. and J.T.W.; methodology, F.J., N.M.L., and S.G.R.; formal analysis, F.J., N.M.L., and S.G.R.; writing—original draft preparation, F.J.; review and editing, F.J., N.M.L., S.G.R., I.M.C.D. and J.T.W. visualization, F.J., N.M.L., and S.G.R.; supervision, I.M.C.D. and J.T.W; project administration, J.T.W.; funding acquisition, I.M.C.D and J.T.W.

Funding: The work in this project was supported by a Canadian Institutes for Health Research (CIHR) operating grant to I.M.C.D. and J.T.W. F.J. was supported by a graduate studentship from Research Manitoba. N.M.L. is supported by a CIHR Frederick Banting and Charles Best Canada Research Scholarship and a Graduate Studentship (PhD) from Research Manitoba.

Acknowledgments: We are also grateful for infrastructure support from The St. Boniface Hospital Foundation as well as Heart and Stroke Foundation of Canada (I.M.C.D.—with salary support for S.G.R.).

Conflicts of Interest: The authors declare no conflict of interest.

Abbreviations

ECM	extracellular matrix
α-SMA	α-smooth muscle actin
SMemb	embryonic smooth muscle myosin heavy chain
ZEB2	zinc finger E box-binding homeobox 2

References

1. Krenning, G.; Zeisberg, E.M.; Kalluri, R. The origin of fibroblasts and mechanism of cardiac fibrosis. *J. Cell. Physiol.* **2010**, *225*, 631–637. [CrossRef] [PubMed]
2. Travers, J.G.; Kamal, F.A.; Robbins, J.; Yutzey, K.E.; Blaxall, B.C. Cardiac Fibrosis: The Fibroblast Awakens. *Circ. Res.* **2016**, *118*, 1021–1040. [CrossRef] [PubMed]
3. Baum, J.; Duffy, H.S. Fibroblasts and myofibroblasts: What are we talking about? *J. Cardiovasc. Pharmacol.* **2011**, *57*, 376–379. [CrossRef] [PubMed]

4. Gabbiani, G. The myofibroblast in wound healing and fibrocontractive diseases. *J. Pathol.* **2003**, *200*, 500–503. [CrossRef] [PubMed]
5. Moore-Morris, T.; Tallquist, M.D.; Evans, S.M. Sorting out where fibroblasts come from. *Circ. Res.* **2014**, *115*, 602–604. [CrossRef] [PubMed]
6. Czubryt, M.P. Common threads in cardiac fibrosis, infarct scar formation, and wound healing. *Fibrogenesis Tissue Repair* **2012**, *5*, 19. [CrossRef] [PubMed]
7. Micallef, L.; Vedrenne, N.; Billet, F.; Coulomb, B.; Darby, I.A.; Desmouliere, A. The myofibroblast, multiple origins for major roles in normal and pathological tissue repair. *Fibrogenesis Tissue Repair* **2012**, *5* (Suppl. 1), S5. [PubMed]
8. Souders, C.A.; Bowers, S.L.; Baudino, T.A. Cardiac fibroblast: The renaissance cell. *Circ. Res.* **2009**, *105*, 1164–1176. [CrossRef] [PubMed]
9. Valiente-Alandi, I.; Potter, S.J.; Salvador, A.M.; Schafer, A.E.; Schips, T.; Carrillo-Salinas, F.; Gibson, A.M.; Nieman, M.L.; Perkins, C.; Sargent, M.A.; et al. Inhibiting Fibronectin Attenuates Fibrosis and Improves Cardiac Function in a Model of Heart Failure. *Circulation* **2018**, *138*, 1236–1252. [CrossRef] [PubMed]
10. Gourdie, R.G.; Dimmeler, S.; Kohl, P. Novel therapeutic strategies targeting fibroblasts and fibrosis in heart disease. *Nat. Rev. Drug Discov.* **2016**, *15*, 620–638. [CrossRef] [PubMed]
11. Weber, K.T.; Sun, Y.; Bhattacharya, S.K.; Ahokas, R.A.; Gerling, I.C. Myofibroblast-mediated mechanisms of pathological remodelling of the heart. *Nat. Rev. Cardiol.* **2013**, *10*, 15–26. [CrossRef] [PubMed]
12. Willems, I.E.; Havenith, M.G.; De Mey, J.G.; Daemen, M.J. The α-smooth muscle actin-positive cells in healing human myocardial scars. *Am. J. Pathol.* **1994**, *145*, 868–875. [PubMed]
13. Kanisicak, O.; Khalil, H.; Ivey, M.J.; Karch, J.; Maliken, B.D.; Correll, R.N.; Brody, M.J.; LIN, S.-C.J.; Aronow, B.J.; Tallquist, M.D.; et al. Genetic lineage tracing defines myofibroblast origin and function in the injured heart. *Nat. Commun.* **2016**, *7*, 12260. [CrossRef] [PubMed]
14. Vasquez, C.; Benamer, N.; Morley, G.E. The cardiac fibroblast: Functional and electrophysiological considerations in healthy and diseased hearts. *J. Cardiovasc. Pharmacol.* **2011**, *57*, 380–388. [CrossRef] [PubMed]
15. Cunnington, R.H.; Northcott, J.M.; Ghavami, S.; Filomeno, K.L.; Jahan, F.; Kavosh, M.S.; Davies, J.J.; Wigle, J.T.; Dixon, I.M. The Ski-Zeb2-Meox2 pathway provides a novel mechanism for regulation of the cardiac myofibroblast phenotype. *J. Cell Sci.* **2014**, *127*, 40–49. [CrossRef] [PubMed]
16. Khalil, H.; Kanisicak, O.; Prasad, V.; Correll, R.N.; Fu, X.; Schips, T.; Vagnozzi, R.J.; Liu, R.; Huynh, T.; Lee, S.J.; et al. Fibroblast-specific TGF-β-Smad2/3 signaling underlies cardiac fibrosis. *J. Clin. Investig.* **2017**, *127*, 3770–3783. [CrossRef] [PubMed]
17. Wang, B.; Hao, J.; Jones, S.C.; Yee, M.S.; Roth, J.C.; Dixon, I.M. Decreased Smad 7 expression contributes to cardiac fibrosis in the infarcted rat heart. *Am. J. Physiol. Heart Circ. Physiol.* **2002**, *282*, H1685–H1696. [CrossRef] [PubMed]
18. Wang, B.; Omar, A.; Angelovska, T.; Drobic, V.; Rattan, S.G.; Jones, S.C.; Dixon, I.M. Regulation of collagen synthesis by inhibitory Smad7 in cardiac myofibroblasts. *Am. J. Physiol. Heart Circ. Physiol.* **2007**, *293*, H1282–H1290. [CrossRef] [PubMed]
19. Alehagen, U.; Benson, L.; Edner, M.; Dahlstrom, U.; Lund, L.H. Association Between Use of Statins and Mortality in Patients With Heart Failure and Ejection Fraction of >/=50. *Circ. Heart Fail.* **2015**, *8*, 862–870. [CrossRef] [PubMed]
20. Brilla, C.G.; Funck, R.C.; Rupp, H. Lisinopril-mediated regression of myocardial fibrosis in patients with hypertensive heart disease. *Circulation* **2000**, *102*, 1388–1393. [CrossRef] [PubMed]
21. Shi, Y.; Li, D.; Tardif, J.C.; Nattel, S. Enalapril effects on atrial remodeling and atrial fibrillation in experimental congestive heart failure. *Cardiovasc Res.* **2002**, *54*, 456–461. [CrossRef]
22. Eclov, J.A.; Qian, Q.; Redetzke, R.; Chen, Q.; Wu, S.C.; Healy, C.L.; Ortmeier, S.B.; Harmon, E.; Shearer, G.C.; O'Connell, T.D. EPA, not DHA, prevents fibrosis in pressure overload-induced heart failure: Potential role of free fatty acid receptor 4. *J. Lipid Res.* **2015**, *56*, 2297–2308. [CrossRef] [PubMed]
23. Thandapilly, S.J.; Louis, X.L.; Behbahani, J.; Movahed, A.; Yu, L.; Fandrich, R.; Zhang, S.; Kardami, E.; Anderson, H.D.; Netticadan, T. Reduced hemodynamic load aids low-dose resveratrol in reversing cardiovascular defects in hypertensive rats. *Hypertens. Res.* **2013**, *36*, 866–872. [CrossRef] [PubMed]

24. Pereira, B.L.B.; Reis, P.P.; Severino, F.E.; Felix, T.F.; Braz, M.G.; Nogueira, F.R.; Silva, R.A.C.; Cardoso, A.C.; Lourenco, M.A.M.; Figueiredo, A.M.; et al. Tomato (*Lycopersicon esculentum*) or lycopene supplementation attenuates ventricular remodeling after myocardial infarction through different mechanistic pathways. *J. Nutr. Biochem.* **2017**, *46*, 117–124. [CrossRef] [PubMed]
25. Aloud, B.M.; Raj, P.; McCallum, J.; Kirby, C.; Louis, X.L.; Jahan, F.; Yu, L.; Hiebert, B.; Duhamel, T.A.; Wigle, J.T.; et al. Cyanidin 3-O-glucoside prevents the development of maladaptive cardiac hypertrophy and diastolic heart dysfunction in 20-week-old spontaneously hypertensive rats. *Food Funct.* **2018**, *9*, 3466–3480. [CrossRef] [PubMed]
26. Chen, T.; Li, J.; Liu, J.; Li, N.; Wang, S.; Liu, H.; Zeng, M.; Zhang, Y.; Bu, P. Activation of SIRT3 by resveratrol ameliorates cardiac fibrosis and improves cardiac function via the TGF-β/Smad3 pathway. *Am. J. Physiol. Heart Circ. Physiol.* **2015**, *308*, H424–H434. [CrossRef] [PubMed]
27. O'Connell, T.D.; Block, R.C.; Huang, S.P.; Shearer, G.C. omega3-Polyunsaturated fatty acids for heart failure: Effects of dose on efficacy and novel signaling through free fatty acid receptor 4. *J. Mol. Cell. Cardiol.* **2017**, *103*, 74–92. [CrossRef] [PubMed]
28. Santiago, J.J.; Dangerfield, A.L.; Rattan, S.G.; Bathe, K.L.; Cunnington, R.H.; Raizman, J.E.; Bedosky, K.M.; Freed, D.H.; Kardami, E.; Dixon, I.M. Cardiac fibroblast to myofibroblast differentiation in vivo and in vitro: Expression of focal adhesion components in neonatal and adult rat ventricular myofibroblasts. *Dev. Dyn.* **2010**, *239*, 1573–1584. [CrossRef] [PubMed]
29. Masur, S.K.; Dewal, H.S.; Dinh, T.T.; Erenburg, I.; Petridou, S. Myofibroblasts differentiate from fibroblasts when plated at low density. *Proc. Natl. Acad. Sci. USA* **1996**, *93*, 4219–4223. [CrossRef] [PubMed]
30. Vandewalle, C.; Comijn, J.; De Craene, B.; Vermassen, P.; Bruyneel, E.; Andersen, H.; Tulchinsky, E.; Van Roy, F.; Berx, G. SIP1/ZEB2 induces EMT by repressing genes of different epithelial cell-cell junctions. *Nucleic Acids Res.* **2005**, *33*, 6566–6578. [CrossRef] [PubMed]
31. Darby, I.; Skalli, O.; Gabbiani, G. A-smooth muscle actin is transiently expressed by myofibroblasts during experimental wound healing. *Lab. Investig.* **1990**, *63*, 21–29. [PubMed]
32. Serini, G.; Bochaton-Piallat, M.L.; Ropraz, P.; Geinoz, A.; Borsi, L.; Zardi, L.; Gabbiani, G. The fibronectin domain ED-A is crucial for myofibroblastic phenotype induction by transforming growth factor-β1. *J. Cell Biol.* **1998**, *142*, 873–881. [CrossRef] [PubMed]
33. Frangogiannis, N.G.; Michael, L.H.; Entman, M.L. Myofibroblasts in reperfused myocardial infarcts express the embryonic form of smooth muscle myosin heavy chain (SMemb). *Cardiovasc. Res.* **2000**, *48*, 89–100. [CrossRef]
34. Nam, E.H.; Lee, Y.; Zhao, X.F.; Park, Y.K.; Lee, J.W.; Kim, S. ZEB2-Sp1 cooperation induces invasion by upregulating cadherin-11 and integrin α5 expression. *Carcinogenesis* **2014**, *35*, 302–314. [CrossRef] [PubMed]
35. Postigo, A.A.; Depp, J.L.; Taylor, J.J.; Kroll, K.L. Regulation of Smad signaling through a differential recruitment of coactivators and corepressors by ZEB proteins. *EMBO J.* **2003**, *22*, 2453–2462. [CrossRef] [PubMed]
36. Verschueren, K.; Remacle, J.E.; Collart, C.; Kraft, H.; Baker, B.S.; Tylzanowski, P.; Nelles, L.; Wuytens, G.; Su, M.T.; Bodmer, R.; et al. SIP1, a novel zinc finger/homeodomain repressor, interacts with Smad proteins and binds to 5′-CACCT sequences in candidate target genes. *J. Biol. Chem.* **1999**, *274*, 20489–20498. [CrossRef] [PubMed]
37. Gabbiani, G.; Ryan, G.B.; Majne, G. Presence of modified fibroblasts in granulation tissue and their possible role in wound contraction. *Experientia* **1971**, *27*, 549–550. [CrossRef] [PubMed]
38. Hinz, B.; Celetta, G.; Tomasek, J.J.; Gabbiani, G.; Chaponnier, C. A-smooth muscle actin expression upregulates fibroblast contractile activity. *Mol. Biol. Cell.* **2001**, *12*, 2730–2741. [CrossRef] [PubMed]
39. Follonier Castella, L.; Gabbiani, G.; McCulloch, C.A.; Hinz, B. Regulation of myofibroblast activities: Calcium pulls some strings behind the scene. *Exp. Cell Res.* **2010**, *316*, 2390–2401. [CrossRef] [PubMed]
40. Espira, L.; Lamoureux, L.; Jones, S.C.; Gerard, R.D.; Dixon, I.M.; Czubryt, M.P. The basic helix-loop-helix transcription factor scleraxis regulates fibroblast collagen synthesis. *J. Mol. Cell. Cardiol.* **2009**, *47*, 188–195. [CrossRef] [PubMed]
41. Chilosi, M.; Calio, A.; Rossi, A.; Gilioli, E.; Pedica, F.; Montagna, L.; Pedron, S.; Confalonieri, M.; Doglioni, C.; Ziesche, R.; et al. Epithelial to mesenchymal transition-related proteins ZEB1, β-catenin, and β-tubulin-III in idiopathic pulmonary fibrosis. *Mod. Pathol.* **2017**, *30*, 26–38. [CrossRef] [PubMed]

42. Gregory, P.A.; Bert, A.G.; Paterson, E.L.; Barry, S.C.; Tsykin, A.; Farshid, G.; Vadas, M.A.; Khew-Goodall, Y.; Goodall, G.J. The miR-200 family and miR-205 regulate epithelial to mesenchymal transition by targeting ZEB1 and SIP1. *Nat. Cell Biol.* **2008**, *10*, 593–601. [CrossRef] [PubMed]
43. Hao, J.; Wang, B.; Jones, S.C.; Jassal, D.S.; Dixon, I.M. Interaction between angiotensin II and Smad proteins in fibroblasts in failing heart and in vitro. *Am. J. Physiol. Heart Circ. Physiol.* **2000**, *279*, H3020–H3030. [CrossRef] [PubMed]
44. Smith, P.K.; Krohn, R.I.; Hermanson, G.T.; Mallia, A.K.; Gartner, F.H.; Provenzano, M.D.; Fujimoto, E.K.; Goeke, N.M.; Olson, B.J.; Klenk, D.C. Measurement of protein using bicinchoninic acid. *Anal. Biochem.* **1985**, *150*, 76–85. [CrossRef]
45. Schneider, C.A.; Rasband, W.S.; Eliceiri, K.W. NIH Image to ImageJ: 25 years of image analysis. *Nat. Methods* **2012**, *9*, 671–675. [CrossRef] [PubMed]
46. Zeglinski, M.R.; Davies, J.J.; Ghavami, S.; Rattan, S.G.; Halayko, A.J.; Dixon, I.M. Chronic expression of Ski induces apoptosis and represses autophagy in cardiac myofibroblasts. *Biochim. Biophys. Acta* **2016**, *1863*, 1261–1268. [CrossRef] [PubMed]

International Journal of
Molecular Sciences

Review

Klotho: A Major Shareholder in Vascular Aging Enterprises

Kenneth Lim [1,2,*], Arvin Halim [3], Tzong-shi Lu [3], Alan Ashworth [4] and Irene Chong [5]

1 Division of Nephrology, Department of Medicine, Massachusetts General Hospital and Harvard Medical School, Boston, MA 02114, USA
2 MGH Renal Associates, 165 Cambridge Street, Suite 302, Boston, MA 02114, USA
3 Renal Division, Department of Medicine, Brigham and Women's Hospital and Harvard Medical School, Boston, MA 02114, USA; arv.halim@gmail.com (A.H.); tlu1@rics.bwh.harvard.edu (T.L.)
4 Helen Diller Family Comprehensive Cancer Center, University of California San Francisco (UCSF), 1450 3rd St, San Francisco, CA 94158, USA; Alan.Ashworth@ucsf.edu
5 The Institute of Cancer Research and The Royal Marsden NHS Foundation Trust, Chester Beatty Laboratories, 237 Fulham Road, London SW3 6JB, UK; Irene.Chong@icr.ac.uk
* Correspondence: kjlim@mgh.harvard.edu

Received: 6 August 2019; Accepted: 13 September 2019; Published: 19 September 2019

Abstract: Accelerated vascular aging is a condition that occurs as a complication of several highly prevalent inflammatory conditions such as chronic kidney disease, cancer, HIV infection and diabetes. Age-associated vascular alterations underlie a continuum of expression toward clinically overt cardiovascular disease. This has contributed to the striking epidemiologic transition whereby such noncommunicable diseases have taken center stage as modern-day global epidemics and public health problems. The identification of α-Klotho, a remarkable protein that confers powerful anti-aging properties has stimulated significant interest. In fact, emerging data have provided fundamental rationale for Klotho-based therapeutic intervention for vascular diseases and multiple other potential indications. However, the application of such discoveries in Klotho research remains fragmented due to significant gaps in our molecular understanding of Klotho biology, as well as hurdles in clinical research and experimental barriers that must first be overcome. These advances will be critical to establish the scientific platform from which future Klotho-based interventional trials and therapeutic enterprises can be successfully launched.

Keywords: vascular aging; vascular calcification; arteriosclerosis; Klotho; chronic kidney disease (CKD), cancer; diabetes

---

1. Introduction

Advancing age is a major risk factor for both subclinical and clinically overt cardiovascular disease (CVD) [1]. Age-associated changes of cardiovascular structure and function occur universally in apparently healthy persons without overt clinical consequences. However, these changes can compromise cardiovascular reserve capacity, alter the threshold for symptoms and signs, and can occur at an accelerated rate in various disease states thereby leading to occult CVD [1]. Features of accelerated or premature cardiovascular aging and reduced lifespan accompany patients with a number of chronic disease states, including chronic kidney disease (CKD), cancer, diabetes, HIV infection and inflammatory arthropathies [2,3]. It is becoming increasingly apparent that premature age-related vascular alterations compound in different vascular beds to cause an exponential increase in disability and function as a major contributor to occult CVD [4]. Models that include population aging, increasing rates of urbanization and globalization and increasing prevalence of these chronic disease states have demonstrated a striking epidemiologic transition whereby noncommunicable diseases such as accelerated CVD, have taken over as a modern-day global public health problem and a leading cause of death [5].

The kidneys are among the organs that are functionally most sensitive to the aging process, and the link between aging and kidney function is well-recognized and appears bidirectional [2,6]. Premature CVD is well-illustrated in CKD patients by the observation that cardiovascular mortality in a 20 year-old dialysis patient is similar to that of an 80 year-old person without significant renal impairment [7]. In fact, CVD remains the leading cause of death in patients with CKD [7–9]. The uremic phenotype in CKD recapitulates many features of aging, such as arteriosclerosis, atherosclerosis, osteoporosis, poor wound healing, sarcopenia, inflammatory and oxidative stress, insulin resistance, frailty, hypogonadism, skin atrophy, cognitive dysfunction and disability; therefore, CKD has been increasingly recognized as a model for premature aging syndrome [2]. Age-associated vascular changes in CKD is a complex process driven by both traditional and CKD-related risk factors, such as uremia, mineral disorders, fibroblast growth factor (FGF)-23, inflammation, post-translational protein modifications, metabolites, advanced glycation end products and pressure and volume overload [10–13]. Similarly, inflammation and oxidative stress beyond traditional risk factors are also key contributors to premature vascular aging in other conditions, such as in cancer and diabetes [14,15]. In survivors of cancer, the direct effects of various chemotherapies and radiation on telomere length, senescent cells, epigenetic modifications and microRNAs have also been linked to accelerated aging [16].

Aging and age-associated changes of the vascular system involve fundamentally different alterations from atherosclerosis. Vascular alterations attributable to aging include arterial dilatation, vascular calcification, increased collagen-to-elastin ratio with fragmentation, endothelial dysfunction and hypertrophy of vascular smooth muscles cells (VSMCs) [17]. These changes are typically described as arteriosclerosis and provide the milieu for the development of overt vascular disease; this interaction between age-associated changes and the development of clinically overt disease has previously been termed the "vascular aging-vascular disease" pathway [18–20]. The terms "arteriosclerosis" and "atherosclerosis" have long been frequently confused, however they represent distinct groups of pathologic processes: atherosclerotic disease primarily affects the intima leading to plaque formation, while arteriosclerotic disease is primarily a disease of the medial layer of arteries.

Medial vascular calcification occurs at an accelerated rate in age-associated conditions such as CKD, cancer and diabetes with the consequence of loss of arterial distensibility and increasing arterial stiffening. As arteries stiffen, this leads to a lack of buffering capabilities during oscillatory changes in blood pressure caused by intermittent ventricular ejection and results in a more rigid aorta that can accommodate less stroke volume; these changes can be clinically detected by greater pressure augmentation in systole and higher pulse pressures [17]. End organs such as the heart, brain and kidneys are consequently exposed to higher systolic pressures and greater pressure fluctuations resulting in microvascular damage, reduced diastolic coronary perfusion and promotion of subendocardial ischemia and myocardial fibrosis. Additionally, age-associated arterial hardening leads to left ventricular hypertrophy (LVH), increased cardiac afterload and characteristic diastolic failure [18–20].

α-Klotho, hereby referred to as Klotho, was first discovered by Kuro-o et al. In 1997 and has made a triumphant entry onto the center stage in the field of aging [21]. The identification of Klotho as a novel anti-aging protein has become a major focal point of aging research and challenged the long-held paradigm of aging as a passive, inevitable process of deteriorating systemic organ function. Klotho knockout mice demonstrate a premature aging syndrome phenotype and exhibit shortened life span [21]. Restoration of Klotho in Klotho-deficient mice however ameliorated these changes and these mice live 30% longer than wild-type [21–23]. The kidney is the primary organ of functional Klotho expression and its expression has been consistently linked with kidney function. Analysis of 2946 participants within the Healthy Aging and Body Composition Study demonstrated that a high soluble Klotho level was independently associated with a lower risk of decline in kidney function [24]. In fact, total Klotho protein levels decline in serum as kidney dysfunction ensues in CKD and with advancing age [25]. Critically, accumulating experimental, clinical and epidemiologic evidence have shown that

Klotho deficiency is associated with a variety of vascular outcomes, such as cardiovascular events and arterial stiffness [26], vascular calcification [13,27] and atherosclerosis [28]. Moreover, therapeutic restoration of Klotho can ameliorate these changes [21–23,27].

Taken together, emerging data suggest that arterial aging and age-associated arterial diseases are active, tightly regulated cell-mediated processes that may be potentially modifiable. Given the implications of these discoveries for human health, Klotho as a potential longevity-modulating therapeutic strategy has attracted widespread attention and as a result, the field of aging research has flourished over the past decade. This review will consider the biology of Klotho and contemporary evidence for its therapeutic applications in the treatment of age-associated vascular diseases [21,29].

2. The Biology of Klotho

A significant advance in the field of aging research came about in 1997 after the serendipitous discovery of the aging-suppressor gene named *Klotho* [21]. The gene was named after Clotho, a figure in ancient Greek mythology responsible for spinning the thread of human life. Mice lacking the *Klotho* gene develop an aging-like phenotype similar to premature human aging, including endothelial dysfunction, vascular calcification, progressive atherosclerosis and myocardial hypertrophy of the cardiovascular system [21,27,30]. These mice develop aging-like changes beginning at three weeks of age at which time their growth is stunted and have an average life span of 15 weeks, compared with wild-type mice that have an average lifespan of approximately 2.5–3 years.

2.1. Structure of Klotho

The *Klotho* gene spans approximately 50 kb and is composed of five exons located on chromosome 13q12 in humans [29]. *Klotho* encodes a type 1 transmembrane glycoprotein that is 1014 and 1012 amino acids long in mouse and human, respectively [21,29]. Full-length Klotho consists of a putative N-terminal signal sequence and two internal repeats (mKL1 and mKL2) constituting the extracellular domain, a single-pass membrane spanning domain, and a short C-terminal intracellular domain of 10 amino acids [29]. Full-length Klotho is an approximately 135 kDa in molecular weight, the size being influenced by N-glycosylation [31]. Klotho also exists in the circulation as both soluble and secreted isoforms [31]: Soluble Klotho is generated from cleavage of full-length Klotho by membrane proteases (ADAM10 and ADAM17) in an α-cut to generate a 130 kDa protein that contains both the KL1 and KL2 domains, but lacks the transmembrane and intracellular components. Following the α-cut, the remaining transmembrane and cytosolic 5 kDa portion then undergoes proteolysis by γ-secretase [32,33]. A 65 kDa isoform is also generated by a β-cut and contains the KL1 domain.

In the mouse an alternatively spliced *Klotho* mRNA is produced which lacks exons 4 and 5, and in humans a premature stop codon leads to the truncation of the Klotho protein [29,34]. Therefore, a 65 kDa fragment consisting of the KL1 domain was also previously thought to be secreted by alternative splicing in humans [31]. However, this paradigm was challenged in a recent study by Mencke et al. that reported that the premature stop codon responsible for the truncated form primes the alternatively spliced mRNA for degradation [35]. Therefore, these data now suggest that the secreted Klotho isoform does not exist in humans. It remains unclear whether a secreted Klotho isoform exists in mice.

The KL1 and KL2 domains have sequence similarity to family 1 β-glycosidases and are most similar to mammalian lactase–phlorizin hydrolase (LPH) [21]. Although the internal repeats lack the prototypical catalytic glutamic acid residues of β-glycosidases, they have substitute catalytic acid-base Asn in KL1 and a nucleophilic Ser residue in KL2 for humans. Fluorescence assays with chimeric 958 amino acid-long mouse Klotho extracellular domain with human immunoglobulin Fc suggested the internal repeats have β-glucouronidase activity [36]. However, in a recent landmark study by Chen et al. using 3.0 Å resolution crystallography of a 1:1:1 ternary complex of soluble Klotho bound to FGF23 and fibroblast growth factor receptor (FGFR)1c, refuted the hypothesis that Klotho has intrinsic catalytic activity. They confirmed the absence of catalytic Glu residues in the putative catalytic pockets of KL1 and KL2, but also showed that soluble Klotho did not hydrolyze the substrates of β-glucouronidase and sialidase, in vitro [37]. With crystallography evidence showing that KL1 and KL2 each appear to take on a $(\beta\alpha)_8$ riosephosphate isomerase (TIM) barrel fold, Chen et al. suggested that Klotho is the first and so-far only known TIM barrel protein that acts only as a non-enzymatic molecular scaffold [37].

2.2. Tissue Expression of Klotho

Studies in the mouse originally identified Klotho expression mainly in distal convoluted tubule cells and to a lesser extent, in the proximal convoluted tubule cells of the kidneys [21,38]. In an elegant study by Lindberg et al. A novel mouse strain with the *Klotho* gene deleted throughout the nephron was generated and this was found to exhibit an 80% reduction in circulating Klotho levels confirming that the kidneys are the primary source of soluble Klotho [39]. These findings are consistent with observations in human patients who show an approximately 30% reduction in circulating Klotho following unilateral nephrectomy [40].

In a study by Lim et al., that characterized systemwide tissue expression of transmembrane Klotho using targeted proteomic analysis in parallel with conventional antibody-based methods, Klotho expression was identified in a variety of organ systems, including arterial (in both endothelial cells and vascular smooth muscle cells (VSMCs)), epithelial, endocrine, reproductive and neuronal tissues [3]. Extra-renal expression of Klotho appears to be less abundant than expression at the kidneys, and this suggests that local expression of Klotho could serve in an autocrine/paracrine fashion to regulate tissue health locally while the kidneys remain the principal source of endocrine Klotho. This situation mirrors the vitamin D hormonal system, whereby endocrine vitamin D is principally sourced from the kidney [41].

2.3. The Klotho–FGF23 Axis

A major but not exclusive role for Klotho is serving as an obligate coreceptor for fibroblast growth factor 23 (FGF23) signaling. FGF23 is a phosphaturic hormone that is secreted by osteocytes and plays a pivotal role in regulating phosphate homeostasis [42,43]. FGF23 knockout mice were found to develop a complex aging-like phenotype very similar to that observed in Klotho-deficient mice [44]. Both Klotho and FGF23 deficient mice develop hyperphosphatemia and high serum levels of active 1,25-dihydroxyvitamin D levels together with premature aging features [44,45]. However, most of these symptoms of premature aging are alleviated by feeding these mice a low phosphate diet to restore phosphate homeostasis, despite this diet stimulating further increases in vitamin D levels [46]. These results suggested that phosphate imbalance, rather than increased serum vitamin D levels, is a major regulator of aging as discussed below.

Klotho forms a complex with fibroblast growth factor receptors (FGFR) 1c, 3c or 4 which converts their canonical functions into specific receptors for FGF23 [47–49]. Both membrane-bound and soluble Klotho can function as a co-receptor for FGF23. Jimbo et al. showed that FGF23 enhanced extracellular signal-regulated kinases (ERK)1/2 phosphorylation in Klotho-overexpressing, but not in naive VSMCs without detectable Klotho [50]. Structural studies suggest soluble Klotho acts to enhance the binding affinity of FGF23 to FGFR1c via proximity, with the assistance of Zn^{2+} prosthetic group [37]. This is speculated to promote heparan sulfate (HS)-induced dimerization of FGF23–FGFR1c, which is likely required for FGF23-mediated FGFR1c activation [37]. Binding of FGF23 to the Klotho–FGFR complex induces the internalization and degradation of sodium-dependent phosphate transport protein 2A (NPT2A) and downregulates expression of *NPT2A* (SLC34AS1) at the local brush border membrane of proximal tubular cells [51]. Additionally, FGF23 suppresses the expression of Cyp27b1 which encodes 1α-hydroxylase, the enzyme that converts 25-hydroxyvitamin D_3 to active 1,25-dihydroxyvitamin D_3 and stimulates its metabolic breakdown by increasing expression of Cyp24a1 which encodes 24-hydroxylase [52].

These effects collectively operate to help maintain phosphate balance with significant implications on vascular health, given that phosphate is a major determinant of CVD in CKD and the aging process. High phosphate concentrations have been widely shown to stimulate endothelial and vascular smooth muscle cell (VSMC) damage and calcification [53,54]. Phosphate toxicity can result in the formation of calcium phosphate (CaPi) product; in the blood, CaPi binds to serum protein fetuin-A and forms colloidal particles termed calciprotein particles (CPP). CPP can induce endothelial damage, VSMC calcification and innate immune responses, thereby contributing to accelerated vascular aging [55]. Notably, phosphate retention, progressive hyperphosphatemia, rising FGF23 levels and low Klotho expression collectively are observed in human patients with advancing CKD and has been associated with progressive age-associated cardiovascular alterations [13,41]. This has prompted the view that CKD is a strategic clinical model to study premature cardiovascular aging [2]. These observations have consistently linked phosphate to the aging process and emphasize the importance of Klotho–FGF23 as a counterregulatory hormonal system. Of note, Klotho–FGF23 also regulates other channels in the kidney, such as members of the transient receptor potential (TRP) vanilloid (V) subgroup, TRPV5 and TRPV6 calcium channels and renal outer medullary potassium channel 1 [56]; these effects are beyond the scope of this review.

2.4. Klotho Exerts Pleiotropic Functions Independent of FGF23

In the absence of FGF23, soluble Klotho can directly exert phosphaturic activity by promoting the endocytosis and degradation of NPT2A, NPT3 and phosphate transporter 1 (PiT1) and 2 (PiT2) [57,58]. However, the role of soluble Klotho in regulating mineral metabolism has been subject to considerable controversy. Chen et al. showed that in vitro soluble and transmembrane Klotho possess similar capacities to facilitate FGF23 signaling [37]. Recombinant soluble Klotho injected into wild-type mice results in a small, but significant increase in urinary phosphate excretion. However, injection of a mutated soluble Klotho isoform lacking the FGF receptor binding arm resulted in a striking downregulation of FGF23 target genes in the kidney with the development of hyperphosphatemia. These findings suggest that the effects of soluble Klotho on mineral metabolism are FGF23 dependent.

Klotho can modulate a number of evolutionary conserved intracellular signaling pathways involved in longevity, including insulin and insulin-like growth factor 1 [59], target of rapamycin [60], cyclic adenosine monophosphate [61], protein kinase C [62], transforming growth factor-β [63], p53/p21 [64] and Wnt signaling [65]. Unfortunately, the mechanisms by which Klotho exerts anti-aging effects and how Klotho overexpressing mice live 20–30% longer compared to wildtype animals is still not well-understood.

3. Vasculo-Protective Effects of Klotho

Accumulating epidemiological and observational studies have linked circulating Klotho levels to cardiovascular risk and outcomes: In a study by Semba et al. that examined 804 community-dwelling adults aged 65 or greater, the investigators found that participants in the lowest tertile of plasma Klotho (<575 pg/mL) had an increased risk of death compared with participants in the highest tertile (>763 pg/mL; hazard ratio 1.78, 95% CI 1.20–2.63) [66]. Arking et al. examined the association of a functional variant of Klotho, termed KL-VS genotyped in 525 Ashkenazi Jews composed of 216 probands (age ≥ 95 years) and 309 unrelated individuals (ages 51 to 94) [67]. KL-VS was associated with cardiovascular disease risk factors including high-density lipoprotein ($p < 0.05$) and systolic blood pressure ($p < 0.008$). Furthermore, homozygous KL-VS individuals had the highest risk of vascular events (odds ratio (OR) 30.65; 95% CI 2.55–368) and a 4.49-fold (95% CI 1.35 to 14.97) relative risk for mortality. Similarly, several other studies have shown that plasma Klotho is inversely associated with CVD [66,68], and macrovascular disease (including coronary artery disease and cerebrovascular accidents) in type 2 diabetics [28,66].

Mounting experimental evidence suggest that the presence of Klotho is critical for vascular health and its administration can exert vasculo-protective effects (Figure 1). Klotho knock-out mice develop striking vascular disease, including widespread vascular calcification, endothelial dysfunction and progressive atherosclerosis together with severe hypervitaminosis D, hypercalcemia and hyperphosophatemia [69]. Conversely, a growing body of evidence suggest that restoration of Klotho ameliorates these changes [21–23,27].

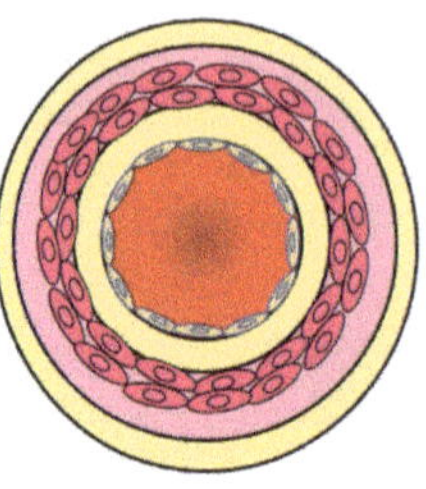

Figure 1. Vasculo-protective effects of Klotho. The presence of Klotho can exert pleiotropic protective effects against age-associated arterial changes. VSMC, Vascular Smooth Muscle Cells.

3.1. Vascular Calcification

Molecularly, age-associated changes of the arterial wall are characterized by osteogenic transformation of vascular smooth muscle cells (VSMCs) and loss of their contractile phenotype, upregulation of transcriptional regulators of osteoblastic differentiation such as Runx2, increased expression of bone markers (alkaline phosphatase, osteopontin, osteocalcin), release of matrix vesicles, apoptosis, extracellular matrix degradation and nuclear changes [70–72]. The continuum of expression of these molecular and cellular alterations underlie the development of vascular calcification, arteriosclerosis and arterial remodeling. Both the extent of vascular calcification and arterial stiffening is a hallmark of vascular aging and has been used as a measure of biological cardiovascular age [73–75]. Accumulation of biomarkers such as senescence associated β-galactosidase (SAβG) is also widely used to assess biological aging of the arterial tree [76].

Emerging data suggest that dysregulation of Klotho is centrally involved in the development of calcification and has provided a case to support potential therapeutic interference by restoration or supplementation of Klotho. Both endogenous tissue Klotho and soluble Klotho have been shown to exert anti-calcific effects: Hu et al. showed that transgenic CKD mice that overexpress Klotho had attenuated development of vascular calcification together with better renal function and enhanced phosphaturia, compared to wild-type mice with CKD [27]. Conversely, Klotho deficient mice with CKD developed severe calcification and worse renal function. Soluble Klotho directly suppressed Na-dependent uptake of phosphate and mineralization induced by high phosphate as well as preserving differentiation in VSMCs, in vitro. Similarly, intraperitoneal administration of recombinant Klotho [77], increased soluble Klotho by vitamin D receptor agonists (VDRA) treatment [69], and Klotho gene delivery were all associated with reduced vascular calcification [23].

Lim et al. were the first investigators to describe endogenous Klotho expression in human arteries and VSMCs [13]. They showed that CKD is a state of vascular Klotho deficiency that can be promoted by inflammatory, uremic and metabolic stressors. Klotho knockdown in VSMCs abrogated FGF23 mediated intracellular signaling and promoted the development of accelerated VSMC calcification, in vitro. Furthermore, restoration of Klotho deficiency by vitamin D receptor activators conferred responsiveness of VSMCs to potential FGF23 anti-calcific effects. Fang et al. showed that mice with early CKD by mild renal ablation developed a reduction in vascular Klotho expression together with vascular osteoblastic transition, increased osteocytic secreted proteins and inhibition of skeletal modeling, characteristic of mineral bone disorder (MBD) [78]. Several other studies have confirmed vascular expression of Klotho in various animal models [79–82], while others have not detected [83,84] or have not found changes in its vascular expression in CKD (Table 1) [50].

Table 1. Arterial Klotho expression in human and animal aortas. CKD, chronic kidney disease; VDRA, vitamin D receptor agonist.

Arterial Klotho Expression	Experimental Observations	Reference
Decreased mRNA and protein in human CKD	Human aorta Klotho deficiency in CKD can be reversed and calcification is attenuated ex vivo with VDRA	Lim et al. [13]
Decreased mRNA and protein in CKD mice	Low aortic Klotho but high circulating Klotho associated with vascular calcification in ldlr $^{-/-}$ CKD mice	Fang et al. [78]
mRNA but no protein in mouse aorta	Aortic Klotho has no role in vascular calcification	Lindberg et al. [79]
mRNA in human aorta, coronary arteries and thrombus	Klotho mRNA detectable in human arteries and thrombi of occlusive coronary disease	Donate-Correa et al. [81]
Increased mRNA and protein in calcified aorta of Enpp1$^{-/-}$ mice	Increased Klotho associated with decreased vascular calcification in CKD mice	Zhu et al. [82]
mRNA and protein expression in rat aorta but not in rat vascular smooth muscle cells	No native VSMC Klotho expression, however overexpression worsens calcification	Jimbo et al. [50]
No mRNA or protein expression in mouse aorta	VDRA in vivo increases plasma αKlotho an decreases vascular calcification in CKD mice	Lau et al. [69]
No mRNA in normal and calcified aortas of CKD mice	No aortic Klotho expression and no Klotho effect in vitro	Scialla et al. [83]

Deficiency of Klotho in VSMCs results in loss of smooth muscle cell contractile phenotype; similarly transformation of VSMCs from a contractile to a secretory phenotype has been associated with vascular Klotho deficiency [13]. These results suggest that endogenous Klotho expression is present only in VSMCs with a contractile phenotype.

3.2. Endothelial Dysfunction

Endothelial dysfunction is an early event in the development of atherosclerosis and encompasses a constellation of maladaptive alterations with a variety of implications, such as dysregulation of local vascular tone via regulation of nitric oxide (NO) availability, redox balance and orchestration of acute and chronic inflammatory reactions within the arterial wall [85]. Studies have shown that soluble Klotho decreases H_2O_2- and etoposide- induced apoptosis in human umbilical vascular endothelial cells (HUVECs). These anti-apoptotic effects occurred through the caspase-3/caspase-9 and p53/p21 pathways [64]. Six et al. showed that treatment of HUVECs with Klotho partially reverts FGF23-induced vasoconstriction, induced relaxation of preconstricted aorta by phosphate exposure and enhanced endothelial NO production [86].

Saito et al. showed that Klotho heterozygous mice exhibited attenuated aortic and arteriolar vasodilatation, however parabiosis between wild-type heterozygous Klotho mice restored endothelial function in heterozygous Klotho mice [87]. Additionally, heterozygous Klotho mice exhibited reduced nitric oxide metabolites (NO-2 and NO-3) in urine compared to wild-type mice, suggesting a decrease in NO production. In a related study, using the Otsuka Long-Evans Tokushima Fatty (OLETF) rat, an animal model with multiple atherogenic risk factors, adenovirus-mediated Klotho gene delivery ameliorated vascular endothelial dysfunction, increased nitric oxide production, reduced elevated blood pressure and prevented medial hypertrophy and perivascular fibrosis [88].

3.3. Oxidative Stress and Inflammation

Several studies have provided evidence that Klotho can suppress oxidative stress and inflammation, central processes firmly established in the development of vascular dysfunction, calcification and atherosclerosis. Klotho deficiency increases endogenous reactive oxygen species (ROS) generation and accentuates oxidative stress [89]. Conversely, overexpression of Klotho can decrease H_2O_2 induced-apoptosis, superoxide anion generation as well as β-galactosidase activity, mitochondrial DNA fragmentation, lipid peroxidation and Bax protein expression [58,89,90]. FOXO3a is a transcription factor that functions as a negative regulator of mitochondrial ROS generation [91]. It upregulates the expression of manganese superoxide dismutase (MnSOD), an important enzyme involved in mitochondrial antioxidant defense [22,59]. Klotho increases FOXO3a phosphorylation, suggesting that Klotho may suppress ROS-related oxidative stress. This is supported by observations that transgenic mice that overexpress Klotho have higher MnSOD expression and lower oxidative stress as evidenced by lower levels of urinary 8-hydroxy-2-deoxyguanosine, a marker of oxidative DNA damage [59,92,93]. Overexpression of Klotho or treatment with recombinant Klotho enhanced MnSOD expression, partially via activation of the cAMP signaling pathway [61]. In a study by Wang et al. The investigators found that Klotho gene transfer decreased nicotinamide adenine dinucleotide phosphate (NADPH) oxidase 2 (Nox2) protein expression, intracellular superoxide production and oxidative stress in rat aortic smooth muscle cells, in vitro [90]. Klotho gene expression also significantly attenuated angiotensin II (AngII)-induced superoxide production, oxidative damage and apoptosis. In another study, Klotho gene delivery in spontaneous hypertensive rats decreased upregulation of NADPH oxidase 2 activity and superoxide production and prevented the progression of spontaneous hypertension [94].

Anti-inflammatory actions of Klotho may underlie some of its vasculo-protective effects. Klotho protein has been shown to suppress the expression of intracellular adhesion molecule-1 (ICAM-1) and vascular cell adhesion molecule-1 (VCAM-1) in HUVECs exposed to tumour necrosis factor (TNF)-alpha [95]. These effects were associated with attenuation of nuclear factor (NF)-kappaB activation, IkappaB phosphorylation and inhibition of TNF-alpha induced monocyte adhesion. Liu et al. showed that intracellular, but not secreted Klotho interacts with retinoic-acid-inducible gene-I (RIG-I) thereby inhibiting RIG-I induced expression of interleukin (IL)-6 and IL-8 in senescent cells [96]. Using a senescence-accelerated mice P1 (SAMP1) aging model that developed aortic valve fibrosis, Chen et al. showed that adenovirus delivery of secreted Klotho inhibited inflammatory processes in aortic valves, including inhibition of monocyte chemoattractant protein-1 (MCP1), intercellualar adhesion molecule 1 (ICAM-1) expression, transforming growth factor (TGF)β upregulation, attenuated upregulation of tartrate-resistant acid phosphatase (TRAP) and matrix metallopeptidase (MMP)-2 expression and suppressed myofibroblastic transition [97].

4. Experimental Challenges and the Future of Klotho-Based Therapies

Despite the growing number of promising basic science discoveries over the past two decades supporting the therapeutic potential of Klotho to modulate CVD and other disease phenotypes, there are currently no clinical trials exploring the efficacy of Klotho-based therapies. However, several different Klotho-based delivery strategies have been explored. These include recombinant Klotho protein, gene therapy delivery of Klotho or small molecules that can enhance Klotho expression (Figure 2) [98,99]. These various strategies are now employed in an emerging landscape of Klotho-based biotechnology start-up companies for various indications, such as "Klotho Therapeutics" (https://www.klotho.com/), "Klogene" (http://www.klogene.com/) and "Unity Biotechnology" (https://unitybiotechnology.com/).

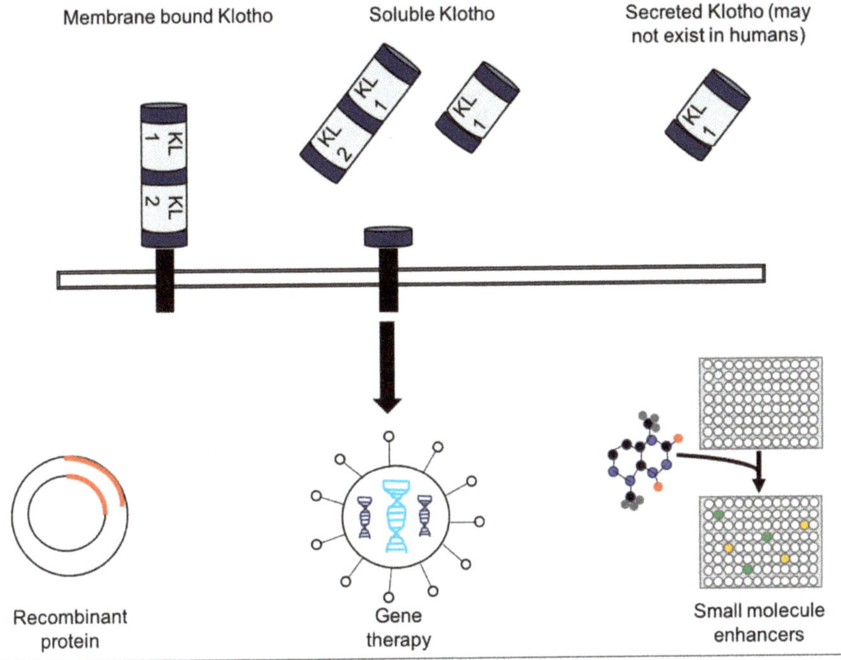

Figure 2. Potential delivery modalities of Klotho-based therapies. Full-length transmembrane Klotho is a ~135 kDa protein. Cleavage of full-length Klotho by membrane proteases (ADAM10 and ADAM17) in an α-cut generates a 130 kDa soluble isoform containing the KL1 and KL2 domains. Cleavage in a β-cut generates a 65 KDa isoform that contains only the KL1 domain. Recent evidence has challenged the existent of secreted Klotho by alternative splicing of Klotho mRNA. Various Klotho-based delivery strategies have been explored as illustrated.

Several critical considerations, gaps and challenges must first be overcome before Klotho-based interventions can be successfully translated into potential therapies for vascular diseases and beyond: Firstly, the precise functional role of the various Klotho isoforms in health and disease, their molecular mechanisms and the identity of the Klotho receptor for FGF23 independent effects remains to be elucidated. Therapeutic strategies will need to carefully target the isoform with most clinical benefit and carefully consider their application from a standpoint of prophylactic treatment (before overt clinical disease appears in high-risk patients) or reversal of existing vascular diseases. Given the pleiotropic effects of Klotho and its inherent large globular form, significant molecular analysis is still needed to identify the mechanisms and active sites of Klotho responsible for its various activities. This knowledge would form a critical platform on which successful pharmaceutical engineering can then be built.

From a clinical viewpoint, the precise concentrations of circulating Klotho and their specific isoforms that would be considered sufficient is still unknown and this is likely influenced by genetic variability between populations. Additionally, it is unknown whether supplementation above levels considered sufficient would be of therapeutic benefit or result in toxicity. In fact, high levels of Klotho may result in hypophosphatemic rickets and hyperparathyroidism [100]. These latter considerations are critical given that emerging concepts in precision medicine suggesting that individual responsiveness to therapeutic intervention is a function of naturally occurring genetic variants [101,102]. This would influence both the dosing and timing of Klotho delivery. Furthermore, studies are required to help guide the selection of patients who would qualify as good candidates for a Klotho-based intervention.

These experimental and clinical studies remain hampered by significant challenges with current antibody-based techniques for assessing Klotho: While commercially available antibodies against Klotho are available, most of them appear to be unspecific and cross-react with other proteins [103]. A novel synthetic anti-Klotho antibody (termed sb106) has been shown to detect Klotho in tissue and soluble Klotho in serum and urine, however this antibody is currently not commercially available [103,104]. Similarly, antibody-based assays for assessing serum Klotho levels have provided inconsistent results in CKD patients. In one study that compared a time-resolved fluorescence immunoassay (TRF, Cusabio, China) to an ELISA (IBL, Japan), surprisingly, no correlation was found between the assays and the levels of serum Klotho differed with by a factor 1000 [105]. At present, assessment of circulating Klotho using an immunoprecipitation-immunblot (IP-IB) assay has been shown to be highly correlated with glomerular filtration rate (GFR) in never-thawed serum samples of humans with varying severity of kidney disease compared to commercial ELISA [106]. Additionally, it is unclear whether commercial assays are detecting the 130 kDa and/or 65 kDa circulating isoforms.

A related technical challenge is that soluble Klotho appears to be highly unstable in blood and urine [104,107]. Prevention of degradation to conserve soluble Klotho, standardization of techniques and rigorous, in-house validation is therefore essential, however these have not been described. Similarly, generation of recombinant Klotho for experimental studies have been challenging and the variability and unpredictable quality of commercially available recombinant proteins may affect the reproducibility of reported effects [99]. Additionally, functional assays to detect Klotho activity are lacking [108]. Leveraging advances in next-generation sequencing technologies and Mendelian randomization studies are therefore imperative in the interim to first help identify genetic variants as instruments for strengthening causal inference in observation studies, while methodological improvements in antibody-based techniques and assays are being made.

5. Summary

The central thesis of this review was that accumulating evidence has stimulated significant interest and provided fundamental rationale for the therapeutic role of Klotho for age-related vascular diseases. As such, there is much reason for optimism toward the development of Klotho-based therapies. However, there are several significant gaps in our molecular and clinical understanding, as well as experimental challenges as discussed earlier. These gaps or hurdles must first be overcome before we can harness the clinical benefits of Klotho-based therapies as an elixir for vascular disease treatment or prevention. However, it seems likely that the pleiotropic nature of Klotho has brought together investigators from multiple different basic science and clinical disciplines that would have otherwise had traditionally disparate research emphases to work together toward a concerted strategy. What is interesting about a potential Klotho-based therapy is the possibility for a single drug to have multiple different disease indications. Transcending traditional barriers between disciplines offers immense opportunities for speeding innovative research that can address the growing burden of non-communicable diseases, in this case age-associated vascular diseases that remain a significant public health burden today.

Funding: This publication was supported by an NIH DK115683-01 award to K.L.

Conflicts of Interest: The authors declare no relevant conflict of interest.

Disclosures: K.L. is the recipient of an NIH career development award (K23 DK115683-01). I.Y.C. has received NHS funding from the RMH/ICR NIHR Biomedical Research Centre and research funding from Merck KGaA, Janssen, Red Trouser Day, the Thornton Foundation and the Royal Marsden Charity. A.A. declares that he is a consultant for AtlasMDX, Third Rock Ventures, Pfizer, Sun Pharma, Bluestar, TopoRx and ProLynx; is a member of the Scientific Advisory Board of Genentech, GLAdiator and Driver; receives grant support from Sun Pharma and AstraZeneca; is co-founder of Tango Therapeutics; and is a shareholder of Syncona.

References

1. Lakatta, E.G.; Levy, D. Arterial and cardiac aging: Major shareholders in cardiovascular disease enterprises: Part II: The aging heart in health: Links to heart disease. *Circulation* **2003**, *107*, 346–354. [CrossRef] [PubMed]
2. Stenvinkel, P.; Larsson, T.E. Chronic kidney disease: A clinical model of premature aging. *Am. J. Kidney Dis.* **2013**, *62*, 339–351. [CrossRef]
3. Lim, K.; Groen, A.; Molostvov, G.; Lu, T.; Lilley, K.S.; Snead, D.; James, S.; Wilkinson, I.B.; Ting, S.; Hsiao, L.L.; et al. alpha-Klotho Expression in Human Tissues. *J. Clin. Endocrinol. Metab.* **2015**, *100*, E1308–E1318. [CrossRef] [PubMed]
4. O'Rourke, M.F. Arterial aging: Pathophysiological principles. *Vasc. Med.* **2007**, *12*, 329–341. [CrossRef] [PubMed]
5. Kwan, G.F.; Benjamin, E.J. Global Health and Cardiovascular Disease. *Circulation* **2015**, *132*, 1217. [CrossRef]
6. Kanasaki, K.; Kitada, M.; Koya, D. Pathophysiology of the aging kidney and therapeutic interventions. *Hypertens Res.* **2012**, *35*, 1121–1128. [CrossRef]
7. Foley, R.N.; Parfrey, P.S.; Sarnak, M.J. Clinical epidemiology of cardiovascular disease in chronic renal disease. *Am. J. Kidney Dis.* **1998**, *32* (Suppl. 3), S112–S119. [CrossRef]
8. Hodes, R.J. *Aging Hearts and Arteries: A Scientific Quest*; National Institute of Aging: Bethesda, MA, USA, 1990.
9. Roderick, P.J.; Atkins, R.J.; Smeeth, L.; Mylne, A.; Nitsch, D.D.; Hubbard, R.B.; Bulpitt, C.J.; Fletcher, A.E. CKD and mortality risk in older people: A community-based population study in the United Kingdom. *Am. J. Kidney Dis.* **2009**, *53*, 950–960. [CrossRef]
10. Kahn, M.R.; Robbins, M.J.; Kim, M.C.; Fuster, V. Management of cardiovascular disease in patients with kidney disease. *Nat. Rev. Cardiol.* **2013**, *10*, 261–273. [CrossRef]
11. Verbrugge, F.H.; Tang, W.H.; Hazen, S.L. Protein carbamylation and cardiovascular disease. *Kidney Int.* **2015**, *88*, 474–478. [CrossRef]
12. Rhee, E.P.; Gerszten, R.E. Metabolomics and cardiovascular biomarker discovery. *Clin. Chem.* **2012**, *58*, 139–147. [CrossRef] [PubMed]
13. Lim, K.; Lu, T.S.; Molostvov, G.; Lee, C.; Lam, F.T.; Zehnder, D.; Hsiao, L.L. Vascular Klotho deficiency potentiates the development of human artery calcification and mediates resistance to fibroblast growth factor 23. *Circulation* **2012**, *125*, 2243–2255. [CrossRef] [PubMed]
14. Guzik, T.J.; Touyz, R.M. Oxidative Stress, Inflammation, and Vascular Aging in Hypertension. *Hypertension* **2017**, *70*, 660–667. [CrossRef] [PubMed]
15. Libby, P.; Kobold, S. Inflammation: A common contributor to cancer, aging, and cardiovascular diseases-expanding the concept of cardio-oncology. *Cardiovasc. Res.* **2019**, *115*, 824–829. [CrossRef] [PubMed]
16. Cupit-Link, M.C.; Kirkland, J.L.; Ness, K.K.; Armstrong, G.T.; Tchkonia, T.; LeBrasseur, N.K.; Armenian, S.H.; Ruddy, K.J.; Hashmi, S.K. Biology of premature ageing in survivors of cancer. *ESMO Open* **2017**, *2*, e000250. [CrossRef] [PubMed]
17. Tolle, M.; Reshetnik, A.; Schuchardt, M.; Hohne, M.; van der Giet, M. Arteriosclerosis and vascular calcification: Causes, clinical assessment and therapy. *Eur. J. Clin. Investig.* **2015**, *45*, 976–985. [CrossRef]
18. London, G.; Covic, A.; Goldsmith, D.; Wiecek, A.; Suleymanlar, G.; Ortiz, A.; Massy, Z.; Lindholm, B.; Martinez-Castelao, A.; Fliser, D.; et al. Arterial aging and arterial disease: Interplay between central hemodynamics, cardiac work, and organ flow-implications for CKD and cardiovascular disease. *Kidney Int. Suppl.* **2011**, *1*, 10–12. [CrossRef]
19. Najjar, S.S.; Scuteri, A.; Lakatta, E.G. Arterial aging: Is it an immutable cardiovascular risk factor? *Hypertension* **2005**, *46*, 454–462. [CrossRef]
20. Lakatta, E.G.; Levy, D. Arterial and cardiac aging: Major shareholders in cardiovascular disease enterprises: Part I: Aging arteries: A "set up" for vascular disease. *Circulation* **2003**, *107*, 139–146. [CrossRef]
21. Kuro-o, M.; Matsumura, Y.; Aizawa, H.; Kawaguchi, H.; Suga, T.; Utsugi, T.; Ohyama, Y.; Kurabayashi, M.; Kaname, T.; Kume, E.; et al. Mutation of the mouse klotho gene leads to a syndrome resembling ageing. *Nature* **1997**, *390*, 45–51. [CrossRef]
22. Kurosu, H.; Yamamoto, M.; Clark, J.D.; Pastor, J.V.; Nandi, A.; Gurnani, P.; McGuinness, O.P.; Chikuda, H.; Yamaguchi, M.; Kawaguchi, H.; et al. Suppression of aging in mice by the hormone Klotho. *Science* **2005**, *309*, 1829–1833. [CrossRef]

23. Masuda, H.; Chikuda, H.; Suga, T.; Kawaguchi, H.; Kuro-o, M. Regulation of multiple ageing-like phenotypes by inducible klotho gene expression in klotho mutant mice. *Mech. Ageing Dev.* **2005**, *126*, 1274–1283. [CrossRef]
24. Drew, D.A.; Katz, R.; Kritchevsky, S.; Ix, J.; Shlipak, M.; Gutierrez, O.M.; Newman, A.; Hoofnagle, A.; Fried, L.; Semba, R.D.; et al. Association between Soluble Klotho and Change in Kidney Function: The Health Aging and Body Composition Study. *J. Am. Soc. Nephrol.* **2017**, *28*, 1859–1866. [CrossRef]
25. Wang, Q.; Su, W.; Shen, Z.; Wang, R. Correlation between Soluble alpha-Klotho and Renal Function in Patients with Chronic Kidney Disease: A Review and Meta-Analysis. *Biomed. Res. Int.* **2018**, *2018*, 9481475.
26. Memmos, E.; Sarafidis, P.; Pateinakis, P.; Tsiantoulas, A.; Faitatzidou, D.; Giamalis, P.; Vasilikos, V.; Papagianni, A. Soluble Klotho is associated with mortality and cardiovascular events in hemodialysis. *BMC Nephrol.* **2019**, *20*, 217. [CrossRef]
27. Hu, M.C.; Shi, M.; Zhang, J.; Quinones, H.; Griffith, C.; Kuro-o, M.; Moe, O.W. Klotho deficiency causes vascular calcification in chronic kidney disease. *J. Am. Soc. Nephrol.* **2011**, *22*, 124–136. [CrossRef]
28. Pan, H.C.; Chou, K.M.; Lee, C.C.; Yang, N.I.; Sun, C.Y. Circulating Klotho levels can predict long-term macrovascular outcomes in type 2 diabetic patients. *Atherosclerosis* **2018**, *276*, 83–90. [CrossRef]
29. Matsumura, Y.; Aizawa, H.; Shiraki-Iida, T.; Nagai, R.; Kuro-o, M.; Nabeshima, Y. Identification of the human klotho gene and its two transcripts encoding membrane and secreted klotho protein. *Biochem. Biophys. Res. Commun.* **1998**, *242*, 626–630. [CrossRef]
30. Hu, M.C.; Shi, M.; Cho, H.J.; Adams-Huet, B.; Paek, J.; Hill, K.; Shelton, J.; Amaral, A.P.; Faul, C.; Taniguchi, M.; et al. Klotho and phosphate are modulators of pathologic uremic cardiac remodeling. *J. Am. Soc. Nephrol.* **2015**, *26*, 1290–1302. [CrossRef]
31. Xu, Y.; Sun, Z. Molecular basis of Klotho: From gene to function in aging. *Endocr. Rev.* **2015**, *36*, 174–193. [CrossRef]
32. Chen, C.D.; Podvin, S.; Gillespie, E.; Leeman, S.E.; Abraham, C.R. Insulin stimulates the cleavage and release of the extracellular domain of Klotho by ADAM10 and ADAM17. *Proc. Natl. Acad. Sci. USA* **2007**, *104*, 19796–19801. [CrossRef]
33. Bloch, L.; Sineshchekova, O.; Reichenbach, D.; Reiss, K.; Saftig, P.; Kuro-o, M.; Kaether, C. Klotho is a substrate for alpha-, beta- and gamma-secretase. *FEBS Lett.* **2009**, *583*, 3221–3224. [CrossRef]
34. Shiraki-Iida, T.; Aizawa, H.; Matsumura, Y.; Sekine, S.; Iida, A.; Anazawa, H.; Nagai, R.; Kuro-o, M.; Nabeshima, Y. Structure of the mouse klotho gene and its two transcripts encoding membrane and secreted protein. *FEBS Lett.* **1998**, *424*, 6–10. [CrossRef]
35. Mencke, R.; Harms, G.; Moser, J.; van Meurs, M.; Diepstra, A.; Leuvenink, H.G.; Hillebrands, J.L. Human alternative Klotho mRNA is a nonsense-mediated mRNA decay target inefficiently spliced in renal disease. *JCI Insight* **2017**, *2*. [CrossRef]
36. Tohyama, O.; Imura, A.; Iwano, A.; Freund, J.N.; Henrissat, B.; Fujimori, T.; Nabeshima, Y. Klotho is a novel beta-glucuronidase capable of hydrolyzing steroid beta-glucuronides. *J. Biol. Chem.* **2004**, *279*, 9777–9784. [CrossRef]
37. Chen, G.; Liu, Y.; Goetz, R.; Fu, L.; Jayaraman, S.; Hu, M.C.; Moe, O.W.; Liang, G.; Li, X.; Mohammadi, M. alpha-Klotho is a non-enzymatic molecular scaffold for FGF23 hormone signalling. *Nature* **2018**, *553*, 461–466. [CrossRef]
38. Hu, M.C.; Shi, M.; Zhang, J.; Quinones, H.; Kuro-o, M.; Moe, O.W. Klotho deficiency is an early biomarker of renal ischemia-reperfusion injury and its replacement is protective. *Kidney Int.* **2010**, *78*, 1240–1251. [CrossRef]
39. Lindberg, K.; Amin, R.; Moe, O.W.; Hu, M.C.; Erben, R.G.; Ostman Wernerson, A.; Lanske, B.; Olauson, H.; Larsson, T.E. The kidney is the principal organ mediating klotho effects. *J. Am. Soc. Nephrol.* **2014**, *25*, 2169–2175. [CrossRef]
40. Kakareko, K.; Rydzewska-Rosolowska, A.; Brzosko, S.; Gozdzikiewicz-Lapinska, J.; Koc-Zorawska, E.; Samocik, P.; Kozlowski, R.; Mysliwiec, M.; Naumnik, B.; Hryszko, T. The effect of nephrectomy on Klotho, FGF-23 and bone metabolism. *Int. Urol. Nephrol.* **2017**, *49*, 681–688. [CrossRef]
41. Lim, K.; Hamano, T.; Thadhani, R. Vitamin D and Calcimimetics in Cardiovascular Disease. *Semin. Nephrol.* **2018**, *38*, 251–266. [CrossRef]
42. Liu, S.; Zhou, J.; Tang, W.; Jiang, X.; Rowe, D.W.; Quarles, L.D. Pathogenic role of Fgf23 in Hyp mice. *Am. J. Physiol. Endocrinol. Metab.* **2006**, *291*, E38–E49. [CrossRef] [PubMed]

43. Feng, J.Q.; Ward, L.M.; Liu, S.; Lu, Y.; Xie, Y.; Yuan, B.; Yu, X.; Rauch, F.; Davis, S.I.; Zhang, S.; et al. Loss of DMP1 causes rickets and osteomalacia and identifies a role for osteocytes in mineral metabolism. *Nat. Genet.* **2006**, *38*, 1310–1315. [CrossRef] [PubMed]
44. Shimada, T.; Kakitani, M.; Yamazaki, Y.; Hasegawa, H.; Takeuchi, Y.; Fujita, T.; Fukumoto, S.; Tomizuka, K.; Yamashita, T. Targeted ablation of Fgf23 demonstrates an essential physiological role of FGF23 in phosphate and vitamin D metabolism. *J. Clin. Investig.* **2004**, *113*, 561–568. [CrossRef] [PubMed]
45. Tsujikawa, H.; Kurotaki, Y.; Fujimori, T.; Fukuda, K.; Nabeshima, Y. Klotho, a gene related to a syndrome resembling human premature aging, functions in a negative regulatory circuit of vitamin D endocrine system. *Mol. Endocrinol.* **2003**, *17*, 2393–2403. [CrossRef] [PubMed]
46. Stubbs, J.R.; Liu, S.; Tang, W.; Zhou, J.; Wang, Y.; Yao, X.; Quarles, L.D. Role of hyperphosphatemia and 1,25-dihydroxyvitamin D in vascular calcification and mortality in fibroblastic growth factor 23 null mice. *J. Am. Soc. Nephrol.* **2007**, *18*, 2116–2124. [CrossRef]
47. Urakawa, I.; Yamazaki, Y.; Shimada, T.; Iijima, K.; Hasegawa, H.; Okawa, K.; Fujita, T.; Fukumoto, S.; Yamashita, T. Klotho converts canonical FGF receptor into a specific receptor for FGF23. *Nature* **2006**, *444*, 770–774. [CrossRef] [PubMed]
48. Goetz, R.; Beenken, A.; Ibrahimi, O.A.; Kalinina, J.; Olsen, S.K.; Eliseenkova, A.V.; Xu, C.; Neubert, T.A.; Zhang, F.; Linhardt, R.J.; et al. Molecular insights into the klotho-dependent, endocrine mode of action of fibroblast growth factor 19 subfamily members. *Mol. Cell Biol.* **2007**, *27*, 3417–3428. [CrossRef]
49. Goetz, R.; Nakada, Y.; Hu, M.C.; Kurosu, H.; Wang, L.; Nakatani, T.; Shi, M.; Eliseenkova, A.V.; Razzaque, M.S.; Moe, O.W.; et al. Isolated C-terminal tail of FGF23 alleviates hypophosphatemia by inhibiting FGF23-FGFR-Klotho complex formation. *Proc. Natl. Acad. Sci. USA* **2010**, *107*, 407–412. [CrossRef]
50. Jimbo, R.; Kawakami-Mori, F.; Mu, S.; Hirohama, D.; Majtan, B.; Shimizu, Y.; Yatomi, Y.; Fukumoto, S.; Fujita, T.; Shimosawa, T. Fibroblast growth factor 23 accelerates phosphate-induced vascular calcification in the absence of Klotho deficiency. *Kidney Int.* **2014**, *85*, 1103–1111. [CrossRef]
51. Murer, H.; Forster, I.; Biber, J. The sodium phosphate cotransporter family SLC34. *Pflugers Arch.* **2004**, *447*, 763–767. [CrossRef]
52. Meyer, M.B.; Benkusky, N.A.; Kaufmann, M.; Lee, S.M.; Onal, M.; Jones, G.; Pike, J.W. A kidney-specific genetic control module in mice governs endocrine regulation of the cytochrome P450 gene Cyp27b1 essential for vitamin D3 activation. *J. Biol. Chem.* **2017**, *292*, 17541–17558. [CrossRef]
53. Di Marco, G.S.; Hausberg, M.; Hillebrand, U.; Rustemeyer, P.; Wittkowski, W.; Lang, D.; Pavenstadt, H. Increased inorganic phosphate induces human endothelial cell apoptosis in vitro. *Am. J. Physiol. Renal Physiol.* **2008**, *294*, F1381–F1387. [CrossRef]
54. Jono, S.; McKee, M.D.; Murry, C.E.; Shioi, A.; Nishizawa, Y.; Mori, K.; Morii, H.; Giachelli, C.M. Phosphate regulation of vascular smooth muscle cell calcification. *Circ. Res.* **2000**, *87*, E10–E17. [CrossRef]
55. Kuro, O.M. Molecular Mechanisms Underlying Accelerated Aging by Defects in the FGF23-Klotho System. *Int. J. Nephrol.* **2018**, *2018*, 9679841. [CrossRef]
56. Kuro, O.M. The Klotho proteins in health and disease. *Nat. Rev. Nephrol.* **2019**, *15*, 27–44. [CrossRef]
57. Reiser, J.; Polu, K.R.; Moller, C.C.; Kenlan, P.; Altintas, M.M.; Wei, C.; Faul, C.; Herbert, S.; Villegas, I.; Avila-Casado, C.; et al. TRPC6 is a glomerular slit diaphragm-associated channel required for normal renal function. *Nat. Genet.* **2005**, *37*, 739–744. [CrossRef]
58. Haruna, Y.; Kashihara, N.; Satoh, M.; Tomita, N.; Namikoshi, T.; Sasaki, T.; Fujimori, T.; Xie, P.; Kanwar, Y.S. Amelioration of progressive renal injury by genetic manipulation of Klotho gene. *Proc. Natl. Acad. Sci. USA* **2007**, *104*, 2331–2336. [CrossRef]
59. Yamamoto, M.; Clark, J.D.; Pastor, J.V.; Gurnani, P.; Nandi, A.; Kurosu, H.; Miyoshi, M.; Ogawa, Y.; Castrillon, D.H.; Rosenblatt, K.P.; et al. Regulation of oxidative stress by the anti-aging hormone klotho. *J. Biol. Chem.* **2005**, *280*, 38029–38034. [CrossRef]
60. Zhao, Y.; Zhao, M.M.; Cai, Y.; Zheng, M.F.; Sun, W.L.; Zhang, S.Y.; Kong, W.; Gu, J.; Wang, X.; Xu, M.J. Mammalian target of rapamycin signaling inhibition ameliorates vascular calcification via Klotho upregulation. *Kidney Int.* **2015**, *88*, 711–721. [CrossRef]
61. Rakugi, H.; Matsukawa, N.; Ishikawa, K.; Yang, J.; Imai, M.; Ikushima, M.; Maekawa, Y.; Kida, I.; Miyazaki, J.; Ogihara, T. Anti-oxidative effect of Klotho on endothelial cells through cAMP activation. *Endocrine* **2007**, *31*, 82–87. [CrossRef]

62. Imai, M.; Ishikawa, K.; Matsukawa, N.; Kida, I.; Ohta, J.; Ikushima, M.; Chihara, Y.; Rui, X.; Rakugi, H.; Ogihara, T. Klotho protein activates the PKC pathway in the kidney and testis and suppresses 25-hydroxyvitamin D3 1alpha-hydroxylase gene expression. *Endocrine* **2004**, *25*, 229–234. [CrossRef]
63. Doi, S.; Zou, Y.; Togao, O.; Pastor, J.V.; John, G.B.; Wang, L.; Shiizaki, K.; Gotschall, R.; Schiavi, S.; Yorioka, N.; et al. Klotho inhibits transforming growth factor-beta1 (TGF-beta1) signaling and suppresses renal fibrosis and cancer metastasis in mice. *J. Biol. Chem.* **2011**, *286*, 8655–8665. [CrossRef] [PubMed]
64. Ikushima, M.; Rakugi, H.; Ishikawa, K.; Maekawa, Y.; Yamamoto, K.; Ohta, J.; Chihara, Y.; Kida, I.; Ogihara, T. Anti-apoptotic and anti-senescence effects of Klotho on vascular endothelial cells. *Biochem. Biophys. Res. Commun.* **2006**, *339*, 827–832. [CrossRef] [PubMed]
65. Liu, H.; Fergusson, M.M.; Castilho, R.M.; Liu, J.; Cao, L.; Chen, J.; Malide, D.; Rovira, I.I.; Schimel, D.; Kuo, C.J.; et al. Augmented Wnt signaling in a mammalian model of accelerated aging. *Science* **2007**, *317*, 803–806. [CrossRef] [PubMed]
66. Semba, R.D.; Cappola, A.R.; Sun, K.; Bandinelli, S.; Dalal, M.; Crasto, C.; Guralnik, J.M.; Ferrucci, L. Plasma klotho and mortality risk in older community-dwelling adults. *J. Gerontol. A Biol. Sci. Med. Sci.* **2011**, *66*, 794–800. [CrossRef] [PubMed]
67. Arking, D.E.; Atzmon, G.; Arking, A.; Barzilai, N.; Dietz, H.C. Association between a functional variant of the KLOTHO gene and high-density lipoprotein cholesterol, blood pressure, stroke, and longevity. *Circ. Res.* **2005**, *96*, 412–418. [CrossRef] [PubMed]
68. Semba, R.D.; Cappola, A.R.; Sun, K.; Bandinelli, S.; Dalal, M.; Crasto, C.; Guralnik, J.M.; Ferrucci, L. Plasma klotho and cardiovascular disease in adults. *J. Am. Geriatr. Soc.* **2011**, *59*, 1596–1601. [CrossRef] [PubMed]
69. Lau, W.L.; Leaf, E.M.; Hu, M.C.; Takeno, M.M.; Kuro-o, M.; Moe, O.W.; Giachelli, C.M. Vitamin D receptor agonists increase klotho and osteopontin while decreasing aortic calcification in mice with chronic kidney disease fed a high phosphate diet. *Kidney Int.* **2012**, *82*, 1261–1270. [CrossRef]
70. Shanahan, C.M.; Crouthamel, M.H.; Kapustin, A.; Giachelli, C.M. Arterial calcification in chronic kidney disease: Key roles for calcium and phosphate. *Circ. Res.* **2011**, *109*, 697–711. [CrossRef]
71. Ragnauth, C.D.; Warren, D.T.; Liu, Y.; McNair, R.; Tajsic, T.; Figg, N.; Shroff, R.; Skepper, J.; Shanahan, C.M. Prelamin A acts to accelerate smooth muscle cell senescence and is a novel biomarker of human vascular aging. *Circulation* **2010**, *121*, 2200–2210. [CrossRef]
72. Reynolds, J.L.; Joannides, A.J.; Skepper, J.N.; McNair, R.; Schurgers, L.J.; Proudfoot, D.; Jahnen-Dechent, W.; Weissberg, P.L.; Shanahan, C.M. Human vascular smooth muscle cells undergo vesicle-mediated calcification in response to changes in extracellular calcium and phosphate concentrations: A potential mechanism for accelerated vascular calcification in ESRD. *J. Am. Soc. Nephrol.* **2004**, *15*, 2857–2867. [CrossRef] [PubMed]
73. Shaw, L.J.; Raggi, P.; Berman, D.S.; Callister, T.Q. Coronary artery calcium as a measure of biologic age. *Atherosclerosis* **2006**, *188*, 112–119. [CrossRef] [PubMed]
74. Girndt, M.; Seibert, E. Premature cardiovascular disease in chronic renal failure (CRF): A model for an advanced ageing process. *Exp. Gerontol.* **2010**, *45*, 797–800. [CrossRef] [PubMed]
75. Blacher, J.; Safar, M.E.; Guerin, A.P.; Pannier, B.; Marchais, S.J.; London, G.M. Aortic pulse wave velocity index and mortality in end-stage renal disease. *Kidney Int.* **2003**, *63*, 1852–1860. [CrossRef] [PubMed]
76. Debacq-Chainiaux, F.; Erusalimsky, J.D.; Campisi, J.; Toussaint, O. Protocols to detect senescence-associated beta-galactosidase (SA-betagal) activity, a biomarker of senescent cells in culture and in vivo. *Nat. Protoc.* **2009**, *4*, 1798–1806. [CrossRef] [PubMed]
77. Chen, T.H.; Kuro, O.M.; Chen, C.H.; Sue, Y.M.; Chen, Y.C.; Wu, H.H.; Cheng, C.Y. The secreted Klotho protein restores phosphate retention and suppresses accelerated aging in Klotho mutant mice. *Eur. J. Pharmacol.* **2013**, *698*, 67–73. [CrossRef] [PubMed]
78. Fang, Y.; Ginsberg, C.; Sugatani, T.; Monier-Faugere, M.C.; Malluche, H.; Hruska, K.A. Early chronic kidney disease-mineral bone disorder stimulates vascular calcification. *Kidney Int.* **2014**, *85*, 142–150. [CrossRef] [PubMed]
79. Lindberg, K.; Olauson, H.; Amin, R.; Ponnusamy, A.; Goetz, R.; Taylor, R.F.; Mohammadi, M.; Canfield, A.; Kublickiene, K.; Larsson, T.E. Arterial klotho expression and FGF23 effects on vascular calcification and function. *PLoS ONE* **2013**, *8*, e60658. [CrossRef]
80. Yamazaki, M.; Ozono, K.; Okada, T.; Tachikawa, K.; Kondou, H.; Ohata, Y.; Michigami, T. Both FGF23 and extracellular phosphate activate Raf/MEK/ERK pathway via FGF receptors in HEK293 cells. *J. Cell. Biochem.* **2010**, *111*, 1210–1221. [CrossRef]

81. Donate-Correa, J.; Mora-Fernandez, C.; Martinez-Sanz, R.; Muros-de-Fuentes, M.; Perez, H.; Meneses-Perez, B.; Cazana-Perez, V.; Navarro-Gonzalez, J.F. Expression of FGF23/KLOTHO system in human vascular tissue. *Int. J. Cardiol.* **2013**, *165*, 179–183. [CrossRef]
82. Zhu, D.; Mackenzie, N.C.; Millan, J.L.; Farquharson, C.; MacRae, V.E. A protective role for FGF-23 in local defence against disrupted arterial wall integrity? *Mol. Cell. Endocrinol.* **2013**, *372*, 1–11. [CrossRef] [PubMed]
83. Scialla, J.J.; Lau, W.L.; Reilly, M.P.; Isakova, T.; Yang, H.Y.; Crouthamel, M.H.; Chavkin, N.W.; Rahman, M.; Wahl, P.; Amaral, A.P.; et al. Fibroblast growth factor 23 is not associated with and does not induce arterial calcification. *Kidney Int.* **2013**, *83*, 1159–1168. [CrossRef] [PubMed]
84. Navarro-Gonzalez, J.F.; Donate-Correa, J.; Muros de Fuentes, M.; Perez-Hernandez, H.; Martinez-Sanz, R.; Mora-Fernandez, C. Reduced Klotho is associated with the presence and severity of coronary artery disease. *Heart* **2014**, *100*, 34–40. [CrossRef] [PubMed]
85. Gimbrone, M.A., Jr.; Garcia-Cardena, G. Endothelial Cell Dysfunction and the Pathobiology of Atherosclerosis. *Circ. Res.* **2016**, *118*, 620–636. [CrossRef] [PubMed]
86. Six, I.; Okazaki, H.; Gross, P.; Cagnard, J.; Boudot, C.; Maizel, J.; Drueke, T.B.; Massy, Z.A. Direct, acute effects of Klotho and FGF23 on vascular smooth muscle and endothelium. *PLoS ONE* **2014**, *9*, e93423. [CrossRef] [PubMed]
87. Saito, Y.; Yamagishi, T.; Nakamura, T.; Ohyama, Y.; Aizawa, H.; Suga, T.; Matsumura, Y.; Masuda, H.; Kurabayashi, M.; Kuro-o, M.; et al. Klotho protein protects against endothelial dysfunction. *Biochem. Biophys. Res. Commun.* **1998**, *248*, 324–329. [CrossRef] [PubMed]
88. Saito, Y.; Nakamura, T.; Ohyama, Y.; Suzuki, T.; Iida, A.; Shiraki-Iida, T.; Kuro-o, M.; Nabeshima, Y.; Kurabayashi, M.; Nagai, R. In vivo klotho gene delivery protects against endothelial dysfunction in multiple risk factor syndrome. *Biochem. Biophys. Res. Commun.* **2000**, *276*, 767–772. [CrossRef] [PubMed]
89. Izbeki, F.; Asuzu, D.T.; Lorincz, A.; Bardsley, M.R.; Popko, L.N.; Choi, K.M.; Young, D.L.; Hayashi, Y.; Linden, D.R.; Kuro-o, M.; et al. Loss of Kitlow progenitors, reduced stem cell factor and high oxidative stress underlie gastric dysfunction in progeric mice. *J. Physiol.* **2010**, *588 Pt 16*, 3101–3117. [CrossRef]
90. Wang, Y.; Kuro-o, M.; Sun, Z. Klotho gene delivery suppresses Nox2 expression and attenuates oxidative stress in rat aortic smooth muscle cells via the cAMP-PKA pathway. *Aging Cell* **2012**, *11*, 410–417. [CrossRef]
91. Emerling, B.M.; Weinberg, F.; Liu, J.L.; Mak, T.W.; Chandel, N.S. PTEN regulates p300-dependent hypoxia-inducible factor 1 transcriptional activity through Forkhead transcription factor 3a (FOXO3a). *Proc. Natl. Acad. Sci. USA* **2008**, *105*, 2622–2627. [CrossRef]
92. Mitobe, M.; Yoshida, T.; Sugiura, H.; Shirota, S.; Tsuchiya, K.; Nihei, H. Oxidative stress decreases klotho expression in a mouse kidney cell line. *Nephron Exp. Nephrol.* **2005**, *101*, e67–e74. [CrossRef] [PubMed]
93. Chen, B.; Wang, X.; Zhao, W.; Wu, J. Klotho inhibits growth and promotes apoptosis in human lung cancer cell line A549. *J. Exp. Clin. Cancer Res.* **2010**, *29*, 99. [CrossRef] [PubMed]
94. Wang, Y.; Sun, Z. Klotho gene delivery prevents the progression of spontaneous hypertension and renal damage. *Hypertension* **2009**, *54*, 810–817. [CrossRef] [PubMed]
95. Maekawa, Y.; Ishikawa, K.; Yasuda, O.; Oguro, R.; Hanasaki, H.; Kida, I.; Takemura, Y.; Ohishi, M.; Katsuya, T.; Rakugi, H. Klotho suppresses TNF-alpha-induced expression of adhesion molecules in the endothelium and attenuates NF-kappaB activation. *Endocrine* **2009**, *35*, 341–346. [CrossRef]
96. Liu, F.; Wu, S.; Ren, H.; Gu, J. Klotho suppresses RIG-I-mediated senescence-associated inflammation. *Nat. Cell Biol.* **2011**, *13*, 254–262. [CrossRef] [PubMed]
97. Chen, J.; Fan, J.; Wang, S.; Sun, Z. Secreted Klotho Attenuates Inflammation-Associated Aortic Valve Fibrosis in Senescence-Accelerated Mice P1. *Hypertension* **2018**, *71*, 877–885. [CrossRef] [PubMed]
98. Abraham, C.R.; Chen, C.; Cuny, G.D.; Glicksman, M.A.; Zeldich, E. Small-molecule Klotho enhancers as novel treatment of neurodegeneration. *Future Med. Chem.* **2012**, *4*, 1671–1679. [CrossRef]
99. Cheikhi, A.; Barchowsky, A.; Sahu, A.; Shinde, S.N.; Pius, A.; Clemens, Z.J.; Li, H.; Kennedy, C.A.; Hoeck, J.D.; Franti, M.; et al. Klotho: An Elephant in Aging Research. *J. Gerontol. A Biol. Sci. Med. Sci.* **2019**, *74*, 1031–1042. [CrossRef] [PubMed]
100. Brownstein, C.A.; Adler, F.; Nelson-Williams, C.; Iijima, J.; Li, P.; Imura, A.; Nabeshima, Y.; Reyes-Mugica, M.; Carpenter, T.O.; Lifton, R.P. A translocation causing increased alpha-klotho level results in hypophosphatemic rickets and hyperparathyroidism. *Proc. Natl. Acad. Sci. USA* **2008**, *105*, 3455–3460. [CrossRef]
101. Smart, A.; Martin, P. The promise of pharmacogenetics: Assessing the prospects for disease and patient stratification. *Stud. Hist. Philos. Biol. Biomed. Sci.* **2006**, *37*, 583–601. [CrossRef]

102. Gat-Viks, I.; Chevrier, N.; Wilentzik, R.; Eisenhaure, T.; Raychowdhury, R.; Steuerman, Y.; Shalek, A.K.; Hacohen, N.; Amit, I.; Regev, A. Deciphering molecular circuits from genetic variation underlying transcriptional responsiveness to stimuli. *Nat. Biotechnol.* **2013**, *31*, 342–349. [CrossRef] [PubMed]
103. Barker, S.L.; Pastor, J.; Carranza, D.; Quinones, H.; Griffith, C.; Goetz, R.; Mohammadi, M.; Ye, J.; Zhang, J.; Hu, M.C.; et al. The demonstration of alphaKlotho deficiency in human chronic kidney disease with a novel synthetic antibody. *Nephrol. Dial. Transplant.* **2015**, *30*, 223–233. [CrossRef] [PubMed]
104. Hu, M.C.; Shi, M.; Zhang, J.; Addo, T.; Cho, H.J.; Barker, S.L.; Ravikumar, P.; Gillings, N.; Bian, A.; Sidhu, S.S.; et al. Renal Production, Uptake, and Handling of Circulating alphaKlotho. *J. Am. Soc. Nephrol.* **2016**, *27*, 79–90. [CrossRef] [PubMed]
105. Pedersen, L.; Pedersen, S.M.; Brasen, C.L.; Rasmussen, L.M. Soluble serum Klotho levels in healthy subjects. Comparison of two different immunoassays. *Clin. Biochem.* **2013**, *46*, 1079–1083. [CrossRef] [PubMed]
106. Neyra, J.A.; Moe, O.W.; Pastor, J.; Gianella, F.; Sidhu, S.S.; Sarnak, M.J.; Ix, J.H.; Drew, D.A. Performance of soluble Klotho assays in clinical samples of kidney disease. *Clin. Kidney J.* **2019**. [CrossRef]
107. Adema, A.Y.; Vervloet, M.G.; Blankenstein, M.A.; Heijboer, A.C. alpha-Klotho is unstable in human urine. *Kidney Int.* **2015**, *88*, 1442–1444. [CrossRef] [PubMed]
108. Grabner, A.; Faul, C. The role of fibroblast growth factor 23 and Klotho in uremic cardiomyopathy. *Curr. Opin. Nephrol. Hypertens* **2016**, *25*, 314–324. [CrossRef]

International Journal of
Molecular Sciences

Review

Cardiotonic Steroids—A Possible Link Between High-Salt Diet and Organ Damage

Aneta Paczula, Andrzej Wiecek and Grzegorz Piecha *

Department of Nephrology, Transplantation and Internal Medicine, Medical University of Silesia, Francuska 20-24, 40-027 Katowice, Poland; paczula@poczta.onet.pl (A.P.); awiecek@sum.edu.pl (A.W.)
* Correspondence: g.piecha@outlook.com; Tel.: +48-32-2552695; Fax: +48-32-2553726

Received: 31 December 2018; Accepted: 28 January 2019; Published: 30 January 2019

Abstract: High dietary salt intake has been listed among the top ten risk factors for disability-adjusted life years. We discuss the role of endogenous cardiotonic steroids in mediating the dietary salt-induced hypertension and organ damage.

Keywords: marinobufagenin; ouabain; salt; hypertension; fibrosis

1. Salt and Blood Pressure

High dietary salt intake has been listed among the top ten risk factors for disability-adjusted life years (DALYs) [1]. The role of dietary salt in the pathogenesis of increased blood pressure has been demonstrated by several large clinical trials, such as the International Study of Salt and Blood Pressure (INTERSALT) [2] and the Dietary Approaches to Stop Hypertension (DASH) study [3]. High salt consumption is associated with increased blood pressure (BP) and vascular stiffening due to altered endothelial and vascular smooth muscle cells (VSMCs) function and extended arterial wall fibrosis [4–6]. Notably, high dietary salt intake correlates positively with a faster pulse wave velocity (PWV), indicating arterial stiffening, which precedes the development of hypertension with aging [7–9]. Conversely, dietary salt restriction is accompanied by a reduction in PWV, indicating less arterial stiffening [10].

Salt/sodium is absolutely necessary for survival. As the availability of salt/sodium is scarce in nature outside the oceans, the mechanisms for salt conservation are very efficient and well known. However, the mechanisms for elimination of excess salt are less understood.

Laragh et al. postulated two forms of essential hypertension: one related to vasoconstriction (largely the result of the renin-angiotensin system activation) and the other form due to volume expansion (excess salt and water) in which plasma renin activity is suppressed [11].

An unresolved issue in the pathogenesis of hypertension is the specific mechanism or "signaling pathway" by which salt retention elevates the blood pressure (BP). Mean arterial BP is a function of cardiac output (CO) and total peripheral vascular resistance (TPR). Cardiac output, which is generated by a heart rate (HR) and stroke volume (SV), is in turn directly related to the extracellular fluid volume, specifically the volume of the venous return to the heart. TPR is regulated by vasoconstriction or vasodilatation of small resistance arteries by three mechanisms: baroreflexes and other neuro-humoral mechanisms, endothelial and myogenic mechanisms. Hypertension has often been associated with structural changes in arterial wall that decrease the wall-to-lumen ratio and increase wall stiffness. It is not clear, however, whether such a vascular remodeling is only a consequence of hypertension or is it also an important factor in the pathogenesis of elevated blood pressure. Recently, it was reported that in some models of hypertension, most of the increase in TPR can be attributed to functional and not structural alterations in small resistance arteries. The contraction of the vascular smooth muscle cells (VSMCs) is activated by a rise in the cytosolic Ca^{2+} concentration [12].

For many years it has been assumed that increased sodium intake is paralleled by increased sodium excretion maintaining steady sodium body content and that sodium is accumulated only with a corresponding volume of extracellular fluid. This assumption places the kidney as a central regulator of sodium handling. In recent years, sodium has been found to be accumulated in osmotic inactive state in the interstitium of the skin and other organs [13]. Regulation or dysregulation of this storage may affect blood pressure. Some data suggests that sodium and potassium may regulate the stiffness of endothelial cells and their nitric oxide release and thus the vessel tone and blood pressure [14]. Central nervous system emerged as another site of salt sensing in cerebrospinal fluid by a novel isoform of Na channels (Na_X), sensing of CSF osmolality by nonselective cation channels (transient receptor potential vanilloid type 1 channels), and osmolarity sensing by volume-regulated anion channels in glial cells of supraoptic and paraventricular nuclei [15].

2. "Humoral Factor" Increases Blood Pressure in Response to Salt Intake

The hypothesis of a circulating "humoral factor" that induces salt-sensitive hypertension came from the study performed by Dahl et al. over half a century ago [16,17]. Later, de Wardener and Clarkson suggested that this unidentified "humoral factor", implicated in the pathogenesis of salt-sensitive hypertension, was an endogenous natriuretic hormone, and had digitalis-like properties [18]. Cardiotonic steroids (CTS, Figure 1) were first found in plants, most notably digitalis in the foxglove plant, and then in the skin of toads like the *Bufo marinus* [19]. They have been used in traditional ancient medicine to treat congestive heart failure [20]. Endogenous CTS have been implicated in sodium homeostasis and blood pressure regulation through their effects on the Na/K-ATPase in renal and cardiovascular tissue [19]. Cardiotonic steroids (CTS) are also called digitalis-like factors. They are a group of steroid hormones that circulate in the blood and are excreted in the urine. CTS synthesis has been demonstrated in the adrenal cortex, placenta and hypothalamus [21]. They belong to two groups with different chemical structure: cardenolides (e.g., ouabain) and bufadienolides (e.g., marinobufagenin). Until recently, their biological role has been linked to their ability to inhibit activity of the ubiquitous transport enzyme called sodium-potassium adenosine triphosphatase (Na/K-ATPase). Over the last several years, their signaling capabilities unrelated to the Na/K-ATPase inhibition have caught the attention of many scientific groups.

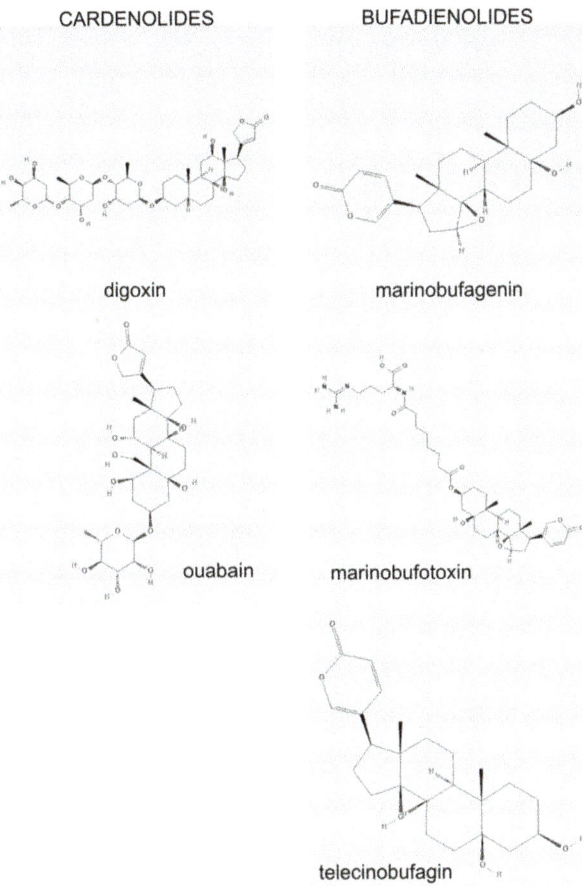

Figure 1. Chemical structure of cardiotonic steroids.

3. Na/K-ATPase: A Pump and a Receptor

The Na/K-ATPase is an ubiquitous enzyme present on the surface of all cells, the primary role of which is to maintain the difference in natrium and potassium concentrations between cytosolic and extracellular compartments. These differences are essential for cell-to-cell communication, contractility, and response to stimuli. The Na/K-ATPase is a heterodimer consisting of alpha and beta subunits. The alpha subunit is the "catalytic subunit" and contains binding sites for ATP, CTS, and other ligands, while the beta subunit is essential for the structural assembly of the enzyme. There are four α and three β isoforms known, thus allowing numerous combinations of $\alpha\beta$ complexes among tissues with different characteristics including different sensitivity to different cardiotonic steroids. The $\alpha1\beta1$ complex is the most common combination and is present in nearly every tissue. The $\alpha2$ isoform is expressed in adult heart, smooth muscle, skeletal muscle, brain, adipocytes, cartilage, and bone. The $\alpha3$ isoform is expressed in the central and peripheral nervous tissues and in the conductive system of the heart. The $\alpha4$ isoform has been found only in testis. The $\beta2$ and $\beta3$ isoforms are expressed in the brain, cartilage and erythrocytes, whereas $\beta2$ can also be found in cardiac tissue and $\beta3$ in lungs. The cardenolides have been determined to have a predilection for the $\alpha2$ and $\alpha3$ isoforms (Table 1 [22,23]), whereas the bufadienolides also inhibit the $\alpha1$ isoform. There are, however, differences between species in terms of the sensitivity of those isoforms to different CTS (e.g., in rats the $\alpha1$ isoform is resistant to ouabain, while in humans it is not). The K_i values of human α1, α2 and α3 isoforms range from 10^{-8} to 10^{-9} M/l [23]. Differences have even been found in different cellular localization of the enzyme: the α1 Na^+/K^+-ATPase, expressed in the renal epithelium, is ouabain-resistant, while the α1 isoform, found in the caveolae of renal tubular cells, exhibits remarkable sensitivity to ouabain [24]. In *rats* and in *humans*, ouabain has been detected in plasma at concentrations between 10^{-9} and 10^{-10} M/l [25]. Marinobufagenin has been reported in *rat* plasma at concentrations 10^{-9} to 10^{-10} M/l and in *human* plasma at concentrations between 0.5×10^{-9} and 10^{-8} M/l [25].

Table 1. Inhibition constant (K_i) of the Na-K-ATPase isozymes [22,23].

Isozyme	Ouabain Inhibition in Rats K_i, M	Ouabain Inhibition in Humans K_i, M
α1β1	4.3×10^{-5}	1.3×10^{-8}
α2β1	1.7×10^{-7}	3.2×10^{-8}
α2β2	1.5×10^{-7}	
α3β1	3.1×10^{-8}	1.7×10^{-8}
α3β2	4.7×10^{-8}	

Apart from the "classic" function of the Na/K-ATPase of maintaining the gradient of sodium and potassium concentrations across the plasmalemmal barrier, an alternative or "signaling" function for the enzyme has been described in recent years. This model proposes that some of plasmalemmal Na/K-ATPase resides in the caveole of the cells and does not seem to actively "pump" sodium and potassium but is closely associated with other key signaling proteins [19]. The Na/K-ATPase has been colocalized with signaling molecules including Src, PLC-γ, PI3K, IP3R, ankyrin, adducin, and caveolin-1 [26].

Activation of this receptor complex by CTS results in stimulation of the protein kinase cascades and generation of second messengers. Binding of ouabain to the caveolar complex of Na/K-ATPase phosphorylates epithelial growth factor receptor (EGFR) via Src and this results in activation of the Ras/Raf/MEK/ERK1/2 cascade [27]. These ouabain-induced signaling events may be specific for a particular cell type. For example, ouabain simulates the Src-dependent activation and translocation of several PKC isoforms in cardiac myocytes, which in turn activate the Ras/Raf/ERK1/2 cascade [28]. Moreover, in cardiac myocytes ouabain is also able to induce phosphorylation of protein kinase B (Akt) [29]. The cumulative effects of Akt, ERK1/2 and calcium signaling results in hypertrophic growth of cardiac myocytes, stimulate proliferation in renal epithelial cells [29], but cause growth inhibition in some cancer cells [30].

4. Marinobufagenin is a Ligand for Na/K ATPase

Marinobufagenin (MBG) by inhibiting the Na/K-ATPase participates in the regulation of renal sodium transport and arterial blood pressure. MBG promotes natriuresis through inhibition of sodium pump in the renal proximal tubules and vasoconstriction through inhibition of the same enzyme in vascular smooth muscle cells [31]. The synthesis of MBG by the adrenocortical cells is stimulated by high salt intake and is observed in volume-expanded states, such as preeclampsia, chronic kidney disease, and resistant arterial hypertension [31]. Elevation of plasma MBG concentration is preceded by transient ouabain increase [32]. Ouabain does not have natriuretic properties and increase of its plasma concentration after increased salt intake is only short-term. Inhibition of Na/K-ATPase in the vascular smooth muscle cells (VSMC) results in an increase of cytosolic Ca^{2+} concentration through Na^+/Ca^{2+} exchanger (NCX) which results in VSMC contraction [33].

Only some of the endogenous cardiotonic steroids may increase natriuresis. This is not only due to inhibition of the Na/K-ATPase and subsequently renal sodium reabsorption, but also due to internalization of the sodium pump in the proximal tubule and decreased expression of the transport protein, Na/H exchanger (NHE3) in apical membrane of the renal proximal tubule [34]. Additional studies have reported that CTS may play an important role in regulation of several pathways, including renal sodium handling and blood pressure regulation through the activation of a Src-EGFR (Epithelial Growth factor receptor) signaling cascade via caveolar Na/K ATPase [35].

5. Marinobufagenin and Fibrosis

It has recently been demonstrated that MBG in concentrations which are insufficient to block the pumping mechanism of the Na/K-ATPase initiates pro-fibrotic signaling by binding to the Na/K-ATPase and activating Src (sarcoma; proto-oncogene tyrosine-protein kinase) and EGFR (epidermal growth factor receptor) signaling, resulting in degradation of Fli-1 (negative nuclear regulator of the *procollagen-1* gene) in the myocardium and induction of collagen-1 synthesis [36]. Cardiac fibrosis was observed in rats administered with MBG by osmotic minipumps, and in a rat models of uremic cardiomyopathy, in which endogenous MBG concentrations were concurrently elevated [37]. High-salt diet increased TGFβ_1 and subsequent fibrosis in the heart and kidney in both normotensive and hypertensive rats [38]. These results suggest that excessive salt intake may be an important direct pathogenic factor for cardiovascular disease. Both clinical and experimental evidence support the development of salt-induced hypertrophy of the arterial wall in the absence of arterial pressure changes [39,40].

In a study performed in normotensive rats, Fedorova et al. demonstrated that high salt intake stimulates MBG production and tissue remodeling in heart and kidney, without affecting BP [41]. In another study, the same authors demonstrated that MBG is essential for the development of aortic fibrosis due to high salt intake. However, immunization against MBG abrogated only the pro-fibrotic effects of a high salt diet without affecting the blood pressure [42]. High salt-intake have been also shown to paradoxically increase the tissue renin-angiotensin system activation in Dahl salt-sensitive *rats*. It was documented that such an increase of tissue angiotensin II stimulates adrenocortical MBG production in this *rat* model. Moreover, AT1 receptor blocker losartan prevented stimulation of MBG biosynthesis both in vivo and in vitro [32]. A strong relationship between high salt intake, activation of the renin-angiotensin system and pro-fibrotic signaling has been demonstrated in this study leading to the damage of cardiovascular and renal tissues. Administration of a highly specific monoclonal antibody against MBG in vivo reduced aortic fibrosis and restored aortic relaxation in animals after prolonged high salt intake. The observed changes in vascular wall morphology in the absence of hemodynamic changes indicate that possible arterial stiffening is independent of blood pressure and that the pro-fibrotic factor MBG is responsible for the development of tissue fibrosis [42].

In normotensive rats, high dietary salt intake have been associated with the activation of TGF-β signaling within the arterial wall and increased aortic stiffness in the presence of elevated levels of the Na/K-ATPase ligand MBG despite unchanged blood pressure [43]. Moreover, the *rats* exposed to a reduced salt diet after the period of high salt intake exhibited a decrease in MBG levels, downregulation of the pro-fibrotic TGF-β pathway, a decrease of aortic wall collagen content and normalization of the pulse wave velocity to control levels. The authors also demonstrated that MBG stimulates collagen production in parallel with activation of TGF-β in cultured VSMCs in vitro, in the absence of hemodynamic effects [43]. Lowering the salt intake can improve vascular elasticity and decrease the cardiovascular risk by reducing the plasma MBG concentration.

In *humans*, dietary sodium restriction has been shown to reduce urinary MBG excretion which correlated with reduction in blood pressure and aortic stiffness [10]. Most importantly, MBG excretion positively correlated with blood pressure and changes in dietary sodium intake typical for a Western diet, extending previous observations in rodents and humans fed with experimentally high-sodium diets [44].

Contrary to the findings for MBG, high doses of ouabain have been demonstrated to inhibit the TGF-β-induced fibrosis in cultured human lung fibroblasts [45,46] suggesting that different CTS may have opposing actions.

6. Marinobufagenin and Cardiovascular Complications

Recently we have shown that plasma marinobufagenin concentration is increased in patients with advanced chronic kidney disease irrespective of their blood pressure [47]. Moreover, the higher the plasma MBG concentration the worse the survival was in this population. Recent data from the African-PREDICT study showed that both high salt intake and elevated plasma MBG concentration were correlated with increased stiffness of large arteries measured by pulse-wave velocity [48,49]. Left ventricular mass is positively and independently associated with marinobufagenin urinary excretion in young healthy adults as well [50]. As these morphological changes also correlated with blood pressure it is not possible to differentiate the direct effects of dietary salt and MBG from the blood pressure-dependent effects. The possibility to diminish or at least postpone arterial stiffness or heart hypertrophy by simple dietary adjustments seems to be very attractive. Experimental data support such a possibility: in normotensive *rats*, low sodium diet resulted in less arterial stiffness, less vascular TGF-β-dependent fibrosis and lower plasma MBG concentration without affecting blood pressure [43]. However, as always, one has to remember that too deep an intervention also has negative effects. In an experimental study both high and low sodium diet resulted in lower nephron number and hypertension in *rat* offspring [51].

The magnitude of systolic blood pressure (SBP) response to acute change in dietary NaCl intake, the salt-sensitivity of blood pressure, increases with advancing age [4]. Specific determinants of the greater blood pressure responsiveness to dietary NaCl observed in older subjects remain to be identified. It has been proposed that salt ingestion results in an increase in plasma volume and natriuresis. It has been postulated for some time that endogenous substances are stimulated by increased Na intake and increase natriuresis by inhibiting renal tubular Na exchangers to lower the renal reabsorption of filtered sodium. The age-associated differences in circulating endogenous Na/K-ATPase inhibitors may be implicated in the age-associated increase in SBP and increased salt sensitivity of SBP in the elderly. Anderson et al. were the first to demonstrate in normotensive *humans* that following a change from a low to a high salt diet, a sustained increase in MBG synthesis occurs, and renal fractional sodium elimination increases and correlates directly with increased urinary MBG excretion. In contrast to the sustained increase in MBG synthesis on high salt diet, ouabain levels in these subjects increased only transiently [52].

7. Endogenous Ouabain and Other CTS

Ouabain is another cardiotonic steroid demonstrated in *human* and animal plasma. In *humans* it does not increase sodium excretion, but it does have a role in the adaptation to both sodium depletion and loading. Although a few studies have shown that high salt loading in normotensive *rats* stimulates the release of ouabain, other experiments performed in *dogs*, *rats* and *humans* did not confirm these findings. In 180 patients with untreated hypertension, plasma levels of endogenous ouabain did not change during 2 weeks of salt loading, but increased following 2 weeks of sodium depletion [53]. Recent studies indicate that endogenous ouabain might act as a central mediator of salt - sensitive hypertension. In Dahl salt-sensitive *rats*, an important interaction seems to occur between brain and peripheral cardiotonic steroids. After acute or chronic salt-loading, a transient increase in circulating endogenous ouabain precedes a sustained increase in circulating marinobufagenin concentration [54]. This observation led to the postulate that endogenous ouabain acts as a neurohormone, triggering release of MBG, which in turn increases in cardiac contractility, peripheral vasoconstriction and natriuresis by inhibiting the Na/K-ATPase. More recently it was demonstrated that, similar to observations in Dahl salt-sensitive *rats*, normotensive *humans* on increased salt intake exhibit a transient increase in urinary endogenous ouabain excretion, which precedes a more sustained increase in renal MBG excretion [52].

Experiments in Milan hypertensive *rats*, which carry a mutation in the cytoskeletal protein adducin gene and exhibit increased circulating levels of endogenous ouabain, administration of the digoxin derivative rostofuroxin antagonized the effects of ouabain and lowered blood pressure [55]. The experimental data are promising and led to a clinical trial aimed to show the hypotensive effects of rostofuroxin in *humans*. The results in *humans*, however, could not demonstrate the blood pressure lowering effects after rostafuroxin administration [56].

There is substantial uncertainty as to whether the "endogenous ouabain" is indeed identical with the plant derived ouabain [57]. Although adrenals are supposed to be the source of the endogenous ouabain, the details of the adrenal biosynthetic pathway remain to be defined. A large portion of the data supporting the presence of "endogenous ouabain" is based on immunodetection. Cross-reactivity with similar compounds is an important issue in these methods. Some authors; however, failed to detect any measurable amount of true ouabain using state-of-the-art mass spectrometry [58]. This suggests that the "endogenous ouabain" may differ slightly from the plant ouabain. Further research is definitely needed in order to determine the exact structure of the compound referred to as "endogenous ouabain". Oubain-like immunoreactivity has been localized mainly to neuronal cells, especially hypothalamus [59]. In contrast, marinobufagenin immunoreactivity has been detected primarily in adrenals. It has been hypothesized that endogenous ouabain in the central nervous system responds to increased sodium load and increases sympathetic nervous activity resulting in hypertension [59].

Other CTS have been identified in mammalian tissues: marinobufotoxin [60], telocinobufagin [61], digoxin [62]. It is not known whether different CTS have different roles or are they different metabolites of a single active compound.

8. Summary

High dietary salt intake is a cause of elevated blood pressure and cardiovascular risk. However, it was demonstrated that even if the blood pressure did not increase on high salt diet, organ damage may still occur. Both effects are mediated (among other mechanisms) by endogenous digitalis-like cardiotonic steroids (Figure 2). They are released in order to maintain body sodium and act on the NaK-ATPase not only blocking the pumping mechanism but also triggering cellular responses leading to fibrosis.

9. Future Perspectives

Interfering with this pathway may present a new therapeutic target for treating hypertension and cardiovascular disease. Much work is needed before drug development is possible. Antibodies that bind cardiotonic steroids are not useful for long-term treatment of hypertension and cardiovascular events, although they could be useful in short-term situations like preeclampsia. Exact molecular mechanisms in CTS biosynthesis and their regulation will be studied further. Finding a way to influence differently the Na/K-ATPase blocking and signaling functions would be a major step forward in developing new medications in this pathway.

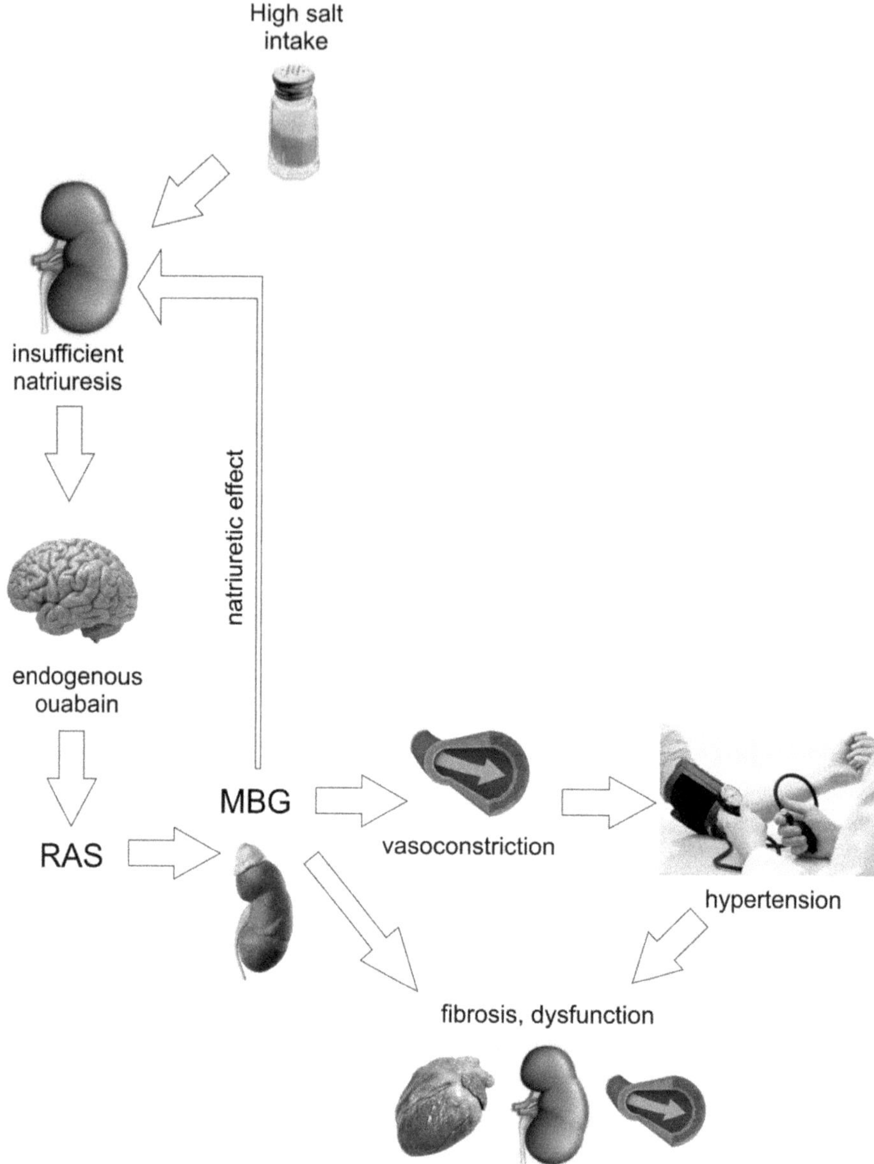

Figure 2. A possible mechanism of salt-induced hypertension and organ damage in humans. NaCl loading stimulates brain endogenous ouabain. Endogenous ouabain in the brain activates the local renin-angiotensin system (RAS) as well as sympathetic nervous system (SNS). These actions stimulate renin-angiotensin system in adrenal cortex and release of adrenocortical marinobufagenin (MBG). MBG is secreted in order to facilitate natriuresis, but at the same time MBG induces vasoconstriction which increases blood pressure and promotes fibrosis leading to permanent heart, kidney, and arterial damage and dysfunction.

Funding: The study was financed by a grant #UMO-2013/09/B/NZ5/02489 from the National Science Centre Poland.

Conflicts of Interest: The authors declare no conflict of interest.

References

1. GBD 2015 Risk Factors Collaborators. Global, regional, and national comparative risk assessment of 79 behavioural, environmental and occupational, and metabolic risks or clusters of risks, 1990-2015: A systematic analysis for the Global Burden of Disease Study 2015. *Lancet* **2016**, *388*, 1659–1724. [CrossRef]
2. Intersalt: An international study of electrolyte excretion and blood pressure. Results for 24 hour urinary sodium and potassium excretion. Intersalt Cooperative Research Group. *BMJ* **1988**, *297*, 319–328. [CrossRef]
3. Appel, L.J.; Moore, T.J.; Obarzanek, E.; Vollmer, W.M.; Svetkey, L.P.; Sacks, F.M.; Bray, G.A.; Vogt, T.M.; Cutler, J.A.; Windhauser, M.M.; et al. A clinical trial of the effects of dietary patterns on blood pressure. DASH Collaborative Research Group. *N. Engl. J. Med.* **1997**, *336*, 1117–1124. [CrossRef] [PubMed]
4. Weinberger, M.H.; Miller, J.Z.; Luft, F.C.; Grim, C.E.; Fineberg, N.S. Definitions and characteristics of sodium sensitivity and blood pressure resistance. *Hypertension* **1986**, *8*, II127–II134. [CrossRef] [PubMed]
5. De Wardener, H.E.; MacGregor, G.A. Sodium and blood pressure. *Curr. Opin. Cardiol.* **2002**, *17*, 360–367. [CrossRef]
6. Bagrov, A.Y.; Lakatta, E.G. The dietary sodium-blood pressure plot "stiffens". *Hypertension* **2004**, *44*, 22–24. [CrossRef] [PubMed]
7. Han, W.; Han, X.; Sun, N.; Chen, Y.; Jiang, S.; Li, M. Relationships between urinary electrolytes excretion and central hemodynamics, and arterial stiffness in hypertensive patients. *Hypertens. Res.* **2017**, *40*, 746–751. [CrossRef]
8. Kaess, B.M.; Rong, J.; Larson, M.G.; Hamburg, N.M.; Vita, J.A.; Levy, D.; Benjamin, E.J.; Vasan, R.S.; Mitchell, G.F. Aortic stiffness, blood pressure progression, and incident hypertension. *JAMA* **2012**, *308*, 875–881. [CrossRef]
9. Sutton-Tyrrell, K.; Najjar, S.S.; Boudreau, R.M.; Venkitachalam, L.; Kupelian, V.; Simonsick, E.M.; Havlik, R.; Lakatta, E.G.; Spurgeon, H.; Kritchevsky, S.; et al. Elevated aortic pulse wave velocity, a marker of arterial stiffness, predicts cardiovascular events in well-functioning older adults. *Circulation* **2005**, *111*, 3384–3390. [CrossRef]
10. Jablonski, K.L.; Fedorova, O.V.; Racine, M.L.; Geolfos, C.J.; Gates, P.E.; Chonchol, M.; Fleenor, B.S.; Lakatta, E.G.; Bagrov, A.Y.; Seals, D.R. Dietary sodium restriction and association with urinary marinobufagenin, blood pressure, and aortic stiffness. *Clin. J. Am. Soc. Nephrol.* **2013**, *8*, 1952–1959. [CrossRef]
11. Laragh, J.H.; Sealey, J.E.; Niarchos, A.P.; Pickering, T.G. The vasoconstriction-volume spectrum in normotension and in the pathogenesis of hypertension. *Fed. Proc.* **1982**, *41*, 2415–2423. [PubMed]
12. Blaustein, M.P.; Hamlyn, J.M. Signaling mechanisms that link salt retention to hypertension: Endogenous ouabain, the Na(+) pump, the Na(+)/Ca(2+) exchanger and TRPC proteins. *Biochim. Biophys. Acta* **2010**, *1802*, 1219–1229. [CrossRef] [PubMed]
13. Titze, J. Sodium balance is not just a renal affair. *Curr. Opin. Nephrol. Hypertens.* **2014**, *23*, 101–105. [CrossRef] [PubMed]
14. Oberleithner, H.; Kusche-Vihrog, K.; Schillers, H. Endothelial cells as vascular salt sensors. *Kidney Int.* **2010**, *77*, 490–494. [CrossRef] [PubMed]
15. Orlov, S.N.; Mongin, A.A. Salt-sensing mechanisms in blood pressure regulation and hypertension. *Am. J. Physiol.* **2007**, *293*, H2039–H2053. [CrossRef] [PubMed]
16. Dahl, L.K. Possible role of chronic excess salt consumption in the pathogenesis of essential hypertension. *Am. J. Cardiol.* **1961**, *8*, 571–575. [CrossRef]
17. Dahl, L.K.; Knudsen, K.D.; Iwai, J. Humoral transmission of hypertension: Evidence from parabiosis. *Circ. Res.* **1969**, *24*, 21–33.
18. de Wardener, H.E.; Clarkson, E.M. Concept of natriuretic hormone. *Physiol. Rev.* **1985**, *65*, 658–759. [CrossRef] [PubMed]
19. Bagrov, A.Y.; Shapiro, J.I.; Fedorova, O.V. Endogenous cardiotonic steroids: Physiology, pharmacology, and novel therapeutic targets. *Pharmacol. Rev.* **2009**, *61*, 9–38. [CrossRef] [PubMed]

20. Fisch, C. William Withering: An account of the foxglove and some of its medical uses 1785-1985. *J. Am. Coll. Cardiol.* **1985**, *5*, 1A–2A. [CrossRef]
21. Schoner, W.; Scheiner-Bobis, G. Endogenous and exogenous cardiac glycosides: Their roles in hypertension, salt metabolism, and cell growth. *Am. J. Physiol. Cell Physiol.* **2007**, *293*, C509–C536. [CrossRef] [PubMed]
22. Blanco, G.; Mercer, R.W. Isozymes of the Na-K-ATPase: Heterogeneity in structure, diversity in function. *Am. J. Physiol.* **1998**, *275*, F633–F650. [CrossRef] [PubMed]
23. Muller-Ehmsen, J.; Juvvadi, P.; Thompson, C.B.; Tumyan, L.; Croyle, M.; Lingrel, J.B.; Schwinger, R.H.; McDonough, A.A.; Farley, R.A. Ouabain and substrate affinities of human Na(+)-K(+)-ATPase alpha(1)beta(1), alpha(2)beta(1), and alpha(3)beta(1) when expressed separately in yeast cells. *Am. J. Physiol. Cell Physiol.* **2001**, *281*, C1355–C1364. [CrossRef] [PubMed]
24. Ferrandi, M.; Molinari, I.; Barassi, P.; Minotti, E.; Bianchi, G.; Ferrari, P. Organ hypertrophic signaling within caveolae membrane subdomains triggered by ouabain and antagonized by PST 2238. *J. Biol. Chem.* **2004**, *279*, 33306–33314. [CrossRef]
25. Khalaf, F.K.; Dube, P.; Mohamed, A.; Tian, J.; Malhotra, D.; Haller, S.T.; Kennedy, D.J. Cardiotonic Steroids and the Sodium Trade Balance: New Insights into Trade-Off Mechanisms Mediated by the Na(+)/K(+)-ATPase. *Int. J. Mol. Sci.* **2018**, *19*, 2576. [CrossRef]
26. Liu, J.; Xie, Z.J. The sodium pump and cardiotonic steroids-induced signal transduction protein kinases and calcium-signaling microdomain in regulation of transporter trafficking. *Biochim. Biophys. Acta* **2010**, *1802*, 1237–1245. [CrossRef]
27. Haas, M.; Wang, H.; Tian, J.; Xie, Z. Src-mediated inter-receptor cross-talk between the Na+/K+-ATPase and the epidermal growth factor receptor relays the signal from ouabain to mitogen-activated protein kinases. *J. Biol. Chem.* **2002**, *277*, 18694–18702. [CrossRef]
28. Mohammadi, K.; Kometiani, P.; Xie, Z.; Askari, A. Role of protein kinase C in the signal pathways that link Na+/K+-ATPase to ERK1/2. *J. Biol. Chem.* **2001**, *276*, 42050–42056. [CrossRef]
29. Liu, L.; Zhao, X.; Pierre, S.V.; Askari, A. Association of PI3K-Akt signaling pathway with digitalis-induced hypertrophy of cardiac myocytes. *Am. J. Physiol. Cell Physiol.* **2007**, *293*, C1489–C1497. [CrossRef]
30. Tian, J.; Li, X.; Liang, M.; Liu, L.; Xie, J.X.; Ye, Q.; Kometiani, P.; Tillekeratne, M.; Jin, R.; Xie, Z. Changes in sodium pump expression dictate the effects of ouabain on cell growth. *J. Biol. Chem.* **2009**, *284*, 14921–14929. [CrossRef]
31. Bagrov, A.Y.; Shapiro, J.I. Endogenous digitalis: Pathophysiologic roles and therapeutic applications. *Nat. Clin. Pract.* **2008**, *4*, 378–392. [CrossRef] [PubMed]
32. Fedorova, O.V.; Agalakova, N.I.; Talan, M.I.; Lakatta, E.G.; Bagrov, A.Y. Brain ouabain stimulates peripheral marinobufagenin via angiotensin II signalling in NaCl-loaded Dahl-S rats. *J. Hypertens.* **2005**, *23*, 1515–1523. [CrossRef] [PubMed]
33. Blaustein, M.P. Sodium ions, calcium ions, blood pressure regulation, and hypertension: A reassessment and a hypothesis. *Am. J. Physiol.* **1977**, *232*, C165–C173. [CrossRef] [PubMed]
34. Periyasamy, S.M.; Liu, J.; Tanta, F.; Kabak, B.; Wakefield, B.; Malhotra, D.; Kennedy, D.J.; Nadoor, A.; Fedorova, O.V.; Gunning, W.; et al. Salt loading induces redistribution of the plasmalemmal Na/K-ATPase in proximal tubule cells. *Kidney Int.* **2005**, *67*, 1868–1877. [CrossRef] [PubMed]
35. Liu, J.; Liang, M.; Liu, L.; Malhotra, D.; Xie, Z.; Shapiro, J.I. Ouabain-induced endocytosis of the plasmalemmal Na/K-ATPase in LLC-PK1 cells requires caveolin-1. *Kidney Int.* **2005**, *67*, 1844–1854. [CrossRef] [PubMed]
36. Elkareh, J.; Periyasamy, S.M.; Shidyak, A.; Vetteth, S.; Schroeder, J.; Raju, V.; Hariri, I.M.; El-Okdi, N.; Gupta, S.; Fedorova, L.; et al. Marinobufagenin induces increases in procollagen expression in a process involving protein kinase C and Fli-1: Implications for uremic cardiomyopathy. *Am. J. Physiol. Renal. Physiol.* **2009**, *296*, F1219–F1226. [CrossRef] [PubMed]
37. Elkareh, J.; Kennedy, D.J.; Yashaswi, B.; Vetteth, S.; Shidyak, A.; Kim, E.G.; Smaili, S.; Periyasamy, S.M.; Hariri, I.M.; Fedorova, L.; et al. Marinobufagenin stimulates fibroblast collagen production and causes fibrosis in experimental uremic cardiomyopathy. *Hypertension* **2007**, *49*, 215–224. [CrossRef] [PubMed]
38. Yu, H.C.; Burrell, L.M.; Black, M.J.; Wu, L.L.; Dilley, R.J.; Cooper, M.E.; Johnston, C.I. Salt induces myocardial and renal fibrosis in normotensive and hypertensive rats. *Circulation* **1998**, *98*, 2621–2628. [CrossRef]
39. Ying, W.Z.; Aaron, K.; Sanders, P.W. Mechanism of dietary salt-mediated increase in intravascular production of TGF-beta1. *Am. J. Physiol. Renal. Physiol.* **2008**, *295*, F406–F414. [CrossRef]

40. Ying, W.Z.; Aaron, K.; Wang, P.X.; Sanders, P.W. Potassium inhibits dietary salt-induced transforming growth factor-beta production. *Hypertension* **2009**, *54*, 1159–1163. [CrossRef]
41. Fedorova, O.V.; Anderson, D.E.; Lakatta, E.G.; Bagrov, A.Y. Interaction of NaCl and behavioral stress on endogenous sodium pump ligands in rats. *Am. J. Physiol. Regul. Integr. Comp. Physiol.* **2001**, *281*, R352–R358. [CrossRef] [PubMed]
42. Grigorova, Y.N.; Juhasz, O.; Zernetkina, V.; Fishbein, K.W.; Lakatta, E.G.; Fedorova, O.V.; Bagrov, A.Y. Aortic Fibrosis, Induced by High Salt Intake in the Absence of Hypertensive Response, is Reduced by a Monoclonal Antibody to Marinobufagenin. *Am. J. Hypertens.* **2016**, *29*, 641–646. [CrossRef] [PubMed]
43. Grigorova, Y.N.; Wei, W.; Petrashevskaya, N.; Zernetkina, V.; Juhasz, O.; Fenner, R.; Gilbert, C.; Lakatta, E.G.; Shapiro, J.I.; Bagrov, A.Y.; et al. Dietary Sodium Restriction Reduces Arterial Stiffness, Vascular TGF-beta-Dependent Fibrosis and Marinobufagenin in Young Normotensive Rats. *Int. J. Mol. Sci.* **2018**, *19*, 3168. [CrossRef] [PubMed]
44. Fedorova, O.V.; Kolodkin, N.I.; Agalakova, N.I.; Namikas, A.R.; Bzhelyansky, A.; St-Louis, J.; Lakatta, E.G.; Bagrov, A.Y. Antibody to marinobufagenin lowers blood pressure in pregnant rats on a high NaCl intake. *J. Hypertens.* **2005**, *23*, 835–842. [CrossRef]
45. La, J.; Reed, E.B.; Koltsova, S.; Akimova, O.; Hamanaka, R.B.; Mutlu, G.M.; Orlov, S.N.; Dulin, N.O. Regulation of myofibroblast differentiation by cardiac glycosides. *Am. J. Physiol. Lung Cell Mol. Physiol.* **2016**, *310*, L815–L823. [CrossRef] [PubMed]
46. La, J.; Reed, E.; Chan, L.; Smolyaninova, L.V.; Akomova, O.A.; Mutlu, G.M.; Orlov, S.N.; Dulin, N.O. Downregulation of TGF-beta Receptor-2 Expression and Signaling through Inhibition of Na/K-ATPase. *PLoS ONE* **2016**, *11*, e0168363. [CrossRef] [PubMed]
47. Piecha, G.; Kujawa-Szewieczek, A.; Kuczera, P.; Skiba, K.; Sikora-Grabka, E.; Wiecek, A. Plasma marinobufagenin immunoreactivity in patients with chronic kidney disease: A case control study. *Am. J. Physiol. Renal. Physiol.* **2018**, *315*, F637–F643. [CrossRef]
48. Strauss, M.; Smith, W.; Wei, W.; Bagrov, A.Y.; Fedorova, O.V.; Schutte, A.E. Large artery stiffness is associated with marinobufagenin in young adults: The African-PREDICT study. *J. Hypertens.* **2018**, *36*, 2333–2339. [CrossRef]
49. Strauss, M.; Smith, W.; Kruger, R.; van der Westhuizen, B.; Schutte, A.E. Large artery stiffness is associated with salt intake in young healthy black but not white adults: The African-PREDICT study. *Eur. J. Nutr.* **2018**, *57*, 2649–2656. [CrossRef]
50. Strauss, M.; Smith, W.; Kruger, R.; Wei, W.; Fedorova, O.V.; Schutte, A.E. Marinobufagenin and left ventricular mass in young adults: The African-PREDICT study. *Eur. J. Prev. Cardiol.* **2018**, *25*, 1587–1595. [CrossRef]
51. Koleganova, N.; Piecha, G.; Ritz, E.; Becker, L.E.; Muller, A.; Weckbach, M.; Nyengaard, J.R.; Schirmacher, P.; Gross-Weissmann, M.L. Both high and low maternal salt intake in pregnancy alter kidney development in the offspring. *Am. J. Physiol. Renal. Physiol.* **2011**, *301*, F344–F354. [CrossRef] [PubMed]
52. Anderson, D.E.; Fedorova, O.V.; Morrell, C.H.; Longo, D.L.; Kashkin, V.A.; Metzler, J.D.; Bagrov, A.Y.; Lakatta, E.G. Endogenous sodium pump inhibitors and age-associated increases in salt sensitivity of blood pressure in normotensives. *Am. J. Physiol. Regul. Integr. Comp. Physiol.* **2008**, *294*, R1248–R1254. [CrossRef] [PubMed]
53. Manunta, P.; Messaggio, E.; Ballabeni, C.; Sciarrone, M.T.; Lanzani, C.; Ferrandi, M.; Hamlyn, J.M.; Cusi, D.; Galletti, F.; Bianchi, G. Plasma ouabain-like factor during acute and chronic changes in sodium balance in essential hypertension. *Hypertension* **2001**, *38*, 198–203. [CrossRef] [PubMed]
54. Fedorova, O.V.; Lakatta, E.G.; Bagrov, A.Y. Endogenous Na,K pump ligands are differentially regulated during acute NaCl loading of Dahl rats. *Circulation* **2000**, *102*, 3009–3014. [CrossRef] [PubMed]
55. Ferrari, P.; Ferrandi, M.; Tripodi, G.; Torielli, L.; Padoani, G.; Minotti, E.; Melloni, P.; Bianchi, G. PST 2238: A new antihypertensive compound that modulates Na,K-ATPase in genetic hypertension. *J. Pharmacol. Exp. Ther.* **1999**, *288*, 1074–1083.
56. Staessen, J.A.; Thijs, L.; Stolarz-Skrzypek, K.; Bacchieri, A.; Barton, J.; Espositi, E.D.; de Leeuw, P.W.; Dluzniewski, M.; Glorioso, N.; Januszewicz, A.; et al. Main results of the ouabain and adducin for Specific Intervention on Sodium in Hypertension Trial (OASIS-HT): A randomized placebo-controlled phase-2 dose-finding study of rostafuroxin. *Trials* **2011**, *12*, 13. [CrossRef] [PubMed]
57. Lewis, L.K.; Yandle, T.G.; Hilton, P.J.; Jensen, B.P.; Begg, E.J.; Nicholls, M.G. Endogenous ouabain is not ouabain. *Hypertension* **2014**, *64*, 680–683. [CrossRef]

58. Baecher, S.; Kroiss, M.; Fassnacht, M.; Vogeser, M. No endogenous ouabain is detectable in human plasma by ultra-sensitive UPLC-MS/MS. *Clin. Chim. Acta* **2014**, *431*, 87–92. [CrossRef]
59. Takahashi, H.; Yoshika, M.; Komiyama, Y.; Nishimura, M. The central mechanism underlying hypertension: A review of the roles of sodium ions, epithelial sodium channels, the renin-angiotensin-aldosterone system, oxidative stress and endogenous digitalis in the brain. *Hypertens. Res.* **2011**, *34*, 1147–1160. [CrossRef]
60. Yoshika, M.; Komiyama, Y.; Takahashi, H. Isolation of marinobufotoxin from the supernatant of cultured PC12 cells. *Clin. Exp. Pharmacol. Physiol.* **2011**, *38*, 334–337. [CrossRef]
61. Komiyama, Y.; Nishimura, N.; Munakata, M.; Mori, T.; Okuda, K.; Nishino, N.; Hirose, S.; Kosaka, C.; Masuda, M.; Takahashi, H. Identification of endogenous ouabain in culture supernatant of PC12 cells. *J. Hypertens.* **2001**, *19*, 229–236. [CrossRef] [PubMed]
62. Goto, A.; Ishiguro, T.; Yamada, K.; Ishii, M.; Yoshioka, M.; Eguchi, C.; Shimora, M.; Sugimoto, T. Isolation of a urinary digitalis-like factor indistinguishable from digoxin. *Biochem. Biophys. Res. Commun.* **1990**, *173*, 1093–1101. [CrossRef]

International Journal of
Molecular Sciences

Review

Relationship of Circulating Irisin with Body Composition, Physical Activity, and Cardiovascular and Metabolic Disorders in the Pediatric Population

Leticia Elizondo-Montemayor [1,2,3,*], Gerardo Mendoza-Lara [1,2], Gustavo Gutierrez-DelBosque [1,2], Mariana Peschard-Franco [1,2], Bianca Nieblas [1,2] and Gerardo Garcia-Rivas [1,3,*]

1 Tecnologico de Monterrey, Escuela de Medicina y Ciencias de la Salud, Ave. Morones Prieto 3000, Monterrey N.L. 64710, Mexico; germendoza14@gmail.com (G.M.-L.); gustavogtz92@gmail.com (G.G.-D.); marianapeschard@gmail.com (M.P.-F.); A00599747@itesm.mx (B.N.)

2 Tecnologico de Monterrey, Center for Research in Clinical Nutrition and Obesity, Ave. Morones Prieto 300, Monterrey N.L. 64710, Mexico

3 Tecnologico de Monterrey, Cardiovascular and Metabolomics Research Group, Hospital Zambrano Hellion, San Pedro Garza Garcia P.C. 66278, Mexico

* Correspondence: lelizond@itesm.mx (L.E.-M.); gdejesus@itesm.mx (G.G.-R.); Tel.: +521-8888-2191 (L.E.-M.)

Received: 24 October 2018; Accepted: 10 November 2018; Published: 23 November 2018

Abstract: Exercise-induced irisin, a recently discovered myokine, has been linked to insulin resistance, obesity, and other diseases in adults; however, information in children is scarce and contradictory. We analyzed the limited evidence of irisin's effects in children and adolescents, and its association with body composition, exercise training, cardiovascular risk factors, and metabolic diseases, as well as the results of dietetic interventions. Both positive and negative correlations between irisin concentrations and body mass index, fat mass, fat-free mass, and other anthropometric parameters were found. Likewise, contradictory evidence was shown associating irisin plasma levels with cardiovascular and metabolic parameters such as glucose, insulin resistance, and cholesterol and other lipid and fatty acid plasma levels in healthy children, as well as in those with obesity and the metabolic syndrome. Gender, puberty, and hormonal differences were also examined. Furthermore, important contradictory findings according to the type and duration of exercise and of dietetic interventions in healthy and unhealthy subjects were demonstrated. In addition, correlations between mother–infant relations and circulating irisin were also identified. This review discusses the potential role of irisin in health and disease in the pediatric population.

Keywords: irisin; pediatric; children; cardiovascular disease; nutrition; diet; body composition; metabolic syndrome; obesity, neonates

1. Introduction

Böstrom et al., 2012 [1] discovered irisin as a myokine derived from fibronectin type III domain-containing protein 5 (FNDC5) that is regulated by PGC1-α in mice. PGC1-α has several effects related to energy metabolism, including the activation of PPAR-γ that regulates the expression of the uncoupling protein 1 (UCP1) and thermogenesis in brown adipose tissue. It also stimulates the expression of messenger RNA (mRNA) of FNDC5, a muscle gene product that encodes a type 1 membrane protein. Irisin is a 112-amino acid polypeptide that is proteolytically cleaved from FNDC5, which undergoes glycosylation and is then secreted into the bloodstream. The original findings of Böstrom et al., 2012 [1] showed that exercise in humans induced circulating irisin, which activated browning and thermogenic genes in white adipose cells through UCP1, while it downregulated genes involved in white fat development. They hypothesized that irisin could play a role in increasing total body energy expenditure, reducing body weight, and improving obesity-related insulin resistance [1].

More recent information indicates that irisin is also an adipokine with endocrine and autocrine functions secreted by white adipose tissue, and to a lesser extent, visceral adipose tissue (VAT) in the subcutaneous adipose tissue (SAT) [2]. Furthermore, irisin has also been found to be secreted by muscle. The expression of FNDC5 in muscle is related to SAT and VAT irisin levels and to the expression of FNDC5 and UCP1 genes in SAT. Muscle expression of the FNDC5 gene was 200-fold higher than that of adipocytes, indicating a relationship between muscle and adipose tissue functions in metabolic diseases. The correlation between irisin levels and FNDC5 expression in adipose tissue demonstrates a positive feedback mechanism for autocrine or paracrine production of irisin by adipose tissue, which further increases its capacity to metabolize glucose and fatty acids [3] (Figure 1).

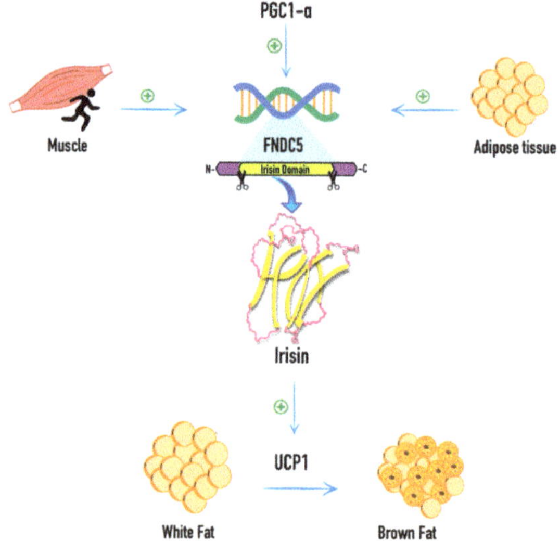

Figure 1. Irisin expression and function. Irisin is a myokine cleaved from fibronectin type III domain-containing protein 5 (FNDC5). Its expression is regulated by PGC1-α and it has been found to be secreted by adipose tissue and muscle. Irisin activates browning and thermogenic genes in white adipose cells through uncoupling protein 1 (UCP1), and downregulates genes involved in white fat development. PGC1-α: Peroxisome proliferator-activated receptor gamma coactivator 1-alpha. +: Positive influence.

Research has focused on finding mechanisms to explain the influence of irisin in the regulation of obesity, cardiovascular risk factors, metabolic syndrome (MS), and other related diseases. Irisin, named after the Greek messenger goddess Iris, has been linked to insulin resistance, obesity, exercise training, and cardiovascular and metabolic diseases in adults; however, the information on its role in children is scarce and contradictory. In this review, we discuss current knowledge in the pediatric population concerning irisin. The analysis also includes findings according to the type and duration of exercise in health and disease, those of dietetic and nutrition education interventions, as well as the influence of gender, puberty, and hormonal status on irisin plasma levels. In addition, the correlations between irisin and mother–newborn relationships are identified, along with the findings of irisin levels in plasma and tissues in other diseases in the pediatric population.

2. Association of Circulating Irisin with Body Mass Index and Body Composition

The associations of circulating irisin with body mass index and body composition parameters are shown in Table 1.

Table 1. Association of circulating irisin with body mass index and body composition.

Author	Sample	BMI%/BMI Z-Score	Body Composition Measurement	FM	MM%/FFM	BFP/BFM	WC/WHR	Others
Elizondo-Montemayor 2017 [4]	n = 40 (20 boys) Mexico 6–12 y-o. UW (n = 5), NW (n = 5), OW (n = 5), OB (n = 5).	+	Fat mass = [(weight-kg) × (body fat%)]/100 Body muscle= (height-cm) [0.264 + (0.0029 × MUAMA-cm^2)] FFM = (weight-kg − (weight-kg × body fat%)) PBF:Bioelectric impedance analysis (TANITA TBF 300)	0	MM (−)	0	WC (+)	N/A
Catli 2016 [5]	n = 66 Turkey 8-15 y-o. OB (n = 20 (20 boys)), NW (n = 30 (16 boys)).	0	Bioelectric impedance analysis (TanitaBC-41)	0	N/A	+	WC (0)	SBP (0), DBP (0)
Palacios-González 2015 [6]	n = 85 (40 boys) —Mexico 8–10 y-o. NW (n = 25), OW (n = 23), OB (n = 37).	+	N/A	N/A	N/A	N/A	N/A	N/A
Binay 2017 [7]	n = 120 Turkey 10–18 y-o. OB (n = 90), NW (n = 30).	+	Bioelectrical impedance analysis (BC-418MA Tanita Segmental Body Composition Analyzer)	+	0	+	WHR (+)	SBP (+)
Jang 2017 [8]	n = 618 (316 boys) Korea 12–15 y-o. NW (n = 370), OB (n = 248).	+	Bioelectrical impedance analysis (BC-418; Tanita)	+	+	+	WC (+)	N/A
Shim 2018 [9]	n = 96 (56 boys) Korea 6–10 y-o. NW (n = 54), OW (n = 16), OB (n = 26).	-	N/A	N/A	N/A	N/A	WC (-)	SBP (+), DBP (+)
Reinehr 2015 [10]	n = 60 Germany 10–15 y-o NW [n = 20 (10 boys)], OB (n = 40 (20 boys)).	0	N/A	N/A	N/A	N/A	N/A	DBP (+)
Löffler 2015 [11]	n = 105 (46 boys) Germany NW (n = 20), OW/OB (n = 64) 8–21 y-o and OB (n = 58 (23 boys)) 7–17 y-o.	BMI (+)	Bioimpedance analyses (Nutriguard-MS) Fat free mass and body cell mass (NutriPlus Software)	0	FFM (+) adults FFM (−) children	0	WHR (+)	SBP (0) DBP (0)
Singhal 2014 [12]	n = 85 women (81 Caucasian and Asian, 11 mixed- race, 5 Black) 14–21 y-o. AA (n = 38), EA (n = 24), NA (n = 23).	N/A	Dual energy x-ray absorptiometry (DXA)	0	(+) in all athletes (+) in AA (0) in EA	N/A	N/A	Spine BMD Z-score (+) Whole body BMD Z-score (+) Total vBMD (+) Trabecular vBMD (+)
Soininen 2018 [13]	n = 472 (245 boys) Finland 6–8 y-o.	N/A	MM, FM, PBF, BMD = Lunar Prodigy Advance DXA device	N/A	N/A	N/A	N/A	BMD (+) all children, not with boys and girls separately
Faienza 2018 [14]	n = 127 Italy 6–16 y-o. DM1 [n = 96 (41 boys)] 8–16 y-o, CO (n = 35 (21 boys)) 6–12 y-o.	(+) in patients with SCII	N/A	N/A	N/A	N/A	N/A	(+) with BTT-Z score, PTH, osteocalcin (−) with serum calcium, 25(OH) vitamin D, DKK-1, and sclerostin

y-o (years old), AA (amenorrheic athletes), BFM (body fat mass), BFP (body fat percentage), BMD (bone mineral density), DBP (Dystolic blood pressure), EA (eumenorrheic athletes), FFM (fat free mass), FM (fat mass), MM (muscle mass), MUAMA (mid-upper arm muscle area) PBF (percent body fat), NA (non-athletes), N/A (not available), NW (normal weight), OB (obese), OW (overweight), SBP (Systolic blood pressure), UW (underweight), vBMD (volumetric bone mineral density), WC (waist circumference), WHR (waist–hip ratio), + (positive correlation), − (negative correlation), 0 (no correlation).

2.1. Body Mass Index And Anthropometric Parameters

The relationships between irisin, body mass index (BMI), and anthropometric parameters are still not completely understood. Literature findings have shown different correlations. Most clinical studies in adults have observed a positive correlation between irisin levels, weight, and BMI at both ends of the body weight spectrum [15–17]. Additionally, positive associations between irisin and body fat, waist circumference (WC), waist-to-hip ratio, and muscle mass have also been found [11,18,19].

However, the findings in children still differ substantially. In a study in Mexican children aged 2–6 years, plasma irisin levels were lower in the underweight group compared with the normal weight and obese groups. Irisin levels correlated positively with WC and BMI percentile, but after multiple linear regression analyses, the correlation remained only for the latter. The lower levels in the underweight children might be explained by the fact that this group presented the lowest total fat mass and body fat percentage, as well as the highest proportion of body muscle mass to fat mass [4].

Irisin levels have also been found to be higher in obese children compared with healthy children [5,20]. Other studies have shown a positive relationship of circulating irisin with BMI and waist-to-hip ratio in Turkish children [7] and with BMI and WC in Korean adolescents [8]. Elevated irisin was independently associated with the risk of obesity even after adjusting for age, sex, physical activity, puberty status, tryglicerides, low density lipoprotein-c (LDL-c), and HOMA-IR [8]. In addition, both before and after a physical activity intervention program, irisin levels have been shown to have a positive association with BMI [6].

In contrast, in prepubescent Korean children, those with obesity tended to exhibit a lower irisin concentration compared with normal-weight children [9]. There was also a significant inverse correlation between irisin and both BMI and WC; although after adjusting for age and gender, this relationship remained significant only for BMI in the normal-weight group, but not in the overweight/obese group. However, in a study that included obese and normal-weight children, irisin was significantly correlated with BMI at baseline, but after a lifestyle intervention program, changes in BMI were not related to changes in irisin levels [10], concluding that during childhood irisin levels are not related to weight status.

2.2. Muscle Mass and Fat Free Mass

The relationship between irisin levels and both muscle mass and fat free mass (FFM) is still contradictory. One study showed no correlation between irisin levels and FFM [7], while in others, a negative association was found between irisin and FFM in German children and adolescents [11], as well as in amenorrheic athletes (AAs), though not in eumenorrheic athletes (EAs) [12]. In contrast, other researchers have demonstrated a positive correlation between irisin and FFM [8]. Furthermore, in another study, the positive correlation between circulating irisin and FFM was lost after multiple regression analyses, while there was no relationship with muscle mass [4].

2.3. Fat Mass

Studies in Korean [8] and Turkish [7] children found a positive correlation between irisin concentration and both percent body fat and total fat mass. Jang et al., 2017 [8] considered body fat mass to be the most important independent factor in this relationship. However, no correlation was found between irisin plasma levels and total fat mass in Mexican children [4], or in athletic and non-athletic lean female adolescents [12].

These contradictory findings in children and adolescents have been attributed to the different body composition of children compared with adults and to the variations that occur during growth development and puberty resulting from the interplay between total fat mass, muscle mass, and the fat/muscle mass ratio. The lower irisin levels shown in underweight children might reflect an adaptive response to conserve energy, while irisin could play an ambiguous role in people who are obese. The higher levels in obese children suggest a compensatory role to increase subcutaneous brown adipose tissue and energy expenditure, and to improve obesity-related insulin resistance. Additionally, the higher irisin levels observed in obese people might be attributed to irisin resistance.

2.4. Bone Mineral Density

Irisin also appears to play a role in metabolic bone health. Several biomarkers secreted by adipose tissue, skeletal muscle, or bone may affect bone metabolism and bone mineral density (BMD). In adults, negative associations between irisin levels and BMD have been found in patients with previous osteoporotic fractures [21] and with an increased risk of hip fractures [22], probably due to the positive influence of irisin in bone quality, but not in bone mass [23].

In contrast, in a study of prepubescent Finnish children, irisin was positively and independently associated with BMD after adjusting for age and sex, and after controlling for lean and fat mass [13]. Similarly, a positive association between irisin levels and spine, femoral neck, and whole body bone density Z-scores, total and trabecular volumetric-BMD, and strength estimates has also been demonstrated in adolescent female athletes [12]. Thus, it is possible for irisin to induce the metabolic effects of brown adipose tissue on bone strength [24] and cortical thickness [25], as previously reported.

3. Association of Circulating Irisin with Physical Activity, Exercise Training and Dietetic Interventions

Table 2 shows the associations of circulating irisin with physical activity, exercise training, and dietetic interventions.

Studies in children concerning the changes in circulating irisin levels induced by physical activity have demonstrated inconsistent results and certain limitations. Regular, moderate physical activity appeared to maintain higher irisin levels in normal-weight adolescents compared with their sedentary counterparts; although this difference was not observed in overweight/obese adolescents [8]. However, another study showed a negative correlation between aerobic exercise and irisin levels in Mexican children alongside the entire BMI spectrum [4].

Varying results have also been demonstrated in dietetic, nutrition education, and physical activity intervention studies. Evidence suggests that acute intervals of aerobic physical activity seem to increase irisin concentrations in subjects. This was confirmed after observing a 2.23-fold increase in irisin levels after as little as 15 min of intense ergometer activity in normal-weight and obese German children [11]. A similar result was observed in the EXIT intervention trial in obese Canadian adolescents [26]. Circulating irisin was observed to increase by 60% after one 45-min session of aerobic exercise, but no change was observed after one 45-min session of weight-training exercise.

Long-term exercise and diet counseling seem to induce contradictory results. Irisin concentrations were found to be higher after one year of diet counseling and a combined endurance and resistance physical activity intervention program in overweight and obese German children, which resulted in decreased BMI [27]. No correlation between changes in BMI and irisin concentration was found, concluding that irisin levels were the direct result of exercise, and they were not influenced by changes in BMI. Similarly, in another one-year nutrition education and physical activity intervention consisting of a weekly low-intensity program, irisin levels were higher in obese children without significant weight reduction, but no changes in irisin levels were observed in obese patients with significant weight loss [10]. The authors suggested that unidentified confounders could explain these findings.

In contrast, after a 4 to 6-week nutritional education, psychological counseling, and daily exercise intervention that resulted in significantly lower BMIs, although inconsistent changes in irisin concentration were observed, most of the patients displayed decreased irisin levels [11]. The same authors evaluated a 3-year longitudinal comparison between children who performed regular, increased, or competitive levels of physical activity and found no difference in irisin levels among the groups. Similarly, a non-significant tendency towards decreased irisin levels was observed in healthy children after an 8-month physical activity program consisting of five weekly sessions of 25 min of moderate intensity exercise compared with overweight/obese children [6].

Table 2. Association of circulating irisin with physical activity and exercise training and dietetic interventions.

Author	Sample	Intervention	Correlation	Results
Jang 2017 [8]	n = 618 (316 boys) Korea 12–15 y-o. NW (n = 370), OB (n = 248)	No Intervention	+	In NW girls
			+	In NW active
Elizondo-Montemayor 2017 [4]	n = 40 (20 boys) Mexico 6–12 y-o. UW (n = 5), NW (n = 5), OW (n = 5), OB (n = 5).	Reported aerobic exercise (days per week and hours per day)	−	With days per week
			-	With hours per day
Löffler 2015 [11]	n = 29 (11 boys) Germany 8–21 y-o. OB (n = 10 (2 boys)).	15-min maximum cycle ergometer	+	In all subjects
	n = 58 OB (23 boys) Germany 7–17 y-o.	4–6 weeks of nutritional and aerobic training	+/−	In boys
	n = 88 Germany 11–12 y-o. CO [n = 29 (12 boys)], IN (n = 34 (20 boys)), CS (n = 25 (16 boys))	Intervention group increased one unit of sports activities for 3 years	0	No difference among groups
Blizzard LeBlanc 2017 [26]	n = 11 OB (6 boys) Canada 15–16 y-o.	Acute bouts of exercise: Aerobic: 45 min at 60% HRR	+	In all subjects
		Acute bouts of exercise: Resistance: 45 min at 60–65% 1RM 12–15 reps	0	In all subjects
Palacios-González 2015 [6]	n = 85 (40 boys) Mexico 8–10 y-o. NW (n = 25), OW (n = 23), OB (n = 37).	5-min warm-up and 25-min aerobic activity at 75% HRMax, 8 months, 5 days/week	−	In all subjects
Blüher 2014 [27]	n = 65 Germany OW/OB (35 boys) 7–18 y-o.	39 session over 1 year of 150 min/week of combined endurance and resistance exercise plus diet counseling	+	In all subjects
Reinehr 2015 [10]	n = 60 Germany 10-15 y-o NW (n = 20 (10 boys)), OB (n = 40 (20 boys)).	Exercise sessions once per week plus nutrition education for 4–6 weeks	0	In OB children who lost weight
			+	In OB children who did not lose weight
Singhal 2014 [12]	n = 85 women (81 Caucasian and Asian, 11 mixed-race, 5 Black) 14–21 y-o. AA (n = 38), EA (n = 24), NA (n = 23).	No intervention	−	In AA
			0	In EA and NA

AA (amenorrheic athletes), CO (Control), CS (Competitive sports), DBP (Dystolic Blood Pressure), EA (eumenorrheic athletes), HRmax (maximum heart rate), HRR (heart rate reserve), IN (Intervention), Ir (irisin), NA (non-athletes), NW (normal weight), OB (obese), OGTT (oral glucose tolerance test), OW (overweight), RM (repetition maximum), UW (underweight) WC (waist circumference), + (positive correlation), − (negative correlation), 0 (no correlation).

Finally, AAs aged 14–21 years showed lower irisin levels than EAs and non-athletes (NAs), even after controlling for age, body fat, and lean mass. Irisin concentration correlated with higher resting energy expenditure in all subjects. No difference was found between EAs and NAs, suggesting that irisin response is an adaptive reaction to preserve energy by decreasing resting energy expenditure and brown adipogenesis in AAs [12].

Interpretation of the results of studies concerning the association between irisin and physical activity is difficult because of the different physical activity regimens in regard to intensity, duration, regularity, number of sessions, the tools used to measure such activity, the combined dietetic interventions, and the studied pediatric populations. Irisin concentrations appear to increase after acute bouts of aerobic exercise, but not after long-term programs, especially in obese populations. Thus, the original hypothesis by Böstrom et al, 2012 [1] that irisin increases energy expenditure and decreases weight status in adults could apply to short bouts of exercise in normal-weight children and adolescents. The possibility that during long-term exercise programs, the body adapts to irisin thermogenesis has to be considered. As irisin seems to not increase in obese children even after weight loss, the loss of muscle mass that occurs during weight reduction might be responsible for the lack of change in irisin concentrations.

4. Association of Circulating Irisin with Cardiovascular and Metabolic Alterations

The association of irisin plasma levels with metabolic changes in insulin resistance, glucose, triglycerides, cholesterol, fatty acid composition, and other variables is shown in Table 3.

4.1. Insulin Resistance and Glucose Regulation

The relationship between circulating irisin and insulin resistance and impaired glucose metabolism in children is not completely understood. In a cross-sectional study in healthy children from Saudi Arabia, a negative correlation between plasma glucose and irisin was observed, but after adjusting for gender, this correlation remained significant in girls only [28]. In contrast, in obese pediatric populations, several studies have demonstrated a positive correlation of irisin levels with insulin resistance and glucose levels. In two Turkish cross-sectional studies, circulating irisin exhibited a positive correlation with glucose, insulin levels, and HOMA-IR in obese children compared with normal-weight ones [5,7].

However, in a cohort of overweight/obese and normal-weight Brazilian children, irisin levels showed a positive correlation with glucose and insulin levels and with HOMA-IR in both groups, though this correlation remained significant only for insulin after multiple logistic regression analyses [29]. Likewise, no significant differences in serum irisin levels between obese children with and without insulin resistance were demonstrated in Italian children [20]. Both studies attributed the lack of observed differences to the small number of subjects analyzed. Contradictory results were shown between two different longitudinal weight loss interventional studies in obese German children. In a study by Reinehr et al, 2015 [10], irisin levels were found to be higher in obese children with impaired glucose tolerance compared with those with normal glucose tolerance at baseline. Additionally, positive correlations were found between irisin concentration and insulin levels, HOMA-IR, and glucose tolerance tests. Although the positive correlations persisted after the intervention, irisin levels were not associated with changes in BMI; rather, a correlation was observed in children entering into puberty, probably due to the effects of insulin resistance related to hormonal changes. In contrast, a study by Bluher et al., 2014 [27] found a 12% increase in irisin levels after weight loss in overweight/obese subjects, but no correlation was found between irisin and insulin levels, HOMA-IR, BMI, or fasting glucose levels either at baseline or after the intervention.

Regarding the role of irisin in type 1 diabetes mellitus (T1DM) in children and adolescents, an Italian cross-sectional study found irisin levels to be higher in T1DM patients than in controls. Circulating irisin was even higher in patients with subcutaneous insulin infusion compared with the ones on multiple daily injection treatments. Furthermore, irisin showed a negative relation with HbA1c%, serum glucose, and years since the diagnosis of T1DM [14].

Table 3. Association of circulating irisin with cardiovascular risk factors and metabolic alterations.

Author	Sample	Insulin	HOMA	Glucose	TG	HDL-c	LDL-c	TC	MS	Leptin	Others
Löffler 2015 [11]	n = 105 (46 boys) Germany NW (n = 20), OW/OB (n = 64) 8–21 y-o and OB (n = 58 (23 boys)) 7–17 y-o.	0	0	0	0	0	0	0	N/A	N/A	N/A
Al-Daghri 2014 [28]	n = 133 (76 boys) Saudi Arabia 9–15 years y-o. OB (n = 30).	0	0	–	0	0	0	0	N/A	0	(+) with ANG II
Reinehr 2015 [10]	n = 60 Germany 10–15 y-o NW (n = 20 (10 boys)), OB (n = 40 (20 boys))	+	+	0	+	–	+	N/A	N/A	N/A	(+) with 2-h OGTT and DBP
Binay 2017 [7]	n = 120 Turkey 10–18 y-o NW (n =30), OB (n = 90).	+	+	+	N/A	N/A	N/A	N/A	N/A	N/A	(+) with SBP in OB
Catli 2016 [5]	n = 66 Turkey 8–15 y-o. OB (n = 20 (20 boys)), NW (n = 30 (16 boys)).	+	+	0	0	–	0	0	N/A	0	N/A
Nigro 2017 [20]	n = 27 (19 boys) OB Italy 4–13 y-o, NW (n = 13 (4 boys)).	N/A	N/A	N/A	N/A	N/A	N/A	N/A	N/A	N/A	(−) with adiponectin
Blüher 2014 [27]	n = 65 OB (35 boys) 7–18 y-o.	0	0	0	N/A	N/A	N/A	N/A	N/A	N/A	No relation with adiponectin, leptin, or resistin
De Meneck 2018 [29]	n = 87 (47 boys) Brazi l6–12 y-o. NW (n = 63), OW/OB (n = 24)	+	+	+	+	–	N/A	0	N/A	N/A	(+) with SBP and DBP in the entire cohort, (+) with EPCs
Viitasalo 2015 [30]	n = 444 (247 boys) Finland 6–9 y-o. NW (n = 388), OW/OB (n = 55).	N/A	N/A	N/A	N/A	N/A	N/A	N/A	N/A	N/A	(+) with unfavorable fatty acid profile
Shim 2018 [9]	n = 96 (56 boys) Korea 6 to 10 y-o. NW (n = 54), OW (n = 16), OB (n = 26).	N/A	N/A	–	(−) in OW/OB	0	0	0	–	N/A	(+) with SBP and DBP in OB/OW
Palacios-González 2015 [6]	n = 85 (40 boys) Mexico 8–10 y-o. NW (n = 25), OW (n = 23), OB (n = 37).	N/A	N/A	N/A	N/A	N/A	N/A	N/A	N/A	+	N/A

DBP (diastolic blood pressure), HDL-c (high-density lipoprotein cholesterol), LDL-c (low-density lipoprotein cholesterol), N/A (not available), NW (normal weight), OB (obese), OGGT (oral glucose tolerance test), OW (overweight), PBF (percent body fat), SBP (systolic blood pressure), TC (total cholesterol, + (positive correlation), – (negative correlation), 0 (no correlation).

Because irisin levels were negatively correlated with glucose levels only in healthy girls, glucose could be an independent predictor of circulating irisin, attributing the difference between genders to dissimilarities in circulating hormone levels, or to the difference in brown adipose tissue quantity. Additionally, entry into puberty rather than BMI might be responsible for increased levels of irisin. Despite the contradictory correlations of irisin concentration with insulin resistance and glucose levels in the obese population, irisin has been proposed as a marker to differentiate obese children from normal-weight children. The increased irisin levels observed in obese subjects may represent the body's compensation mechanism for the insulin resistance observed in this population by increasing insulin sensitivity. On the other hand, the increased irisin concentration may reflect a state of irisin resistance. The findings in T1DM children suggest that better metabolic control is related to higher irisin levels in this pediatric population.

4.2. Cardiovascular Risk Factors and the Metabolic Syndrome

Previous studies have reported a link between irisin and the MS in adults, but only a few studies have evaluated this relationship in the pediatric population. A significant positive correlation exists between irisin levels and BMI, WC, triglyceride levels, systolic blood pressure (SBP), and diastolic blood pressure (DBP); but an inverse correlation with HDL-c levels were observed in a study by De Meneck et al., 2018 [29]. However, after multiple regression analyses, the relationship remained significant only for WC. Similarly, a cross-sectional study by Jang et al., 2017 [8] found irisin levels to be positively correlated with SBP, WC, triglycerides, fasting plasma glucose, HOMA-IR, and LDL-c. Higher irisin levels increased the risk for obesity and MS by two-fold, even after adjusting for age, sex, physical activity, and puberty, but after adjusting for BMI, this odds ratio was lost. A positive correlation between circulating irisin levels and branched and aromatic amino acids was also found. The authors suggested that the metabolic actions of irisin start during childhood and that beta cell dysfunction and evolution towards metabolic diseases are driven by the interplay between circulating irisin and branched-chain amino acids, which highly influence adiposity, lipids, and glucose.

In contrast, a cross-sectional study in prepubescent children found circulating irisin to be positively correlated with SBP and DBP only, but an inverse correlation with other components of the MS was demonstrated. After adjustments, irisin concentrations were found to be significantly lower in overweight and obese children with the MS compared with those without the MS. The authors proposed an irisin concentration of 15.43 ng/mL as a cutoff value for MS distinction, suggesting that irisin might be used as a biomarker for the MS [9].

The explanations for the contradictory role of irisin in metabolic diseases in children have yet to be clarified. Although myocytes are responsible for exercise-induced irisin secretion, in the context of increased adiposity, fat cells may be the primary source of high circulating irisin observed in some obese individuals with the MS. Increased fat mass may stimulate irisin production as a means to counteract new set points in energy balance. Conversely, a decrease in adipose tissue browning in obese individuals with the MS may be related to lower circulating irisin. Notably, in studies that showed positive associations between the MS and irisin, only anthropometric parameters such as BMI and WC, but not metabolic parameters, remained significant after adjustments, supporting the role of adipose tissue and its association with irisin in children. Furthermore, although it has been suggested that circulating irisin might regulate energy expenditure in adults with altered glucose metabolism, this compensatory mechanism could be limited in children, particularly before puberty, as a consequence of a proportionally lower muscle mass compared with adult populations.

4.3. Adipocytokines

Contradictory results have also been demonstrated regarding the relationship between irisin levels and adipocytokines. Circulating irisin has been positively correlated with leptin, but negatively associated with adiponectin in both obese and normal-weight Korean children [8]. A negative correlation between irisin and adiponectin was also observed in obese and normal-weight Italian children [20]. In an interventional study, a positive correlation between irisin and leptin was observed before and after an 8-month physical activity program in normal weight and obese children. The decrease in leptin after the exercise intervention showed a strong association with the decrease in irisin levels across all BMI subgroups [6]. In contrast, two studies found no correlation between circulating irisin and leptin, adiponectin, or resistin levels, either at baseline or after lifestyle interventions [27,30]. Concerning inflammatory cytokines, positive correlations between irisin and TNFα and IL-6 were observed in a cross-sectional analysis of the PANIC study [27,30].

The negative association between irisin and adiponectin levels may indicate that in states of low energy expenditure, decreased adiponectin might stimulate a compensatory increase in irisin in order to increase energy expenditure. However, the positive association between irisin and leptin levels, which both decrease after a physical exercise intervention, supports the hypothesis that irisin is produced by adipose tissue. Since there are some inconsistent results, the interplay between irisin and adipocytokines needs to be further investigated. The association of irisin levels with TNFα and IL-6 suggests that irisin levels could be related to a proinflammatory profile.

4.4. Fatty Acids Composition

Higher irisin levels have recently been associated with metabolically unfavorable fatty acid profiles in overweight and obese children, compared with normal weight ones, in a cross-sectional analysis from the ongoing PANIC study. In a subset of children, higher irisin levels were associated with polymorphism linked to an increased accumulation of hepatic triglycerides, suggesting that increased irisin levels may be intended to prevent lipid accumulation and progressive steatosis and fibrosis. Additionally, higher proportions of oleic acid, adrenic acid, and docosapentaenoic acid in plasma were associated with higher plasma irisin levels among overweight/obese children, which suggests that irisin is directly associated with increased activity of elongation and desaturation steps following the desaturation of linoleic acid. A possible association between plasma irisin and delta-6-desaturase activity in plasma cholesteryl esters was also found [31], which has previously been reported to be associated with insulin-resistant states [32]. Excess body fat could modulate these relationships through fatty acid-mediated cross-talk between metabolically active tissues.

4.5. Energy Intake and Expenditure

Recent studies have elucidated the interplay between new energy intake and expenditure regulators, such as oxytocin, which is involved in food intake regulation in the central nervous system. A positive correlation between circulating irisin and oxytocin levels was observed in AAs compared with EAs and NAs [33]. In contrast, in an obese population, irisin and oxytocin appeared to have opposite roles, as the first was found to be higher in obese and overweight pubescent children and adolescents compared with controls, while the opposite was found for oxytocin levels. Yet, a correlation between irisin and oxytocin was not studied by Binay et al., 2017 [7]. Thus, oxytocin signaling and the regulation of food intake may be more significant in high energy consumption situations, such as in AAs, while an inverse association is found in obesity.

5. Association of Circulating Irisin with Gender, Puberty, and Hormonal Status

Regarding the differences in irisin concentrations between genders, some studies in normal-weight subjects have concluded that circulating irisin levels were higher in lean girls than in lean boys. Noteworthy this difference between genders has not been observed in overweight or obese children [5,6,8,11,27,28]. As for puberty, a significant difference in irisin concentration between prepubescent and pubescent subjects was also observed in obese children. After one-year follow-up, an increase in irisin levels was found in five obese subjects that had begun puberty, compared with those who had not [10].

Exploring the relationship between irisin concentrations and hormonal levels, in a study of adolescent athletes, Singhal et al., 2014 [12] found no significant difference in sexual hormones between groups. However, irisin was found to correlate with a higher free androgen index in AAs, while a positive correlation between irisin and estradiol was observed in NAs. The meaning of these findings is not yet understood.

Most studies have concluded that gender differences in irisin concentrations in children and adolescents are not due to puberty status, but rather to total adipose tissue, brown adipose tissue, and metabolic activity, with girls having a higher body fat mass than boys. However, a contradictory conclusion was obtained in one of the previous studies stating that in the obese population, entry into puberty was the main factor involved in the increased irisin levels.

6. Association of Circulating Irisin with Mother–Offspring Relationship and Gestational Age in Neonates

Circulating levels of irisin in newborns are believed to be maternally inherited. In a cross-sectional study in Arab families, circulating irisin was found to be an inheritable trait between mother and offspring [34]. Others have found significantly lower irisin levels in newborns than in mothers. In a study of Mexican mother/newborn pairs with single, non-complicated pregnancies, only newborns from cesarean sections presented lower irisin concentrations than their mothers compared with labor-born neonates [35]. Arterial cord blood total antioxidant capacity, IL-1β, and IL1-RA levels positively predicted newborn irisin concentrations. Maternal IL-13 negatively predicted offspring irisin levels, while IL-1β positively predicted newborn irisin concentrations.

In another cross-sectional study in 70 pairs of newly-delivered Greek neonates and their mothers, irisin levels were also lower in the neonates [36]. A possible explanation for the higher irisin levels in labor-born neonates compared with cesarean-born ones could be related to physical stress during labor possibly mimicking physical activity. Additionally, differences in neonate and adult muscle mass could explain the lower irisin concentrations found in neonates compared with their mothers.

Associations between irisin concentrations and gestational age in newborns have been demonstrated. Comparing newborns from different gestational ages, many studies have found higher irisin levels in term infants compared with preterm newborns, as well as positive correlations with birth weight Z-scores. Measurement of irisin in umbilical cord blood has shown that irisin levels are lower in small for gestational age (SGA) newborns compared with appropriate for gestational age (AGA) and large for gestational age (LGA) neonates [36–38]. In contrast, in a study of Caucasian women, a higher irisin level in preterm infants compared with term infants was found [39]. Irisin levels were also significantly higher in maternal blood compared with umbilical cord blood. In intrauterine growth restriction (IUGR) subjects, fetal irisin concentrations were found to be significantly lower, compared with AGA controls and LGA neonates [40]. In the LGA group, fetal irisin concentrations were positively correlated with fetal insulin levels.

The lower irisin concentration in SGA and IUGR newborns might be attributable to their smaller muscle mass. It has been previously shown that IUGR neonates have impaired skeletal muscle growth [41], while SGA neonates have a smaller skeletal muscle mass [42] and total body fat percentage [43] compared with AGA newborns. Since irisin is needed for non-shivering thermogenesis, which is crucial for the adaptation of the newborn to the postnatal environment IUGR, and SGA neonates with low irisin levels might be predisposed to hypothermia at birth [40]. In addition, IUGR, together with LGA neonates, are at high risk for obesity and metabolic disorders, as well as for alterations in fetal adipose tissue development and hormonal dysfunctions [44]. These facts might explain the positive correlation between fetal irisin and insulin levels observed in the LGA group. Therefore, irisin might be an important metabolic factor during very early stages of life that may render these neonates susceptible to insulin resistance and the MS later in life.

7. Association of Circulating Irisin and Other Diseases

Although irisin has been studied in a few other organs and diseases in the pediatric population, its role remains unclear. High irisin immunoreactivity evaluated through immunohistochemistry was demonstrated in acute appendicitis biopsies from Turkish children. Positive correlations of irisin levels with urine, saliva, and serum, as well as with total white blood count, neutrophil percentage, and reactive C-protein have also been found. Irisin was higher in all samples from patients with appendicitis, compared with controls. In the post-operative period, irisin concentration reached baseline levels 72 h after surgery, suggesting that irisin secretion could be increased in response to the acute inflammation of appendix tissues [45].

Irisin was also studied in Egyptian children with epilepsy. Children with idiopathic epilepsy presented higher irisin levels compared with controls [46]. Plasma irisin showed a positive correlation with the severity of seizures and the duration of the disease, probably playing a role as a predictor of uncontrolled seizures. The authors hypothesized that the elevated irisin levels may play a protective role against the hypoxic effects of seizures, as has been demonstrated by Zhao et al., 2016 [47] and Mazur-Bialy et al., 2017 [48] in animal models.

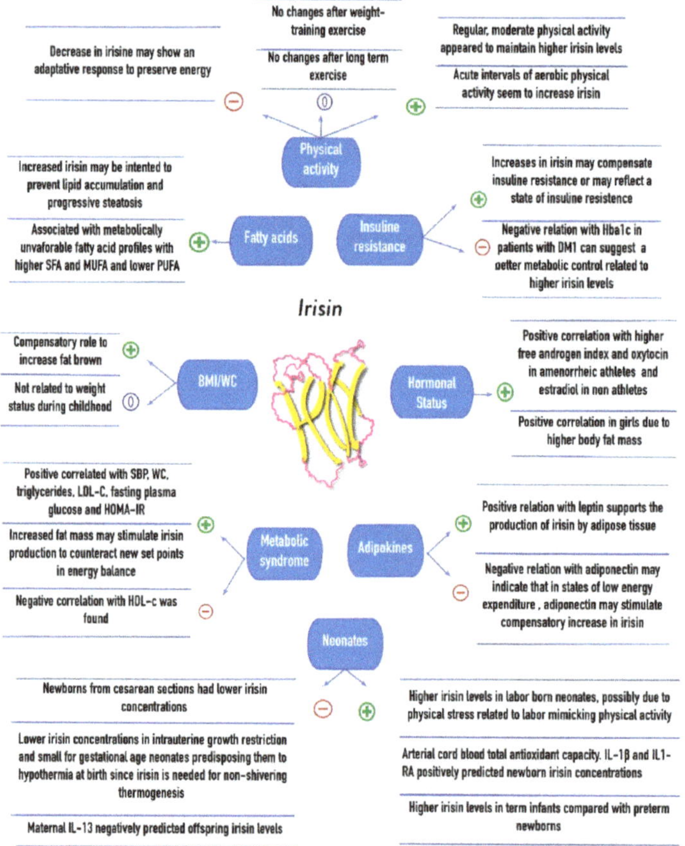

Figure 2. Association of irisin concentrations with cardiovascular, metabolic, and anthropometric parameters, physical activity, and mother-infant relations in children. +: Positive correlation; −: Negative correlation; 0: No correlation; SFA: Saturated fatty acids; MUFA: Monounsaturated fatty acids; PUFA: Polyunsaturated fatty acids; SBP: Systolic blood pressure; WC: Waist circumference; LDL-c: Low density lipoprotein cholesterol; HDL-c: High density lipoprotein cholesterol; HOMA-IR: Homeostatic model assessment of insulin resistance; DM1: Diabetes Mellitus type 1; IL13: Interleukin 13; IL-1β: Interleukin 1 beta; IL1-RA: Interleukin 1 receptor antagonist.

Finally, in a study of prepubescent patients with Turner syndrome who underwent recombinant human growth hormone (rhGH) therapy, a significant increase in irisin levels was found after treatment, although no relationship between irisin and IGF-1 was observed before or after therapy [49]. Negative associations between irisin levels and metabolic parameters were found, while a positive association of irisin with HbA1c was identified. These results suggest that in treatment-naïve Turner syndrome patients, who are predisposed to the MS, the physiological role of irisin may be disrupted and that rhGH therapy may restore it.

Figure 2 summarizes the associations of plasma irisin levels with body composition, cardiovascular risk factors, the metabolic syndrome, diet and physical activity interventions, as well as neonates and infant–mother correlations in the pediatric population.

8. Conclusions

The role of irisin as a regulator of body composition, and cardiovascular and metabolic diseases, as well as its correlation with physical activity and dietetic interventions has been understudied in the pediatric population, and it is still poorly understood. Contradictory findings have been found regarding the association of irisin with BMI, WC, fat mass, muscle mass, cardiovascular risk factors, insulin resistance, fasting glucose, and lipid levels, as well as its role in obesity and the MS. Irisin levels have been found to differ by gender in lean children, but not in obese ones. Controversial hypotheses to explain these findings have also been explored.

Irisin might represent an adaptive response to preserve energy in children with decreased muscle and fat mass, such as those who are underweight, and in SGA and IUGR neonates. Meanwhile, in children and adolescents with obesity, cardiovascular risk factors and the MS, irisin might either increase energy expenditure through thermogenesis, or it may represent an insulin-resistant state, especially considering the negative association of irisin with adiponectin and its positive association with leptin. Furthermore, irisin has even been associated with a proinflammatory profile.

Increased irisin levels during short bouts of aerobic exercise may only represent increased energy expenditure, but the lack of response during long term regimens, which even include nutrition and diet counseling, may be attributed to an adaptive thermogenesis. Further research in the pediatric population is required to confirm the association of irisin concentration with body composition parameters, physical activity, and cardiometabolic diseases, as well as to elucidate the role of irisin and its underlying mechanisms as a regulator of the metabolic state.

Funding: This research was funded by the Center for Research in Obesity and Clinical Nutrition, Tecnologico de Monterrey (LEM), and by CONACYT grant number 256577, a1-s-48883 (GGR).

Acknowledgments: The authors wish to thank Luisa Perez-Villarreal for her support in the literature search as well as to the Center for Research in Clinical Nutrition and Obesity (Leticia Elizondo-Montemayor), to the Cardiovascular and Metabolomics Research Group (Gerardo Garcia-Rivas) and to CONACYT, grant number 256577, a1-s-48883.

Conflicts of Interest: The authors declare no conflict of interest.

Abbreviations

UCP-1	uncoupling protein 1
FNDC5	fibronectin type III domain-containing protein 5
VAT	visceral adipose tissue
SAT	subcutaneous adipose tissue
MS	metabolic syndrome
BMI	body mass index
WC	waist circumference
LDL-c	lipoprotein-c
FFM	fat free mass
AAs	amenorrheic athletes
EAs	eumenorrheic athletes

BMD	bone mineral density
NAs	non-athletes
T1DM	type 1 diabetes mellitus
SBP	systolic blood pressure
DBP	diastolic blood pressure
SGA	small for gestational age
AGA	appropriate for gestational age
LGA	large for gestational age
IUGR	intrauterine growth restriction
rhGH	recombinant human growth hormone

References

1. Boström, P.; Wu, J.; Jedrychowski, M.P.; Korde, A.; Ye, L.; Lo, J.C.; Rasbach, K.A.; Boström, E.A.; Choi, J.H.; Long, J.Z.; et al. A PGC1-α-dependent myokine that drives brown-fat-like development of white fat and thermogenesis. *Nature* **2012**, *481*, 463–468. [CrossRef] [PubMed]
2. Roca-Rivada, A.; Castelao, C.; Senin, L.L.; Landrove, M.O.; Baltar, J.; Belen Crujeiras, A.; Seoane, L.M.; Casanueva, F.F.; Pardo, M. FNDC5/irisin is not only a myokine but also an adipokine. *PLoS ONE* **2013**, *8*, e60563. [CrossRef] [PubMed]
3. Moreno-Navarrete, J.M.; Ortega, F.; Serrano, M.; Guerra, E.; Pardo, G.; Tinahones, F.; Ricart, W.; Fernandez-Real, J.M. Irisin is expressed and produced by human muscle and adipose tissue in association with obesity and insulin resistance. *J. Clin. Endocrinol. Metab.* **2013**, *98*, E769–E778. [CrossRef] [PubMed]
4. Elizondo-Montemayor, L.; Silva-Platas, C.; Torres-Quintanilla, A.; Rodríguez-López, C.; Ruiz-Esparza, G.U.; Reyes-Mendoza, E.; Garcia-Rivas, G. Association of Irisin Plasma Levels with Anthropometric Parameters in Children with Underweight, Normal Weight, Overweight, and Obesity. *Biomed. Res. Int.* **2017**, *2017*, 2628968. [CrossRef] [PubMed]
5. Çatlı, G.; Küme, T.; Tuhan, H.; Anık, A.; Çalan, Ö.; Böber, E.; Abacı, A. Relation of serum irisin level with metabolic and antropometric parameters in obese children. *J. Diabetes Complicat.* **2016**, *30*, 1560–1565. [CrossRef] [PubMed]
6. Palacios-González, B.; Vadillo-Ortega, F.; Polo-Oteyza, E.; Sánchez, T.; Ancira-Moreno, M.; Romero-Hidalgo, S.; Meráz, N.; Antuna-Puente, B. Irisin levels before and after physical activity among school-age children with different BMI: A direct relation with leptin. *Obesity* **2015**, *23*, 729–732. [CrossRef] [PubMed]
7. Binay, Ç.; Paketçi, C.; Güzel, S.; Samancı, N. Serum Irisin and Oxytocin Levels as Predictors of Metabolic Parameters in Obese Children. *J. Clin. Res. Pediatr. Endocrinol.* **2017**, *9*, 124–131. [CrossRef] [PubMed]
8. Jang, H.B.; Kim, H.J.; Kang, J.H.; Park, S.I.; Park, K.H.; Lee, H.J. Association of circulating irisin levels with metabolic and metabolite profiles of Korean adolescents. *Metabolism* **2017**, *73*, 100–108. [CrossRef] [PubMed]
9. Shim, Y.S.; Kang, M.J.; Yang, S.; Hwang, I.T. Irisin is a biomarker for metabolic syndrome in prepubertal children. *Endocr. J.* **2018**, *65*, 23–31. [CrossRef] [PubMed]
10. Reinehr, T.; Elfers, C.; Lass, N.; Roth, C.L. Irisin and its relation to insulin resistance and puberty in obese children: A longitudinal analysis. *J. Clin. Endocrinol. Metab.* **2015**, *100*, 2123–2130. [CrossRef] [PubMed]
11. Löffler, D.; Müller, U.; Scheuermann, K.; Friebe, D.; Gesing, J.; Bielitz, J.; Erbs, S.; Landgraf, K.; Wagner, I.V.; Kiess, W.; et al. Serum irisin levels are regulated by acute strenuous exercise. *J. Clin. Endocrinol. Metab.* **2015**, *100*, 1289–1299. [CrossRef] [PubMed]
12. Singhal, V.; Lawson, E.A.; Ackerman, K.E.; Fazeli, P.K.; Clarke, H.; Lee, H.; Eddy, K.; Marengi, D.A.; Derrico, N.P.; Bouxsein, M.L.; et al. Irisin levels are lower in young amenorrheic athletes compared with eumenorrheic athletes and non-athletes and are associated with bone density and strength estimates. *PLoS ONE* **2014**, *9*, e100218. [CrossRef] [PubMed]
13. Soininen, S.; Sidoroff, V.; Lindi, V.; Mahonen, A.; Kröger, L.; Kröger, H.; Jääskeläinen, J.; Atalay, M.; Laaksonen, D.E.; Laitinen, T.; et al. Body fat mass, lean body mass and associated biomarkers as determinants of bone mineral density in children 6–8 years of age—The Physical Activity and Nutrition in Children (PANIC) study. *Bone* **2018**, *108*, 106–114. [CrossRef] [PubMed]

14. Faienza, M.F.; Brunetti, G.; Sanesi, L.; Colaianni, G.; Celi, M.; Piacente, L.; D'Amato, G.; Schipani, E.; Colucci, S.; Grano, M. High irisin levels are associated with better glycemic control and bone health in children with Type 1 diabetes. *Diabetes Res. Clin. Pract.* **2018**, *141*, 10–17. [CrossRef] [PubMed]
15. Perakakis, N.; Triantafyllou, G.A.; Fernandez-Real, J.M.; Huh, J.Y.; Park, K.H.; Seufert, J.; Mantzoros, C.S. Physiology and role of irisin in glucose homeostasis. *Nat. Rev. Endocrinol.* **2017**, *13*, 324–337. [CrossRef] [PubMed]
16. Huh, J.Y.; Panagiotou, G.; Mougios, V.; Brinkoetter, M.; Vamvini, M.T.; Schneider, B.E.; Mantzoros, C.S. FNDC5 and irisin in humans: I. Predictors of circulating concentrations in serum and plasma and II. mRNA expression and circulating concentrations in response to weight loss and exercise. *Metabolism* **2012**, *61*, 1725–1738. [CrossRef] [PubMed]
17. Pardo, M.; Crujeiras, A.B.; Amil, M.; Aguera, Z.; Jimenez-Murcia, S.; Banos, R.; Botella, C.; de la Torre, R.; Estivill, X.; Fagundo, A.B.; et al. Association of irisin with fat mass, resting energy expenditure, and daily activity in conditions of extreme body mass index. *Int. J. Endocrinol.* **2014**, *2014*, 857270. [CrossRef] [PubMed]
18. Crujeiras, A.B.; Zulet, M.A.; Lopez-Legarrea, P.; de la Iglesia, R.; Pardo, M.; Carreira, M.C.; Martínez, J.A.; Casanueva, F.F. Association between circulating irisin levels and the promotion of insulin resistance during the weight maintenance period after a dietary weight-lowering program in obese patients. *Metabolism* **2014**, *63*, 520–531. [CrossRef] [PubMed]
19. Stengel, A.; Hofmann, T.; Goebel-Stengel, M.; Elbelt, U.; Kobelt, P.; Klapp, B.F. Circulating levels of irisin in patients with anorexia nervosa and different stages of obesity-correlation with body mass index. *Peptides* **2013**, *39*, 125–130. [CrossRef] [PubMed]
20. Nigro, E.; Scudiero, O.; Ludovica Monaco, M.; Polito, R.; Schettino, P.; Grandone, A.; Perrone, L.; Miraglia Del Giudice, E.; Daniele, A. Adiponectin profile and Irisin expression in Italian obese children: Association with insulin-resistance. *Cytokine* **2017**, *94*, 8–13. [CrossRef] [PubMed]
21. Anastasilakis, A.D.; Polyzos, S.A.; Makras, P.; Gkiomisi, A.; Bisbinas, I.; Katsarou, A.; Filippaios, A.; Mantzoros, C.S. Circulating irisin is associated with osteoporotic fractures in postmenopausal women with low bone mass but is not affected by either teriparatide or denosumab treatment for 3 months. *Osteoporos. Int.* **2014**, *25*, 1633–1642. [CrossRef] [PubMed]
22. Yan, J.; Liu, H.J.; Guo, W.C.; Yang, J. Low serum concentrations of Irisin are associated with increased risk of hip fracture in Chinese older women. *Jt. Bone Spine* **2018**, *85*, 353–358. [CrossRef] [PubMed]
23. Palermo, A.; Strollo, R.; Maddaloni, E.; Tuccinardi, D.; D'Onofrio, L.; Briganti, S.I.; Defeudis, G.; De Pascalis, M.; Lazzaro, M.C.; Colleluori, G.; et al. Irisin is associated with osteoporotic fractures independently of bone mineral density, body composition or daily physical activity. *Clin. Endocrinol.* **2015**, *82*, 615–619. [CrossRef] [PubMed]
24. Lee, P.; Brychta, R.J.; Collins, M.T.; Linderman, J.; Smith, S.; Herscovitch, P.; Millo, C.; Chen, K.Y.; Celi, F.S. Cold-activated brown adipose tissue is an independent predictor of higher bone mineral density in women. *Osteoporos. Int.* **2013**, *24*, 1513–1518. [CrossRef] [PubMed]
25. Ponrartana, S.; Aggabao, P.C.; Hu, H.H.; Aldrovandi, G.M.; Wren, T.A.; Gilsanz, V. Brown adipose tissue and its relationship to bone structure in pediatric patients. *J. Clin. Endocrinol. Metab.* **2012**, *97*, 2693–2698. [CrossRef] [PubMed]
26. Blizzard LeBlanc, D.R.; Rioux, B.V.; Pelech, C.; Moffatt, T.L.; Kimber, D.E.; Duhamel, T.A.; Dolinsky, V.W.; McGavock, J.M.; Senechal, M. Exercise-induced irisin release as a determinant of the metabolic response to exercise training in obese youth: The EXIT trial. *Physiol. Rep.* **2017**, *5*. [CrossRef] [PubMed]
27. Bluher, S.; Panagiotou, G.; Petroff, D.; Markert, J.; Wagner, A.; Klemm, T.; Filippaios, A.; Keller, A.; Mantzoros, C.S. Effects of a 1-year exercise and lifestyle intervention on irisin, adipokines, and inflammatory markers in obese children. *Obesity* **2014**, *22*, 1701–1708. [CrossRef] [PubMed]
28. Al-Daghri, N.M.; Alkharfy, K.M.; Rahman, S.; Amer, O.E.; Vinodson, B.; Sabico, S.; Piya, M.K.; Harte, A.L.; McTernan, P.G.; Alokail, M.S.; et al. Irisin as a predictor of glucose metabolism in children: Sexually dimorphic effects. *Eur. J. Clin. Investig.* **2014**, *44*, 119–124. [CrossRef] [PubMed]
29. De Meneck, F.; Victorino de Souza, L.; Oliveira, V.; do Franco, M.C. High irisin levels in overweight/obese children and its positive correlation with metabolic profile, blood pressure, and endothelial progenitor cells. *Nutr. Metab. Cardiovasc. Dis.* **2018**, *28*, 756–764. [CrossRef] [PubMed]

30. Viitasalo, A.; Atalay, M.; Pihlajamaki, J.; Jaaskelainen, J.; Korkmaz, A.; Kaminska, D.; Lindi, V.; Lakka, T.A. The 148 M allele of the PNPLA3 is associated with plasma irisin levels in a population sample of Caucasian children: The PANIC Study. *Metabolism* **2015**, *64*, 793–796. [CrossRef] [PubMed]
31. Viitasalo, A.; Agren, J.; Venalainen, T.; Pihlajamaki, J.; Jaaskelainen, J.; Korkmaz, A.; Atalay, M.; Lakka, T.A. Association of plasma fatty acid composition with plasma irisin levels in normal weight and overweight/obese children. *Pediatr. Obes.* **2016**, *11*, 299–305. [CrossRef] [PubMed]
32. Steffen, L.M.; Vessby, B.; Jacobs, D.R., Jr.; Steinberger, J.; Moran, A.; Hong, C.P.; Sinaiko, A.R. Serum phospholipid and cholesteryl ester fatty acids and estimated desaturase activities are related to overweight and cardiovascular risk factors in adolescents. *Int. J. Obes.* **2008**, *32*, 1297–1304. [CrossRef] [PubMed]
33. Lawson, E.A.; Ackerman, K.E.; Slattery, M.; Marengi, D.A.; Clarke, H.; Misra, M. Oxytocin secretion is related to measures of energy homeostasis in young amenorrheic athletes. *J. Clin. Endocrinol. Metab.* **2014**, *99*, E881–E885. [CrossRef] [PubMed]
34. Al-Daghri, N.M.; Al-Attas, O.S.; Alokail, M.S.; Alkharfy, K.M.; Yousef, M.; Vinodson, B.; Amer, O.E.; Alnaami, A.M.; Sabico, S.; Tripathi, G.; et al. Maternal inheritance of circulating irisin in humans. *Clin. Sci.* **2014**, *126*, 837–844. [CrossRef] [PubMed]
35. Hernandez-Trejo, M.; Garcia-Rivas, G.; Torres-Quintanilla, A.; Laresgoiti-Servitje, E. Relationship between Irisin Concentration and Serum Cytokines in Mother and Newborn. *PLoS ONE* **2016**, *11*, e0165229. [CrossRef] [PubMed]
36. Karras, S.N.; Polyzos, S.A.; Newton, D.A.; Wagner, C.L.; Hollis, B.W.; Ouweland, J.V.D.; Dursun, E.; Gezen-Ak, D.; Kotsa, K.; Annweiler, C.; et al. Adiponectin and vitamin D-binding protein are independently associated at birth in both mothers and neonates. *Endocrine* **2018**, *59*, 164–174. [CrossRef] [PubMed]
37. Joung, K.E.; Park, K.H.; Filippaios, A.; Dincer, F.; Christou, H.; Mantzoros, C.S. Cord blood irisin levels are positively correlated with birth weight in newborn infants. *Metabolism* **2015**, *64*, 1507–1514. [CrossRef] [PubMed]
38. Keles, E.; Turan, F.F. Evaluation of cord blood irisin levels in term newborns with small gestational age and appropriate gestational age. *SpringerPlus* **2016**, *5*, 1757. [CrossRef] [PubMed]
39. Pavlova, T.; Zlamal, F.; Tomandl, J.; Hodicka, Z.; Gulati, S.; Bienertova-Vasku, J. Irisin Maternal Plasma and Cord Blood Levels in Mothers with Spontaneous Preterm and Term Delivery. *Dis. Mark.* **2018**, *2018*, 7628957. [CrossRef] [PubMed]
40. Baka, S.; Malamitsi-Puchner, A.; Boutsikou, T.; Boutsikou, M.; Marmarinos, A.; Hassiakos, D.; Gourgiotis, D.; Briana, D.D. Cord blood irisin at the extremes of fetal growth. *Metabolism* **2015**, *64*, 1515–1520. [CrossRef] [PubMed]
41. Padoan, A.; Rigano, S.; Ferrazzi, E.; Beaty, B.L.; Battaglia, F.C.; Galan, H.L. Differences in fat and lean mass proportions in normal and growth-restricted fetuses. *Am. J. Obstet. Gynecol.* **2004**, *191*, 1459–1464. [CrossRef] [PubMed]
42. Van de Lagemaat, M.; Rotteveel, J.; Lafeber, H.N.; van Weissenbruch, M.M. Lean mass and fat mass accretion between term age and 6 months post-term in growth-restricted preterm infants. *Eur. J. Clin. Nutr.* **2014**, *68*, 1261–1263. [CrossRef] [PubMed]
43. Verkauskiene, R.; Beltrand, J.; Claris, O.; Chevenne, D.; Deghmoun, S.; Dorgeret, S.; Alison, M.; Gaucherand, P.; Sibony, O.; Levy-Marchal, C. Impact of fetal growth restriction on body composition and hormonal status at birth in infants of small and appropriate weight for gestational age. *Eur. J. Endocrinol.* **2007**, *157*, 605–612. [CrossRef] [PubMed]
44. Ornoy, A. Prenatal origin of obesity and their complications: Gestational diabetes, maternal overweight and the paradoxical effects of fetal growth restriction and macrosomia. *Reprod. Toxicol.* **2011**, *32*, 205–212. [CrossRef] [PubMed]
45. Bakal, U.; Aydin, S.; Sarac, M.; Kuloglu, T.; Kalayci, M.; Artas, G.; Yardim, M.; Kazez, A. Serum, Saliva, and Urine Irisin with and Without Acute Appendicitis and Abdominal Pain. *Biochem. Insights* **2016**, *9*, 11–17. [CrossRef] [PubMed]
46. Elhady, M.; Youness, E.R.; Gafar, H.S.; Abdel Aziz, A.; Mostafa, R.S.I. Circulating irisin and chemerin levels as predictors of seizure control in children with idiopathic epilepsy. *Neurol. Sci.* **2018**, *39*, 1453–1458. [CrossRef] [PubMed]

47. Zhao, Y.T.; Wang, H.; Zhang, S.; Du, J.; Zhuang, S.; Zhao, T.C. Irisin Ameliorates Hypoxia/Reoxygenation-Induced Injury through Modulation of Histone Deacetylase 4. *PLoS ONE* **2016**, *11*, e0166182. [CrossRef] [PubMed]
48. Mazur-Bialy, A.I.; Pochec, E.; Zarawski, M. Anti-Inflammatory Properties of Irisin, Mediator of Physical Activity, Are Connected with TLR4/MyD88 Signaling Pathway Activation. *Int. J. Mol. Sci.* **2017**, *18*, 701. [CrossRef] [PubMed]
49. Wikiera, B.; Zawadzka, K.; Laczmanski, L.; Sloka, N.; Bolanowski, M.; Basiak, A.; Noczynska, A.; Daroszewski, J. Growth Hormone Treatment Increases Plasma Irisin Concentration in Patients with Turner Syndrome. *Horm. Metab. Res.* **2017**, *49*, 122–128. [CrossRef] [PubMed]

International Journal of
Molecular Sciences

Article

Ginseng Berry Extract Rich in Phenolic Compounds Attenuates Oxidative Stress but not Cardiac Remodeling post Myocardial Infarction

Mihir Parikh [1,2,3,†], Pema Raj [1,3,†], Liping Yu [1,4], Jo-Ann Stebbing [1,4], Suvira Prashar [1,4], Jay C. Petkau [1,4], Paramjit S. Tappia [5], Grant N. Pierce [1,2,3], Yaw L. Siow [1,3,4], Dan Brown [1,4], Heather Blewett [1,3,4,6] and Thomas Netticadan [1,3,4,*]

1 Canadian Centre for Agri-Food Research in Health and Medicine, St. Boniface Hospital Albrechtsen Research Centre, Winnipeg, MB R2H2A6, Canada; mparikh@sbrc.ca (M.P.); praj@sbrc.ca (P.R.); lyu@sbrc.ca (L.Y.); jstebbing@sbrc.ca (J.-A.S.); sprashar@sbrc.ca (S.P.); jpetkau@sbrc.ca (J.C.P.); gpierce@sbrc.ca (G.N.P.); csiow@sbrc.ca (Y.L.S.); danielbrown500@gmail.com (D.B.); hblewett@sbrc.ca (H.B.)
2 Institute of Cardiovascular Sciences, St. Boniface Hospital Albrechtsen Research Centre, Max Rady College of Medicine, Rady Faculty of Health Sciences, University of Manitoba, Winnipeg, MB R2H2A6, Canada
3 Department of Physiology and Pathophysiology, University of Manitoba, Winnipeg, MB R3E0J9, Canada
4 Agriculture and Agri-Food Canada, Winnipeg, MB R2H2A6, Canada
5 Asper Clinical Research Institute & Office of Clinical Research, St. Boniface Hospital, Winnipeg, MB R2H2A6, Canada; ptappia@sbrc.ca
6 Department of Food and Human Nutritional Sciences, University of Manitoba, Winnipeg, MB R3T2N2, Canada
* Correspondence: tnetticadan@sbrc.ca
† These authors contributed equally to this work.

Received: 22 December 2018; Accepted: 20 February 2019; Published: 24 February 2019

Abstract: The cardioprotective effects of ginseng root extracts have been reported. However, nothing is known about the myocardial actions of the phenolic compounds enriched in ginseng berry. Therefore, this study was undertaken to investigate the effects of American ginseng berry extract (GBE) in an experimental model of myocardial infarction (MI). Coronary artery ligation was performed on Sprague–Dawley male rats to induce MI after which animals were randomized into groups receiving either distilled water or GBE intragastrically for 8 weeks. Echocardiography and assays for malondialdehyde (MDA) and TNF-α were conducted. Flow cytometry was used to test the effects of GBE on T cell phenotypes and cytokine production. Although GBE did not improve the cardiac functional parameters, it significantly attenuated oxidative stress in post-MI rat hearts. GBE treatment also resulted in lower than control levels of TNF-α in post-MI rat hearts indicating a strong neutralizing effect of GBE on this cytokine. However, there was no effect of GBE on the proportion of different T cell subsets or ex-vivo cytokine production. Taken together, the present study demonstrates GBE reduces oxidative stress, however no effect on cardiac structure and function in post-MI rats. Moreover, reduction of TNF-α levels below baseline raises concern regarding its use as prophylactic or preventive adjunct therapy in cardiovascular disease.

Keywords: *Panax quinquefolius*; ginseng berry; myocardial infarction; phenolic compounds

1. Introduction

Ginseng has been used for centuries in the traditional medicines in Asia. It is a herb derived from genus *Panax*, of family *Araliaceae* and has thirteen different species which are indigenous to Asia and North America [1]. Active components of ginseng include ginsenosides, saponins, polysaccharides, alkaloids, peptides, polyacetylenes, phenolics, and fatty acids [2,3].

Out of all the bioactive compounds of ginseng, ginsenosides have been more extensively studied compared to the phenolic compounds. However, phenolic compounds are now being investigated in several studies for their diverse biological actions [3]. Salicylic acid, p-coumaric acid, ferulic acid, cinnamic acid, and quercetin are some of the phenolic compounds identified in ginseng [3]. Unlike ginseng roots, the chemical composition of ginseng berry is less known. A comprehensive profile of the phenolic compounds found in Korean ginseng berry, root, and leaf has been reported [3]. In the Korean ginseng berry, chlorogenic acid was reported to be the predominant compound present, followed by gentisic acid and rutin [3]. In the North American ginseng berry, caffeic acid and chlorogenic acid were reported to be the active polyphenolic constituents in a study of the protective effect of ginseng berry extract against oxidant injury in cardiomyocytes [4]. In spite of reports of the presence of bioactives in ginseng berry, it is not commercially used and is often discarded as a 'useless by-products' [5,6].

A recent study showed that ginseng berry has higher total phenol content (including quercetin, rutin, and resveratrol) than the root [5]. Although both ginseng root and berry have pharmacological actions, in some instances berry has been found to be more effective. Ginseng berry has been reported to have a more potent antihyperglycemic action than the root at the same dose [7]. Ginseng berry has been shown to reduce coagulation of blood [8], improve insulin sensitivity [9], and regulate glucose metabolism [10]. Ginseng berry extract (GBE) was found to protect cardiomyocyte against oxidative stress by activating the antioxidant Nrf2 pathway [11]. An echocardiography study using ginseng root extract demonstrated a significant improvement in left ventricular function [12]. However, ginseng berry with its high phenolic content has not been evaluated for its effect on cardiac structure and function. Accordingly, the present study investigated the effect of a phenolic rich GBE on cardiac structure and function. Furthermore, the damage to the heart muscle resultant from a myocardial infarction (MI) triggers an immune response [13]. When this immune response is uncontrolled it can cause more damage to the heart. Phenolic compounds have been shown to modulate immune responses [14]; but there is a paucity of information on immunomodulatory effects of phenolic extracts from ginseng berry. Thus, our study also assessed immunomodulatory activity of GBE in the myocardial infarction (MI) model induced in rats by coronary artery ligation.

2. Results

2.1. Phenolic Content and Antioxidant Capacity of GBE

The total phenolic content of the GBE was 3586 ± 04 mg gallic acid equivalents/100 g dry weight using the Folin–Ciocalteu assay. Oxygen radical absorbance capacity (ORAC) assay was performed to assess the oxygen radical scavenging activity of the extract. As expected, GBE exhibited a strong antioxidant capacity with a value of 151,864 ± 883 μmol Trolox equivalents/100 g dry weight. The proximate analysis is presented in Table 1.

Table 1. Proximate analysis of ginseng berry extract.

Constituent	Value (%)
Moisture	8.19
Dry matter	91.81
Crude protein	8.61
Crude fibre	1.54
Fat	0.90
Ash	12.50
Non-fibre carbohydrates	68.26
Total digestible nutrients	69.77

2.2. Body Weight and Heart Weight Characteristics after MI

Biometrical characteristics were assessed to check whether the coronary artery ligation or treatment with GBE was able to produce any alterations. No significant changes in body weight and heart weight were observed between groups 8 weeks post-surgery. The percentage of scar weight was 21.31% less in GBE-treated MI animals, but not statistically significant when compared to the water-treated MI animals (Table 2). No differences were observed in the heart weight to tibia length ratio values between groups. Similarly, liver and lung weights also did not show significant differences between groups (values not shown).

Table 2. Biometrical characteristics of sham and MI animals with and without ginseng berry extract treatment for 8 weeks after coronary artery ligation.

Parameter	Sham-W	Sham-G	MI-W	MI-G
Body weight (g)	573.1 ± 11	579.9 ± 21	538.4 ± 12	549 ± 16
Heart weight (g)	1.433 ± 0.05	1.355 ± 0.05	1.458 ± 0.03	1.415 ± 0.05
LV weight (g)	0.70 ± 0.060	0.86 ± 0.051	0.73 ± 0.03	0.82 ± 0.04
% infarct	-	-	32.37 ± 2.8	25.47 ± 2.3
Heart weight/Tibia length (g/cm)	0.32 ± 0.009	0.30 ± 0.01	0.33 ± 0.0	0.31 ± 0.0
LV weight/Tibia length (g/cm)	0.15 ± 0.013	0.19 ± 0.011	0.16 ± 0.01	0.18 ± 0.01

Values presented are mean ± SEM. MI: Myocardial infarction; LV: Left Ventricle; Sham-W: Sham MI treated with distilled water (n = 8); Sham-G: Sham MI treated with GBE 150 mg/kg/body weight/day (n = 8); MI-W: MI treated with distilled water (n = 12–14); MI-G: MI treated with GBE 150 mg/kg/body weight/day (n = 12–14).

2.3. Lack of Improvement in Cardiac Structure and Function with GBE Treatment

M-mode echocardiography was carried out to assess the effect of GBE on the left ventricular remodeling at 4 and 8 weeks post-MI. At 4 weeks, left ventricle (LV) internal diameter (LVID) values at systole and diastole were comparable between water- and GBE-treated sham animals (Figure 1A,C). In contrast, the water-treated and GBE-treated MI groups had significantly higher LVID values when compared to the water-treated sham group indicating left ventricular dilatation (Figure 1A,C). A similar trend was observed at 8 weeks with significantly high LVID values at systole and diastole in the water-treated and GBE-treated MI groups in comparison to water-treated sham group (Figure 1B,D). However, no changes were observed in LVID after GBE treatment in MI group (vs. water treated MI group) at 4 and 8 weeks (Figure 1A–D). LV volumes were also observed to be altered. MI animals treated with water and GBE showed significantly higher end-diastolic volume (EDV) and end-systolic volume (ESV) when compared to water treated sham animals (Figure 1E–H). There was no difference observed between sham animals treated with water and GBE, for EDV and ESV. GBE treatment of MI animals did not produce significant changes in EDV and ESV when compared to the water-treated MI animals (Figure 1E–H).

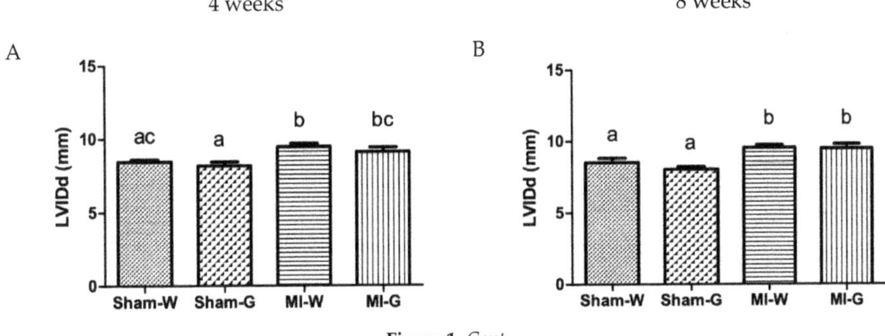

Figure 1. *Cont.*

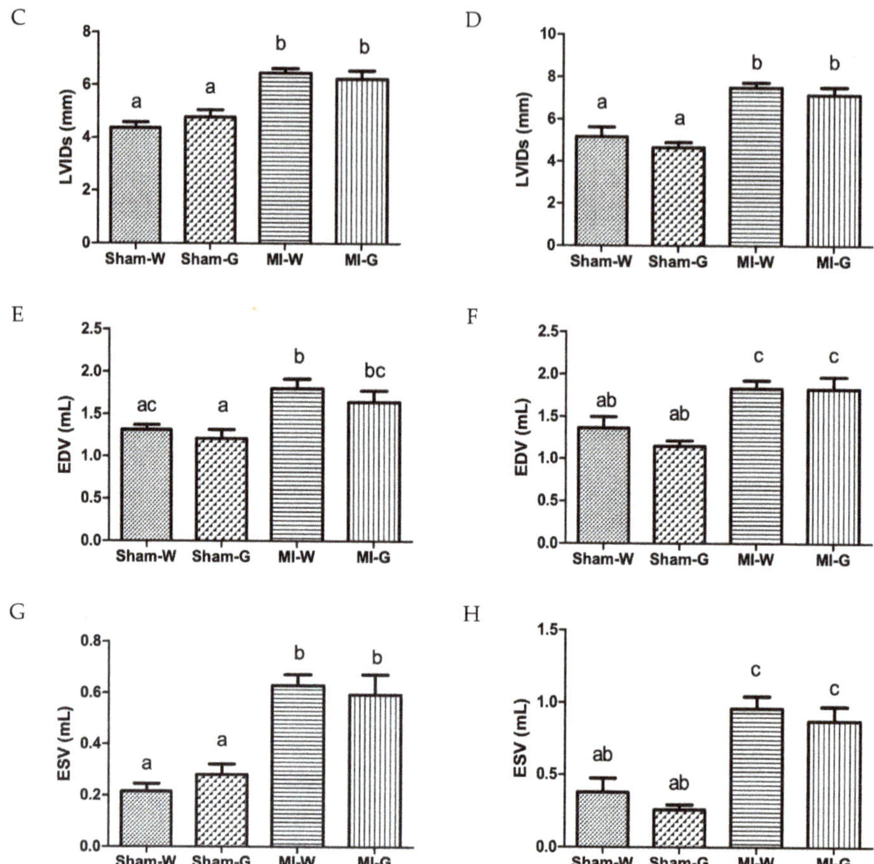

Figure 1. Effect of ginseng berry extract on ventricular chamber dilatation and remodeling. (**A**) LV internal diameter at diastole (LVIDd) at 4 weeks; (**B**) LVIDd at 8 weeks; (**C**) LV internal diameter at systole (LVIDs) at 4 weeks; (**D**) LVIDs at 8 weeks; (**E**) end diastolic volume (EDV) at 4 weeks; (**F**) EDV at 8 weeks; (**G**) end systolic volume (ESV) at 4 weeks; (**H**) ESV at 8 weeks. Values presented are mean ± SEM. Bars with differing letter are significantly different, $p < 0.05$. Sham-W: Sham MI treated with distilled water ($n = 8$); Sham-G: Sham MI treated with GBE 150 mg/kg/body weight/day ($n = 8$); MI-W: MI treated with distilled water ($n = 12–14$); MI-G: MI treated with GBE 150 mg/kg/body weight/day ($n = 12–14$).

To check for left ventricular hypertrophy along with the increased LV dilatation, the thickness of interventricular septal thickness (IVS) and LV posterior wall (LVPW) was measured. IVS thickness at systole was observed to be significantly lower at 4 weeks (Figure 2C), and at both systole and diastole at 8 weeks (Figure 2B,D), in the water- and GBE-treated MI animals when compared to water treated sham animals. LVPW thickness values were also found to be significantly low at systole in the water-treated MI group when compared to water treated sham group (Figure 2H). At both 4 and 8 weeks, values for IVS and LVPW thickness were found to be comparable between water and GBE treated MI groups (Figure 2A–H).

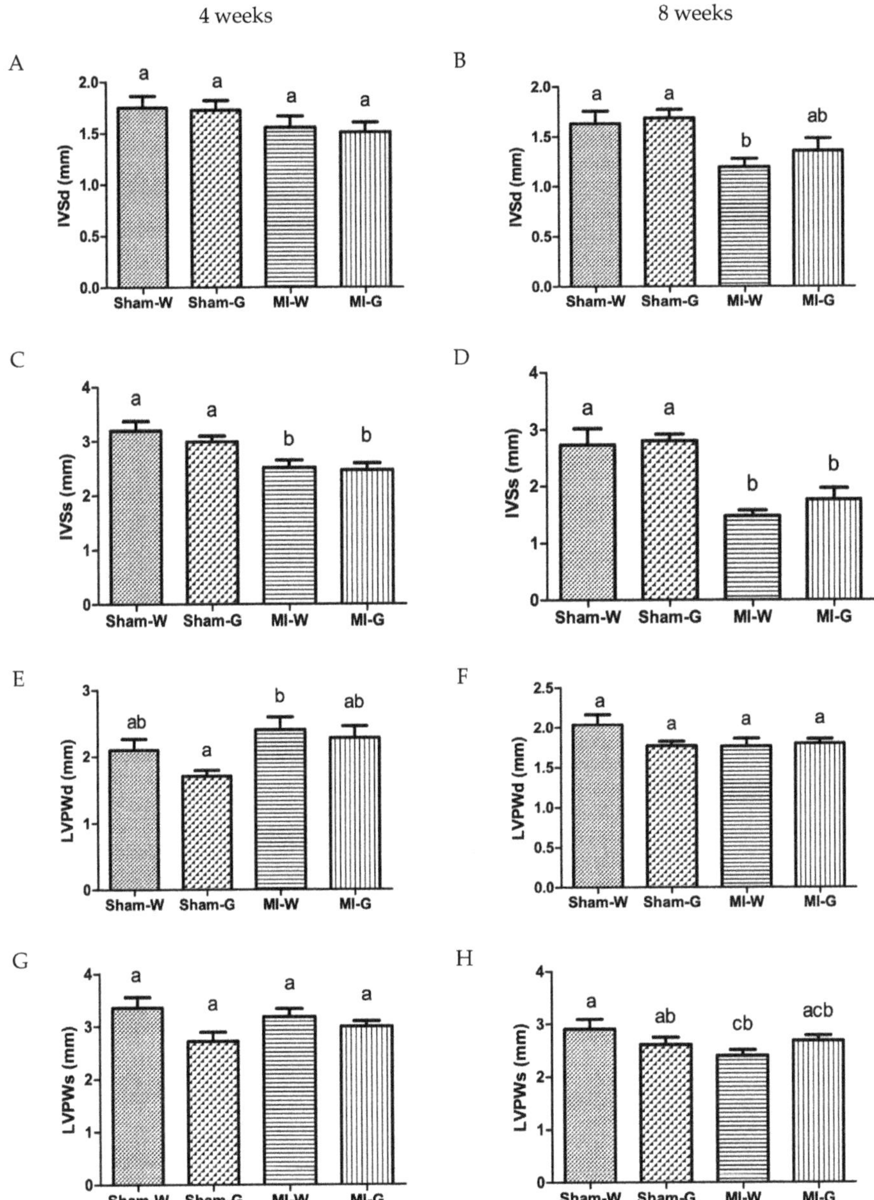

Figure 2. Effect of ginseng berry extract on cardiac hypertrophy. (**A**) Interventricular septal thickness at diastole (IVSd) at 4 weeks; (**B**) IVSd at 8 weeks; (**C**) IVS at systole (IVSs) at 4 weeks; (**D**) IVSs at 8 weeks; (**E**) LV posterior wall thickness at diastole (LVPWd) at 4 weeks; (**F**) LVPWd at 8 weeks; (**G**) LV posterior wall thickness at systole (LVPWs) at 4 weeks; (**H**) LVPWs at 8 weeks. Values presented are mean ± SEM. Bars with differing letter are significantly different, $p < 0.05$. Sham-W: Sham MI treated with distilled water ($n = 8$); Sham-G: Sham MI treated with GBE 150 mg/kg/body weight/day ($n = 8$); MI-W: MI treated with distilled water ($n = 12$–14); MI-G: MI treated with GBE 150 mg/kg/body weight/day ($n = 12$–14).

Cardiac function was assessed by using M-mode echocardiography. Ejection fraction (EF), a measure of cardiac contraction (systolic heart function), was identical between water- and GBE-treated sham animals. The water-treated and GBE-treated MI groups showed significantly lower EF at 4 and 8 weeks when compared to water treated sham group (Figure 3A,B). EF values were not significantly different between the water- and GBE-treated MI groups (Figure 3A,B). Another parameter of cardiac contraction, fractional shortening (FS), was also significantly reduced in the water- and GBE-treated MI groups when compared to water treated Sham controls (Figure 3C,D). No significant differences were noted for FS values between water and GBE treated MI groups (Figure 3C,D).

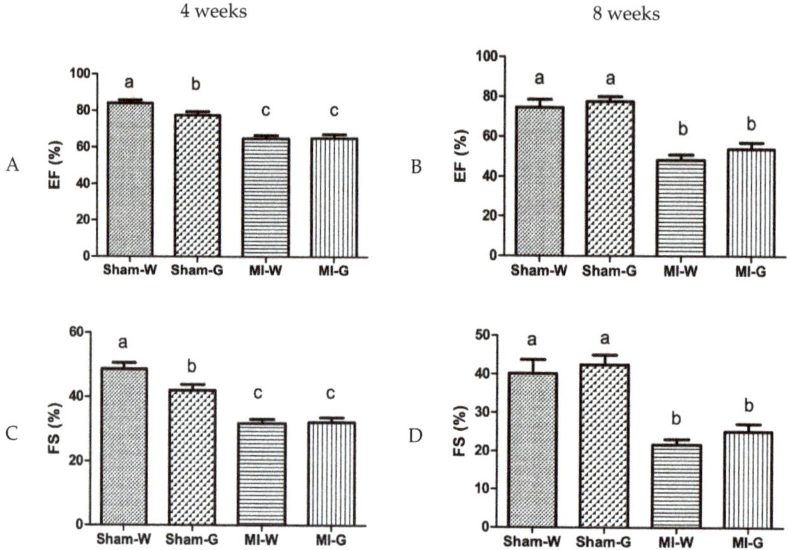

Figure 3. Effect of ginseng berry extract on systolic heart function. (**A**) Ejection fraction (EF) at 4 weeks; (**B**) EF at 8 weeks; (**C**) fractional shortening (FS) at 4 weeks; (**D**) FS at 8 weeks. Values presented are mean ± SEM. Bars with differing letter are significantly different, $p < 0.05$. Sham-W: Sham MI treated with distilled water (n = 8); Sham-G: Sham MI treated with GBE 150 mg/kg/body weight/day (n = 8); MI-W: MI treated with distilled water (n = 12–14); MI-G: MI treated with GBE 150 mg/kg/body weight/day (n = 12–14).

Doppler echocardiography was used to examine the changes in cardiac relaxation (diastolic heart function), post-MI. At 4 and 8 weeks, no changes were observed in the parameters assessing diastolic heart function, mitral valve (MV) E and A wave velocity in all groups (Figure 4A,B). At 4 weeks, the values of another diastolic heart function parameter, Isovolumic relaxation time (IVRT) was also found to be similar in all the groups (Figure 4C). However, IVRT was observed to be significantly increased in the water- and GBE-treated MI groups when compared to water treated sham controls at 8 weeks (Figure 4D). GBE-treated MI animals had values comparable to water-treated MI animals (Figure 4D).

2.4. Reduction in Oxidative Stress and Inflammation with Ginseng Berry Extract Treatment

The 8-week water treated MI animals demonstrated significantly increased MDA levels in the heart tissue when compared to water-treated sham animals (Figure 5A). The GBE treated MI group showed a significant decrease (29%) in MDA levels when compared to the water-treated MI group (Figure 5A). Although, values for TNF-α were comparable between 8-week water treated sham and MI groups, a significant reduction of ~94% was observed in 8-week GBE-treated MI and sham animals when compared to water-treated sham and MI animals, respectively (Figure 5B).

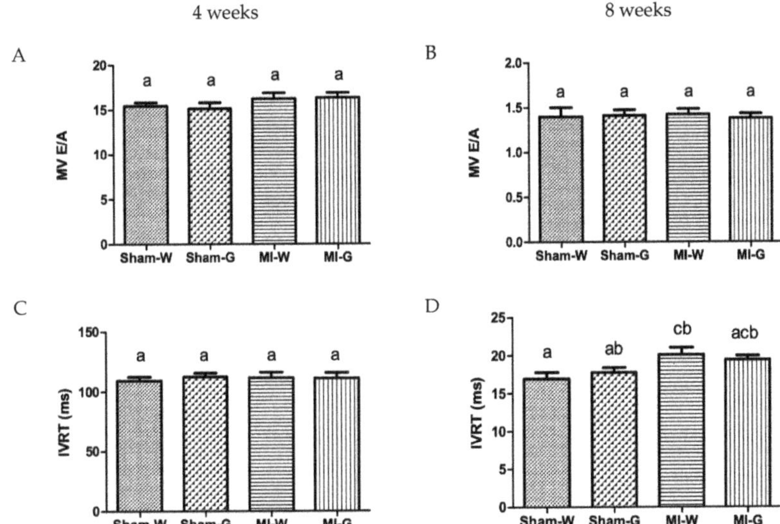

Figure 4. Effect of ginseng berry extract on diastolic heart function. (**A**) Mitral valve E- to A- wave ratio (MV E/A) at 4 weeks; (**B**) MV E/A at 8 weeks; (**C**) Isovolumic relaxation time (IVRT) at 4 weeks; (**D**) IVRT at 8 weeks. Values presented are mean ± SEM. Bars with differing letter are significantly different, $p < 0.05$. Sham-W: Sham MI treated with distilled water ($n = 8$); Sham-G: Sham MI treated with GBE 150 mg/kg/body weight/day ($n = 8$); MI-W: MI treated with distilled water ($n = 12$–14); MI-G: MI treated with GBE 150 mg/kg/body weight/day ($n = 12$–14).

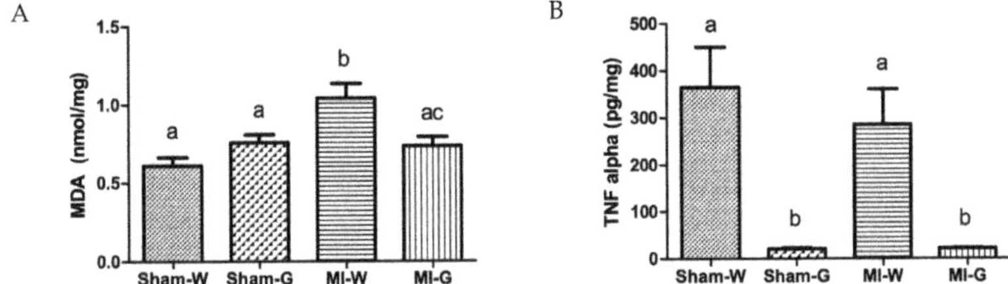

Figure 5. Effect of ginseng berry extract on oxidative stress and inflammation. (**A**) Malondialdehyde (MDA); (**B**) tumor necrosis factor-α (TNF-α). Values presented are mean ± SEM. Bars with differing letter are significantly different, $p < 0.05$. Sham-W: Sham MI treated with distilled water ($n = 8$); Sham-G: Sham MI treated with GBE 150 mg/kg/body weight/day ($n = 8$); MI-W: MI treated with distilled water ($n = 12$–14); MI-G: MI treated with GBE 150 mg/kg/body weight/day ($n = 12$–14).

2.5. Immune Cell Phenotypes

Using forward vs. side scatter we can identify viable lymphoid cells. Cell shrinkage, one of the first indicators of apoptosis, can be identified by a decrease in forward light scatter. There was no difference among groups in the proportion of cells that appeared in the non-viable lymphoid region.

In the viable lymphoid region (Figure 6), MI had no effect on the proportion of total T-cells (identified as $CD3^+$), helper T-cells (identified as $CD3^+CD4^+CD8^-$), cytotoxic T-cells (identified as $CD3^+CD4^-CD8^+$) or activated T-cells (identified as $CD3^+CD25^+$) and activated cytotoxic T-cells (identified as $CD3^+CD4^-CD8^+CD25^+$). MI did result in an 8.0% lower proportion of activated helper T-cells (identified as $CD3^+CD4^+CD8^-CD25^+$) and 9.8% lower proportion of T-regulatory cells (identified as $CD3^+CD4^+CD8^-CD25^+foxp3^+$) compared to sham-operated rats. GBE treated rats had a 9.4% higher proportion of T-cells and 13.2% higher proportion of cytotoxic T-cells, but a 7.4% lower proportion of helper T-cells compared to water treated rats. There was no effect of GBE on the proportion of activated T-cells, activated helper T-cells, activated cytotoxic T-cells and T-regulatory cells (Table 3).

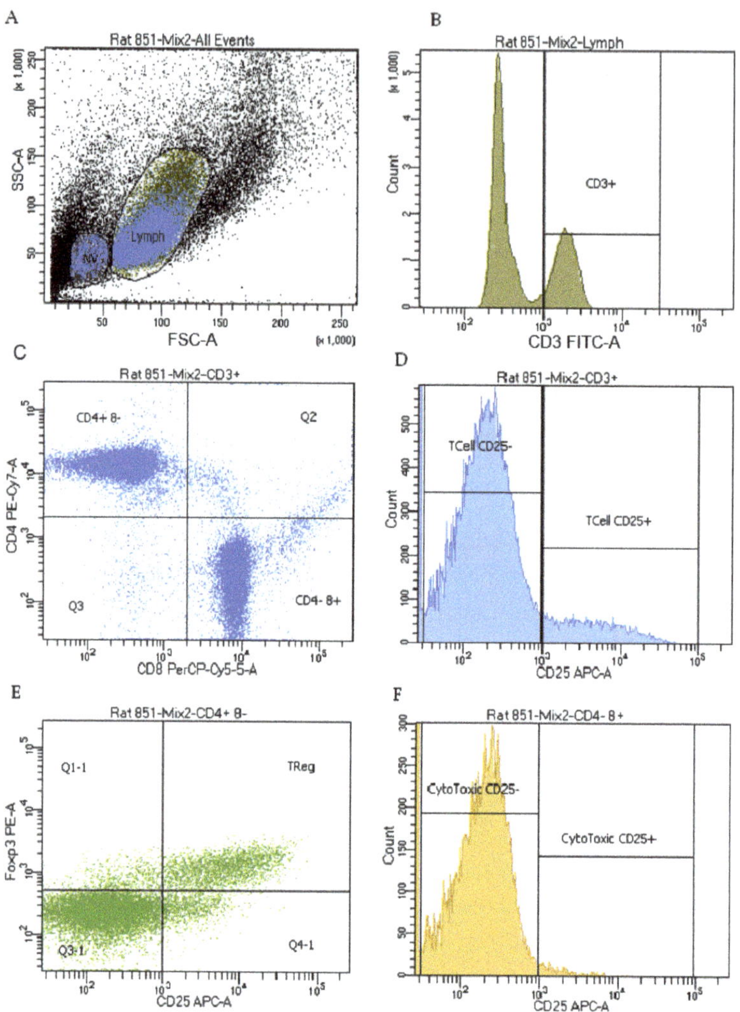

Figure 6. Immune cell phenotyping. (**A**) Definition of viable lymphocytes (Lymph) and non-viable lymphocytes (NV); (**B**) Definition of CD3 binding after gating on Lymph; (**C**) Definition of CD4 and CD8 after gating on $CD3^+$ Lymph; (**D**) Definition of CD25 binding after gating on $CD3^+$ Lymph; (**E**) Definition of Foxp3 and CD25 after gating on $CD3^+CD4^+CD8^-$ Lymph; (**F**) Definition of CD25 binding after gating on $CD3^+CD4^-CD8^+$ Lymph.

In the non-viable lymphoid region compared to the viable lymphoid region, there were lower proportions of all T-cell phenotypes, except for helper T-cells (Figure 7). Rats that had an MI had a 30% lower proportion of activated cytotoxic T-cells compared to sham-operated rats in the non-viable region. There was a trend towards a lower proportion of T-regulatory cells after MI compared to sham ($p = 0.06$), but no significant difference in the proportion of total T-cells, activated T-cells, helper T-cells, activated helper T-cells, or cytotoxic T-cells in the non-viable region. Compared to water-treated controls, rats treated with GBE had a 29.4% higher proportion of T-cells and a 10% lower proportion of activated Helper T-cells in this region. There was no difference in the proportion of activated T-cells, helper T-cells, T-regulatory cells, cytotoxic T-cells and activated cytotoxic T-cells (Table 3).

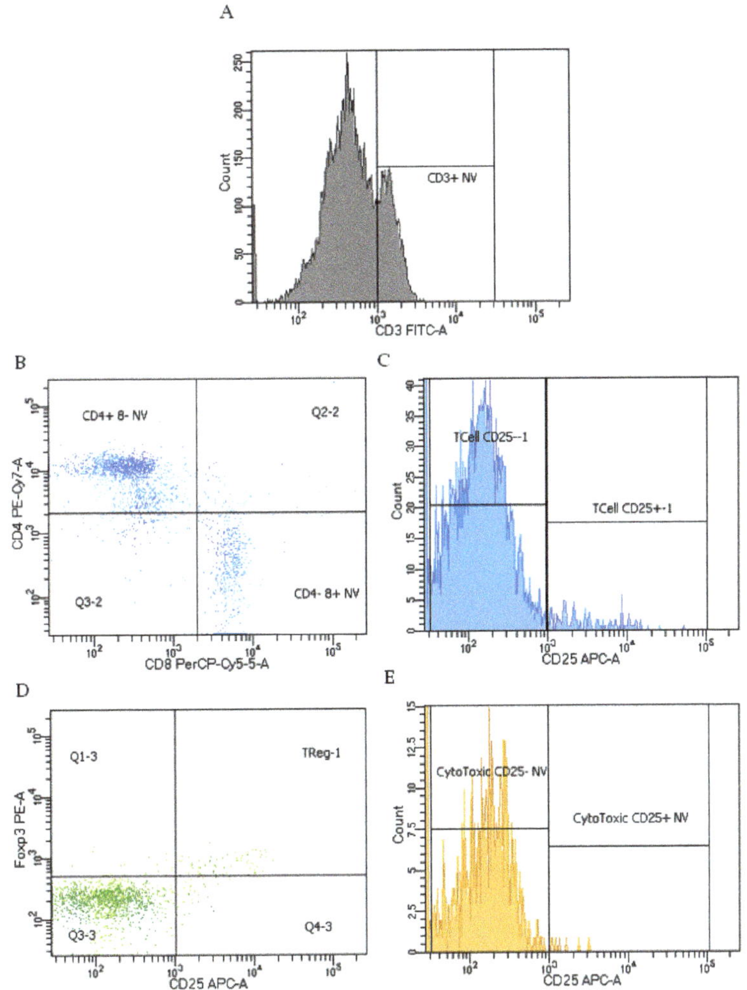

Figure 7. Representative flow cytometry plot for non-viable lymphocytes. (**A**) Definition of CD3 binding after gating on non-viable lymphocytes (NV); (**B**) Definition of CD4 and CD8 after gating on $CD3^+$ NV; (**C**) Definition of CD25 binding after gating on $CD3^+$ NV; (**D**) Definition of Foxp3 and CD25 after gating on $CD3^+CD4^+CD8^-$ NV; (**E**) Definition of CD25 binding after gating on $CD3^+CD4^-CD8^+$ NV.

2.6. Cytokines Production

ConA stimulated splenocytes from rats that had an MI when compared to controls, did not show a significant difference in IL-2, IFNγ, IL-10 or TNFα concentrations produced or between GBE treatment and control (Table 4).

Unstimulated samples were below the lower limit of detection for IFNγ (6.8 pg/mL) and TNFα (27.7 pg/mL). Unstimulated concentrations of IL-2 and IL-10 were not affected by MI or GBE treatment (Table 4).

Table 3. T-cell phenotypes in viable and non-viable gates.

Lymphocyte Phenotypes	Model			Treatment			Interaction				
	Sham $n = 16$	MI $n = 23$	p-Value	Water $n = 20$	GBE $n = 19$	p-Value	Sham Water $n = 8$	Sham GBE $n = 8$	MI Water $n = 12$	MI GBE $n = 11$	p-Value
Viable Lymphocytes											
$CD3^+$ % lymphocytes	30.0 ± 0.9	30.0 ± 0.8	0.99	28.7 ± 0.8	31.4 ± 0.9	0.03	29.2 ± 1.3 ^{ab}	30.9 ± 1.3 ^{ab}	28.2 ± 1.0 ^{b}	31.9 ± 1.1 ^{a}	0.39
[1] $CD25^+$ % of $CD3^+$	11.5 ± 0.3	10.9 ± 0.3	0.14	11.5 ± 0.3	10.9 ± 0.3	0.15	11.7 ± 0.5	11.3 ± 0.4	11.3 ± 0.4	10.4 ± 0.4	0.65
$CD4^+CD8^-$ % of $CD3^+$	57.1 ± 1.2	57.5 ± 1.0	0.80	59.5 ± 1.1	55.1 ± 1.1	0.01	58.7 ± 1.7 ^{a}	55.5 ± 1.7 ^{ab}	60.4 ± 1.4 ^{a}	54.6 ± 1.5 ^{b}	0.41
[2] $CD25^+$ % of $CD3^+CD4^+CD8^-$	16.3 ± 0.4	15.0 ± 0.4	0.04	15.9 ± 0.4	15.4 ± 0.4	0.40	16.7 ± 0.6 ^{a}	15.9 ± 0.6 ^{ab}	15.1 ± 0.6 ^{ab}	14.9 ± 0.6 ^{b}	0.54
[3] $Foxp3^+CD25^+$ % of $CD3^+CD4^+CD8^-$	12.3 ± 0.4	11.1 ± 0.3	0.03	11.9 ± 0.4	11.6 ± 0.4	0.60	12.8 ± 0.5 ^{a}	11.8 ± 0.5 ^{ab}	10.9 ± 0.5 ^{b}	11.4 ± 0.5 ^{ab}	0.17
$CD4^-CD8^+$ % of $CD3^+$	39.6 ± 1.2	39.2 ± 1.0	0.80	37.0 ± 1.1	41.9 ± 1.1	0.004	37.8 ± 1.7 ^{ab}	41.5 ± 1.7 ^{a}	36.2 ± 1.4 ^{b}	42.3 ± 1.5 ^{a}	0.46
[4] $CD25^+$ % of $CD3^+CD4^-CD8^+$	2.8 ± 0.2	2.7 ± 0.1	0.66	2.7 ± 0.2	2.7 ± 0.1	0.97	2.7 ± 0.2	2.9 ± 0.2	2.8 ± 0.2	2.6 ± 0.2	0.30
Non-Viable lymphocytes											
$CD3^+$ % non-viable cells	27.2 ± 1.1	25.7 ± 0.9	0.31	23.1 ± 1.0	29.9 ± 1.0	<0.0001	22.9 ± 1.5 ^{b}	31.5 ± 1.5 ^{a}	23.2 ± 1.3 ^{b}	28.2 ± 1.4 ^{a}	0.22
[5] $CD25^+$ % of $CD3^+$	4.9 ± 0.2	4.3 ± 0.2	0.07	4.9 ± 0.2	4.3 ± 0.2	0.06	4.9 ± 0.3 ^{a}	4.8 ± 0.3 ^{a}	4.8 ± 0.3 ^{a}	3.9 ± 0.3 ^{b}	0.17
$CD4^+CD8^-$ % of $CD3^+$	75.6 ± 1.3	74.4 ± 1.1	0.49	74.9 ± 1.1	75.1 ± 1.2	0.94	75.7 ± 1.8	75.5 ± 1.8	74.2 ± 1.5	74.7 ± 1.6	0.83
[6] $CD25^+$ % of $CD3^+CD4^+CD8^-$	4.9 ± 0.2	4.5 ± 0.2	0.13	4.9 ± 0.2	4.4 ± 0.2	0.05	4.9 ± 0.3 ^{a}	4.9 ± 0.3 ^{a}	5.0 ± 0.2 ^{a}	4.0 ± 0.2 ^{b}	0.08
[7] $Foxp3^+CD25^+$ % of $CD3^+CD4^+CD8^-$	3.1 ± 0.1	2.7 ± 0.1	0.06	3.0 ± 0.1	2.8 ± 0.1	0.18	3.2 ± 0.2 ^{a}	3.0 ± 0.2 ^{ab}	2.9 ± 0.2 ^{ab}	2.6 ± 0.2 ^{b}	0.59
$CD4^-CD8^+$ % of CD3+	17.5 ± 0.9	18.3 ± 0.8	0.56	17.3 ± 0.9	18.5 ± 0.9	0.36	16.7 ± 1.3	18.4 ± 1.3	18.0 ± 1.1	18.6 ± 1.2	0.66
[8] $CD25^+$ % of $CD3^+CD4^-CD8^+$	2.3 ± 0.3	1.6 ± 0.2	0.05	2.2 ± 0.2	1.7 ± 0.2	0.17	2.4 ± 0.4 ^{a}	2.2 ± 0.4 ^{ab}	1.9 ± 0.3 ^{ab}	1.2 ± 0.3 ^{b}	0.54

Values are means (pg/mL) ± standard error. Means with different superscript letters are significantly different ($p < 0.05$). GBE: Ginseng berry extract. [1,4]: Outliers (value >3× standard deviation of mean) were removed: 2 from MI water and 1 from sham water group; [2,3,7]: Outliers (value >3× standard deviation of mean) were removed: 2 from MI water; [5,6,8]: Outlier (value >3× standard deviation of mean) were removed: 1 from MI water.

Table 4. Supernatant cytokine concentrations from concanvalin A-stimulated and unstimulated splenocytes.

Cytokines	Model			Treatment			Interaction				
	Sham $n = 16$	MI $n = 23$	p-Value	Water $n = 20$	GBE $n = 19$	p-Value	Sham Water $n = 8$	Sham GBE $n = 8$	MI Water $n = 12$	MI GBE $n = 11$	p-Value
IL-2 ConA	2111 ± 181	1890 ± 79	0.24	1910 ± 133	2055 ± 114	0.47	2067 ± 279	2155 ± 248	1805 ± 125	1983 ± 90	0.80
[1] IL-2 UNS	3.4 ± 0.2	4.0 ± 0.4	0.35	3.9 ± 0.4	3.4 ± 0.3	0.52	3.5 ± 0.6	3.3 ± 0.2	4.2 ± 0.6	3.6 ± 0.5	0.72
[2] IFNγ ConA	383.0 ± 75	267.0 ± 35	0.13	312.9 ± 59	312.7 ± 47	0.79	427.8 ± 131	343.7 ± 89	245.9 ± 49	290.1 ± 51	0.40
IFNγ UNS	ND	ND	/	ND	ND	/	ND	ND	ND	ND	/
[3] IL-10 ConA	49.2 ± 3.3	47.3 ± 3.1	0.65	48.4 ± 4.1	47.9 ± 2.3	0.81	51.9 ± 5.4	46.9 ± 4.2	45.9 ± 5.9	48.7 ± 2.7	0.41
[4] IL-10 UNS	33.1 ± 4.2	29.1 ± 1.8	0.49	28.3 ± 2.4	33.1 ± 3.1	0.30	29.2 ± 6.8	35.4 ± 5.6	27.9 ± 2.4	30.8 ± 2.8	0.70
TNFα ConA	773.6 ± 57	727.4 ± 31	0.43	779.1 ± 42	711.8 ± 41	0.30	792.9 ± 87	754.3 ± 79	770.0 ± 42	680.9 ± 43	0.68
TNFα UNS	ND	ND	/	ND	ND	/	ND	ND	ND	ND	/

Values are means (pg/mL) ± standard error. Abbreviations: Ginseng berry extract (GBE); Interleukin-2 (IL-2); concanvalin A (ConA); unstimulated (UNS); interferon-γ (IFNγ); interleukin-10 (IL-10); tumor necrosis factor α (TNFα); ND = not detectable. [1] IL-2 UNS 4 samples from sham/water, 2 from sham/GBE, 4 from MI/water, 6 from MI/GBE were below detection limit of 0.46 pg/Ml; [2] IFNγ ConA 1 outlier removed from sham/water group; [3] IL-10 ConA 2 samples from MI/water and 1 from sham/water were below detection limit of 19.4 pg/mL; [4] IL-10 UNS 5 samples from sham/water, 3 from sham/GBE, 5 from MI/water, 6 from MI/GBE were below detection limit of 19.4 pg/mL.

3. Discussion

The present study is the first to report that ginseng berry phenolic extract has a strong antioxidant effect despite no effect on cardiac structure and function in the condition of a MI.

Although there is no information in the literature on the effects of ginseng berry on heart structure and function in any settings of cardiovascular disease, some studies conducted with Korean ginseng (*Panax ginseng*) have shown that GBE exerts anti-atherosclerotic [15], anti-diabetic, and anti-obesity effects [16]. Unlike ginseng root extract, GBE has been found to inhibit the mRNA expression of interleukins 1β and 6 inflammatory markers and it is likely through this mechanism that it exerts its antihyperglycemic and anti-obesity effects [17]. GBE also increases prothrombin time, indicating potential atherothrombotic effects [8].

Ginseng consists of phenolic compounds which include phenolic acids and flavonoids with strong antioxidant activity, but these are less characterized as compared to the ginsenosides. Chung et al. reported 4–9-fold higher phenolic compounds in the 3–6 year-old Korean ginseng berry (*Panax ginseng*) compared to the same age roots [3]. A higher phenolic content has been observed in the older Korean ginseng berries when compared to younger ones; the total amount increasing by 20–48% with an increase in the cultivation year. DPPH free-radical-scavenging activity (DPPH activity) of the 3–6 years old Korean ginseng berry is 3–5 fold higher than the root, suggesting the strong antioxidant activity of the berry [3]. In this context, an extract derived from American ginseng berries protected cardiomyocytes from oxidative stress induced by H_2O_2 and antimycin A [8]. An approximately 60% reduction in cardiomyocyte death was observed with the American GBE, clearly demonstrating that it can salvage cardiomyocytes from the oxidative injury [18]. The proposed mechanisms for the American GBE antioxidant actions included direct free radical scavenging activity and stimulation of NO synthesis [18]. This study, therefore, examined the antioxidant potential of North American GBE (*Panax quinquefolius*) in addition to examining its cardioprotective potential in an experimental model of MI.

The present study demonstrated a significant reduction in ventricular function after a MI. The reasons behind this adverse outcome include an impaired contractile function as well as infarct expansion leading to ventricular dilatation. Two-dimensional echocardiography performed on patients with an acute transmural MI has demonstrated dilatation of the infarct zone. This infarct expansion phase, which occurs due to the slippage of necrotic myocardial fibers, starts as early as 3 days after infarction [19]. This early regional dilation occurring in the infarcted zone results in an overall left ventricular dilatation. The ventricular dilatation may continue until 30 months after infarction. Unlike the early expansion phase, both infarcted and non-infarcted segments are affected during the chronic phase dilation [20]. This disproportionate cardiac dilatation alters left ventricular topography and is associated with poor prognosis for long-term survival of MI patients [19]. Left ventricular dilatation has been associated with the progression of cardiac dysfunction as cardiac and stroke indices decrease. Of note, ventricular dilatation results in and is not the result of deteriorating cardiac function [21]. Left ventricular function is usually assessed by ejection fraction and end diastolic and systolic volumes. Ejection fraction has been observed to have a strong effect on mortality. With the increase in left ventricular internal diameter, there are substantial increases in end systolic and diastolic volumes with consequently increased stress on ventricular wall. Earlier studies with gated scintiphotography revealed an increase in left ventricular end-systolic volume and decrease in ejection fraction in MI patients compared to normal subjects [22]. Three months follow-up demonstrated increased values of LV ejection fraction in patients with clinical improvement and decreased values with clinically worsened cases [22]. Moller et al demonstrated that echocardiographically determined ejection fraction was a powerful predictor of mortality during a median follow up of 19 months [23]. A progressive increase was noted in the cardiac mortality in patients with ejection fraction below 40% [24]. Another important determinant of the ventricular function post-MI is the end systolic volume (ESV). Consideration of ESV along with EF adds more predictive power for mortality risk stratification of MI patients [25]. Echocardiographic analysis revealed that GBE treatment does not prevent infarct-related impairment in cardiac structure and function.

Reactive oxygen species produced during oxidative stress damage membrane lipids, proteins, DNA thereby causing apoptosis of cardiomyocytes and eventually resulting in cardiac dysfunction [26]. Infiltration of inflammatory cells in the myocardium after an ischemic event initiates an exaggerated inflammatory response, this further accelerates and worsens ventricular remodeling by increasing myocardial injury [27]. Higher levels of the inflammatory marker TNF-α have been associated with ventricular dilation and cardiac fibrosis. Our results demonstrate a significant decrease in the LV levels of TNF-α and MDA due to GBE treatment, indicating that it can reduce the inflammatory and oxidative response in MI. The substantial attenuation of cardiac oxidative stress observed in MI rats could be associated with the robust antioxidant capacity of GBE, as determined by its oxygen radical scavenging activity. Despite exerting strong antioxidant effects, however, GBE was unable to ameliorate cardiac remodeling and rescue cardiac function in MI rats.

The large GBE mediated decline in cardiac inflammatory response observed in the present study is consistent with a previous report which showed that treatment with Korean ginseng berry (*Panax ginseng*) suppressed the expression of TNF-α and thereby reduced atherosclerotic lesions [15]. Chronic use of TNF-α blockers has nevertheless been associated with increased risk of cardiovascular complications. A deficiency in TNF signaling for an extended period of time can cause immune system defects. The cardiac effects of TNF-α are biphasic. While high levels of TNF-α is associated with apoptosis, the basal level of TNF-α is required to maintain cytoprotective Nrf2 signaling [28]. Kurrelmeyer et al reported the importance of TNF signaling in protecting cardiomyocytes against ischemic injury [29]. The protective mechanism against hypoxic damage could be via activation of protein kinase A which stimulates SERCA2a thereby reducing intracellular calcium concentration during calcium overload [30]. Thus, TNF-α signaling helps in pumping calcium ions in the sarcoplasmic reticulum and restoring cytosolic calcium back to baseline levels [30]. It is possible that the reduction of TNF-α levels due to GBE observed in the present study may be as a consequence of: (a) reduced expression of the TNF-α gene and a lessened production of TNF-α protein, or, (b) increased production of soluble TNF-α receptors, which could bind and inactivate TNF-α [31]. However, the results of our examination of T cell phenotypes and function and ex-vivo cytokine production showed no differences in the immune response to GBE.

Limitations and Future Opportunities

This study has certain limitations that should be taken into account while interpreting the observed findings. Despite the study being sufficiently powered for analysis, the sample size may have been too small as the observed effect size was small. A preliminary pharmacokinetic study would have assisted in appropriate dose selection and treatment regimen. In addition, more than once daily administration might have resulted in a sufficient plasma steady state concentration that could have produced a different outcome. Furthermore, pretreatment with GBE may have resulted in an adequate cardiac tissue distribution of polyphenols necessary for potentiating the endogenous antioxidant system during an ischemic insult. The results of our study open up opportunities for further investigation including a comparative study of a polyphenol rich extract from ginseng berries vs. polyphenol rich extract from roots, and other parts of the plant.

4. Materials and Methods

4.1. Ginseng Berry Pulp Extract

Ginseng berries from three-year-old *Panax quinquefolius* L. (North American ginseng) were provided by C & R Atkinson Farms Ltd, St. Williams, ON, Canada. The berries were stored at −20 °C until they could be freeze-dried in smaller batches. After freeze-drying berries, the seeds were removed and the pulp was ground to a fine powder. Pulp extracts were prepared in batches at a ratio of 10 g ground pulp per 200 mL of 80% methanol. The slurry was mixed at room temperature using a rotary shaker, at 200 rpm for 1.5 h. The mixture was centrifuged at room temperature for 15 min at 10,000× g, saving the pellet for further extraction. The supernatant was collected and stored at −20 °C and the remaining pulp went through two additional rounds of extraction with 80% methanol. All supernatants were pooled together and stored at −20 °C. The solvent was removed by rotary evaporation and the remaining aqueous extract was freeze-dried. Multiple extracts were performed to produce sufficient ginseng berry pulp extract for the study. Dried extracts were further ground to a fine powder and combined to form a homogeneous product for the study. The pH of the dried extract was 4.5 and the final product was stored at −80 °C

4.2. Quantification of Phenolic Content and Antioxidant Activity

To ensure complete mixing of dried samples three biological replicates were prepared. A quantity of 0.125 g dried powder of each of these replicates (a, b, and c) was dissolved in 3 mL of autoclaved 18 'Ω Milli-Q water. Tubes were vortexed, sonicated for 10 min and then centrifuged at 10,000× g. The supernatant was used for subsequent measurement. The assay was performed in a 96 well plate by combining reagent and sample in sodium carbonate buffer. By using a protocol modified from Ainsworth and Gillespie [32], the total phenolic content was determined by spectrometry using SpectraMax M5 (Molecular Devices, San Jose, CA, USA) microplate reader at 765 nm wavelength. A method modified from Gillespie et al [33] was used to determine the oxygen radical absorbance capacity (ORAC) using 2,2-azobis(2-amidinopropane) dihydrochloride (Wako Chemicals, Richmond, VA, USA) as a peroxyl generator. The decline in fluorescence of fluorescein was measured kinetically for 60 min using a SpectraMax M5 microplate reader and the area under the curve was calculated for each sample.

4.3. Proximate Analysis

The core proximate and dietary chemical composition of the sample was determined using standard AOAC methods for ash (AOAC 923.03), fat (AOAC Am5-04), crude fiber (AOAC Ba 6a-05), crude protein modification (AOAC 990.03) and moisture (AOAC 930.15). All tests were conducted by the Central Testing Laboratory (Winnipeg, MB, Canada).

4.4. Animal Study

This study protocol was approved by the University of Manitoba Office of Research Ethics and Compliance and Animal Care Committee and was done in accordance with the guidelines by the Canadian Council for Animal Care. Male Sprague Dawley rats (150–175 g; Charles River Laboratories, Quebec, Canada) were housed in a temperature and humidity controlled room with a 12 h light/dark cycle. Rats were anesthetized with 1–5% isoflurane with oxygen at a flow rate of 2 L min^{-1} and kept in a surgical plane of anesthesia with 2% isoflurane during surgery and subjected to permanent ligation of the left anterior descending artery (LAD) to induce MI or to sham surgery. A left thoracotomy was done, and the heart was gently exposed from the pericardial sac through the incision. The left anterior descending coronary artery (LAD) was located and occluded with 6-0 polypropylene silk suture at about 2 mm from aortic root. The suture was tied and the ligation was estimated to be successful when the anterior wall of the left ventricle turned pale. The heart was repositioned, then chest compressed to remove any air from the cavity and the incision was closed using a purse string suture. Sham-operated animals that served as normal control were subjected to similar surgical procedures except that the LAD was not ligated. Buprenorphine 0.05 mg kg^{-1} was administered pre- and post-surgery (2 times a day for 2 days) subcutaneously as an analgesic agent to all rats. All surviving sham and MI rats were assigned to the following 4 treatment groups: (1) Sham MI—distilled water as vehicle (Sham-W); (2) MI—distilled water as vehicle; (MI-W); (3) Sham MI—ginseng berry extract 150 mg/kg/body weight/day (Sham-G); (4) MI—ginseng berry extract 150 mg/kg/body weight/day (MI-G). The 4 groups consisted of sham (n = 8) and MI (n = 12–14) rats and received the treatments by oral gavage for 8 weeks. The sample size was calculated using G*Power statistical software with the power of study kept at 80%. Animals were regularly monitored for well-being.

4.4.1. Transthoracic Echocardiography

All experimental rats were weighed and anesthetized with 3% isoflurane in a chamber, and then kept under 1.5–2% isoflurane throughout the procedure. An echocardiogram was obtained at 4 and 8 weeks post-surgery by 2D guided M-mode and Doppler modalities with a 13 MHz probe (Vivid E9; GE Medical Systems, Milwaukee, WI, USA) as described by us earlier [34]. 2D M-mode parasternal short-axis view images were obtained to determine systolic functional parameters such as the percentage of left ventricle (LV) ejection fraction (EF), fractional shortening (FS), and end-systolic volume (ESV) and end-diastolic volume (EDV). Doppler measurements included isovolumic relaxation time (IVRT), mitral valve (MV) E wave, A wave, and E wave deceleration time. The cardiac structural parameters such as interventricular septal thickness (IVS), LV posterior wall thickness (LVPW), and LV internal diameter (LVID) at diastole and systole were determined from parasternal short-axis view images. All echocardiographic images were analyzed to calculate the listed parameters using EchoPAC software (GE Medical Systems, Milwaukee, WI, USA). All measurements were performed and averaged from three cardiac cycles to account for interbeat variability.

4.4.2. Biological Sample Collection and Analysis

All animals were anesthetized with ketamine/xylazine (9.0 mg per 100 g and 0.9 mg per 100 g IM). The depth of anesthesia was assessed by pedal withdrawal reflex. The blood sample was collected from the inferior vena cava by opening the thoracic cavity and the heart was immediately excised. The whole heart was rinsed in PBS, LVs, septum, and fibrotic scar tissues were separated, weighed and flash frozen in liquid nitrogen.

Percentage of infarct (scarred/fibrotic) LV tissue was calculated by dividing the weight of scarred LV tissue by whole weight of LV tissue as described previously [35]. Evidence of overt heart failure was assessed by determining the presence of ascites and by calculating the lung and liver wet-to-dry weight ratio in all rats.

To determine MI-associated oxidative stress and inflammation, the levels of the lipid peroxidation product, malondialdehyde (MDA), and tumor necrosis factor-α (TNF α) as a proinflammatory marker were assessed in the heart tissue using commercial kits.

4.5. T-cell Phenotypes and Function

To assess the effect of MI and GBE treatment on T-cell phenotypes and activation state and T-cell function, the following experiments were conducted.

4.5.1. Isolation of T-cells

Single cell suspensions were obtained by pressing spleens through a 100µm cell strainer using the barrel of a sterile syringe into a sterile Petri dish containing Hank's buffered saline supplemented with 10 mM-HEPES, 4% fetal bovine serum, and 1% antibiotic/antimycotic at pH 7.4. Erythrocytes were lysed with ammonium chloride buffer. Cells were subsequently washed and re-suspended in label buffer (PBS containing 23 mM-sodium azide and 2% fetal bovine serum). Cell count and viability was completed using a Nexcelom AutoT4Plus. Cell concentration was adjusted to 1×10^7 cells/mL in label buffer [36].

4.5.2. Cytokine Determination

Splenocytes were resuspended in RPMI-1640 supplemented with 10mM-HEPES, 10 mM-sodium bicarbonate, 1 mM-sodium pyruvate, 2 mM-gluatmine, 0.1 mM-non-essential amino acids, 50 µM-2-mercatpoethanol, 1% antibiotic/antimycotic and 5% fetal bovine serum at pH 7.2 at a concentration of 1×10^6 splenocytes/mL and incubated at 37 °C and 5% CO_2 with 2.5 μg/mL Concanavalin A (ConA), or unstimulated for 48 hours. After incubation, samples were centrifuged at 400 g at 4 °C for 5 min to pellet cells. Supernatants were stored at −80 °C until analysis. The concentration of IL-2 (lower limit of detection (LLOD) 0.46 pg/mL), IFNγ (LLOD 6.8 pg/mL), IL-10 (LLOD 19.4 pg/mL), and TNFα (LLOD 27.7 pg/mL) were measured in cell culture supernatants using a cytometric bead array on a FACSCanto II flow cytometer. All samples were analyzed in duplicate with CV <10% according to the BD Cytometric Bead Array Mouse/Rat Soluble Protein Master Buffer Kit Instruction Manual [37].

4.5.3. Phenotyping

T-cell phenotypes were determined by flow cytometry using isolated splenocytes. Monoclonal antibodies against rat CD3 (FITC label, clone 1F4, isotype mouse IgM_{κ}), CD4 (PE-Cy7 label, clone OX-35, isotype mouse $IgG_{2a,\kappa}$), CD8 (PerCP label, clone OX-8, isotype mouse $IgG_{1,\kappa}$), CD25 (APC label, clone OX39, isotype Mouse $IgG_{1,\kappa}$), foxp3 (PE label, clone FJK-16s, isotype rat $IgG_{2a,\kappa}$) were obtained from BD BioSciences (Mississauga, ON, Canada). Antibodies were incubated with 1×10^6 cells/mL for 30 min at 4 °C in the dark. Cells were washed, and then incubated in Foxp3 fixation/permeabilization working solution at 4 °C for 18 h. Following incubation cells were washed with the permeabilization buffer and incubated with mouse CD16/CD32 antibody at 4 °C for 5 min to block non-specific binding. Foxp3 was added to cells and incubated at 4 °C in the dark for at least 30 min. Finally, treated cells were washed using the permeabilization buffer and re-suspended in PBS. Data were acquired on a FACSCanto II flow cytometer using the 488 nm and 633 nm lasers. Figure 6 shows representative flow cytometry plots. Forward versus side-scatter plots were used to gate on intact lymphoid cells and non-viable cells. The data were collected in list-mode format with the analyses based on 100,000 cells satisfying the light scatter gate for lymphocytes using Cell Diva software (v8.0.1). Unstained cells were used to assess auto-fluorescence, isotype controls to assess background staining, and single color samples were employed to adjust color compensation [36].

4.6. Statistical Analysis

All values are represented as means ± SEM. Two-way analysis of variance (ANOVA) test was used to determine the effect of surgery (factor 1) and treatment (factor 2) and their interaction (SAS, version 9.4, SAS Institute, Cary NC, USA). Unpaired t-test was utilized for comparison between 2 groups Significant differences among means were determined using LSmeans. Differences were considered significant at $p \leq 0.05$.

5. Conclusions

The results of this study suggest that although GBE exerted potent antioxidant activity, it was unable to recover cardiac function in post-MI rats.

Author Contributions: The authors were responsible for the following aspects of this paper: conceptualization, T.N.; methodology, T.N., Y.S., and H.B.; formal analysis, M.P.; investigation, P.R., J.S., S.P., J.C.P., and L.Y.; resources, D.B., T.N.; data curation, L.Y.; writing—original draft preparation, M.P., P.R, H.B; writing—review and editing, M.P, P.R., G.N.P., Y.S., H.B., P.S.T., J.S., S.P., J.C.P., and T.N.; supervision, T.N.; project administration, L.Y., and T.N.; funding acquisition, D.B, Y.S., T.N., and H.B.

Funding: This research was funded by Agriculture and Agri-Food Canada.

Acknowledgments: We would like to thank the R.O. Burrell Laboratory for assistance with animal studies.

Conflicts of Interest: The authors declare no conflict of interest.

References

1. Vuksan, V.; Xu, Z.Z.; Jovanovski, E.; Jenkins, A.L.; Beljan-Zdravkovic, U.; Sievenpiper, J.L.; Mark Stavro, P.; Zurbau, A.; Duvnjak, L.; Li, M.Z.C. Efficacy and safety of American ginseng (*Panax quinquefolius* L.) extract on glycemic control and cardiovascular risk factors in individuals with type 2 diabetes: A double-blind, randomized, cross-over clinical trial. *Eur. J. Nutr.* **2018**. [CrossRef] [PubMed]
2. Attele, A.S.; Wu, J.A.; Yuan, C.-S. Ginseng pharmacology: Multiple constituents and multiple actions. *Biochem. Pharmacol.* **1999**, *58*, 1685–1693. [CrossRef]
3. Chung, I.M.; Lim, J.J.; Ahn, M.S.; Jeong, H.N.; An, T.J.; Kim, S.H. Comparative phenolic compound profiles and antioxidative activity of the fruit, leaves, and roots of Korean ginseng (*Panax ginseng* Meyer) according to cultivation years. *J. Ginseng Res.* **2016**, *40*, 68–75. [CrossRef] [PubMed]
4. Mehendale, S.R.; Wang, C.Z.; Shao, Z.H.; Li, C.Q.; Xie, J.T.; Aung, H.H.; Yuan, C.S. Chronic pretreatment with American ginseng berry and its polyphenolic constituents attenuate oxidant stress in cardiomyocytes. *Eur. J. Pharmacol.* **2006**, *553*, 209–214. [CrossRef] [PubMed]
5. Choi, S.Y.; Cho, C.W.; Lee, Y.; Kim, S.S.; Lee, S.H.; Kim, K.T. Comparison of Ginsenoside and Phenolic Ingredient Contents in Hydroponically-cultivated Ginseng Leaves, Fruits, and Roots. *J. Ginseng Res.* **2012**, *36*, 425–429. [CrossRef] [PubMed]
6. Hyun, T.K.; Jang, K.I. Are berries useless by-products of ginseng? Recent research on the potential health benefits of ginseng berry. *Excli J.* **2017**, *16*, 780–784. [PubMed]
7. Dey, L.; Xie, J.T.; Wang, A.; Wu, J.; Maleckar, S.A.; Yuan, C.S. Anti-hyperglycemic effects of ginseng: Comparison between root and berry. *Phytomedicine* **2003**, *10*, 600–605. [CrossRef] [PubMed]
8. Kim, M.H.; Lee, J.; Jung, S.; Kim, J.W.; Shin, J.H.; Lee, H.J. The involvement of ginseng berry extract in blood flow via regulation of blood coagulation in rats fed a high-fat diet. *J. Ginseng Res.* **2017**, *41*, 120–126. [CrossRef] [PubMed]
9. Seo, E.; Kim, S.; Lee, S.J.; Oh, B.C.; Jun, H.S. Ginseng berry extract supplementation improves age-related decline of insulin signaling in mice. *Nutrients* **2015**, *7*, 3038–3053. [CrossRef] [PubMed]
10. Choi, H.S.; Kim, S.; Kim, M.J.; Kim, M.S.; Kim, J.; Park, C.W.; Seo, D.; Shin, S.S.; Oh, S.W. Efficacy and safety of Panax ginseng berry extract on glycemic control: A 12-wk randomized, double-blind, and placebo-controlled clinical trial. *J. Ginseng Res.* **2018**, *42*, 90–97. [CrossRef] [PubMed]
11. Li, J.; Ichikawa, T.; Jin, Y.; Hofseth, L.J.; Nagarkatti, P.; Nagarkatti, M.; Windust, A.; Cui, T. An essential role of Nrf2 in American ginseng-mediated anti-oxidative actions in cardiomyocytes. *J. Ethnopharmacol.* **2010**, *130*, 222–230. [CrossRef] [PubMed]
12. Moey, M.; Gan, X.T.; Huang, C.X.; Rajapurohitam, V.; Martinez-Abundis, E.; Lui, E.M.; Karmazyn, M. Ginseng reverses established cardiomyocyte hypertrophy and postmyocardial infarction-induced hypertrophy and heart failure. *Circ. Heart Fail.* **2012**, *5*, 504–514. [CrossRef] [PubMed]
13. Frangogiannis, N.G. Regulation of the inflammatory response in cardiac repair. *Circ. Res.* **2012**, *110*, 159–173. [CrossRef] [PubMed]
14. Magrone, T.; Jirillo, E. Polyphenols from red wine are potent modulators of innate and adaptive immune responsiveness. *Proc. Nutr. Soc.* **2010**, *69*, 279–285. [CrossRef] [PubMed]
15. Kim, C.-K.; Cho, D.H.; Lee, K.-S.; Lee, D.-K.; Park, C.-W.; Kim, W.G.; Lee, S.J.; Ha, K.-S.; Goo Taeg, O.; Kwon, Y.-G.; Kim, Y.-M. Ginseng Berry Extract Prevents Atherogenesis via Anti-Inflammatory Action by Upregulating Phase II Gene Expression. *Evid. -Based Complement. Altern. Med.* **2012**, *2012*. [CrossRef] [PubMed]
16. Attele, A.S.; Zhou, Y.P.; Xie, J.T.; Wu, J.A.; Zhang, L.; Dey, L.; Pugh, W.; Rue, P.A.; Polonsky, K.S.; Yuan, C.S. Antidiabetic effects of Panax ginseng berry extract and the identification of an effective component. *Diabetes* **2002**, *51*, 1851–1858. [CrossRef] [PubMed]
17. Li, Z.; Kim, H.J.; Park, M.S.; Ji, G.E. Effects of fermented ginseng root and ginseng berry on obesity and lipid metabolism in mice fed a high-fat diet. *J. Ginseng Res.* **2018**, *42*, 312–319. [CrossRef] [PubMed]
18. Shao, Z.-H.; Xie, J.-T.; Vanden Hoek, T.L.; Mehendale, S.; Aung, H.; Li, C.-Q.; Qin, Y.; Schumacker, P.T.; Becker, L.B.; Yuan, C.-S. Antioxidant effects of American ginseng berry extract in cardiomyocytes exposed to acute oxidant stress. *Biochim. Biophys. Acta Gen. Subj.* **2004**, *1670*, 165–171. [CrossRef] [PubMed]
19. Eaton, L.W.; Weiss, J.L.; Bulkley, B.H.; Garrison, J.B.; Weisfeldt, M.L. Regional Cardiac Dilatation after Acute Myocardial Infarction. *N. Engl. J. Med.* **1979**, *300*, 57–62. [CrossRef] [PubMed]

20. Erlebacher, J.A.; Weiss, J.L.; Eaton, L.W.; Kallman, C.; Weisfeldt, M.L.; Bulkley, B.H. Late effects of acute infarct dilation on heart size: A two dimensional echocardiographic study. *Am. J. Card.* **1982**, *49*, 1120–1126. [CrossRef]
21. Gaudron, P.; Eilles, C.; Kugler, I.; Ertl, G. Progressive left ventricular dysfunction and remodeling after myocardial infarction. Potential mechanisms and early predictors. *Circulation* **1993**, *87*, 755–763. [CrossRef] [PubMed]
22. Rigo, P.; Murray, M.; Strauss, H.W.; Taylor, D.; Kelly, D.; Weisfeldt, M.; Pitt, B. Left Ventricular Function in Acute Myocardial Infarction Evaluated by Gated Scintiphotography. *Circulation* **1974**, *50*, 678–684. [CrossRef] [PubMed]
23. Moller, J.E.; Hillis, G.S.; Oh, J.K.; Reeder, G.S.; Gersh, B.J.; Pellikka, P.A. Wall motion score index and ejection fraction for risk stratification after acute myocardial infarction. *Am. Heart J.* **2006**, *151*, 419–425. [CrossRef] [PubMed]
24. Multicenter Postinfarction Research Group. Risk Stratification and Survival after Myocardial Infarction. *N. Engl. J. Med.* **1983**, *309*, 331–336. [CrossRef] [PubMed]
25. White, H.D.; Norris, R.M.; Brown, M.A.; Brandt, P.W.; Whitlock, R.M.; Wild, C.J. Left ventricular end-systolic volume as the major determinant of survival after recovery from myocardial infarction. *Circulation* **1987**, *76*, 44–51. [CrossRef] [PubMed]
26. Janahmadi, Z.; Nekooeian, A.A.; Moaref, A.R.; Emamghoreishi, M. Oleuropein Offers Cardioprotection in Rats with Acute Myocardial Infarction. *Cardiovasc. Toxicol.* **2015**, *15*, 61–68. [CrossRef] [PubMed]
27. Zhao, H.; Zhang, J.; Hong, G. Minocycline improves cardiac function after myocardial infarction in rats by inhibiting activation of PARP-1. *Biomed. Pharmacother.* **2018**, *97*, 1119–1124. [CrossRef] [PubMed]
28. Shanmugam, G.; Narasimhan, M.; Sakthivel, R.; Kumar, R.R.; Davidson, C.; Palaniappan, S.; Claycomb, W.W.; Hoidal, J.R.; Darley-Usmar, V.M.; Rajasekaran, N.S. A biphasic effect of TNF-alpha in regulation of the Keap1/Nrf2 pathway in cardiomyocytes. *Redox Biol.* **2016**, *9*, 77–89. [CrossRef] [PubMed]
29. Kurrelmeyer, K.M.; Michael, L.H.; Baumgarten, G.; Taffet, G.E.; Peschon, J.J.; Sivasubramanian, N.; Entman, M.L.; Mann, D.L. Endogenous tumor necrosis factor protects the adult cardiac myocyte against ischemic-induced apoptosis in a murine model of acute myocardial infarction. *Proc. Natl. Acad. Sci. USA* **2000**, *97*, 5456–5461. [CrossRef] [PubMed]
30. El-Ani, D.; Philipchik, I.; Stav, H.; Levi, M.; Zerbib, J.; Shainberg, A. Tumor necrosis factor alpha protects heart cultures against hypoxic damage via activation of PKA and phospholamban to prevent calcium overload. *Can. J. Physiol. Pharmacol.* **2014**, *92*, 917–925. [CrossRef] [PubMed]
31. Utheim, T.P. Chapter 7: Why Test BCG in Sjögren's Syndrome? In *The Value of BCG and TNF in Autoimmunity*; Faustman, D.L., Ed.; Academic Press: Amsterdam, The Netherlands, 2014; pp. 105–125.
32. Ainsworth, E.A.; Gillespie, K.M. Estimation of total phenolic content and other oxidation substrates in plant tissues using Folin-Ciocalteu reagent. *Nat. Protoc.* **2007**, *2*, 875–877. [CrossRef] [PubMed]
33. Gillespie, K.M.; Chae, J.M.; Ainsworth, E.A. Rapid measurement of total antioxidant capacity in plants. *Nat. Protoc.* **2007**, *2*, 867–870. [CrossRef] [PubMed]
34. Raj, P.; McCallum, J.L.; Kirby, C.; Grewal, G.; Yu, L.; Wigle, J.T.; Netticadan, T. Effects of cyanidin 3-0-glucoside on cardiac structure and function in an animal model of myocardial infarction. *Food Funct.* **2017**, *8*, 4089–4099. [CrossRef] [PubMed]
35. Ju, H.; Zhao, S.; Jassal, D.S.; Dixon, I.M. Effect of AT1 receptor blockade on cardiac collagen remodeling after myocardial infarction. *Cardiovasc. Res.* **1997**, *35*, 223–232. [CrossRef]
36. Blewett, H.J.; Mohankumar, S.K.; Rech, L.; Rector, E.S.; Taylor, C.G. The reduced proportion of New splenic T-cells in the zinc-deficient growing rat is not due to increased susceptibility to apoptosis. *Immunobiology* **2014**, *219*, 602–610. [CrossRef] [PubMed]
37. Blewett, H.J.; Gerdung, C.A.; Ruth, M.R.; Proctor, S.D.; Field, C.J. Vaccenic acid favourably alters immune function in obese JCR:LA-cp rats. *Br. J. Nutr.* **2009**, *102*, 526–536. [CrossRef] [PubMed]

MDPI
St. Alban-Anlage 66
4052 Basel
Switzerland
Tel. +41 61 683 77 34
Fax +41 61 302 89 18
www.mdpi.com

International Journal of Molecular Sciences Editorial Office
E-mail: ijms@mdpi.com
www.mdpi.com/journal/ijms

www.ingramcontent.com/pod-product-compliance
Lightning Source LLC
LaVergne TN
LVHW070153170726
843515LV00007B/1911

* 9 7 8 3 0 3 9 2 8 8 8 7 8 *